Environmental Geology and Sustainability

This book explains the role of geology as the basis of sustainability. It discusses how humans have altered natural balances and the unique dimensions that geology brings to understanding sustainability. Focused on humans' activities in shaping urban areas, this book helps readers identify natural geologic risks created, identify human actions that reduce or increase those risks, or create new risks with unintended negative environmental consequences. It provides sustainability-oriented solutions so that humans can live in harmony with nature.

Features:

- The first book to identify and describe geology as the foundation of sustainability.
- Provides the history and reasoning why geology is important to achieve sustainability and environmental stewardship.
- Goes beyond identifying natural geological and anthropogenic-induced risks by providing numerous case studies and potential solutions.
- Includes an overview of natural geologic and anthropogenic-induced impacts in major cities across the world.
- Examines where environmental regulations in many countries of the world have succeeded or failed and lists those areas where new sustainability-oriented environmental regulations are needed worldwide.

This textbook is for senior undergraduate and graduate students taking courses in environmental geology, Earth science and sustainability, urban planning, and environmental risk analysis. It also serves as an insightful reference for professionals, researchers, and academics in these fields.

Environmental Geology and Sustainability

Risk Mitigation in Urban Areas

Daniel T. Rogers

CRC Press
Taylor & Francis Group
Boca Raton London New York

CRC Press is an imprint of the
Taylor & Francis Group, an **informa** business

Designed cover image: Getty Images

First edition published 2025
by CRC Press
2385 NW Executive Center Drive, Suite 320, Boca Raton, FL 33431

and by CRC Press
4 Park Square, Milton Park, Abingdon, Oxon, OX14 4RN

CRC Press is an imprint of Taylor & Francis Group, LLC

Library of Congress Cataloging-in-Publication Data
Names: Rogers, Daniel T. (Daniel Timothy), 1959– editor.
Title: Environmental geology and sustainability : risk mitigation in urban areas / Daniel T. Rogers.
Other titles: Risk mitigation in urban areas
Description: First edition. | Boca Raton, FL : CRC Press, 2025. |
Includes bibliographical references and index.
Identifiers: LCCN 2024057169 (print) | LCCN 2024057170 (ebook) | ISBN 9781032732169 (hbk) |
ISBN 9781032732206 (pbk) | ISBN 9781003463153 (ebk)
Subjects: LCSH: Human ecology. | Environmental geology. | Sustainable development. | Natural disasters–Environmental aspects. | Cities and towns–Growth–Environmental aspects. | Human beings–Effect of environment on.
Classification: LCC GF43 .E68 2025 (print) | LCC GF43 (ebook)
LC record available at https://lccn.loc.gov/2024057169
LC ebook record available at https://lccn.loc.gov/2024057170

ISBN: 9781032732169 (hbk)
ISBN: 9781032732206 (pbk)
ISBN: 9781003463153 (ebk)

DOI: 10.1201/9781003463153

Typeset in Times
by Newgen Publishing UK

Dedication

To Jennie

Contents

Preface

The science of geology is the study of the composition, structure, and history of Earth. Environmental geology is an applied branch of geology that addresses the interaction between geology and humans. Environmental geology is fundamentally important because it directly impacts every living individual every second of every day at every location on Earth where life exists. This is where the study of environmental geology directly connects with sustainability. Sustainability involves creating a balance between humans and nature that at this point does not exist.

This book focuses on how Earth's geology impacts humans and how human behavior impacts the geology and natural environment of Earth. The outcomes are integrated with sustainability objectives to identify what actions are necessary to create a balance with nature instead of harming nature and perhaps ourselves. The first section of this book addresses geologic hazards, which is a measure of how geology impacts humans. The second section addresses environmental hazards and how they impact humans. The third section examines anthropogenic-induced hazards. Anthropogenic-induced hazards originate from a human activity and affect nature and geology.

As humans expanded the use of agriculture thousands of years ago, they created communities that changed social behavior. This laid the foundation for kinship, community organization, and conflict. Currently, the area of land that humans have developed is staggering. In addition, as cities grew and populations increased, some of the locations where humans settled were within areas of geologic hazards. The outcome was that when the geological hazard was triggered, loss of human life was often significant. The eruption of Mount Vesuvius is an example where thousands lost their lives when the volcano suddenly erupted in 79 AD.

Culture and civilization have unleashed opportunity of imaginative thought. Instead of focusing our efforts on survival, technological advancements have given us the opportunity and time to explore how life on Earth has evolved since its origin billions of years ago. It has also allowed us to explore fundamental subatomic particles, explore and learn about our universe, and travel into space. Our curious nature and the power of our brains provide us with awareness and self-reflection, consciousness, and allow us to question our existence. Now we find ourselves at a crossroad where our success as a species may no longer be an advantage for our future survival but may be leading us down a path where it threatens the existence of many living species on Earth, including our own.

The modern Earth narrative, the scientific story of the 4.54-billion-year natural history of Earth generally does not include significant discussion of humans because our species has been in existence for such a short time. Scientific research centers on the natural history of Earth, on geologic time, and biological evolution. Discussing Earth in this fashion is important as we begin the third millennium of our current calendar. It provides a clear viewpoint and understanding of the increasing impact of human population growth and associated stresses placed on the Earth system. Humans have significantly impacted the air, water, land, and biosphere of the entire planet in such a short period of time that Earth has simply not been able to keep pace and adapt. In fact, many scientists now believe we have moved the needle of geologic time to a new time period termed Anthropocene, meaning the age of humans. Many scientists believe that there is widespread consensus for this new designation. The idea that human activity has become the dominant force on Earth during recent recorded history is well established. What is debated, across scientific disciplines, is not whether the Anthropocene exists, but at what point did our actions as a species become a climate-driving force equal to major geologic processes. An additional debated point is, as that it occurred in such a short period of time, will it truly be a new geologic period or will humans simply go extinct and these impacts will be recorded as no more than a short-term event in Earth history?

We find ourselves on Earth after billions of years of Earth's existence by chance, not by design. Humans are one of approximately 4,500 species of mammals and one species in more than perhaps a 100 million living species that form Earth's biosphere. All life on Earth is dependent on the biosphere. Our species has altered the biosphere so quickly that many other life forms have been overcome and dominated by our success mainly through technological and cultural innovations.

To achieve sustainability and exist with nature will require cooperation not often experienced in our history. Humans must respond and must act individually, as a community, and globally to more fully understand impacts to our planet and act diligently to limit the negative effects and restore ecosystems where possible. We are the first generation experiencing the effects of climate change and the last generation that can do something meaningful about it. These actions must be undertaken before it is too late for the survival of numerous species, including our own.

After more than 50 years of attempting to protect the environment, humans are learning that altering the biosphere and nature to suit our perceived needs has not worked as well as we may have hoped for ourselves, for Earth, or for the living organisms with which we share this planet.

About the Author

FIGURE 0.0 Photograph of the author taken by Jennie Macaluso at Mount Rainier National Park in August 2024 showing hazy skies from wildfires and the glacial front of the receding Nisqually glacier.

Daniel T. Rogers is the Director – Environmental Affairs at Amsted Industries Incorporated in Chicago, IL. Amsted Industries is a diversified manufacturing company of industrial components serving railroad, vehicular, construction and building markets. Amsted Industries has more than 75 manufacturing locations in more than 15 countries. Mr. Rogers participates in environmental due diligence for acquisitions and divestitures, investigation and remediation of contamination, creating and evaluating environmental compliance and sustainability programs, and providing environmental advice, oversight, training, and negotiation strategies at various levels within the organization.

Mr. Rogers has published more than 100 research papers over the last 40 years in professional and academic publications and peer-reviewed journals on subjects including environmental geology, hydrogeology, geologic sensitivity and mapping, contaminant fate and transport, urban geology, environmental site investigations, contaminant risk, brownfield re-development, remediation, pollution prevention, environmental compliance, and sustainability. He has authored Fundamentals of Environmental Law and Compliance in 2024, Environmental Compliance Handbook in 2023, which is a four-volume set that includes Air, Water, Land, and Sustainability. He has also authored *Urban Watersheds: Geology, Contamination, Environmental Regulations, and Sustainability*, 2nd Edition (2020), *Environmental Compliance and Sustainability: Global Challenges and Perspectives*

(2019), and *Environmental Geology of Metropolitan Detroit* (1996). Mr. Rogers has published surficial geologic and contaminant sensitivity maps of the Rouge River watershed in southeastern Michigan. In addition, he is a contributing author of *The Encyclopedia of Global Social Issues* (2013), *Urban Groundwater* (2007), *Geoenvironmental Mapping* (2002), and *Groundwater in the Urban Environment* (1997).

He has taught geology and environmental chemistry at Eastern Michigan University and The University of Michigan and has presented guest lectures at numerous colleges and universities both in the United States and internationally.

Acknowledgments

I wish to thank many regulatory agencies and scientific research organizations who are dedicated to the protection of human health and the environment and have contributed to the development of this book. International regulatory agencies and countries that I wish to thank include the World Health Organization, the United Nations, European Environmental Agency, Environment Canada, Mexico Environmental and Natural Resource Ministry, Norwegian Environment Agency, Turkey Ministry of Environment and Urbanization, India Ministry of Environment, Forestry, and Climate Change, Indonesia Ministry of Environment, Japan Ministry of the Environment, Korea Ministry of the Environment, Malaysia Department of Environment, Egyptian Environmental Affairs Agency, Kenya National Environment Management Authority, South Africa Department of Environmental Affairs, Tanzania Department of Environment, Brazil Ministry of the Environment, Chile Ministry of the Environment, Peru Ministry of the Environment, Argentine Secretariat for the Environment, Australia Department of Environment and Energy, and New Zealand Ministry for the Environment.

Other organizations and state agencies include International Union of Geological Sciences, Geological Society of America, International Association of Hydrogeologists (IAH), United States Environmental Protection Agency, Intergovernmental Panel on Climate Change (IPCC), United States Geological Survey (USGS), British Geological Survey, National Oceanic and Atmospheric Administration, Illinois State Geological Survey, California Environmental Protection Agency, California Department of Toxic Substances and Control, Delaware Department of Natural Resources and Environmental Control, Illinois Environmental Protection Agency, Indiana Department of Environmental Management, Iowa Department of Natural Resources, Kansas Department of Health and Environment, Maryland Department of Environment, Michigan Department of Environment, Great Lakes, and Energy, New Jersey Department of Environmental Protection, Ohio Environmental Protection Agency, Oregon Department of Environmental Quality, and the Wisconsin Department of Natural Resources.

Individuals who have provided guidance and advice from the research and academic field over the years include my colleagues at The University of Michigan, Dr. Martin Kaufman and Dr. Kent Murray and Dr. Ken Howard at The University of Toronto, Dr. Jack Sharp at The University of Texas, Dr. Richard Berg at the Illinois State Geological Survey, Dr. Robert Bobrowsky at the University of British Columbia, Dr. Peter Kolesar at Utah States University, Dr. Bill Farrand at The University of Michigan, Dr. Robert Oaks at Utah State University, Dr. Krause formerly at Henry Ford Community College, Dr. Derek Wong at APEX Environmental, Dr. Fred Payne with Arcadis, Dr. Colin Booth at Northern Illinois University, Dr. Mike Barcelona at Western Michigan University, Dr. Rebecca Spearot formerly with Clayton Environmental, Dr. Mary Ann Thomas at USGS, Dr. Garth van der Kamp at Environment Canada, Dr. Vladimir Kovalevsky at the Russian Academy of Sciences, Dr. John Moore with IAH, Dr. Hugo Loaiciga with the University of California, and Dr. John Chilton and Dr. Craig Foster with the British Geological Survey.

When dealing with environmental regulations of the United States and the world, one must also have guidance from the legal field. I wish to thank those special individuals who have been a source of inspiration, counsel, and debate concerning environmental regulations who include Chris Athas, Esq., Ed Brosius Esq., James Enright Esq., Rick Glick Esq., Geneva Halliday, Esq., Brian Houghton Esq., Michael Krautner Esq., Stephen Lewis, Esq., Michael Maher, Esq., Chris McNevin Esq., Granta Y. Nakayama, P. C., Esq., Jeryl Olsen Esq., Ron Paterson, Esq., Tom Petermann Esq., Doug Schleicher, Esq., Stephen Smith Esq., and Tom Wilczak, Esq.

The author wishes to thank the following professional colleagues for their assistance, cooperation, and ideas over the many years that it took to prepare this book: Lauren Alkadis, Cathee Andrews, Tony Anthony, Ann Barry, Gary Blinkiewicz, Tom Buggey, Kristine Casper, Chris Christensen Russ Chadwick, Bob Cigale, Tom Cok, Marc, DeLoecker, Sheryl Doxtader, Ian Drost, Rob Ellis, Rose

Ellison, Rob Ferree, Jennifer Formoso, Orin Gelderloos, Steve Hoin, Heather Hopkins, George Karalus, Steve Kitler, Shannon Lian, Rick Linnville, Daniel J. Lombardi, Bill Looney, Kim Myers, Ernie Nimister, David O'Donnell, Tom O'Hara, Ray Ostrowski, Paul Owens, Bruce Patterson, Scott Pearson, Sara Pearson, Mark Penzkover, Robert Ribbing, Eric Ross, Jeanne Schlaufman, Dave Slayton, Dave Smith, Ed Stewart, Patricia Thornton, Mary Vanderlaan, Chris Venezia, Nick Welty, Cheryl Wilson, and Tom Wenzel.

Acronyms

ACGIH American Conference of Governmental Industrial Hygienists
ADP atmospheric decontamination plan
APCP Air Prevention and Control of Pollution Act
APP atmospheric prevention plan
AST aboveground storage tank
ASTM American Society for Testing Materials
ATSDR Agency for Toxic Substances and Disease Registry
BACT Best Available Control Technology
BLM Bureau of Land Management
BMP best management practice
BPA bisphenol A
BTEX benzene, toluene, ethyl benzene, and xylenes
CAA Clean Air Act
CAAC Clean Air Alliance of China
CalEPA California Environmental Protection Agency
CAS Chemical Abstract Service
CEPA Canadian Environmental Protection Act
CERCLA Comprehensive Environmental Response, Compensation and Liability Act
CESQG conditionally exempt small quantity generator
CFC chlorofluorocarbon
CFR Code of Federal Regulation
CMEE China Ministry of Ecology and Environment
CONAGUA Mexico National Water Commission
CRF contaminant or pollution risk factor
CRFair contaminant or pollution risk factor for air
CRFgw contaminant or pollution risk factor for groundwater
CRFsoil contaminant or pollution risk factor for soil
CSO combined sewer overflow
CVOC chlorinated volatile organic hydrocarbon
CWA Clean Water Act
DBP di-*n*-butyl phthalate
DCE dichloroethylene, dichloroethane
DCL dioxin-like compounds
DDT dichlorodiphenyltrichloroethane
DEHP bi or bis(2-ethylhexyl) phthalate
DEP diethyl phthalate
DIV Dutch Intervention Value
DNAPL dense nonaqueous phase liquid
DOT Department of Transportation
EEA European Environment Agency
EEC European Economic Community
EH&S environmental health and safety
EHS extremely hazardous substance
EIA Environmental Impact Assessment
EINETICS European Information and Observation Network
EIS environmental impact statement
EPA Environmental Protection Agency

EPCRA	Emergency Planning and Community Right-to-Know Act
ESA	environmental site assessment
ESPM	elimination, substitution, prevention, and minimization
ETS	emission trading system
EU	European Union
FDA	Food and Drug Administration
FEPCA	Federal Environmental Pesticide Control Act
FFDCA	Federal Feed, Drug, and Cosmetic Act
FIFRA	Federal Insecticide, Fungicide, and Rodenticide Act
GACT	Generally Available Control Technology
GHG	greenhouse gas
GWP	global warming potential
HAP	hazardous air pollutant
HAZMAT	hazardous materials
HCFC	hydrochlorofluorocarbons
HFC	hydrofluorocarbon
HMTA	Hazardous Materials and Transportation Act
HRS	Hazard Ranking System
IAEA	International Atomic Energy Agency
IED	Industrial Emissions Directive
IGS	Italian Geological Survey
IPCC	Intergovernmental Panel on Climate Change
KDNREP	Kentucky Department of Natural Resources and Environmental Protection
LEPC	Local Emergency Planning Committee
LNAPL	light nonaqueous phase liquid
LOAEL	lowest-observed-adverse-effect level
LQG	large quantity generator
MACT	Maximum Achievable Control Technology
MCL	maximum contaminant level
MCLG	maximum contaminant level goal
MDEP	Massachusetts Department of Environmental Protection
MDEQ	Michigan Department of Environmental Quality
MEK	methyl ethyl ketone
MEP	China Ministry of Environmental Protection
MNR	Russian Ministry of Natural Resources
MS4S	Large Municipal Separate Storm Sewer Systems
MSDS	material safety data sheet
MTBE	methyl tertiary butyl ether
NAAQS	National Ambient Air Quality Standards
NAFTA	North American Free Trade Agreement
NAICS	North American Industry Classification System
NASA	National Aeronautics and Space Administration
NATO–	North Atlantic Treaty Organization
NCZMP	National Coastal Zone Management Program
NEA	Norwegian Environment Agency
NEPA	National Environmental Protection Act
NEPC	Australian National Environmental Protection Council
NERRS	National Estuarine Research Reserve System
NESHAPS	National Emission Standards for Hazardous Air Pollutants
NGS	National Geographic Society

NHTSA	National Highway Traffic Safety Administration
NJDEP	New Jersey Department of Environmental Protection
NOAA	National Oceanic and Atmospheric Administration
NOM	Mexico Standards or Normas
NOV	notice of violation
NPDES	National Pollutant Discharge Elimination System
NPDWS	National Primary Drinking Water Standards
NPL	National Priority List
NPS	non-point source
NRC	National Response Center
NRC	Nuclear Regulatory Commission
ODS	ozone-depleting substance
OECD	Organization for the Economic Co-Operation and Development
OPP	office of pesticide programs (USEPA)
OSHA	Occupational Safety and Health Administration
OSPA	Oil Spill Prevention Act
PAH	polycyclic aromatic hydrocarbon
PBT	persistent bioaccumulative toxic
PCB	polychlorinated biphenyl
PCE or PERC	tetrachloroethylene, tetrachloroethene, perchloroethylene, perchloroethene
PCP	pentachlorophenol
PEL	permissible exposure limit
PFC	perfluorocarbon
PFOA or PFAS	per and polyfluoroalkyl substances
PM	particulate matter
PNA	polynuclear aromatic hydrocarbon
POP	persistent organic pollutants
POTW	publicly owned treatment works
PPA	Pollution Prevention Act
PPE	personal protective equipment
PPMv	parts per million by volume
PROFEPA	Mexico Federal Attorney Generalship of Environmental Protection
PRP	potential responsible party
PS	point source
PSD	Potential of Significant Deterioration
PTE	potential to emit
PVC	polyvinyl chloride
RCI	reactivity, corrosivity, and ignitability
RCRA	Resource Conservation and Recovery Act
REC	recognized environmental condition
RQ	reportable quantity
SARA	Superfund Amendments and Reauthorization Act
SDS	safety data sheet
SDWA	Safe Drinking Water Act
SEMARNAT	Mexico Secretariat of the Environment and Natural Resources
SEP	supplemental environmental project
SERC	State Emergency Response Commission
SIC	Standard Industrial Code
SPCC	Spill Prevention Control and Countermeasures Plan
SPLP	synthetic precipitant leaching procedure

SQG small quantity generator
SRP spill response plan
SVOC semi-volatile organic compound
SWPPP stormwater pollution prevention plan
TCA trichloroethane
TCDD 2,3,7,8-tetrachlorodibenzodioxin
TCE trichloroethylene, trichloroethene
TCLP toxic characteristic leaching procedure
THM trihalomethane
TPQ Threshold Planning Quantity
TSCA Toxic Substance and Control Act
TSP total suspended particulates
UK United Kingdom
UN United Nations
UNESCO United Nations Educational, Scientific and Cultural Organization
UNFCCC United Nations Framework Convention on Climate Change
USCB United States Census Bureau
USCIA United States Central Intelligence Agency
USDA United States Department of Agriculture
USDC United States Department of Commerce
USDHH United States Department of Health and Human Services
USDOE United States Department of Energy
USDOT United States Department of Transportation
USEPA United States Environmental Protection Agency
USFS United States Forest Service
USGS United States Geological Survey
USNRC United States Nuclear Regulatory Commission
UST underground storage tank
VC vinyl chloride
VEI volcanic eruption index
VOC volatile organic compound
VSQG very small quantity generator
WDNR Wisconsin Department of Natural Resources
WHO World Health Organization

Elements

Ag	silver
As	arsenic
Ba	barium
Br	bromine
C	carbon
Cd	cadmium
Cl	chlorine
Cr	chromium
Cr III	trivalent chromium
Cr VI	hexavalent chromium
Cr^{3+}	trivalent chromium
Cr^{6+}	hexavalent chromium
Cu	copper
F	fluorine
H	hydrogen
Hg	mercury
I	iodine
K	potassium
N	nitrogen
O	oxygen
P	phosphorus
Pb	lead
S	sulfur
Se	selenium
Zn	zinc

Symbols

η	effective porosity
ρ_b	bulk density
Ws	water solubility
ATM	atmosphere
°C	degrees Celsius
°F	degrees Fahrenheit
F_{oc}	organic carbon partition coefficient
H	Henry's Law constant
K_d	distribution coefficient
K_{oc}	fraction of total organic carbon
M	mobility
M_W	molecular weight
P	persistence
P	pressure
R	retardation factor
T	toxicity
VP	vapor pressure
V	volume

Molecules/Minerals

($CaMg[CO_3]_2$)	magnesium–calcium carbonate/dolostone
($CaSO_4.2H_2O$)	gypsum
$(CH_2Cl)_2$	1,2-dichloroethene
$C_{12}H_{10}$-$x$$Cl_x$	polychlorinated biphenyl
C_2H_3Cl	vinyl chloride
C_2HCl_3	trichloroethene
C_6H_6	benzene
$CaCO_3$	calcium carbonate/limestone
CH_4	methane
$Cl_2 = CCl_2$	tetrachloroethene
CO	carbon monoxide
CO_2	carbon dioxide
H_2CO_3	carbonic acid
H_2O	water
N_2O	nitrous oxide
NaCl	sodium chloride
NO_x	nitrogen oxides
O_3	ozone
SF_6	sulfur hexafluoride
SiO_2	silica
SO_2	sulfur dioxide

Units of Measure

BTU British thermal unit
g/l gram per liter
mg/kg milligram per kilogram
mg/l milligram per liter
mg/m^3 milligram per cubic meter
PM2.5 particulate matter less than 2.5 microns
PM10 particulate matter less than 10 microns
ppb part per billion
ppm part per million
ppt part per trillion
TPY tons per year
µg/kg microgram per kilogram
µg/l microgram per liter
µg/m^3 microgram per cubic meter

1 Introduction and Overview

1.1 INTRODUCTION

Geology is a field of science that is defined as the study of the composition, structure, and history of Earth (Flint and Skinner 1974; Coble et al. 1987). Environmental Geology is an applied branch of geology that addresses the interaction between geology and humans (United States Geological Survey (USGS) 2024; Bennett and Doyle 1997).

Environmental geology is fundamentally important because it directly impacts every living individual every second of every day at every location on Earth where life exists. This is where the study of environmental geology directly connects with sustainability. Sustainability involves creating a balance between humans and nature that at this point does not exist (United States Environmental Protection Agency (USEPA) 2024). An important factor that adds complexity to this desired relationship is that the Earth is dynamic and changes constantly.

This book focuses on how Earth's geology impacts humans and how human behavior impacts the geology and natural environment of Earth. This three-way interactive relationship creates both negative and positive outcomes for Earth, humans, and nature. These outcomes are integrated with sustainability objectives to identify what actions are necessary to create a balance with nature instead of harming nature, if a balance actually exists. Table 1.1 lists three types of hazards that will be discussed in detail in this book. The first are geologic hazards, which are a measure of how geology impacts humans. The second are environmental hazards, which include atmospheric-related events and phenomena that impact humans. The third set are anthropogenic-induced hazards, which are those that affect nature and geology and originate from a human activity.

The first section of this book (Chapters 2 through 5) is dedicated to learning about geology and how geology impacts humans. Understanding geology is critical because the geologic environment immediately beneath our feet literally forms the foundation of human civilization and sustainability and influences all aspects of life on Earth.

The second section of the book (Chapters 6 through 10) is dedicated to learning how human behavior has negatively or positively influenced the geologic environment and Earth. This includes impacts to the air, water, and land through many anthropogenic activities that include agriculture, urban development, and industrial activities, including many forms of contaminants that humans have released into the environment and their behavior in nature.

This brings us to the final section of the book (Chapters 11 through 16), which examines how humans can improve and develop methods to protect human health and the environment through sustainability with a greater understanding of environmental geology. Over the past 50 years of learning and attempting to protect the environment, humans have learned a hard lesson that changing nature

DOI: 10.1201/9781003463153-1

TABLE 1.1
Types of Hazards Addressed

Geologic Hazards	Environmental Hazards	Anthropogenic-Induced Hazards
Earthquake	Hurricanes	Habitat destruction
Tsunami	Tornados	Natural resource exploitation
Volcanic eruption	Severe storms	Water cycle disruption
Landslide	Extreme precipitation	Deforestation
Coastal hazard	Flooding	Overfishing
Subsidence and sink holes	Drought	Overhunting and poaching
Erosion	Extreme heat	Invasive Species
Aeolian processes (dust)	Extreme cold	Species extinction
	Wildfire	Light pollution
	Lightning	Noise pollution
	Solar radiation	Ocean pollution
	Hail	Air pollution
	High winds	Water pollution
	Severe winter weather	Land pollution
	Marine heat wave	Chemical contaminants
	High-tide flooding	Agriculture
	Derecho	Land development
		Endangered species
		Pathogens
		Erosion
nl1		Particulate matter or dust
		Nuclear radiation
		Certain buildings
		Certain wind turbines
		Climate change

to suit our perceived needs has not worked out very well for ourselves, for Earth, and the living organisms with which we share this planet.

1.2 MAJOR THEMES OF THIS BOOK

The following sections provide a brief introduction to the central themes that are covered in this book and include the following:

1.2.1 Theme 1: Geology as the Foundation of Sustainability
1.2.2 Theme 2: Environmental Geology in the Urban Environment
1.2.3 Theme 3: Anthropogenic Impacts to the Natural World
1.2.4 Theme 4: Risk Mitigation Methods
1.2.5 Theme 5: Working with Nature Instead of Against Nature
1.2.6 Theme 6: Effectiveness of and Need for Improved Environmental Regulations
1.2.7 Theme 7: Measuring Risk and Progress through Sustainability Modeling

The themes of this book examine the importance of geology and its connections to sustainability. The themes also examine anthropogenic impacts and how these impacts affect all living organisms on Earth. Finally, the themes examine and explore risk mitigation alternatives that can be employed using geology to improve our relationship with nature and move toward a more sustainable planet. Each of the seven major themes of this book is briefly described below.

1.2.1 Theme 1: Geology as the Foundation of Sustainability

Each day when we walk outside, we are walking on a history book, yet most of us are unaware of this and are unaware of how this history book influences our lives. The arrangement, thickness, and composition of the soil and sediment layers just centimeters beneath our feet have a profound influence on our lives and all life on Earth. Locked away in those layers of strata beneath our feet is the history book that Earth scientists unlock and tell the story of Earth. The story may include evidence of earthquakes, volcanic eruptions, landslides, tsunamis, floods, erosion, paleoclimate, extreme atmospheric events, mass extinctions, contaminant pathways, and still other phenomena.

In addition, the soil and sediment layers beneath urban regions dictate where and how cities, roads, and structures are built. Most significantly, these soil and sediment layers are the meeting place between human civilization and the natural world. This contact is the proving ground and point where modifications to the natural world are most noticeable and impact not just humans, but all life on Earth. This contact also serves as the location where contaminants released into the environment begin their destructive journey. We have learned all too well that contaminants in their many forms have now impacted the entire biosphere and all living things on Earth.

Understanding the geology of urban regions is of the utmost importance, not just because of how geologic hazards affect humans but also how the geologic environment influences the behavior of contaminants. Therefore, we must understand our natural geologic world in at least the same scientific detail that we need to understand pollution. The reason is simple: they interact with each other.

For the Earth to be able to sustain a global human population of more than 8 billion people, which continues to increase daily, it is essential that we appreciate and comprehend environmental geology, especially in urban regions, at a level that is greater than most, if not all, scientific levels of inquiry and understanding. A high level of scientific inquiry is necessary because this is where we live, and immediately beneath our feet is the contact between civilization and the natural geologic world.

Much of our human history has been spent in developing ways to conquer nature in numerous ways, which has not turned out well for nature and the organisms with which we share the planet. To underscore this point, Earth scientists are now convinced that we have now moved the needle of geologic time into a new period called the **Anthropocene**. This is significant because the definition of a Geologic Age is a time period that affects the entire Earth and will be recorded in Earth history and cannot be erased. To support the human population on Earth, we must not continue to develop land, consume natural resources, and pollute at current levels. To achieve a sustainable living, vibrant, and diverse world instead of a world without humans, we need to diligently work cooperatively toward living in productive harmony with nature instead of trying to change nature. This means that we must understand and protect the air, the water, and the land. By protecting these three media, we can also protect the biosphere.

We shall examine our natural environment in urban areas and explore how nature influences human life. We will learn that we have control over the chemicals we use, and where and how we use them. Control over the geologic environment, however, is beyond our means. For example, we cannot stop earthquake or volcanic eruption from occurring. Therefore, we must understand the geologic environment where our urban areas are located and develop methods to minimize or eliminate the potential harmful effects of geologic hazards and contaminants upon human health and the environment.

A logical first step is through an *understanding* of urban geology, followed by an *evaluation* of the extent that a given urban area's geology may be at risk of a geologic hazard and influences the migration of contaminants. In addition, as water plays a critical role in assessing a region's sensitivity to natural hazard and contamination, the analyses performed require an understanding of water occurring at the Earth's surface and beneath.

1.2.2 Theme 2: Environmental Geology in Urban Areas

Throughout the world, the largest cities share a geologic environment upon which they are built that is dominated by unconsolidated sedimentary deposits and are located near water. Most of those sedimentary deposits are saturated with water very near the surface and function as sources of drinking water and/or as hydraulic connections to surface water and ecosystems, and ultimately, to the oceans. For nearly 11,000 years, humans have applied themselves toward changing their environment for their own benefit, protection, and enjoyment. In almost every situation, these activities have been conducted with a disregard and lack of understanding of the natural geological environment in which we live. Over the past few decades, and in some isolated instances in the last few hundred years, there has been a realization that perhaps a lack of understanding of environmental geology and attempts to modify or change our environment may not have been a wise choice.

Despite the availability of specific methods and procedures, the natural geological environment of most urban areas is not well understood. Some reasons include that they are (Rogers 2020):

1. developed and covered
2. unconsolidated sediments and are not rock
3. considered overburden
4. recent deposits and not deep in time
5. not of significant economic value (e.g., containing oil or minerals)
6. located on private land
7. complex
8. may be disturbed or modified by humans since deposition

The arrangement, thickness, and composition of the soil and sediment layers just centimeters deep beneath our feet have a profound influence on our lives and all life on Earth. Perhaps most significant to our civilization at this juncture is that this is the meeting place between human civilization and the natural world. This contact is the point at which pollutants released into the environment begin their destructive journey and could eventually negatively impact all living things on Earth. This is why we must understand our natural world in at least the same scientific detail that we need to understand pollution, because they interact with each other.

Scientific factors that control the severity of harm to the environment from contaminant releases are as follows: (1) the geologic and hydrogeologic environment, (2) the physical chemistry of the contaminants and amounts released, and (3) the mechanism in which the release occurs (Rogers 1996; Murray and Rogers 1999; Kaufman et al. 2005; Rogers 2018). There are several techniques available for investigating and managing the complexity of urban geology and water pollution (Rogers 2016a, 2016b). One important and critical concept to keep in mind when evaluating groundwater in urban areas is that the hydrologic cycle is modified and altered by human development (Table 1.2).

Despite the availability of specific methods and procedures, the environmental assessment of many urban areas can become a daunting task. This situation arises because the near-surface geologic deposits in urban areas are poorly understood because they are difficult to study, require in-depth study, are complex, have been anthropogenically disturbed, and exhibit high variability over short distances. Therefore, to achieve any level of success in mitigating environmental contamination and other hazards, it becomes a prerequisite to understand the geology, hydrology, and the fate and migration of contaminants within its specific geology (Rogers 2014).

Environmental problems faced in the typical urban setting can, in part, be attributed to the poor understanding of their subsurface geology. Lack of attention to this area of inquiry by the scientific community is one factor contributing to this knowledge gap. Geologists generally have not spent

TABLE 1.2
Water Bodies and Near-Surface Geology Near the Major Cities of the World

City	Location	Estimated Metropolitan Population (Millions)	Dominant Surficial Geology	Water Body
Tokyo	Japan	37.8	Unconsolidated: fluvial	Pacific Ocean
Shanghai	China	34.8	Unconsolidated: fluvial	Pacific Ocean
Jakarta	Indonesia	31.7	Unconsolidated: fluvial	Pacific Ocean
Delhi	India	26.4	Unconsolidated: fluvial	Yamuna River
Seoul	Korea	25.5	Unconsolidated: fluvial, lacustrine	Han River
Beijing	China	24.9	Unconsolidated: fluvial, lacustrine	Yongding River
New York City	United States	23.8	Unconsolidated: glacial, alluvial	Atlantic Ocean
Mexico City	Mexico	21.6	Unconsolidated: lacustrine	Lerma River
Sao Paulo	Brazil	21.2	Unconsolidated: fluvial, alluvial	Atlantic Ocean
Cairo	Egypt	20.5	Unconsolidated: fluvial, eolian	Nile River
Los Angeles	United States	18.7	Unconsolidated: fluvial, alluvial	Pacific Ocean
Moscow	Russia	16.9	Unconsolidated: fluvial, glacial	Moskva River
Istanbul	Turkey	15.2	Unconsolidated: fluvial, eolian, alluvial	Turkish Straits
London	England	14.2	Unconsolidated: fluvial	Thames River
Buenos Aires	Argentina	13.1	Unconsolidated: fluvial, alluvial	Atlantic Ocean
Paris	France	12.6	Unconsolidated: fluvial	Seine River
Rio de Janeiro	Brazil	12.3	Unconsolidated: fluvial, alluvial	Atlantic Ocean
Chicago	United States	9.8	Unconsolidated: fluvial, glacial, lacustrine	Lake Michigan
Johannesburg	South Africa	9.6	Unconsolidated: fluvial, lacustrine	Jukskei River
Riyadh	Saudi Arabia	7.7	Unconsolidated: eolian, alluvial	Wadi Hanifa River
Santiago	Chile	6.7	Unconsolidated: fluvial, alluvial, lacustrine	Pacific Ocean
Berlin	Germany	6.1	Unconsolidated: fluvial	Spree River
Toronto	Canada	5.9	Unconsolidated: fluvial, glacial, lacustrine	Lake Ontario
Sydney	Australia	5.0	Unconsolidated: fluvial, alluvial	Pacific Ocean

Source: United Nations. 2024a. *World Population Prospects. United Nations Department of Economic and Social Affairs. Population Division*. https://esa.un.org/unpd/wpp/data. (Accessed May 4, 2024).

very much effort or time studying shallow, unconsolidated geologic deposits beneath urban areas for environmental reasons because they (Rogers 2020):

- have only become important in the last 50 years with the onset of significant environmental regulations
- are covered by buildings and roads and other human-constructed surfaces
- are difficult to access because many desirable study locations are on private property
- have been disturbed and modified due to development
- are typically located on unconsolidated sediments of Quaternary age and were historically considered overburden and not as scientifically valuable as the bedrock located beneath the unconsolidated deposits
- are very complex and time-consuming to study
- involve issues that may take decades to be noticed by the general community as compared with other more sudden types of geologic hazards such as earthquakes, tsunamis, and volcanic eruptions.

In addition, the geology of most urban regions generally goes unnoticed by most residents and even the scientific community because it is not visually spectacular like the geology of many national parks. Most residents would be more inclined to learn about a region's geology and its underlying processes if their area was prone to earthquakes, such as parts of California, or volcanic eruptions, such as the Pacific Northwest or Hawaii. The geology beneath most urban regions affects our day-to-day lives in more subtle ways than an earthquake or volcanic eruption; however, its ultimate impact on the quality of our lives could be just as significant (Rogers 2014).

Urban areas functioned well before humans began to alter their natural flows of water, and now many are in a distressed state. Understanding what was once at a given location (site) and how it came into existence are therefore prerequisites for correctly analyzing current problems and prescribing an effective and sustainable remediation strategy. As more and more sites of human interference are analyzed, the overall picture of the linkages between the geology, biology, geography, and human interference becomes clearer. As more of these linkages are made, the geographic extent and impacts of contamination and other hazards in different geological environments become more evident, appropriate mitigation efforts can be more effectively implemented, and the methods for sustainable redevelopment and new development refined (Rogers 2018).

The focus of this theme is to learn how to identify areas within urban regions that are especially sensitive to geologic hazards and contamination and those areas where widespread hazard impact is less likely. To accomplish this task, an additional level of interpretation consisting of a comprehensive sensitivity analysis is added to the near-surface geologic maps of urban areas.

1.2.3 Theme 3: Anthropogenic Impacts to the Natural World

Now with more than 8 billion people on the planet, we are feeling the effects of our own actions by running out of habitable land, polluting the entire planet, and perhaps being responsible for the sixth major extinction in Earth's 4.54-billion-year history. Modifications to the environment have been conducted to suit our needs and to make our lives safer and more comfortable. To avoid feeling guilty and taking responsibility for our own actions, we have fooled ourselves and tried to justify the destruction of the natural environment by creating clever labels such as taming nature, Western expansion, progress, sustainable development, and falsely claiming that Earth was created just for us.

Although there is much we can be proud of for limiting pollutant effects through environmental regulation in the last 50 years, it clearly has not been enough, and our species is now at a critical crossroad. What we choose to do in the next few years to address not only climate change but also other adverse impacts to water, land, and other living species will have a profound impact on how we and other species survive in the future. One significant aspect of our technological advances in the last few centuries, which has introduced a multitude of pollutants and other effects, is that they have greatly outpaced nature's ability to evolve and adapt to human's destructive behavior. Creating culture and civilization has unleashed the opportunity of imaginative thought. Instead of focusing our efforts on individual survival, technological advancements have given us the opportunity and time so that we can explore how life on Earth has evolved since its origin billions of years ago. It has also made it possible for us to explore fundamental subatomic particles, explore and learn about our universe, and travel into space. Our curious nature and the power of our brains permit us the ability of awareness and self-reflection, consciousness, and questioning our existence. Now we find ourselves at a crossroad where our success as a species may no longer be an advantage for our future survival but may be leading us down a path where it threatens the existence of many living species on Earth, including our own.

We are the first generation feeling the effects of climate change and the last generation that can do something meaningful about it. Unfortunately, climate change is just one of many examples of how humans have impacted the environment on a global scale. Other impacts may not be in the public eye but are just as significant and include impacts to the land, oceans, surface water, and groundwater and have affected most all living species on Earth. This book examines how human behavior has negatively impacted the environment, especially in urban areas. This book also addresses the fundamentals of environmental geology in the context of how the natural environment has influenced and impacted humans through our brief history. Finally, by combining what we have learned, this book addresses how employing sustainability measures, with the assistance and knowledge of environmental geology, can greatly improve our relationship with Earth and other species with which we share the planet.

Many believe that we do not purposely treat our planet with malice and that much of the damage inflicted has been out of lack of knowledge or ignorance and that our labels for justifying our destructive behavior have an origin to avoid guilt and responsibility. Perhaps human impacts were best framed by a comment made in the 1970s, during the environmental movement in the United States, which stated, "we have met the enemy, and he is us" (Ohio State University 1972). If we do not learn for our past and continue our attempts to conquer and change nature, we will face the harsh reality that our actions will continue to have significant negative impact on Earth's atmosphere and climate, water, land, and threaten the survival of many species including our own.

Air, water, and land are essential for life on Earth. Life becomes threatened when any of these becomes contaminated or is modified with disregard for other species. In many instances, these actions result in the introduction of pollutants into the environment that degrade not just the local environment but can now be observed on a global scale and affect all life on Earth. Therefore, we must evaluate how these actions are impacting our planet and either eliminate, modify, or reduce the harmful impacts if we and other organisms are to survive. Understanding and protecting these three essentials for life and protecting living organisms form the fundamentals of environmental geology and sustainability.

We have control over the chemicals and other substances we use, and where and how we use them. We also have control on where and how to develop land so that we can live and grow food to sustain our population. Control over the geologic environment, however, is generally beyond our means. Therefore, we must understand the geologic environment where our urban areas are located and develop methods to minimize or eliminate natural risks (e.g., earthquakes, landslides, tsunamis) and the potential harmful effects from pollutants, which threaten human health and the environment.

Anthropogenic impacts on the natural world go beyond introducing pollution into the environment. Other significant impacts from agriculture, habitat destruction, natural resource exploitation, overpopulation, ocean heating, and still others. One of the most alarming and disturbing aspects of anthropogenic impacts on the natural world is the conclusion from the scientific community that biodiversity loss poses a global threat at the same level as the risks posed by climate change (United Nations 2024b). According to the World Wildlife Fund (2025) and the International Union on the Conservation of Nature (IUCN, 2024), hundreds of species of animals are now extinct directly due to humans in several different ways, including hunting, habitat destruction, poisoning, development, and pollution. According to the United Nations, more than 1 million species of animals and plants are threatened with extinction as a direct result of human activities worldwide (United Nations 2024a). The more than 200 scientists who participated in the study concluded that biodiversity loss poses a global threat at the same level as the risks posed by climate change (United Nations 2024c).

Such a collapse in biodiversity would wreak havoc on interconnected ecosystems of the planet, putting human communities at risk by compromising food sources, contaminating clean water and air, and eroding natural defense against extreme weather events such as hurricanes and floods. In addition, the study concluded that species extinctions will become increasingly common in the coming decades if some of the driving factors include continuing urban and agricultural development, deforestation, pollution, overfishing, plastic pollution, and poaching (United Nations 2024c), are not addressed.

1.2.4 Theme 4: Risk Mitigation Methods

The near-surface geologic environment has become very important in characterizing and remediating sites of environmental contamination and in evaluating risk to the environment and human health. The reason is simple: the near-surface geologic environment frequently acts as the migration pathway for contaminants to travel from a specific point of release to a point of ecologic or human exposure. In many cases, this pathway goes undetected because the contaminants migrate beneath the surface of the ground and cannot be observed or detected until it is perhaps too late. Risk mitigation methods for geologic hazards, along with extreme atmospheric hazards and environmental hazards generally confined to urban areas, all have a common mitigation chronology that begins with understanding the geology first. Then each hazard may have a unique set of risk mitigation methods depending on the hazard, specific geology, and the degree of human and environmental exposure to the risk.

The study of the near-surface geologic environment will become increasingly important in the future. The world's human population continues to increase and now exceeds 8 billion people and is increasing in urban areas at a rate greater than the human population as a whole. Nearly 3 billion people currently live in urban areas. By the year 2030, the human population in urban areas is projected to be nearly 5 billion or 60% of the 8.6 billion people on the planet (United Nations 2024a). The current level of population has placed enormous pressure on our environment, including the geologic environment, and the projected increases are likely to cause additional stress. Therefore, a thorough understanding of the subsurface geologic environment is critical to understand and protect the cities and places we live.

Mitigation strategies for geologic and anthropogenic-induced hazards follow a pattern that generally begins with constructing a geologic map that identifies and outlines the hazard. The process of developing a mitigation strategy to address the hazard typically follows a step-by-step process that includes the following (United Nations 2024d):

1. Collecting and Analyzing Date
2. Risk Assessment and History
3. Analyzing the Range of Geological Hazard Risk
4. Creating the Conceptual Site Model or Risk Map
5. Identifying Areas of Highest Risk
6. Developing an Early Warning System
7. Monitoring
8. Communicating Alerts and Warnings
9. Emergency Planning and Preparedness
10. Establishing an Evacuation Plan
11. Education and Awareness
12. Testing and Training
13. Emergency Supplies
14. Critical Response Effectiveness Review

A **hazard** refers to a natural, geologic, environmental, or anthropogenic-induced event. **Risk** is a measure of the magnitude of an event and the frequency of occurrence. Risk can be applied by evaluating the probability of a negative outcome or impact from a geologic or environmental hazard, an extreme atmospheric event, or exposure to a contaminant from an anthropogenic source. **Sensitivity** is a measure of how resilient a target population or ecological sector is to the hazard. To put it in terms that perhaps can be better understood, we do not always have control over the hazard, but we can lower our sensitivity to a negative outcome as a result of knowing and quantifying the risk. Equation 1.1 presents this concept in more mathematical terms whereby a geologic hazard may exist, but the hazard must be realized and must occur within a certain magnitude for there to be risk realized. The next variable is the sensitivity of anthropogenic or environmental exposure to the risk. The outcome is then a measure of the impact severity.

$$\text{Hazard} \times \text{Magnitude and Risk of Occurrence} \times \text{Sensitivity} = \text{Impact Severity} \quad (1.1)$$

Natural geological, environmental, and anthropogenic-induced hazards and associated impacts are of key concern to life on Earth. Understanding and monitoring these hazards through scientific study to identify and quantify risk potential are critical to improve risk reduction methods so that sustainability potential for survival is greatly improved (British Geological Survey (BGS) 2024).

Modifications to current practices combined with the way we choose to live can have a significant impact on improving the environment and considerably increase the effectiveness of sustainability measures. In addition, repurposing the land we already have developed can significantly reduce the need for future development, even with an expanding human population. Several examples and case studies demonstrate that once a high level of scientific understanding of environmental geology is achieved, sustainability measures are much more effective to the point where environmental stewardship can be pursued.

1.2.5 Theme 5: Working with Nature Instead of Against Nature

At least seven policy-oriented actions at the global level are required if working with nature in a sustainable way will be effective. These actions include (Rogers 2019, 2020):

- building a global community
- cohesive environmental policy, regulations, and enforcement
- controlling the human population
- modifying the human diet
- financial fairness and equity distribution
- modifying business models
- improving our relationship with nature

We will cover several subjects within this book that can be pulled together into a cohesive framework that can be used to implement a comprehensive sustainability program at any level with the end goal of environmental stewardship. The pursuit of sustainability involves many subject areas and levels. We will address actions required at the global level, national level and the individual and local level. Those subject areas that will be addressed include the following (Rogers 2019, 2020):

- Cohesive action from the global community
  - Global environmental regulations
  - Obtaining a sustainable human population
  - Modifying the human diet
  - Financial fairness and equity distribution
- Improving environmental regulation at national levels
- Understanding the natural setting of our urban and industrial areas
  - Protecting sensitive ecological areas
  - Smart urban development
  - Improving urban transportation
  - Improving building design
  - Incentivizing multi-family housing

- Knowing the chemicals that are used and how they might cause harm to the environment and humans if they are released
- Knowing what alternative chemicals are available
- How to reduce chemical use
- Preventing chemical releases
- Reducing energy use
- Using alternative sources of energy
- Minimizing solid waste, air emissions, and water discharges
- Reducing water use
- Minimizing stormwater runoff
- Repairing environmental damage to the extent possible
- Attaining environmental stewardship

1.2.6 Theme 6: Effectiveness of and Need for Improved Environmental Regulations

To know where we are heading into the future as a species, we often must reflect back at our history for perspective. From a historical point of view, our treatment of the environment has not been kind. As recently as the early portion of the 20th century, environmental awareness and legislation were generally lacking not only in the United States but also worldwide. There were a few international agreements that primarily focused on boundary waters, navigation, and fishing rights along shared waterways between countries. However, they ignored pollution and ecological issues. In fact, the most convenient and least costly method of waste disposal up until the middle of the 20th century was "up the stack or down the river" (Haynes 1954).

The publishing of the Rachel Carson book, *Silent Spring* in 1962 described the environmental effects of polychlorinated biphenyls (PCBs) on birds and catapulted the environmental movement in the United States into the public eye (Carson 1962). The 1960s and 1970s are generally described as the age of environmentalism in the United States. This is when the United States realized that environmental degradation had significantly affected the air and water quality of many urban areas, especially the air in Los Angeles and the water in Lake Erie, which is one of the Great Lakes. Starting on Earth Day in 1971, the Ad Council and Keep America Beautiful campaign aired public awareness commercials on television depicting a Native American shedding a tear when looking out over the polluted landscape of the United States. The commercial stated, "people start pollution, people can stop pollution" (United States Advisory Council 2025).

Although laws focused on protecting human health or the environment can be traced back more than 2,000 years, it took the United States in the 1970s with the passage of the most significant environmental regulations and laws to set the example for the rest of the world (Hahn 1994).

Environmental regulations of the United States and the world are largely based on reactions to an incident that caused enormous harm to human health and the environment. A few of the many examples of incidents that resulted in the enactment of environmental regulations include Love Canal in New York, Times Beach in Missouri, Bhopal, India, the fire on the Cuyahoga River near Cleveland, the Exxon Valdez oil spill, and the recent incident on the Deep Water Horizon oil drilling platform in the Gulf of Mexico.

We have to accept that contamination does not respect boundaries. Contamination released in one country often will affect another country and sometimes the whole planet if given enough time. Therefore, we must always pay attention to the environmental fundamentals, which are as follows (Rogers 2019):

1. Protect the air
2. Protect the water
3. Protect the land

Factors that influence the effectiveness of environmental regulations vary from country to country and within each country as well. The United States is no different. Factors that influence the degree to which a country has effective environmental regulations include the following (Bates and Ciment 2013):

- Political will
- Environmental tragedies, harm to the environment, and lessons learned
- Cost
- Geography and climate, which is influenced by geology
- Enforcement
- Incentives
- Lack of overriding social factors such as war, political structure, greed, poverty, hunger, lack of infrastructure, educational awareness, and corruption

A few of the most important conclusions concerning environmental regulations include the following (Rogers 2019):

- Each country has environmental regulations
- No two countries are exactly the same
- Each country has a unique set of attributes that influence the effectiveness of environmental protection that include climate, geography, geology, social and economic concerns, and politics
- The platform of environmental regulations of each country is similar to that of the United States or the European Union
- Although some countries are doing a better job than others, each country struggles with protecting human health and the environment for various reasons, two of which are universal and include a growing population and urban expansion
- Every country has been significantly affected by pollution of the air, water, and land

On Earth, the largest polluting activity is agriculture. The largest polluters of air are motor vehicles that include automobiles, diesel trucks, trains, buses, marine vessels, and aircraft. The largest polluters of water worldwide are biological pollutants, largely from human waste and from agricultural activities that include pesticides and herbicides, fertilizers, and erosion (United Nations 2024e).

The final resting place for the vast majority of the pollution of the world is the oceans.

International organizations such as the World Health Organization (WHO) and the United Nations have begun to assess on a global scale, the quality of freshwater and the air on Earth. According to WHO (2013, 2025a) and the United Nations (2024f, 2024g), approximately 2.8 billion humans lack access to basic sanitation and improved drinking water. While this number is improving, it still represents nearly 30% of the planet's human population. In addition, according to WHO (2025b), 9 out of 10 people on Earth breathe air containing high levels of pollution. The United Nations estimates that most land pollution is caused by agricultural activities, such as grazing, use of pesticides, fertilizers, irrigation, plowing, and confined animal facilities (United Nations 2024f).

Some of these actions have affected an individual site or location, and others have affected the entire planet. This book explores how these actions have molded and influenced the development of environmental regulations worldwide and identifies and explores many new approaches that can be undertaken to improve our environment. These improvements and modifications range from the individual to the global level through several avenues that include education, science-based assessments, improving existing environmental regulations, and enacting new environmental regulations with a goal of achieving a human population that is sustainable with Earth.

1.2.7 Theme 7: Measuring Risk and Progress Through Sustainability Modeling

Sustainability is complex and will require significant effort at the global, national, local, and individual level. The rest of this chapter will be dedicated to what actions are necessary at the local and individual level. Many believe that local and individual-level actions are most important because, in essence, it will take cooperation and effort from every individual on Earth to achieve a global community that is sustainable.

Sustainability involves much more than addressing and reducing climate change impacts. As Chapter 9 discusses, many scientists believe that species extinction is of greater concern at this moment. Challenges that must be addressed are impacts and include the following (Rogers 2023):

- Climate change
- Species extinction
- Decline in species abundance
- Invasive species
- Pollution and contamination of the air, land, and water
- Ocean pollution
- Air, water, and land pollution
- Deforestation
- Agriculture practices
- Erosion
- Habitat destruction
- Land development

Challenges exist between countries concerning sustainability and include differences in power and responsibility, environmental and economic concerns, population, and cultural and religious differences that must be addressed before sustainability can proceed on a path to become an effective reality on a global level (United Nations 2024h, 2025).

A sustainability model for any location in the world may seem impossible because of the sheer number of potential variables that greatly increase the complexity and because of the many differences between countries. However, this is just not the case. Sustainability is defined as creating and maintaining conditions under which humans can exist in productive harmony to support present and future generations (United Nations 2025; USEPA 2025). The term "productive harmony" applies to the environment. Therefore, sustainability can be reduced to three variables that include (1) understanding the environment, (2) understanding how aspects of human activities impact the environment, and (3) developing and implementing measures to mitigate unacceptable risk (Rogers 2019, 2020).

Understanding the geological, hydrological, and ecological environment is the first step in building a sustainability model and is represented by the general term **impact severity**. Impact severity is obtained from Equation 1.1. The second step is evaluating aspects of human behavior that affect the environment through either or positive outcomes. The third variable is evaluating the effectiveness of risk reduction measures. An equation for sustainability is created by combining these fundamental concepts, through which a sustainability index is the output (see Equation 1.2). The Sustainability Index represents a measure of sustainability for any particular location with a higher value representing increased risk and potential harm to human health or the environment and, therefore, less sustainable (Rogers 2020):

$$\frac{\text{Impact Severtiy} \times \text{Behavioral Aspects}}{\text{Risk Reduction Measures}} = \text{Sustainability Index} \tag{1.2}$$

The sustainability index can be scaled to evaluate the environmental health of a single location, household, or a whole ecosystem depending on system inputs and evaluation criteria described in the following sections.

As the focus of this book is evaluating geologic and environmental hazards primarily affecting urban areas, the focus of the sustainability model example presented is on exposure to environmental contaminants that are released into the environment from anthropogenic sources. Exposure to contaminants in urban areas was chosen because, according to WHO (2025c, 2025d) and the United Nations (2024e), exposure to environmental contaminants affects every human and is responsible for the premature deaths of millions of humans every year.

1.3 SUMMARY AND CONCLUSION

To be successful at achieving a sustainability goal of existing with nature will require cooperation not often experienced in our history. Humans must respond and must act individually, as a community, and globally to more fully understand the impacts made to our planet and act diligently to limit the negative effects and restore ecosystems where possible. We are the first generation to experience the effects of climate change and the last generation that can do something meaningful about it. These actions must be undertaken before it is too late for the survival of many species, including our own.

Over the past 50 years of learning and attempting to protect the environment, humans have learned a hard lesson that changing nature to suit our perceived needs has not worked out very well for ourselves, for Earth, and the living organisms with which we share this planet.

REFERENCES

Bates, C. G., and Ciment, J. 2013. *Encyclopedia of Global Social Issues*. M.E. Sharpe Publishers. New York, NY. 1450pp.

Bennett, M. R., and Doyle, P. 1997. *Environmental Geology; Geology and the Human Environment*. John Wiley & Sons. New York, NY.

British Geological Survey (BGS). 2024. *Global Geological Risk*. www.bgs.ac.uk/geology-projects/global-geological-risk. (Accessed March 11, 2024).

Carson, R. 1962. *Silent Spring*. Houghton Mifflin. Boston, MA.

Coble, C. R., Murray, E. G., and Rice, D. R. 1987. *Earth Science*. Prentice-Hall Publishers. Englewood Cliffs, NJ. 502pp.

Flint, F. F., and Skinner, B. J. 1974. *Physical Geology*. John Wiley & Sons. New York, NY. 426pp.

Hahn, R. W. 1994. United States Environmental Policy: Past, Present and Future. *Natural Resource Journal*. Vol. 34, no. 1, pp. 306–348.

Haynes, W. 1954. *American Chemical Industry – A History*. Vols. I–IV. Van Nostrand Publishers. New York, NY.

International Union on the Conservation of Nature (IUCN). 2024. *Extinct Species in North America*. www.iucn.org/extinct-species. (Accessed April 8, 2024).

Kaufman, M. M., Rogers, D. T., and Murray, K. S. 2005. An Empirical Model for Estimating Remediation Costs at Contaminated Sites. *Journal of Water, Air and Soil Pollution*. Vol. 167, pp. 365–386.

Murray, K. S., and Rogers, D. T. 1999. Groundwater Vulnerability, Brownfield Redevelopment and Land Use Planning. *Journal of Environmental Planning and Management*. Vol. 42, no. 6, pp. 801–810.

Ohio State University. 1972. *University Libraries. We Have Met the Enemy and He Is Us*. www.library.osu.edu/sitewe=have-met-the-enemy-and-he-is-us. (Accessed May 19, 2024).

Rogers, D. T. 1996. *Environmental Geology of Metropolitan Detroit*. Clayton Environmental Consultants. Novi, MI. 152pp.

Rogers, D. T. 2014. Scientists Call for a Renewed Emphasis on Urban Geology. *American Geophysical Union*. Vol. 95, no. 47, pp. 431–432.

Rogers, D. T. 2016a. Next Generation of Urban Hydrogeologic Investigations. *Journal of the Italian Geological Society*. Vol. 39, no 1, pp. 349–353.

Rogers, D. T. 2016b. Scientific Advancements that Improve the Conceptual Site Model in Urban Hydrogeological Site Investigations. In: *35th International Geological Congress*. Paper 3019. Cape Town, South Africa. London, United Kingdom: International Association of Hydrogeologists.

Rogers, D. T. 2018. Derivation of a Comprehensive Environmental Risk Model for Urban Groundwater Protection. In: *International Association of Hydrogeologists Congress*. Vol. 1. Daejeon, Korea. London, United Kingdom: International Association of Hydrogeologists.

Rogers, D. T. 2019. *Environmental Compliance and Sustainability: Global Challenges and Perspectives*. CRC Press. Boca Raton, FL. 543pp.

Rogers, D. T. 2020. *Urban Watersheds: Geology, Contamination, Environmental Regulations, and Sustainability*. CRC Press. Boca Raton, FL. 606pp.

Rogers, D. T. 2023. *Handbook of Environmental Compliance: Sustainability and Future Environmental Regulations*. Vol. 4. CRC Press. Boca Raton, FL. 548pp.

United Nations. 2024a. *World Population Prospects. United Nations Department of Economic and Social Affairs. Population Division*. https://esa.un.org/unpd/wpp/data. (Accessed May 4, 2024).

United Nations. 2024b. *UN Report: One million Species at Risk of Extinction*. www.un.org/sustainable/one-million-species-at-risk-of-extinction. (Accessed April 7, 2024).

United Nations. 2024c. *UN Report: Nature's Dangerous Decline "Unprecedented"; Species Extinction Rates Accelerating*. www.un.org/sustainabledevelopment/species/extinction. (Accessed April 7, 2024).

United Nations. 2024d. *Computer Modeling for Disaster Readiness*. www.un.org/disaster-preparedness. (Accessed April 21, 2024).

United Nations. 2024e. *Human Development Report*. United Nations Development Programme. www.hdr.undr.org/sites/2017. (Accessed May 4, 2024).

United Nations. 2024f. *Global Drinking Water Quality Index and Development and Sensitivity Analysis Report*. UNEP Water Programme Office. Burlington, ON, Canada. 58pp.

United Nations. 2024g. *World Air Pollution Status*. United Nations News Center. New York, NY. www.un.org/sustainabledevelopment/2016/09. (Accessed May 4, 2024).

United Nations. 2024h. *What Is Sustainable Development*. www.un.org/sustainabledevelopment. (Accessed April 15, 2024).

United Nations. 2025. Global Sustainability Goals. www.undp.org/content/undp/en/home/sustainable-development-goals.html. (Accessed January 28, 2025).

United States Advisory Council. 2025. Pollution: It's a Crying Shame. www.adcouncil.org/ur-campaigns/the-classics/pollution-keep-america-beautiful-iron-eyes-cody. (Accessed January 28, 2025).

United States Environmental Protection Agency (USEPA). 2025. Sustainability. www.epa.gov/sustainability (Accessed January 28, 2025).

United States Geological Survey (USGS). 2025. Glossary of Terms. www.usgs.gov/scientific-collections/glossary-terms. (Accessed January 28, 2025).

World Health Organization. 2013. Water Quality and Health Strategy 2013–2020. www.who.int/water_sanitation_health/dwq/en/. (Accessed January 28, 2025).

World Health Organization. 2025a. Ambient Air Pollution Database. www.who.org/ambient-air-database. (Accessed January 28, 2025).

World Health Organization. 2025b. *Nine Out of Ten People Breathe Unhealthy Air*. Geneva. Switzerland. www.who.int/news-room/detail/02-02-2018. (Accessed January 28, 2025).

World Health Organization (WHO). 2025c. Drinking Water and Pollution. www.who.int/news-room/fact-sheets/drinking-water-pollution. (Accessed January 28, 2025).

World Health Organization (WHO). 2025d. Land Pollution. www.who.int/news-room/fact-sheets/land-pollution. (Accessed January 28, 2025).

World Wildlife Fund (WWF). 2025. Average Decline in Wildlife Populations Since 1970. Living Planet Report 2024. www.worldwildlife.org/livingplanetreport. (Accessed January 28, 2025).

Part I

Fundamentals of Environmental Geology

2 Introduction to Natural Earth Processes and Human Development

2.1 INTRODUCTION

Geology is the science dedicated to the study of the history, structure, and composition of the Earth (Flint and Skinner 1974). The study of anthropogenic or human impacts on the geologic environment is termed **environmental geology** (Bennett and Doyle 1997). For the most part, the study of geology has been conducted for purely scientific and academic purposes, natural resource exploitation, and geologic hazard evaluation and avoidance. Only recently has the geologic environment in urban areas become of interest to geologists and other scientists in evaluating and documenting the extreme upper portion of the lithosphere defined as that portion affected by human activities.

Our efforts in this chapter focus on natural geologic processes. We begin with the age of the Earth and our solar system, then discuss the three aspects that encompass the definition of geology, which are the composition, history, and structure of the Earth. Included in these three aspects of geology, we will also examine important topics such as the formation of rocks, geologic principles, and plate tectonics. The last portion of this chapter will address human development which will introduce the reader to our examination of the increasing impact humans have placed on the Earth system.

2.2 GEOLOGY OF URBAN AREAS

The near-surface geologic environment of urban areas has become very important in characterizing and remediating sites of environmental contamination. Exposure to chemical contamination may place increased risk to our health and adversely affect the natural environment as well. In addition, natural geologic and environmental hazards may exist that may cause extreme harm but go unnoticed because of the infrequent nature of an event. In addition, chemical contamination may also go unnoticed because the migration pathway many contaminants take are located beneath the surface of the ground, and it cannot be observed or detected until it is perhaps too late.

The geology of most urban regions goes unnoticed by most residents and even the scientific community. One reason is that it is not visually spectacular and draws our attention like the geology of many national parks. Most residents would be more inclined to learn about a region's geology and its underlying processes if it were prone to earthquakes, such as parts of California, or volcanic eruptions, such as the Pacific Northwest, Hawaii, Iceland, Italy, Japan, Indonesia, and other parts of the world. The geology beneath most urban regions may not affect our day-to-day lives the majority of the time, but its ultimate impact on the quality of our lives could be just as significant (Rogers 2020).

DOI: 10.1201/9781003463153-3

Water has played a crucial role in shaping our urban areas. Understanding what was once at a given location or site and how it came into existence are prerequisites for correctly analyzing current problems, and prescribing an effective and sustainable strategy for the future. As more and more sites of human impacts are analyzed, the overall picture of the linkages between the geology, hydrology, biology, geography, and human impact becomes clear. As more of these linkages are made, the geographic extent and impacts in different geological environments become evident, and appropriate mitigation methods can be effectively implemented. These actions can have positive inputs for sustainable redevelopment and new development so that environmental stewardship can be pursued (Rogers 2020, 2023; Rogers and Rogers 2022).

2.3 AGE OF THE EARTH

To date, geologists and other scientists have not been able to determine the exact age of the Earth. This is because Earth's oldest rocks may have been destroyed and recycled by a process termed **plate tectonics**, which will be discussed later in this chapter. If there are any rocks that have survived from the time Earth formed, they have not been discovered. Nevertheless, geologists have been able to determine the probable age of the solar system and calculate an age for the Earth by assuming that the Earth formed at the same time as the rest of the solar system (United States Geological Survey (USGS) 2024a).

The age of the Earth is estimated to be 4.54 billion years (USGS 2024a; NASA 2024a). The age of the Earth is estimated through measurements of decay rates of radioactive isotopes, such as potassium and argon, which occur naturally in rocks and minerals. Decay rates are measured as half-lives which may be greater than 700 million years. These measurements of decay rates, which are firmly grounded in physics and are known collectively as radiometric decay rates, are used to measure the last time that a particular rock being assessed was either melted or disturbed sufficiently to re-homogenize its radioactive elements (USGS 2024a). The value of 4.54 billion years for the age of the Earth is based on measurements from iron meteorites, specifically the Canyon Diablo meteorite. However, more than 70 iron meteorites have been measured, and they present an age range from 4.53 to 4.58 billion years, with the preferred age at 4.54 billion, as estimated by measuring the decay rate of lead-207 to lead-206 and the radioactive decay rate of uranium-235 and uranium-238 (USGS 2024a).

The oldest rocks found on the moon have been dated from between 4.4 and 4.5 billion years. The oldest rocks found on Earth thus far have exceeded 3.5 billion years. Rocks at least as old as 3.5 billion years have been discovered on each continent. The oldest rocks found on Earth date to 4.03 billion years and have been termed the **Arcasta Gneisses** and are located in northwestern Canada near Great Slave Lake (USGS 2024a). Other rocks of substantial age have been discovered in western Greenland (3.7 to 3.8 billion years old), Africa (3.4 to 3.5 billion years old), western Australia (3.4 to 3.6 billion years old), and in the United States in Minnesota and northern Michigan (3.5 to 3.7 billion years). In addition, mineral grains belonging to the mineral zircon have been radiometrically estimated to be 4.4 billion years old (USGS 2024a).

An interesting fact concerning the oldest rocks found on Earth is that none of the oldest rocks originated from a primordial crust but are either lava flows or sediments deposited in shallow water. This has been used to support that the age of the Earth must be older (USGS 2024a).

The estimated age of the Earth at 4.54 billion years and our solar system is consistent with current estimates of 11 to 13 billion years for the Milky Way galaxy, which is based on age dating of globular cluster stars and the estimated age of the universe at 12 to 15 billion years, based on the recession of distant galaxies (USGS 2024a; NASA 2024b). Based on cosmic microwave background measurements, the age of the universe comes out to be 13.77 billion years (NASA 2024b).

2.4 COMPOSITION OF THE EARTH

In this section, we will discuss the composition of the Earth from an Earth science point of view. First, we will discuss a few overall facts of the Earth to provide context. Second, we shall discuss minerals and rocks and how rocks form. Then we move to the history of the Earth and, finally, we will discuss the structure of the Earth's crust termed **lithosphere**.

2.4.1 The Solid Earth and Its Composition

The Earth is approximately 12,750 km in diameter and is composed of three main layers that include an outer portion termed the **crust** *or* **lithosphere**, the next layer termed **mantle**, and an inner layer termed the **core**. The weight of the Earth is 5,974,000,000,000,000,000,000,000 kg. The weight of the Earth can also be expressed as 5.972×10^{24} kg. The density of the Earth is 1.217 kg/m^3. The volume of the Earth is 5.9722×10^{10} km^3 (NASA 2024a).

A good example to diagrammatically depict the Earth is to compare it to a boiled egg. An outer crust, next is the egg white which can represent the mantle, and a yoke that represents the core. The crust or lithosphere is the rigid outermost layer and is very thin compared to the other two layers. Beneath the oceans, the crust varies very little in thickness, generally only about 5 km in thickness. The thickness of the crust beneath the continents is much more variable and may average 30 km in thickness. Beneath large mountain ranges, such as the Alps or the Sierra Nevada, the thickness of the crust can be as thick as 100 km (USGS 2024b).

Below the crust is the **mantle**, a dense, hot layer of semi-solid rock approximately 2,900 km thick. The mantle is denser than the crust and is composed of more iron, magnesium, and calcium than the crust. The mantle is hotter and denser because the temperature and pressure increase with depth. At the center of the Earth is the core, and the core is twice as dense as the mantle because its composition is more metallic and made up of iron and nickel alloy. The Earth's core is made up of two parts, one part being a liquid outer core that is 2,200 km thick and an inner core that is approximately 1,250 km thick. As the Earth rotates, the liquid outer core spins, which creates the Earth's magnetic field (USGS 2024b). Figure 2.1 shows a vertical section of the Earth showing each major layer.

2.4.2 Composition of the Atmosphere

Above the ground surface is the atmosphere. The atmosphere is a layer of gas and suspended solid particles extending from the surface up many thousands of miles, becoming increasingly thinner with height, and is held in place by the Earth's gravity (National Oceanic and Atmospheric Administration (NOAA 2024a)). The atmosphere surrounds the Earth and holds the air we breathe. The atmosphere protects us and all living things from outer space, and holds moisture, gases, and solid dust particles. In short, the atmosphere is a protective bubble. The atmosphere is composed of several gases, with the top four being nitrogen, oxygen, argon, and carbon dioxide (see Table 2.1). Nitrogen is the most abundant element in the atmosphere and is required by life forms to produce proteins. Oxygen is required by life forms for respiration. Carbon dioxide is used by plants which produce oxygen. Carbon dioxide also acts as a blanket and prevents the escape of heat into outer space (NOAA 2024a).

There are five basic layers of the atmosphere. Each of the layers is bounded by "pauses" where the greatest changes in thermal characteristics, chemical composition, movement, and density occur. The **exosphere** is the outermost layer and extends from approximately 600 to 10,000 km (375 to 6,200 miles) above the surface of the Earth. In this layer, atoms can escape from Earth's gravity and can drift off into space (NOAA 2024b). The thermosphere is located immediately below the exosphere and is located from 85 to 600 km (53 to 375 miles) above the surface of Earth. Incoming

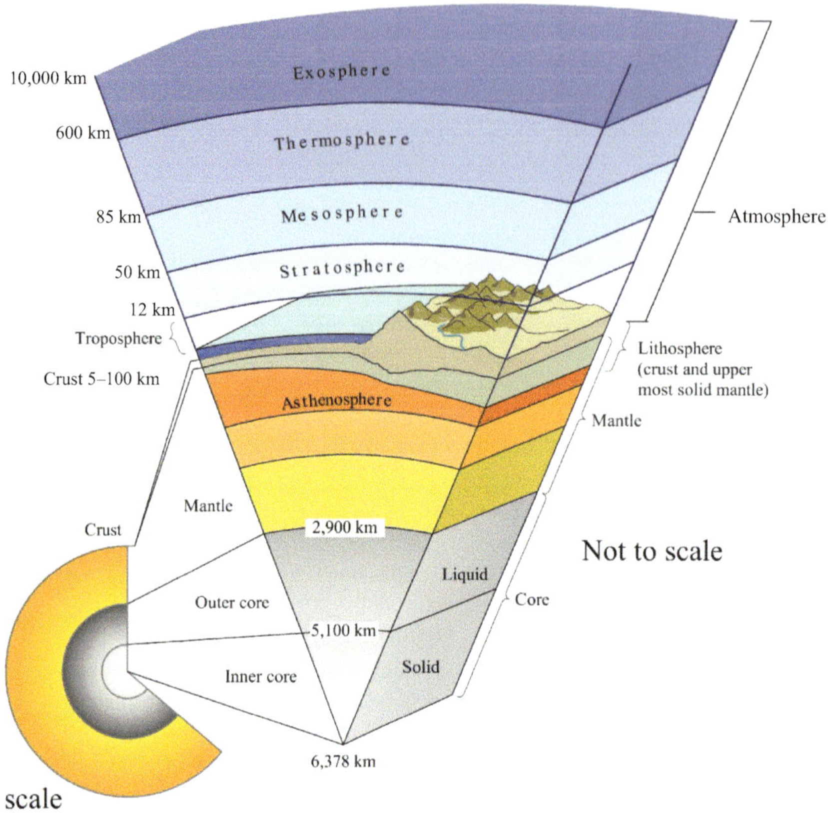

FIGURE 2.1 Cross section of the Earth showing interior and exterior. (Modified from Kious, W. J., and Tilling, R. I. 1996. *The Dynamic Earth: The Story of Plate Tectonics*. United States Geological Survey, United States Government Printing Office. Washington, DC.)

TABLE 2.1
Composition of the Atmosphere

Gas	Chemical Symbol	Mean Molecular Weight (m/mol)	Concentration Parts per Million by Volume(PPMv)
Nitrogen	N_2	28.013	780,840
Oxygen	O_2	31.999	209,460
Argon	Ar	39.948	9,340
Carbon dioxide	CO_2	44.010	426.4
Neon	Ne	20.180	18.18
Helium	He	4.003	5.24
Methane	CH_4	16.043	1.774
Krypton	Kr	83.798	1.14
Hydrogen	H_2	2.106	0.56
Nitrous oxide	N_2O	44.012	0.32
Xenon	Xe	131.293	0.09
Ozone	O_3	47.998	0.01 to 0.10

Source: National Oceanic and Atmospheric Administration (NOAA). 2024a. *Introduction to the Atmosphere*. www.noaa.gov/jetstream/atmisphere. (Accessed September 29, 2024).

high-energy ultraviolet and X-ray radiation from the sun begins to be absorbed by the molecules in the thermosphere. Because of this absorption, the temperature of the thermosphere increases with height, from as low as −120°C (−184°F) to as high as 2,000°C (3,600°F). However, despite these high-temperature levels, this layer of the atmosphere would still feel very cold to our skin. The high temperature indicates the amount of energy absorbed by the molecules, but with so few molecules in this layer, the total number would not be sufficient to heat our skin (NOAA 2024b).

The next layer is called the **mesosphere**. The mesosphere extends from approximately 50 to 85 km (31 to 53 miles) above the surface of the Earth. As the altitude decreases, the gases in this layer become more dense due to the influence of gravity. The next layer of the atmosphere is termed the **stratosphere**. The stratosphere extends from 7 to 20 km (4 to 12 miles) above the surface of Earth. The stratosphere is composed of approximately 19% of the total gases of the atmosphere but has very little water vapor. In the stratosphere, the temperature increases with height. Heat is produced in the process of the formation of ozone in this layer. The temperature increases from approximately −51 to −15°C (−60 to 5°F). The increase in temperature with altitude means that convection is prevented. The bottom of the stratosphere is easily observed as the anvil-shaped tops of cumulonimbus clouds during summer storms (NOAA 2024b).

The lowest layer of the atmosphere is termed the **troposphere**. Nearly all the weather phenomena we experience occur in this region. The troposphere begins at the surface of the Earth and extends to the bottom of the stratosphere. Figure 2.1 shows the different layers of the atmosphere.

2.4.3 Minerals and Rocks

The fundamental building blocks of the Earth's crust are rocks. A **rock** is simply composed of minerals (Flint and Skinner 1974). A mineral is a naturally occurring solid formed through geological processes that has a characteristic chemical composition, a highly ordered atomic structure, and specific physical properties (Dana 1985). Minerals range in composition from pure elements to very complex silicates with thousands of known forms. Rocks do not require a specific chemical composition but have ranges of compositions of typically two or more different minerals. For instance, igneous rocks differ in composition based on the chemical makeup of their parent magma and on their rate of cooling. The composition of sedimentary rocks differs based on the parent material and the methods of transport and deposition.

Understanding and evaluating rocks require knowledge of their composition, and how they were deposited at a given location. On Earth, the dominant physical force that builds and destroys rock is plate tectonics. The Earth is composed of what are termed **lithospheric plates**. These lithospheric plates (crust and uppermost mantle) move on top of the upper mantle called the **asthenosphere** (see Figure 2.1). Their movement is responsible for most earthquakes, volcanoes, sea-floor spreading, mountain building, and ocean trench formation (Stanley 1999). The energy source driving the movement of the plates is believed to be heat from convection currents originating deep within the interior of the Earth. We shall return to the concept of plate tectonics later in this chapter when we discuss the structure of the Earth in greater detail.

Plate tectonics is responsible for, or plays a significant role in, the formation, deformation, and weathering of most rocks on Earth. The three main processes that lead to the formation of rocks include (1) the cooling and solidification of magma, (2) the deposition of sediment, and (3) extreme temperatures and pressure. Igneous, metamorphic, and sedimentary rock, as described below, all originate from these processes. The primary classification of rocks recognizes three categories (igneous, metamorphic, and sedimentary) based on their method of formation.

2.4.3.1 Igneous Rocks

Igneous rocks are formed by the solidification of molten material either beneath the Earth's surface (plutonic igneous rocks) or at the surface (volcanic igneous rocks) (American Geological Institute

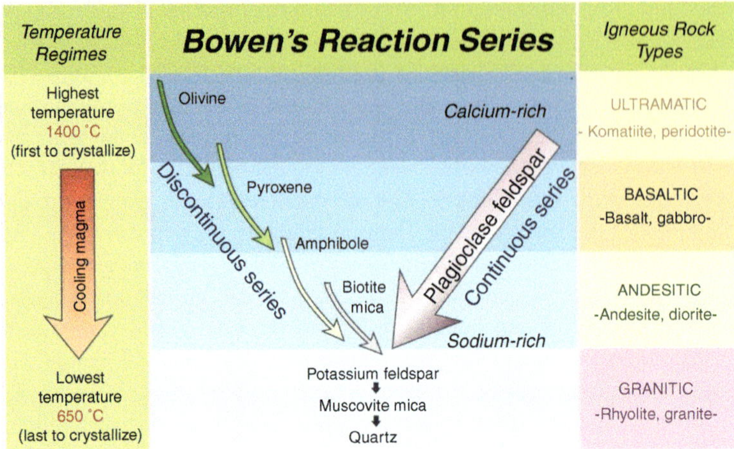

FIGURE 2.2 Bowen's Reaction Sequence showing intrusive rock formation.(Modified from United States Geological Survey (USGS). 2023. *Igneous Rocks.* www.usgs.gov/factswhat-are-igneaous-rocks. GMNA. (Accessed September 26, 2024).)

(AGI) 1962). Examples of plutonic (intrusive) igneous rocks include granite, pegmatite, diorite, gabbro, and peridotite; volcanic (extrusive) igneous rocks include andesite, basalt, obsidian, and rhyolite. Each type of igneous rock is named based on its mineralogic and chemical composition.

A fundamental concept of igneous rock formation and its mineralogy is known as **Bowen's Reaction Sequence**. Bowen's Reaction Sequence describes the crystallization of minerals from molten magma as the temperature of the magma decreases. In basic terms, as magma cools and minerals crystalize, the composition of the remaining magma changes. Two main branches of mineralization, the discontinuous branch and the continuous branch, are described in Bowen's Reaction Sequence. The discontinuous branch represents the minerals that crystallize at distinct temperature intervals from olivine to pyroxene, amphibole, and biotite mica. The continuous branch represents minerals that form continuously as temperature decreases and include plagioclase feldspar anorthite to oligoclase. The last minerals to crystallize are potassium feldspar, muscovite mica, and quartz (AGI 1962). Figure 2.2 depicts the formation of igneous rocks through Bowen's Reaction Sequence.

2.4.3.2 Metamorphic Rocks

Metamorphic rocks are formed by applying extreme temperature and pressure to existing rocks that re-arrange their structure and mineralogy to create a new rock (AGI 1962). Examples of metamorphic rocks include gneiss (formed from the metamorphosis of granite); quartzite (metamorphosed from sandstone); and marble (metamorphosed limestone). Other examples include amphibolite, hornfels, and anthracite.

2.4.3.3 Sedimentary Rocks

Sedimentary rocks are formed from the mechanical weathering or erosion of pre-existing rocks or from dissolved material that precipitates from solution. They are formed on or near the surface of the Earth. The process of sedimentary rock formation contrasts with how intrusive igneous or metamorphic rocks are formed, which are much deeper within the Earth and generally occur at much lower temperatures. The most important geological processes that lead to the creation of sedimentary rocks are erosion weathering, dissolution, precipitation, and lithification. Erosion and weathering include the effects of wind and rain, which slowly break down existing rocks into smaller rocks. Erosion and weathering transform boulders and even whole mountains into sediments, such as sand

and clay. Dissolution is a form of weathering termed **chemical weathering**. Chemical weathering or dissolution is achieved by water that is slightly acidic and slowly dissolves rock over relatively long periods of time. These processes create the raw material for a new sedimentary rock formation.

Precipitation and lithification are two sedimentary processes that build new rocks or minerals. Precipitation is the formation of rocks and minerals from chemicals that precipitate from water. For example, as a lake dries up over many thousands of years, it leaves behind mineral deposits There are two types of sedimentary rocks (Pettijohn 1975):

- *Clastic:* This is composed of grains of fragments of rocks and minerals that have been eroded, transported, deposited, and eventually become cemented together to form a rock. Examples of clastic sedimentary rocks include breccia, conglomerate, sandstone, siltstone, and shale.
- *Chemical or biochemical:* This is composed of minerals that have been precipitated from water by inorganic processes or biological processes. For example, coal is formed from plant debris. Other examples of chemical sedimentary rocks include limestone, dolomite, halite, and gypsum.

Common examples of clastic sedimentary rocks include conglomerate, sandstone, siltstone, and shale. Some of the more common chemical and biochemical sedimentary rocks include limestone, dolostone, coal, evaporite halite, and gypsum. When coal undergoes significant changes through exposure to high temperatures and pressures it becomes a high-grade form of coal called **anthracite**, considered to be a metamorphic rock (Pettijohn 1975).

2.4.4 Rock Formations and Geologic Maps

A geological formation is a body of rock having a consistent set of physical characteristics, which occurs in a particular position in the layers of rock exposed in a geographical region. A rock formation makes up a fundamental unit of what is termed **lithostratigraphy**, which is the study of strata or rock layers. In addition, to be considered a rock formation, it must be both thick enough and extensive enough to plot on a map (Utah Geologic Survey 2024). Figure 2.3 presents a generalized bedrock geologic map of the United States showing the dominant rock types and formations (USGS 2024c).

2.5 HISTORY OF THE EARTH

Now that we know the age of the Earth, which is estimated to be 4.54 billion years, questions arise that include how did the Earth form? Where did all the water come from? What has happened since the Earth formed? Where did the moon come from? These questions will briefly be addressed in this section. Scientists have divided Earth history into subdivisions of time that have corresponded to major historical events, which is termed the **Geologic Time Scale**. Figure 2.4 presents a greatly simplified Geologic Time Scale of Earth (United States National Park Service 2024).

2.5.1 Precambrian

The Precambrian is divided into three Eons. The first is the Hadean, then the Archean, and finally the Proterozoic. The Hadean Eon encompasses the first 500 million years of Earth's history from the formation of the Earth to approximately 4 billion years ago. Not much is known about the Hadean Eon compared to the Archean and Proterozoic Eons. The Hadean Eon is sometimes referred to as *The Great Bombardment* when Earth was bombarded with asteroids and comets. At about this same time, the moon formed. The hypothesis for the formation of the moon is called the **Giant Impact Hypothesis**. The hypothesis centers on the formation of the moon when a Mars-sized

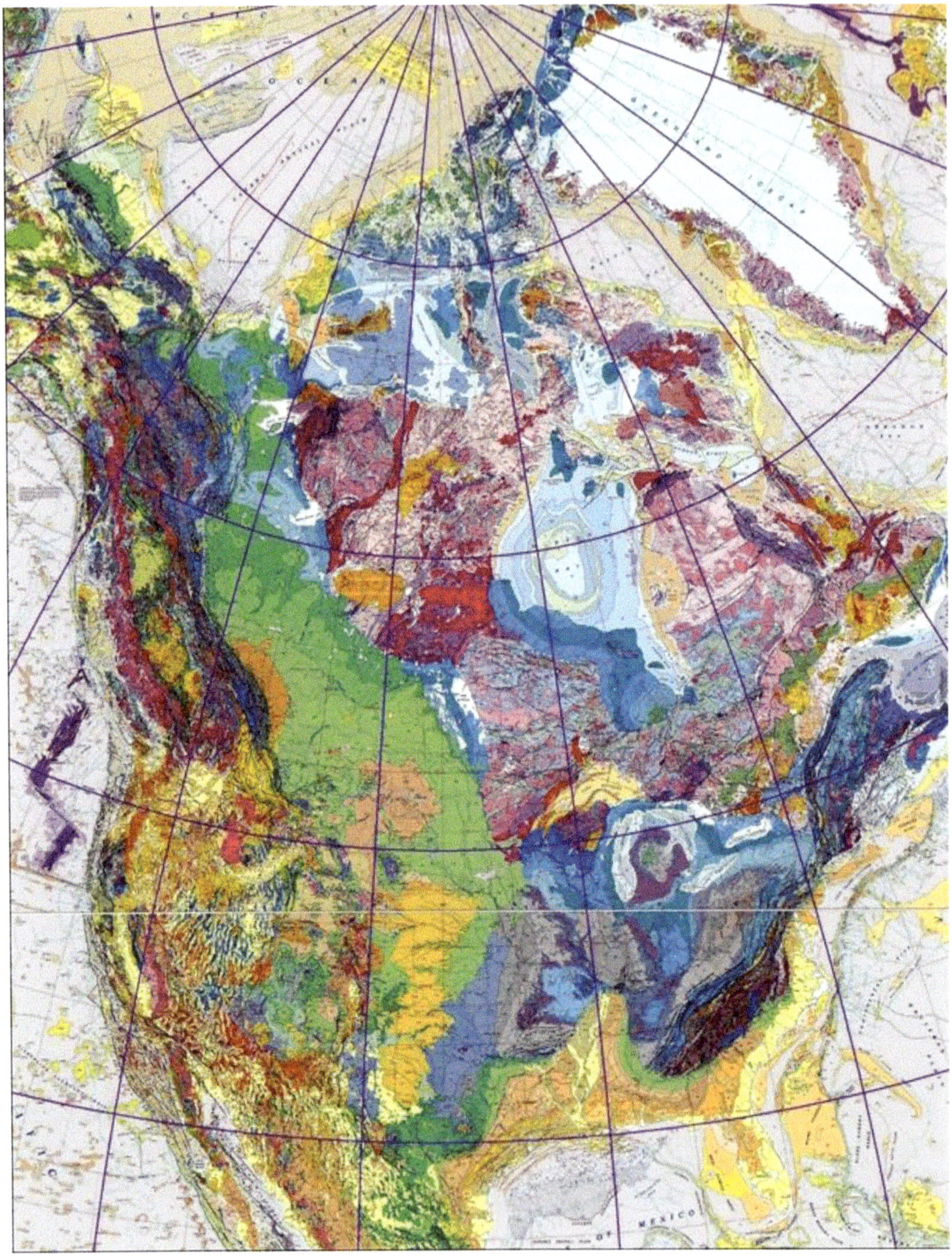

FIGURE 2.3 Geologic map of North America. (United States Geological Survey (USGS). 2024c. *Geologic Map of North America*. www.usgs.gov/media/images/geologic-map-of-north-america. (Accessed May 3, 2024).)

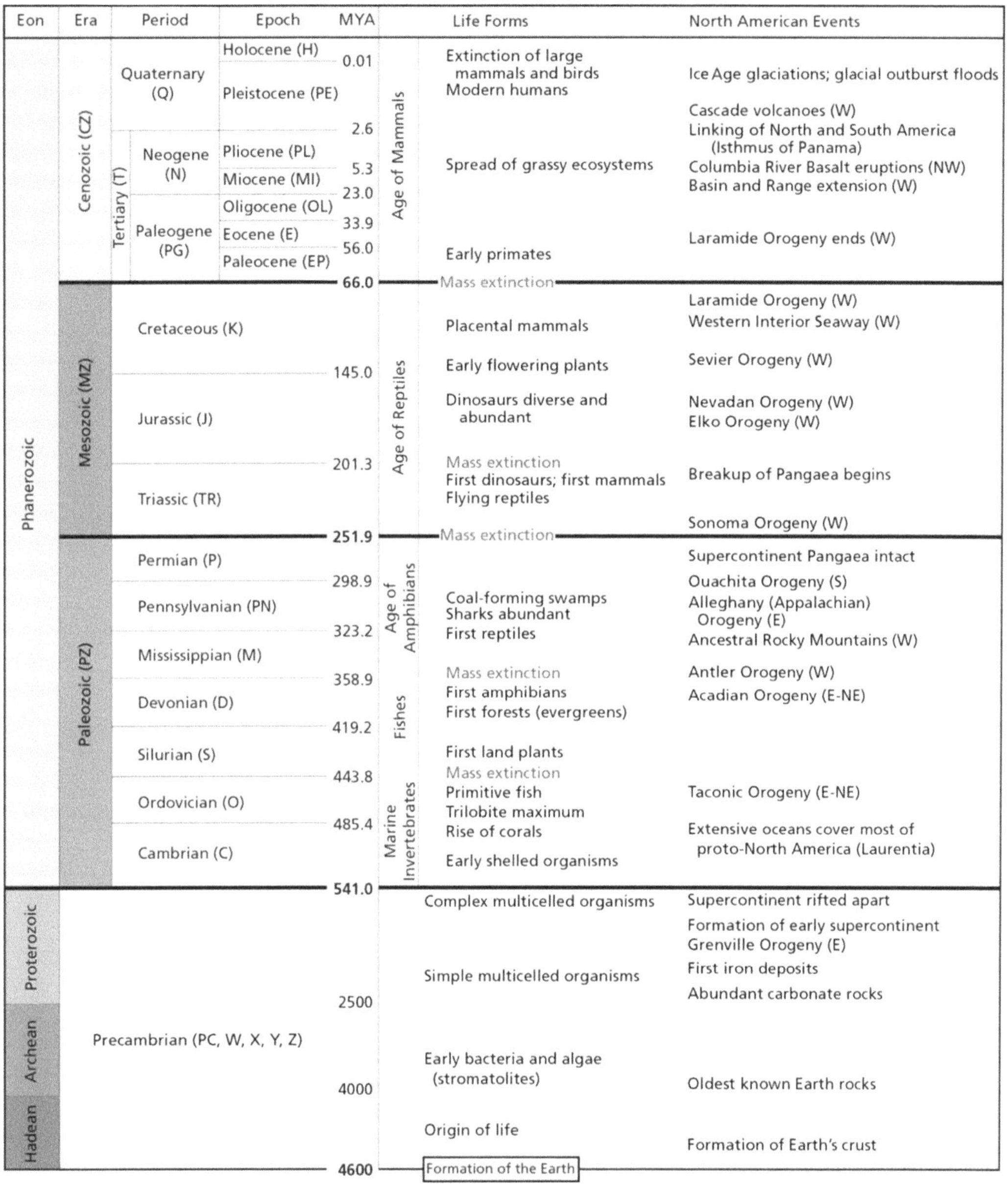

FIGURE 2.4 United States National Park Service.(Geologic Time Scale. 2024. www.nps.gov/subjects/geology/time-scale.htm. (Accessed March 4, 2024).)

planetoid collided with the Earth during the early Hadean Eon (approximately 4.5 billion years ago). Scientists call the imagined planetoid Theia after the deity who gave birth to the moon goddess in ancient Greek mythology (Lawrence Livermore National Laboratory 2024). As the hypothesis states, the Earth–moon system formed together from the impact of two large bodies very early in the solar system history. As the moon lacks plate tectonics (described later in the chapter), and weathering, which are processes that tend to erase or obscure evidence on Earth, the moon is an ideal place to look for evidence of Earth's early history, including the origin of water on Earth (Lawrence Livermore National Laboratory 2024).

Even though close to 70% of the Earth's surface is covered by water, Earth is a relatively dry place compared to many other objects in the solar system. The moon is even drier. Conventional thought is that the lack of volatile substances (such as water) on Earth and the moon was due to the violent impact that caused the depletion of volatile substances. Therefore, Earth was either born with water or was hit by something very early in its history that was nearly all frozen water (Lawrence Livermore National Laboratory 2024).

The Archean Eon began approximately 4 billion years ago and lasted for approximately 1.5 billion years and ended about 2.5 billion years ago (USGS 2024d). Highlights of the Archean Eon include the following (USGS 2024d):

- Formation of early continents (more than 3.5 billion years ago)
- Beginning stages of plate tectonics (more than 3 billion years ago)
- The first evidence of glaciers (approximately 2.8 billion years ago)
- Earliest life (approximately 3.5 billion years ago)

Following the Archean Eon came the Proterozoic Eon. The Proterozoic Eon began at around 2.5 billion years ago and lasted until about 540 million years ago. Highlights of the Proterozoic Eon include the following (USGS 2024d):

- The *Great Oxidation Event* – where great numbers of cyanobacteria produced oxygen in the oceans and atmosphere starting in the early portion of the Proterozoic Eon
- The first Ice Age (approximately 2.3 billion years ago)
- Development of nuclei in cells (perhaps as early as 2 billion years ago)
- Stromatolite formation from blue-green bacteria (more than a billion years ago) (see Figure 2.5)
- Appearance of multi-cellular organisms (around 1.2 billion years ago)
- The second Ice Age (approximately 750 million years ago)
- Appearance of Ediacaran fauna (more complex life forms) late in the Proterozoic

2.5.2 Phanerozoic

The end of the Proterozoic Eon marks the end of the Precambrian. The history of Earth then entered the Phanerozoic Eon approximately 540 million years ago. The Phanerozoic is divided into three

FIGURE 2.5 Top portion of a circular-patterned stromatolite. (Photograph by Daniel T. Rogers.)

Eras and further divided into 12 periods. The first Era of the Phanerozoic is called the **Paleozoic**, the second is termed the **Mesozoic**, and the third is the **Cenozoic** (USGS 2024e).

2.5.2.1 Paleozoic

Paleozoic means ancient life. Life during the Paleozoic was rather primitive and included many invertebrates (animals without backbones) and the earliest fish and amphibians. The Paleozoic Era spanned Earth's history from 540 million years ago to 252 million years ago and is divided into periods which include the following (see Figure 2.4) (USGS 2024e):

- Permian (299 to 252 million years ago)
- Pennsylvanian (323 to 299 million years ago)
- Mississippian (359 to 323 million years ago)
- Devonian (419 to 359 million years ago)
- Silurian (444 to 419 million years ago)
- Ordovician 485 to 444 million years ago)
- Cambrian (540 to 485 million years ago

During the Cambrian Period, the Earth had no polar ice caps and the Earth was warm and wet without distinctive seasons. Marine life flourished but land had not evolved yet. The first vertebrates appeared in the late Cambrian and were jawless fish (USGS 2024e).

The Ordovician Period started out warm and wet like the Cambrian but grew much colder near the end of the Ordovician and an Ice Age began. Many new invertebrates with shells evolved during the Ordovician. The first corals appeared in the Ordovician and jawless armored fish were common. The first land plants appeared in the Ordovician. The first mass extinction occurred at end of the Ordovician period when approximately 80% of species living in the shallow seas became extinct (USGS 2024e).

During the Silurian Period, the Ice Age ended and the climate of Earth was similar to that of today. Sea levels increased due to the melting of ice. The first fish with jaws appeared in the Silurian Period. The first evidence of animals on land appeared in the Silurian and were scorpion and millipede-like animals. The first vascular plants appeared in the Silurian Period and it did not take long for them to cover most of the land surface (USGS 2024e).

During the Devonian Period, three main landmasses were positioned near the equator. The climate was warm and wet along the edges of the continents and was dry in the interiors where salt and gypsum deposits formed. The shallow, tropical seas had abundant sea life. Sharks had evolved and became common. The first lobe-finned fish evolved early in the Devonian, and by the end of the Devonian, had evolved into the first amphibian-like animals that were the first vertebrates to inhabit the land. The oldest preserved insects and centipedes appeared in the Devonian. By the end of the Devonian, the first trees and forests appeared. Ferns and seed-producing plants appeared in the Devonian, but flowering plants had not evolved yet. However, at the end of the Devonian, a mass extinction occurred where approximately 75% of species went extinct. One possible explanation for the mass extinction is that the rapid expansion, diversification, and rapid growth of land plants caused rapid and severe global cooling (USGS 2024e).

The Mississippian and Pennsylvanian periods are commonly referred to as the Carboniferous Period. The Carboniferous Period has a clue in its name, which is essentially the age of carbon. This is when much of the world's coal was formed and many locations on Earth during the time were swampy. The early portion of the Carboniferous had a uniform, tropical, and wet climate with little seasonal changes. Later in the Carboniferous the climate changed and became much cooler, which triggered an Ice Age (USGS 2024e).

During the Permian Period, all the continents were together and formed what is termed a supercontinent called **Pangea** and stretched from the North Pole to the South Pole (see Figure 2.6). This supercontinent experienced huge climatic changes with vast deserts within the continent's interior. During the Permian, reptiles diversified, and precursors to mammals first evolved. Swamps were replaced by temperate forests that contained abundant conifers. At the close of the Permian, the largest mass extinction occurred where approximately 96% of species went extinct. Evidence suggests that the cause of the mass extinction was a massive volcanic eruption in what is now known as Siberia, a meteor impact, and/or a large amount of methane released into the atmosphere from the ocean bottom caused or contributed to the mass extinction. The Permian Period marks the end of the Phanerozoic (USGS 2024e).

2.5.2.2 Mesozoic

The Mesozoic Era followed the Phanerozoic Era approximately 252 million years ago and lasted until 66 million years ago. The Mesozoic Era means **middle life.** The Mesozoic includes the time of the dinosaurs. The Mesozoic Era is divided into three Periods which include the following (USGS 2024f):

- Cretaceous (145 to 66 million years ago)
- Jurassic (201 to 145 million years ago)
- Triassic (252 to 201 million years ago)

During the Triassic Period, Pangea began to split apart and rotate. The resulting tension caused a series of rips within Pangea and portions began to subside forming rifts (see Figure 2.6). These rifts developed between North America and Europe and between Africa and South America. Other rifts later split off Antarctica, India, and Australia. During the early Triassic, many new kinds of animals evolved. The first dinosaurs, marine reptiles, lizards, and tortoises appeared. Mammals also appeared during the Triassic Period but were not significant. Crocodiles were abundant. At the end of the Triassic, the fourth mass extinction occurred where up to 80% of species went extinct. The cause of the mass extinction is believed to have been caused by underwater volcanic eruptions which drastically changed the chemistry of the oceans and caused a sudden warming of the climate (USGS 2024f).

The Jurassic Period experienced a continued breakup of the supercontinent Pangea. Shallow inland seas covered parts of the western United States. Warm tropical greenhouse conditions dominated the worldwide climate. Gian plant-eating dinosaurs roamed the Earth with smaller carnivores stalking them. Flying reptiles and the first birds appeared in the Triassic Period. The shallow oceans contained abundant life from plankton to whale-sized reptiles. Conifers were the most diverse large trees. Cycads (evergreen, cone-bearing, palm, and palm-like plants) became so abundant and diverse that the Jurassic Period is sometimes referred to as the Age of Cycads (USGS 2024f).

The Cretaceous Period is when India separated and became an island continent. North America, Europe, Asia, and Africa became a series of islands (see Figure 2.6). The climate during the Cretaceous was mild with ice-free poles caused by rising sea levels and separated continents, which permitted warm shallow ocean currents to dominate. Many new species of dinosaurs appeared. Birds diversified and grew in numbers. New mammal species also appeared but remained small. Flowering plants first appeared during the Cretaceous Period and, because of this, many modern insect species appeared and began to diversify, including ants, termites, bees, butterflies, aphids, and grasshoppers. The end of the Cretaceous Period marks the fifth mass extinction (USGS 2024f).

The mass extinction at the end of the Cretaceous was caused by a meteorite, with an estimated diameter of 10 km, which crashed into the Gulf of Mexico near the northern tip of the Yucatan Peninsula. The impact crater, termed **Chicxulub Crater**, measures 200 km wide and 1 km deep. The impact of the meteorite generated a massive tsunami perhaps 300 m high, winds as high as 1,000 km/h, and released as much energy as all the earthquakes on Earth combined within the last

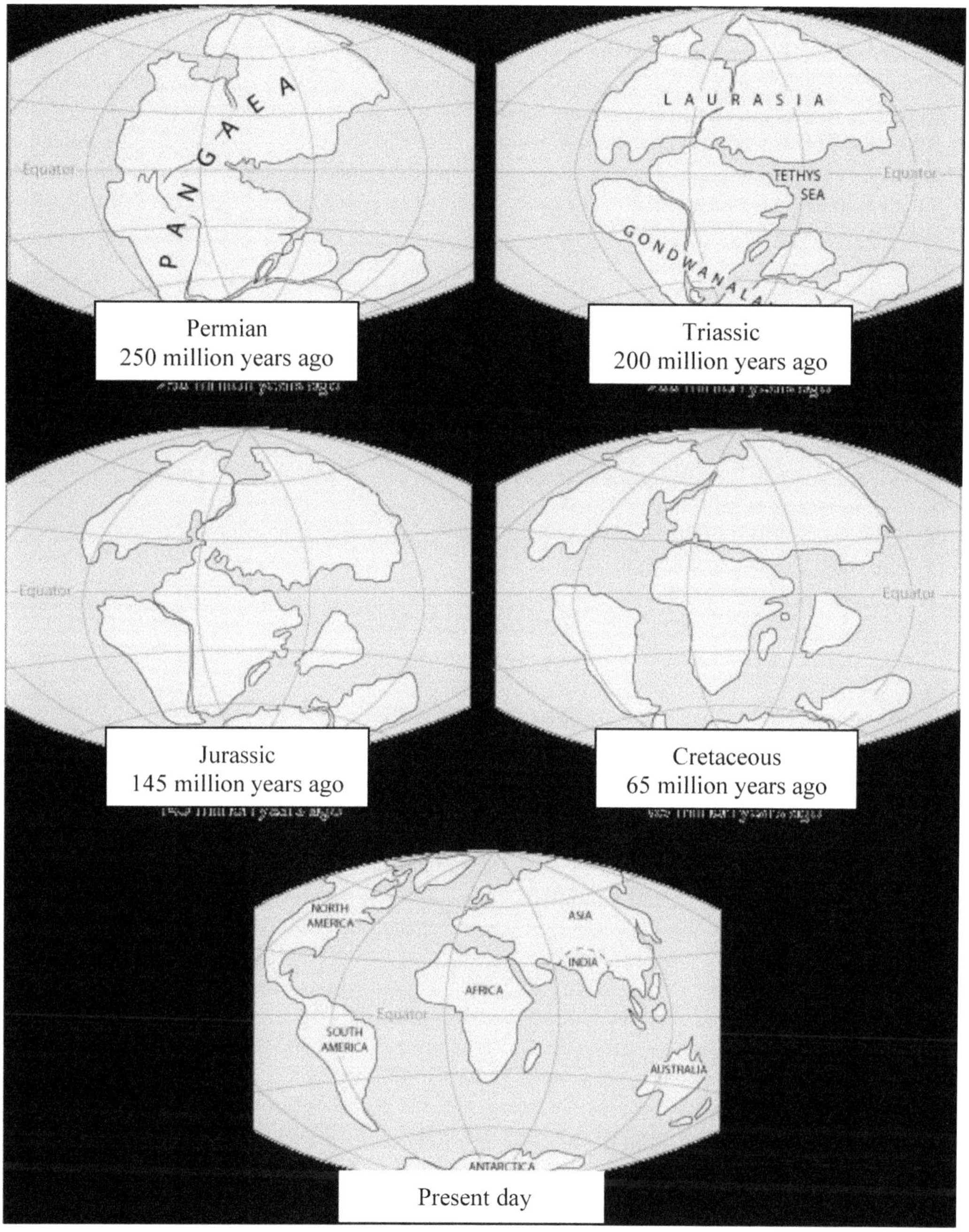

FIGURE 2.6 Position of the continents over the last 250 million years.(United States Geological Survey (USGS). 2024c. *History of the Earth*. https://pubs.usgs.gov/gip/dynamic/historical.html. (Accessed March 3, 2024).)

160 years. The impact also generated massive fires and ejected enough dust and debris into the atmosphere that it blocked much of the sunlight from reaching Earth and caused global cooling and acid rain that killed plant life and almost all large terrestrial animal life. The result was that 80% of species went extinct (Black 2022; Brusatte 2018).

The only land animal with an average weight of 25 kg or more to survive was the crocodile. The only dinosaurs that survived were the small avian dinosaurs. Estimates on the average weight of a land animal before the meteor impact was greater than 1,000 kg. The average weight of a land animal after the meteor impact is estimated at less than 3 kg (Black 2022; Brusatte 2018).

2.5.2.3 Cenozoic

The Cenozoic Era followed the Mesozoic Era from 66 million years ago to the present. The Cenozoic Era means recent life. During this era, plants and animals looked similar to those on Earth today. The Cenozoic is divided into two periods: the Tertiary and Quaternary (USGS 2024g). The Tertiary Period is further divided into what is termed **Epochs** and include the following:

- Paleocene (66 to 56 million years ago)
- Eocene (56 to 34 million years ago)
- Oligocene (34 to 23 million years ago)
- Miocene (23 to 5 million years ago
- Pliocene (5 to 2.6 million years ago)

The Quaternary Period is also further divided into two Epochs:

- Pleistocene (2.6 million to 10,000 years ago)
- Holocene (10,000 years ago to the present)

The Paleocene Period saw rapid evolution of many new species since the mass extinction at the end of the Cretaceous Period. The climate was much warmer than today. South America, Antarctica Australia, India, and Africa were all separate continents. With the extinction of the dinosaurs, mammals evolved, took over, and dominated the Earth. New plants also evolved quickly, and the first pines, cacti, and palm trees appeared. At the end of the Paleocene, the Earth experienced rapid warming (USGS 2024g).

At the beginning of the Eocene, India, which had been moving slowly northward, collided with Asia. As a result of this collision, the Himalayan Mountains were born. Australia drifted away from Antarctica and began to move northward. By the end of the Eocene, the gap between these two continents was large enough that the Circum-Antarctic Current emerged. This event significantly changed ocean currents around the world and resulted in global cooling at the end of the Eocene. During the early Eocene, temperatures were warmer than at any time during the Cenozoic. The land experienced abundant rainfall with no seasons, and no glaciers. Palm trees and alligators were able to live within the Arctic Circle. Many animals and plants continued to diversify during the Eocene (USGS 2024g).

The climate of the Oligocene was drier than the Eocene. India continued to collide with Asia and the Himalayan Mountains grew larger and larger. There was a significant increase in volcanic activity during the Oligocene in Europe and North America. Yellowstone National Park resulted from this volcanic activity. Antarctica glaciers appeared and grew extensively during the Oligocene. During the later portions of the Oligocene, extensive grasslands and savannas appeared. Mammals grew large and a hornless rhinoceros from Asia became the largest land mammal to ever live. Early forms of monkeys, dogs, cats, pigs, horses, and camels appeared (USGS 2024g).

Continental plate movement in the Miocene resulted in new mountain building in North America, South America, Europe, and Africa, and a land bridge existed between Asia and North America (what is now Russia and Alaska). In addition, the Himalayan Mountains continued to be pushed upward. Evidence also suggested that there was a renewed uplift in the Appalachian Mountains. The polar ice cap in Antarctica was still present and continued to grow. Later in the Miocene, due to mountain building combined with changing ocean currents and the Antarctic ice sheet, the global climate began to cool, rain decreased, and seasonal temperatures appeared. True prairies, grasslands, and Savannahs expanded with the cooler temperatures, and forests grew smaller. Hoofed mammals, with multiple stomachs to digest the grasses, came to dominate the plains. There were great mammal migrations from continent to continent. The diversity of mammals was at its highest during the Miocene. Perhaps most important to humans was the appearance of the first anthropoid apes (USGS 2024g).

The continents were in much the same position that are today during the Pliocene. The Cascade Range, many mountain ranges in the Rocky Mountains, and the Colorado Plateau were formed during the Miocene. Not only did the ice cap in Antarctica grow but also a polar ice cap on the North Pole appeared. The emergence of the land bridge between North and South America in the late Pliocene made it possible for many animals to migrate into new regions. Many animals became extinct because of new competition. Armadillos, ground sloths, opossums, and porcupines moved into North America, and dogs, cats, bears, and horses moved into South America. Early hominids in Africa evolved into several distinctly different species with only one of them surviving to the present day as humans (USGS 2024g).

The Quaternary Period began 2.6 million years ago with the Pleistocene Epoch that lasted from 2.6 million to 10,000 years ago. The position of the continents was essentially the way they are today. Global cooling continued and an Ice Age with interglacial periods resulted, which changed the appearance of the coastal regions of the world due to falling sea levels. During the glacial periods, the north 25% of the Earth was covered in ice, and sea levels dropped by as much as 135 m (430 feet). However, the Antarctic Ice Cap variation was much more subtle from glacial to interglacial periods. There was a significant number of large animals including mammoths, mastodons, Smilodon (saber-toothed cats), giant sloths, and cave bears that went extinct. The extinction of these large animals was likely due to stresses induced by a rapidly fluctuating climate and perhaps the hunting of these animals by humans, which may have lowered their numbers enough to endanger their survival. By the end of the Pleistocene, modern humans had spread throughout the world with the exception of Antarctica (USGS 2024g).

The Holocene is the second Epoch of the Quaternary Period and began approximately 10,000 years ago. However, many scholars state that the Holocene began 11,700 years ago and is coincident with the final retreat of glaciers from the last Ice Age (USGS 2024g). The Holocene Epoch also corresponds with the rapid proliferation of the human species worldwide, including our species' written history, technological advancements, significant transition to urban living, and development of our civilization (USGS 2024g).

Many scientists are now convinced that Earth has transitioned into a new Geologic Epoch termed the **Anthropocene** or the age of humans. The date of the beginning of this new Epoch has been debated and centers around two dates. One proposed date is 1753 which coincides with the start of the Industrial Revolution and the other ranges between 1945 and 1952 and coincides with the development and experiments with nuclear detonations. The significance of these actions means that many scientists are now convinced that humans have impacted the entire Earth in a way that will be recorded in Earth's history and it cannot be erased. However, members of the International Union of Geologic Science (IUGS) recently voted down the proposal. The scientists concede that humans have impacted the entire Earth but that it is more of an event than an Epoch (USGS 2024g).

Now that we have completed the brief section on the history of the Earth, it should be apparent that the Earth is a dynamic place and undergoes constant change. This change usually occurs gradually over long periods of change but occasionally occurs suddenly, and when it does occur suddenly or over a short period of time, the outcome can be devastating and result in mass extinction.

The next section addresses a third and final aspect of the definition of geology and that is the structure of the Earth.

2.6 STRUCTURE OF THE EARTH

We have already described the basic structure of the interior of the Earth. However, we have not discussed the structure of the crust of the Earth, which is where we live and is complex. We will first discuss plate tectonics which provides much of the driving force creating the structure of the crust, where and how land masses are made and destroyed, and the oceans. We will then examine some of the fundamental principles of geology and move on to discuss mountain building and erosion.

2.6.1 Plate Tectonics

Plate tectonics was first introduced as the **Theory of Continental Drift** by Alfred Wegener in 1912 and is based on the concept that the Earth's crust or lithosphere is composed of moving plates (Seyfert and Sirkin 1973). In developing his theory, Wegener noticed that some sedimentary rocks known only to form in tropical environments were located in Arctic and Antarctic regions. Wegener then compared these sedimentary deposits with their approximate relative ages and distances from the equator and concluded continental drift had taken place.

These lithospheric plates (crust and uppermost mantle) move on top of the upper mantle called the **asthenosphere** (see Figure 2.1). Their movement is responsible for most earthquakes, volcanoes, sea-floor spreading, mountain building, and ocean trench formation (Stanley 1999). Figure 2.7 shows the arrangement of the tectonic plates and plate boundaries (USGS 2024h). The energy source driving the movement of the plates is believed to be heat from convection currents originating deep within the interior of the Earth and is depicted in Figure 2.8. We shall return to the concept of plate tectonics later in this chapter when we discuss the structure of the Earth

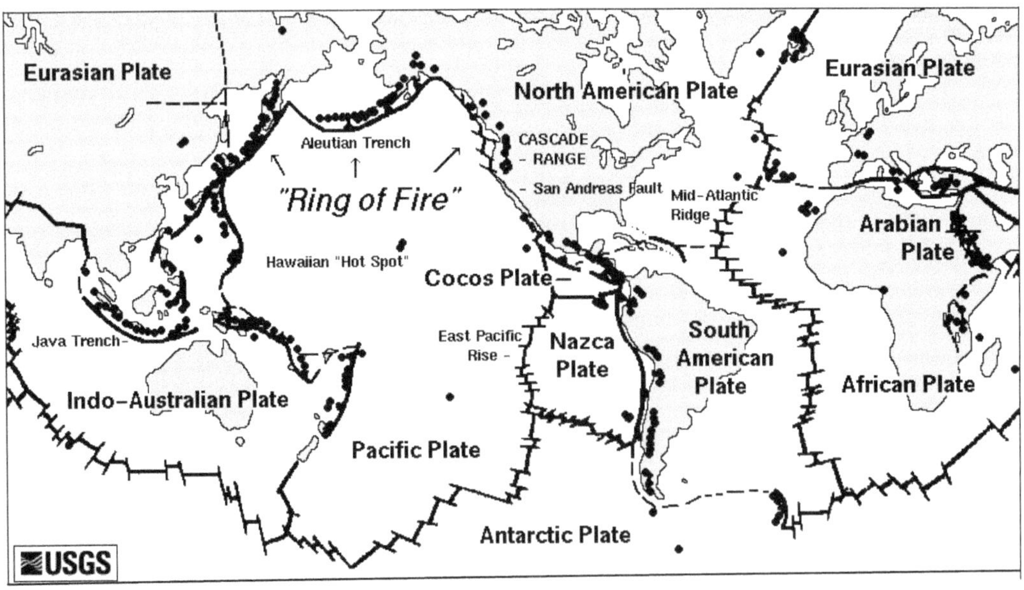

FIGURE 2.7 The major tectonic plates of the world. (From United States Geological Survey. 2024i. *World Map Showing Plate Boundaries and Active Volcanoes.* http://vulcan.wr.usgs.gov/Glossary/PlateTectonics/Maps/map_plate_tectonics_world.html (Accessed March 5, 2024).)

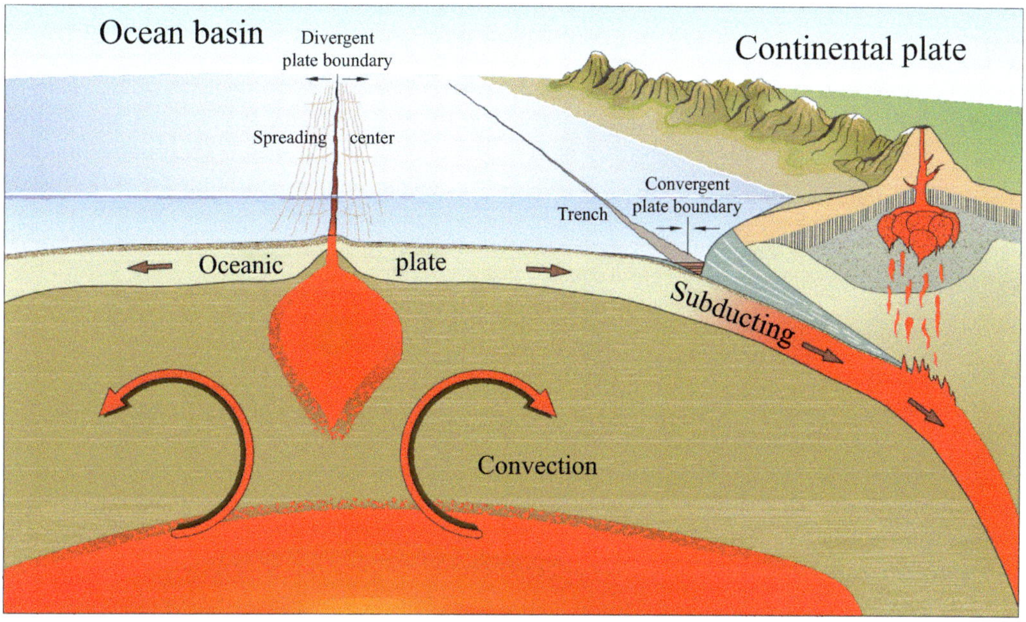

FIGURE 2.8 Concepts of the driver for plate tectonics and continental drift.

in greater detail. Plate tectonics is responsible for, or plays a significant role in, the formation, deformation, and weathering of most rocks on Earth. The three main processes that lead to the formation of rocks include (1) the cooling and solidification of magma, (2) the deposition of sediment, and (3) extreme temperatures and pressure. Igneous, metamorphic, and sedimentary rocks, further described below, all originate from these processes. The primary classification of rocks recognizes three categories (igneous, metamorphic, and sedimentary) based on their method of formation.

Geological processes generally take place at or very near the surface and are primarily sedimentary-type processes, which include processes involving liquid and solid water (ice), gravity, and tectonics. **Sedimentary deposits** are defined as the accumulation of natural materials and sediments formed at or near the surface of the Earth at ordinary temperatures and pressures (Flint and Skinner 1974). Geological processes of sedimentary origin usually occur gradually over long periods of time. On rare occasions, however, sudden events affect sedimentary processes and urban environments. These short-duration events are typically labeled as geologic hazards and may include tectonic episodes such as earthquakes, tsunamis, volcanic eruptions, and landslides. Some short-term geologic processes are also triggered by a weather event such as a flood caused by extensive rains, sudden melting of snowpacks, or a hurricane.

Before we focus on sedimentary rocks, a brief overview of some basic geological principles is required. These principles (uniformity, stratigraphy, and unconformity) provide the necessary foundation for understanding the general characteristics of sedimentary rocks in urban environments.

2.6.2 Principle of Uniformity

During the late 1700s, James Hutton, a Scottish geologist, doctor, and farmer was studying the rock outcrops around Scotland. Hutton, influenced by the Age of Enlightenment, cast aside the existing religious dogma of a very young Earth and developed the **principle of uniformity,** which stated that the external and internal processes on and within the Earth observed today have been operating unchanged, and at the same approximate rates, for a very long time throughout most of the Earth's history (Hutton 1788). Essentially the meaning of the principle of uniformity is that the present is the key to the past.

The significance of this principle cannot be understated, because the recognition of this principle presents the geologist with a great tool for exploring the history of the Earth (Flint and Skinner 1974). Geologists today look at features within observable rocks and search for insights into how they formed long ago. For example, in some urban areas, the presence of sand found near the surface today might indicate the presence of a beach from a lake thousands of years ago. As we will soon learn, the shape of the sand grains and their orientation can reveal even more information about the past and the current physical and anthropogenic impacts.

2.6.3 Principles of Stratigraphy

Stratigraphy is the study of rock layers or unconsolidated sediment and strata, particularly their ages, composition, and relationship with other layers (AGI 1962; Christopherson 2008). In geology, the term **strata** refers to the layers of rock or sediment with internally consistent characteristics that distinguish it from contiguous layers (Krumbein and Sloss 1963). Each individual layer or stratum is generally one of several parallel and originally horizontal layers that were deposited by natural forces and lay one upon the other. Therefore, the study of the stratigraphy of a particular area provides important clues concerning its geologic history. The origins of stratigraphy date back to Italy in the mid-1600s. A Danish anatomist named Nicolas Steno was studying shark teeth and noticed the similarity to fossils of shark teeth in Italy. He deduced that the fossils were remnants of past life. He later deduced that based on his observations that deep rock layers in the Earth were older and that successive layers above gradually became younger. This later was expounded by Hutton and became known as the Principle of Superposition described below (Hongren 2004).

Stratigraphy relies on four basic principles (Anderton 1985):

1. The *principle of original horizontality* states that the sediments formed from sedimentary deposits are generally deposited as horizontal sheets. An example of very fine horizontal layering of a sedimentary deposit composed of layers of clay and fine silt is shown in Figure 2.9. This principle is important because it allows the geologist to distinguish between disturbed or deformed layers. In urban areas, this principle is especially important because so much of the near-surface sediments and soil layers have undergone significant disturbance due to development or deformation from tectonic activity. Some examples of near-surface

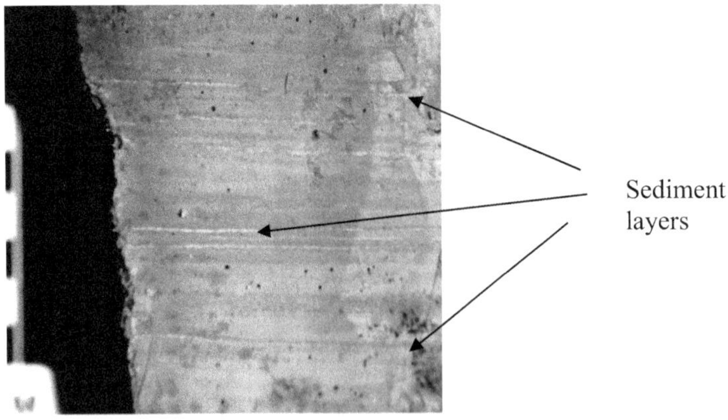

FIGURE 2.9 Example of very fine horizontal layering within a clastic sedimentary deposit. Sample is approximately 8 cm vertical by 6 cm horizontal. (Photograph by Daniel T. Rogers.)

disturbance include farming, constructing buildings and foundations, road building, mining, and landscaping. The term deformation generally refers to both brittle and ductile processes such as faulting or folding caused by tectonic activity. This type of deformation explains why, for example, sedimentary rock deposited thousands of feet below sea level may now be found thousands of feet above sea level. This type of uplift is usually the result of plate interaction. Erosion or removal of a portion of a sedimentary layer through the action of water and wind can also significantly erode or rework a layer of sediment after it has been deposited.

2. The *principle of superposition* defines the depositional sequence in undeformed strata of rock or sediment. Strata are sheets or individual layers of sedimentary deposits, and as a group are visibly distinct from those above and below. Figure 2.10 demonstrates the principle of superposition. The principle of superposition results from the simple observation that new layers of sediment are deposited on top of older layers of sediment. Therefore, strata are deposited sequentially, layer after layer. It is important in areas where tectonic activity has occurred to establish the correct stratigraphic sequence to ensure that the entire sequence of layers has not been overturned.
3. The *principle of cross-cutting relations* states when a layer of sediment is cut, perhaps by faulting, or the intrusion of an igneous dike or sill; the sedimentary layer is older than the event or feature that cuts the layer. In urban environments, it is common, for example, to have a glacial moraine or a river terrace cut by development that obscures the correct interpretation of age and stratigraphic relationships.
4. The *principle of faunal and floral succession* explains the relationship between biological evolution and sedimentary deposits. Because plants and animals evolve over long time periods into new species, sedimentary deposits of different ages will contain fossils of different species. Knowing the age of a fossil will assist in dating a particular sedimentary deposit containing the fossil. This relationship may help when evaluating two sedimentary deposits at different locations.

The work and principles initially conducted by Steno and Hutton began the scientific quest to evaluate the age of Earth which has involved thousands of dedicated scientists from many different

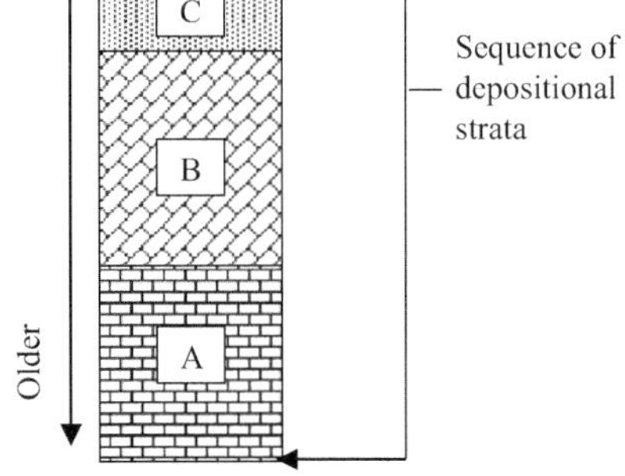

FIGURE 2.10 Example of a stratigraphic sequence demonstrating the principle of superposition.

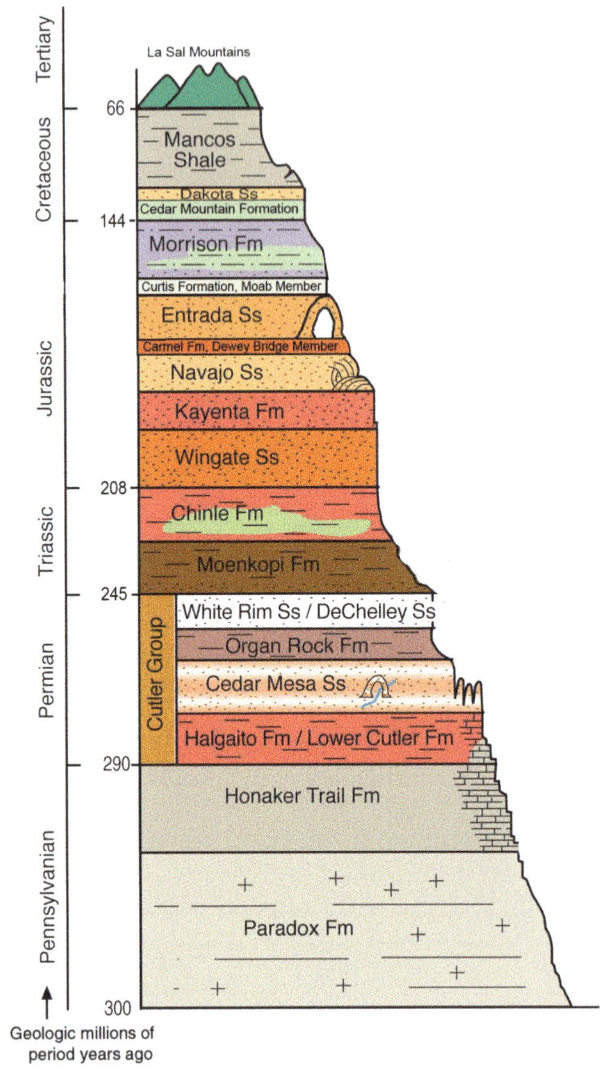

FIGURE 2.11 Geologic cross section of Canyonlands National Park. (Canyonlands Natural History Association. 2024. *Geologic Cross Section of Canyonlands National Park*. www.cnha.org. (Accessed March 7, 2024).)

fields of science using more and more sophisticated methods for more than 200 years. Figure 2.11 is a geologic cross section that depicts many of the concepts of Earth history and stratigraphy of the Earth. Depicted in Figure 2.11 is a cross section of Canyonlands National Park in Utah (Canyonlands Natural Historical Association 2024).

2.6.4 Unconformities

When examining the vertical succession of sedimentary layers, sometimes the layering may not represent a complete timeline of deposition. There may be present within the succession of layers of strata a period of time when there was no deposition, or evidence that erosion removed some

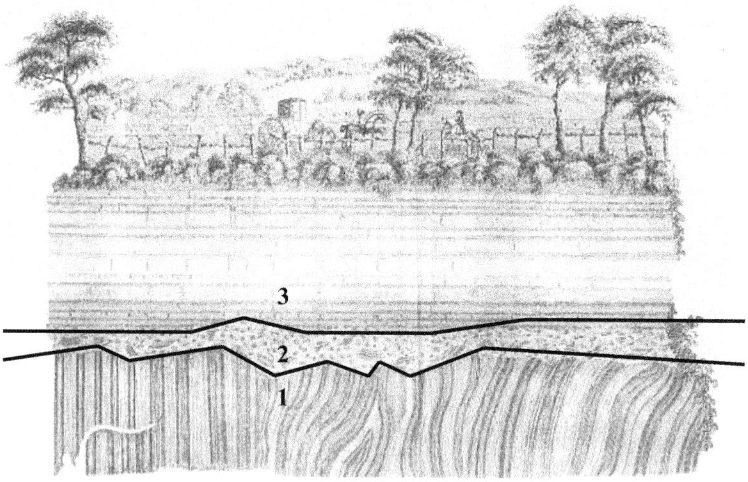

FIGURE 2.12 Concept of unconformities. Unconformities identified by black lines. (Modified from Hutton, J. 1788. Theory of the Earth; or an Investigation of the Laws Observable in the Composition, Dissolution, and Restoration of the Land Upon the Globe. *Transactions of the Royal Society of Edinburgh*. Vol. 1, Part 2, pp. 209–304.)

pre-existing layers. When this occurs and is documented in a depositional succession it is termed an unconformity, and the missing period of time is called a hiatus. An **unconformity** is defined as a buried erosional surface separating two rock masses or sedimentary strata of different ages, indicating that sediment deposition was not continuous (Flint and Skinner 1974).

There are several types of unconformities (Wicander and Monroe 2007):

- Disconformity: an unconformity that exists between parallel layers of sedimentary rocks
- Nonconformity: an unconformity where horizontal strata or layers of sedimentary rock overlie crystalline rock, either igneous or metamorphic.
- Angular unconformity: an unconformity where horizontal strata or layers of sedimentary rock are deposited on tilted and eroded strata. The result produces an angular discordance with the overlying horizontal layers.

Figure 2.12 shows several types of unconformities by Hutton (1788) who was the first geologist to describe them. Examination of Figure 2.12 reveals the sequence of major events creating the unconformities:

- The layers identified as 1 at the bottom portion of the figure were deposited and then metamorphosed, uplifted, folded, and tilted vertically; they represent the oldest layers.
- After folding and tilting of the rocks, they were exposed and eroded.
- Following erosion, sediments identified between the arrows within the zone labeled as 2 were deposited on top of the folded and tilted rock layers creating a nonconformity (lower arrow and line).
- After the layers identified as 2 were deposited, they were uplifted, exposed, and eroded to create a second erosion surface.
- Following a rise in sea level, sequence 3, a marine limestone was deposited, creating a disconformity (the upper arrow and line)

2.6.5 Mountain Building

Mountains form from a variety of processes, with most of the underlying factors of mountain building being associated with the movement of the tectonic plates. The most significant mountain ranges on land result from the collision of continents. A good example of this is the Himalayan Mountains located in northern India, Pakistan, Nepal, Tibet, and southwestern China. The Himalayan Mountains were gradually built over millions of years when the Indian subcontinent slammed into Asia. Mountain ranges also form when one tectonic plate rides up and over another tectonic plate. A good example of this type of mountain-building process is the Andes Mountains of South America. When one tectonic plate rides up and over another tectonic plate, the bottom plate is labeled a subducted plate (USGS 2024i).

Another method of mountain building is when thick continental crust cracks, generally through faulting, and as the crust thins, a series of mountain ranges emerges. These mountain ranges form roughly parallel mountain ranges. A good example of this is the mountain ranges within the basin and range province from the east portion of the Sierra Nevada to western Utah (USGS 2024i).

Mountains and mountain building affect the Earth in numerous ways, some of which include the following (USGS 2024i):

- Wind and precipitation
- Atmospheric circulation
- Weathering and erosion
- Carbon cycle
- Temperature
- Species ranges, populations, and diversity and interactions

Figure 2.13 is a rock outcrop in Kings Canyon National Park in California that demonstrates many of the aspects presented by Hutton in 1788, namely an unconformity and greatly deformed bedding layers.

FIGURE 2.13 Unconformity with deformed and angular bedding layers in Kings Canyon National Park in California. (Photograph by Daniel T. Rogers.)

2.6.5.1 Folding

When rock formations experience stress, such as compression, tension, or shear, they respond to it by deforming in some manner. Rocks tend to give a little, especially at the beginning of the deformation process. If the pressure is eased during this initial stage, the rocks may respond in an elastic manner and will return to their original shape and volume. However, if the stress increases beyond the elastic limit of the strata, the deformation may become permanent (USGS 2024i).

Rocks will permanently deform in one of a couple of ways, depending upon their properties and the confining pressure they experience. Rock strata can break if the rocks behave with a brittle tendency or if the confining pressure is low. Brittle deformation is particularly common in rocks that contain hard and non-elastic minerals such as quartz arenite sandstones, limestones or crystalline dolomites with a high calcite content that are buried relatively shallow and have a low confining pressure. Under these conditions, rocks typically fracture and display joints or displacement or fault-like features (USGS 2024i).

If the rocks are more ductile in nature or if the confining pressure is high, the deformation is more likely to result in bending into folds as is depicted in Figure 2.14. Rocks containing pliable minerals such as gypsum (evaporites), high clay content (shales), or are buried deeply, typically 10 km (6 miles) or more beneath the surface, become folded when subjected to stress (USGS 2024i).

There are several types of folds in structural geology. The three most common types are termed anticline, syncline, and monocline. Figure 2.14 shows an example of each. Figures 2.15 and 2.16 each show photographs of folds as observed in nature.

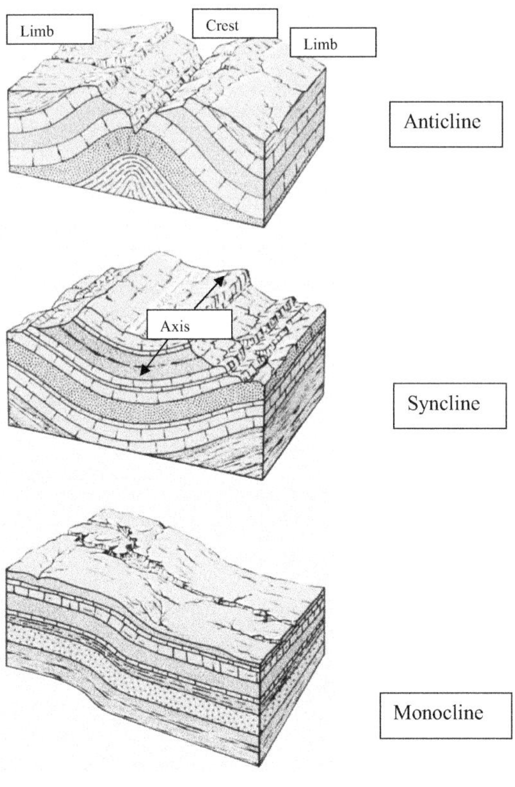

FIGURE 2.14 Diagram showing common types of folds, anticline, syncline, and monocline. (United States Geological Survey (USGS). 2024i. *Geological Survey Bulletin 1393. The Geologic Story of Arches National Park.* www.usgs.gov/publications/geologic-story-arches-national-park). (Accessed March 7, 2024).)

FIGURE 2.15 Photograph showing a shallow angle synclinal fold.(Photograph by Daniel T. Rogers.)

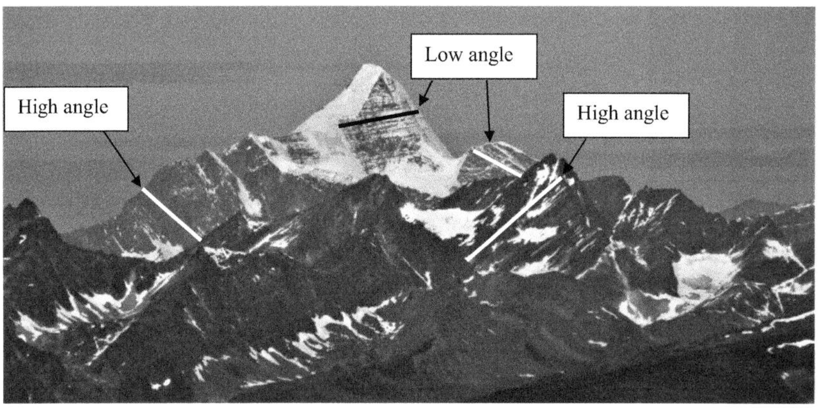

FIGURE 2.16 Example of high and low angle folding and complex geology, Mount Robson, British Columbia, Canada. (Photograph by Daniel T. Rogers.)

The symmetry or orientation of fold structures varies widely. They can be symmetrical, asymmetrical, or overturned fold, which is also called a recumbent fold. An overturned or recumbent fold is a severe type where the fold actually is greater than 90 degrees. Folds also can be inclined relative to horizontal. This type of fold is referred to as plunging. Two other types of folds are domes and basins. Domes consist of strata that have been folded upward. Domes can be very large and they may encompass a whole mountain range. Basins are formed through a much different mechanism more related to sedimentation than to rock deformation. A basin is a down-warping, typically from subsidence which could be as a result of tectonic activity. The significant point is that the down-warping produces a depression that gradually fills in with sediment (USGS 2024i).

2.6.5.2 Faulting

A fault is a fracture or zone of fractures between two blocks of rock. Faults allow the blocks to move relative to each other. This movement may occur rapidly in the form of an earthquake, or it may occur slowly, in the form of what is termed **creep**. Faults may range in length from a few millimeters

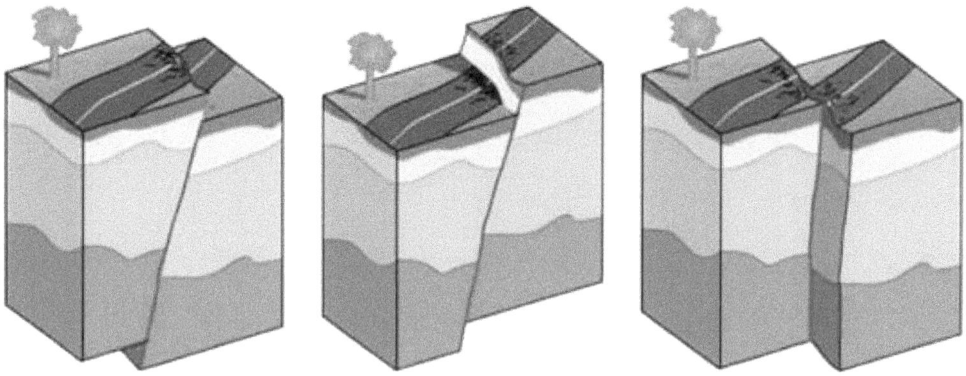

FIGURE 2.17 Strike-slip fault (right), normal fault (center), thrust fault (left). (United States Geological Survey (USGS). 2024k. Diagrams showing fault types. www.usgs.gov/tools/interactive-us-fault-map. (Accessed March 5, 2024).)

to thousands of kilometers. Most faults produce repeated displacements over geologic time. During an earthquake, the rock on one side of the fault suddenly slips with respect to the other. The fault surface can be horizontal or vertical or some arbitrary angle in between (USGS 2024j).

Geologists use the angle of the fault with respect to the surface (known as the dip) and the direction of slip along the fault to classify faults. Faults that move along the direction of the dip plane and dip-slip faults and described as either normal or reverse (thrust), depending on their motion. Faults that move horizontally are known as strike-slip faults and are classified as either right-lateral or left-lateral. Faults that show both dip-slip and strike-slip motion are known as oblique-slip faults (USGS 2024j).

As a review, a **normal fault** is a dip-slip fault in which the block above the fault has moved downward relative to the block below. This type of fault occurs in response to extension and is often observed in the western portions of the United States Basin and Range Province and along ocean ridge systems. A **reverse (thrust) fault** is defined as a dip-slip fault in which the upper block, above the fault plane, moves up and over the lower block. This type of faulting is common in areas of compression, such as regions where one tectonic plate is being subducted under another as is the case in Japan. When the dip angle is shallow, a reverse fault is often described as a thrust fault. A **strike-slip fault** is a fault on which the two blocks slide past one another. The San Andreas Fault in California is an example of a right lateral fault (USGS 2024g). See Figure 2.17 which shows each type of fault. Figure 2.18 shows a photograph of the Sand Andreas Fault in California. Figure 2.19 shows a photograph of a normal fault in Grand Teton National Park in Wyoming (USGS 2024j).

2.6.5.3 Volcanoes

Most mountains form from folding, faulting, uplift, and erosion of the Earth's crust. A volcano is a less common type of mountain that is built by the slow accumulation of erupted lava. The vent may be visible as a small bowl-shaped depression at the summit of a cone or shield-shaped mountain. Through a series of cracks within and beneath the volcano, the vent connects to one or more linked storage areas of molten or partially molten rock called magma. This connection to fresh magma allows the volcano to erupt over and over again in the same location. This allows the volcano to grow and, in some cases, they can grow very large (USGS 2024l).

Molten rock below the surface of the Earth that rises in volcanic vents is called **magma**. After the volcano erupts, what once was magma is now called lava. Magma is typically composed of

FIGURE 2.18 Example of the San Andreas strike-slip fault in California. (Photograph by Daniel T. Rogers.)

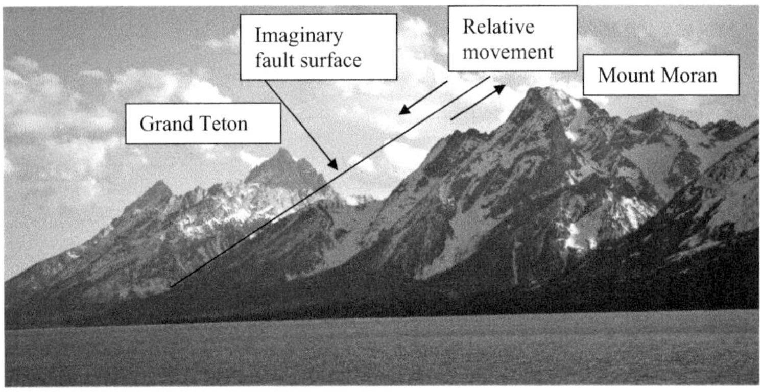

FIGURE 2.19 Normal fault (Grand Teton National Park, Wyoming). (Photograph by Daniel T. Rogers.)

molten rock, crystals, and dissolved gas. The molten rock is typically composed of oxygen, silicon, aluminum, iron, magnesium, calcium, sodium, potassium, titanium, and manganese. After cooling, liquid magma may form crystals of various minerals until it becomes completely solid and forms an igneous rock (USGS 2024l). Originating many tens of kilometers beneath the surface of the ground, magma is lighter than the surrounding solid rock. It is driven toward the Earth's surface by buoyancy, being lighter than the surrounding solid rock, and by pressure from the gases within the magma. The magma forces its way upward and may ultimately break through weaker areas of the crust and then a volcano is born (USGS 2024l). Figure 2.20 is a diagram showing a cross section of a volcano and its creation.

Volcanic eruptions are measured by the size and intensity of explosive volcanic eruptions on a logarithmic scale termed the Volcano Explosivity Index (VEI). The VEI is calculated using the volume of material ejected, the height of the eruption cloud and ejected material, and qualitative observations. The VEI is similar in nature to the Richter scale used for measuring earthquakes (USGS 2024l).

There are five main types of volcanoes, which include cinder cone, composite volcano (strato-volcano), shield volcano, lava dome, and super volcano. They are each described below (USGS 2024l).

2.6.5.3.1 Cinder Cones

Cinder cones are the simplest type of volcano. They are made of small pieces of solid lava, called cinder, which are erupted from a vent. Typically, an eruption from a cinder cone is a powerful blast

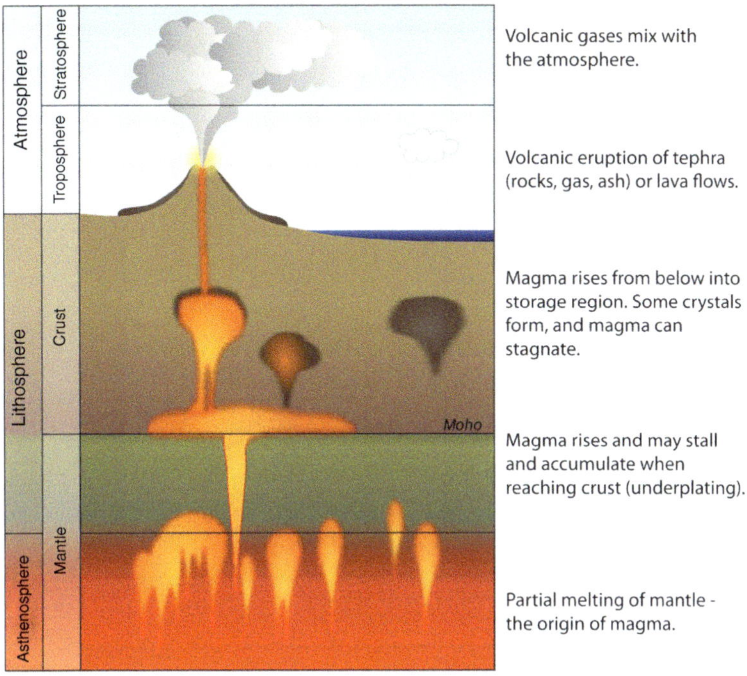

FIGURE 2.20 Cross section of volcano formation. (United States Geological Survey (USGS). 2024l. *Volcanoes*. www.usgs.gov/volcanoes. (Accessed March, 2024).)

that throws molten rock, ash, and gas into the air. The rocks cool quickly in the air, fall to the ground, and may break into smaller pieces that accumulate around and near the vent. As more eruptions occur, the material builds up around the vent and may reach a height of up to 300 m. Ciner cones are numerous, exceeding well more than a thousand, in the western United States.

2.6.5.3.2 Composite Volcano (Stratovolcano)

Some of the Earth's grandest mountains are composite volcanoes. They are usually tall with steep even sides and are made of repeating layers of lava flows, volcanic ash, cinders, blocks, and volcanic bombs. Some composite volcanoes rise over 3,000 m above the surrounding area but reach much higher. For instance, Mount Rainier is over 4,500 m (see Figure 2.21) and Ojos del Salado in Chile is the tallest composite volcano on Earth with a summit elevation greater than 7,000 m.

Some of the most famous and beautiful mountains of the world are composite volcanoes, including Mount Fuji in Japan, Mount Cotopaxi in Ecuador, Mount Shasta in California, Mount Hood in Oregon, Mount St. Helens and Mount Rainier in Washington, and Mount Vesuvius and Mount Etna in Italy.

Mount Etna is located on the island of Sicily off the southern coast of mainland Italy. Mount Etna is one of the most active volcanoes on Earth with six separate eruptions in 2024, with the most eruptions in one year being a total of 40 (Italian Geological Survey (IGS) 2024). Figure 2.22 is a photograph of Mount Etna showing the ash from the most recent eruption which took place in August 2024. This eruption is so recent that the ground is warm, and steam and hydrogen sulfide gases are being emitted. Figure 2.23 is a photograph of the crater of Mount Etna formed in an eruption in 2002. Figure 2.23 also shows part of the metropolitan area of Cantania which lies in the shadow of Mount Etna.

We will discuss these and many other volcanoes later in this book when we discuss the environmental risks they pose to humans.

FIGURE 2.21 Photograph of Mount Rainier, a large composite volcano. (Photograph by Daniel T. Rogers.)

FIGURE 2.22 Mount Etna located in Sicily showing ash, steam, and gases from an August 2024 eruption. (Photograph by Daniel T. Rogers.)

FIGURE 2.23 Crater of Mount Etna formed is the 2002 eruption and showing potions of the city of Catania in the distance. (Photograph by Daniel T. Rogers.)

2.6.5.3.3 Shield Volcano

Shield volcanoes are built almost entirely of fluid lava flows. Lava pours out of vents in all directions, either from the summit or along rift zones (fractures or fissures) that radiate out from the summit like spokes on a bicycle wheel. As lava flows overlap one another from successive eruptions, they construct a broad and gently sloped dome shape. Shield volcanoes build up slowly from the succession of lava flows over time and can spread over rather great distances, sometimes exceeding 10 km. In northern California and Oregon, many shield volcanoes are up to 5 km or more wide and as tall as 500 m or more. The Hawaiian Islands are composed of a chain of shield volcanoes including Kilauea and the world's largest shield volcano, Mauna Loa.

2.6.5.3.4 Lava Dome

A volcano can host many lava domes over a long period of time, so these are technically not a "volcano type" but are certainly considered an eruption phenomenon. Lava domes are lava flows made up of lava too thick to flow away from the vent. Lava squeezes out of the vent and accumulates as a giant pile over and around the vent. Some domes form pointy spines, while others appear as giant muffins. One such lava dome appeared in 1912 during an eruption of Katmai Volcano that measured more than 250 m wide and 100 m high.

2.6.5.3.5 Super Volcanos

A super volcano is a type of volcano that is not characterized by structure but by intensity of explosive eruptions. A volcano that has erupted with a Volcanic Eruption Index (VEI) of 8, the largest recorded value on the VEI index. VEI 8 eruptions are also sometimes referred to as super eruptions (USGS 2022). A super volcano is not a scientific term but has become popular and is now generally accepted by the scientific community.

Super volcanos are characterized as follows (USGS 2022):

- Ejecting more than 1,000 km^3 (240 cubic miles) of material
- Producing eruption columns that reach more than 25 km (16 miles) into the atmosphere, which includes the total height of the troposphere and into the lower portion of the stratosphere
- Causing ash to be carried across the entire Earth by wind
- Creating a circular-shaped collapse feature called a caldera above the evacuated magma chamber

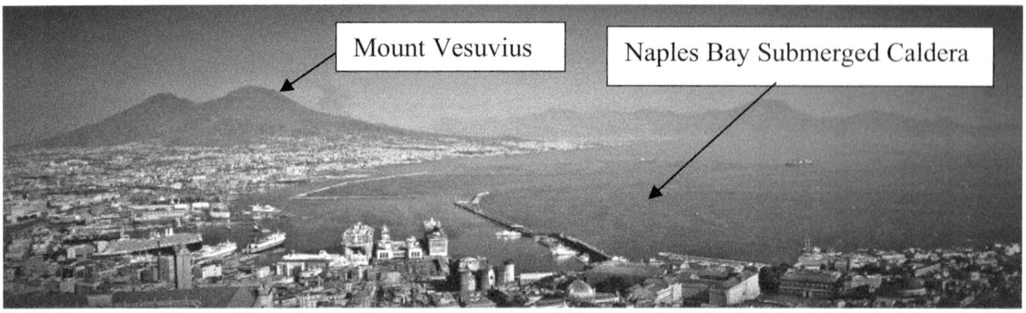

FIGURE 2.24 Photograph showing Naples, Italy, and Mount Vesuvius and a potential super volcano caldera beneath Naples Bay in the center and left portions of the photograph. (Photograph by Daniel T. Rogers.)

Examples of super volcano eruptions include the following (USGS 2022):

- *Taupo, New Zealand:* The most recent eruption was 22,600 years ago.
- *Toba, Indonesia:* The Toba Super Volcano last erupted 74,000 years ago and is believed to be the largest in the geologic record.
- *Yellowstone, United States:* The largest Yellowstone eruption occurred 2.1 million years ago.

The submerged volcanic caldera just offshore of Naples, Italy, is also considered a possible super volcano. Figure 2.24 is a photograph of a portion of Naples Bay along the eastern side of the underwater caldera. If confirmed as a super volcano, the proximity to nearly 3 million residents of Naples and the surrounding area is within just a few kilometers as shown in the photograph (Figure 2.25).

2.6.6 Erosion

Erosion is a geological process in which earth materials are worn away and transported by natural forces such as wind or water. A similar process, termed **weathering**, breaks down rock or dissolves rock, but does not involve movement. Erosion is the opposite of **deposition**, which is the geological process where materials are accumulated, or built up, on a landform. More erosion is performed by liquid water, wind, or ice (typically as a glacier). If the air appears hazy or dusty, or if water or glacial ice appears cloudy or muddy, erosion and transport are taking place. The discoloration is an indicator that rock and soil are suspended in the fluid (air or water) and are undergoing transport from one place to another. The transported material is termed **sediment** (USGS 2024m; National Geographic Society 2024).

Some natural factors impacting erosion include climate, geography, geology, topography, vegetation, and tectonic activity. Climate is perhaps the most influential force impacting the effect of erosion. Climate includes precipitation and wind. Climate also includes seasonal variability, which influences the likelihood of weathered sediments being transported during a weather event such as snowmelt, hurricane, or even a tornado. Topography can contribute to how erosion impacts a given location. Floodplains of river valleys are much more prone to erosion than rocky floodplains because of the energy required to initiate an erosional occurrence (USGS 2024m).

Vegetation can slow the impact of erosion. Plant roots adhere to soil particles and prevent their transport during a rain or wind event. Trees, shrubs, and other plants can even limit the impact of erosional and mass-wasting events such as landslides and other natural hazards such as hurricanes. Deserts generally lack significant vegetation and are, therefore, the most eroded types of landscapes on Earth (USGS 2024m).

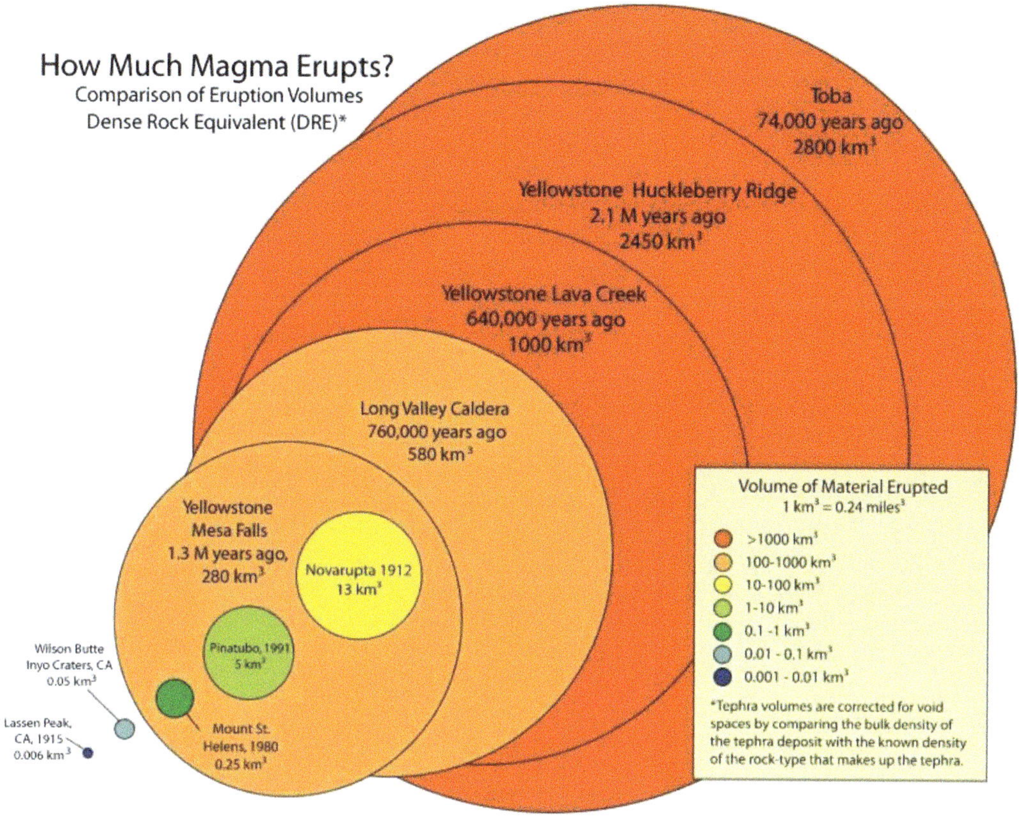

FIGURE 2.25 Comparison of eruption volumes.(From United States Geological Survey (USGS). 2022. Volume of Material Erupted from volcanoes. www.usgs.gov/media/images/comparison-erutption-sizes-using-volume-magma-erupted. (Accessed October 21, 2024).)

Tectonic activity shapes the landscapes themselves and thus greatly influences the way erosion impacts any given area on Earth. Tectonic uplift causes one part of the landscape to rise higher, therefore providing added energy to the system to erode through the force of gravity (USGS 2024m).

2.6.6.1 Physical Erosion

The following sections describe the types of physical erosion, which include physical erosion, water erosion, ocean erosion, wind erosion, ice erosion, thermal erosion, mass-wasting, solar erosion, and eroding animals.

Physical erosion describes the process of rock changing physical properties without changing its basic chemical composition. Physical erosion often causes rocks to get smaller or smoother. Rocks eroded through physical erosion often form deposits termed **clastic sediments**. Clastic sediments are composed of fragments of older rocks that have been transported from their place of origin.

Landslides and other forms of mass-wasting are associated with physical weathering. These processes cause rocks to dislodge from hillside slopes and crumble as they move and tumble down a slope (USGS 2024m). Plants can also contribute to physical erosion in a process called **bioerosion**.

Plants break up earthen materials as their roots grow and expand or create cracks and crevices in rocks. Ice and liquid also contribute to physical erosion as their movement forces rocks to crash together and chip or crash apart. Some rocks shatter and crumble, while others are worn away. Rocks in rivers are often much smoother than rocks found elsewhere (USGS 2024m).

2.6.6.1.1 Water Erosion

Liquid water is a major mechanism of erosion. Rain, rivers, floods, lakes, and the ocean carry away rock particles and soil sometimes long distances as sediment. Rainfall produces four types of soil erosion and each is described below in a succession of increased energy (USGS 2024m; National Geographic Society 2024):

- **Splash erosion** describes the impact of a falling drop of rain, which can scatter small particles of soil and other earthen materials a distance of slightly greater than 0.5 m.
- **Sheet erosion** describes erosion caused by runoff typically during or just after a storm event when the soil becomes saturated and can no longer hold more water.
- **Rill erosion** describes erosion that takes place as runoff develops into discrete streams of water called rills.
- **Gully erosion** is the stage in which soil particles are transported through large and typically rapid-flowing channels.

Gullies carry water for brief periods of time during a rainfall event or rapid snow melt and appear as small valley-like structures or crevasses during dry periods. **Valley erosion** is the process in which rushing water in streams and rivers where away stream or river banks, especially when the stream or river changes direction. In this instance, the water flowing within the outer portion of the directional change travels faster and, as a result of this added energy, erodes the stream or river bank faster (USGS 2024m).

2.6.6.1.2 Ocean Erosion

The ocean is also a major mechanism of erosion. **Coastal erosion** is defined as the wearing away of rocks, earthen material, or sand on the beach. Erosion caused by activity of the ocean can and does change the shape of entire coastlines. During the process of coastal erosion, energy from wave action can change rocks into pebbles, and pebbles into sand. Waves and currents can transport sand away from beaches, moving the coastline farther inland.

The action of erosion caused by the ocean can erode seaside cliffs and create sea stacks (USGS 2024m). Figure 2.26 is a photograph showing the effects of splash and wave erosion by polishing rock on a cliff along the shoreline. Figure 2.27 is a photograph showing sea stacks created from ocean erosion along a shoreline.

2.6.6.1.3 Wind Erosion and Aeolian Processes

Wind is a significant mechanism of erosion. **Aeolian** is a term used to describe wind-driven dust transport that can also include ash and sand particles from one location to another. Wind-blown particles can be transported a few feet or around the world if the conditions are favorable. An example of long-distance transport is a volcanic eruption in which small particles of ash are ejected high into the atmosphere. Another example of long-distance transport of wind-driven dust particles from desert regions is in Africa where they are transported to South and North America (USGS 2024m).

Wind-blown sand can bombard surfaces and polish rocks and cliffs until they are smooth. Wind is responsible for eroded features of some national parks such as Arches National Park in Utah (USGS 2024m). Some of the most destructive examples of wind erosion are dust storms (see Figure 2.28).

2.6.6.1.4 Ice Erosion

Ice, usually as glaciers, can erode the Earth and create dramatic landforms. Figure 2.29 shows the floor of Yosemite Valley, which was once buried in ice less than 20,000 years before the present. The hanging valley in the center of the photograph shows where a glacier once combined with the main

FIGURE 2.26 Effects of splash erosion from wave action along the coast of South Africa at the Cape of Good Hope. (Photograph by Daniel T. Rogers.)

FIGURE 2.27 Photograph showing creation of sea stack along the northern Oregon coast. (Photograph by Daniel T. Rogers.)

FIGURE 2.28 Dust storm at Death Valley National Park in California. (Photograph by Daniel T. Rogers.)

FIGURE 2.29 Photograph of the floor of Yosemite National Park showing how glaciers can carve out a valley. (Photograph by Daniel T. Rogers.)

glacier where Bridalveil Falls now is located. Areas near mountaintops where glaciers are present can transport earthen material downslope in their direction of travel. As they move, they transport material from sand to huge boulders. Rocks carried by glaciers scrape against the ground surface below and erode the ground surface. Glaciers also grind up earthen material within them as the glaciers move downslope. Moving glaciers can gouge out basins and form steep mountain valleys. Eroded sediment that was once carried by glaciers is often visible on and around glaciers (USGS 2024m).

Several times in Earth's history, vast glaciers have covered parts of the Northern Hemisphere. These glacial periods, described earlier in this chapter, are known as ice ages. Recent Ice Age glaciers have carved out much of the modern North American and European landscape.

On a smaller scale, water can erode rock when water infiltrates fracture in rock and then when the water freezes and expands, the rack weakens and can fracture in a process called **freeze–thaw action** (USGS 2024m).

2.6.6.1.5 Thermal Erosion

Thermal erosion typically describes the erosion of permafrost along a river or coastline in arctic regions of the world. Warming temperatures can cause ice-rich permafrost to break from coastlines in large chunks and often carry valuable topsoil and vegetation with them. The eroded floating islands can disintegrate as they thaw in the ocean or even crash into another portion of land which may assist the spread of new life species into new habitats (USGS 2024m).

2.6.6.1.6 Mass-Wasting

Mass-wasting is a type of erosion that involves the downward movement of rock, soil, and vegetation. Mass-wasting incidents include landslides, rockslides, and avalanches. Mass-wasting can erode and transport millions of metric tons of earth materials. Mass-wasting can reshape hills and mountains and often devastate communities in its path (USGS 2024m).

2.6.6.1.7 Solar Erosion

The sun itself can cause erosion. This occurs when rocks heat up when exposed to direct sunlight and, when this occurs, the rocks heat up and expand. The expansion can sometimes be enough to cause the rocks to crack (USGS 2024m).

2.6.6.1.8 Eroding Animals

Burrowing animals such as prairie dogs, chipmunks, and even bears and insects such as some species of beetles, worms, and ants can contribute to erosion by displacing soil (USGS 2024m).

2.6.6.2 Chemical Erosion

Another type of erosion is chemical erosion. All materials that are exposed to the outdoor environment are subject to degradation caused by natural chemical weathering and erosional processes. All rocks that have calcium carbonate in their structure ($CaCO_3$) are subject to chemical erosion through a process called dissolution.

The chemistry of rock composed of some fraction of carbonate has the potential to undergo chemical dissolution, predominantly when in contact with rainwater or groundwater. Under equilibrium conditions, the incremental impact of hydrogen ions in pH 4.0 rainwater on carbonate rocks is expected to be small because the equilibrium solubility of calcite is dominated by its reaction with carbonic acid derived from atmospheric carbon dioxide that has dissolved in rainwater (USGS 2024n). Once the calcium is dissolved, it can migrate vertically and travel in groundwater and precipitate calcium carbonate or other minerals in the form of stalagmites (form from the floor up) and stalactites (form from the ceiling down) in caves. Stalagmites and stalactites are icicle-shaped deposits that form when the water re-deposited calcium carbonate along the ceiling and floors of caves (USGS 2024o).

2.7 HUMAN DEVELOPMENT

You may ask yourself what does human evolution and civilization have to do with environmental geology. Well, the answer is everything. Humans and all living organisms are a product of their environment and the environment at any given location is a product of geological factors. Some of those factors include other planets, planetoids, and local stars. For instance, the sun provides Earth with its solar energy, the moon gifts Earth with tidal influences, and the position of Earth with respect to the sun such that water is present in each of its three fundamental forms, the stability of our solar system, and our place in the Milky Way galaxy.

Additional factors on Earth itself include the tilt and rotation of Earth, plate tectonics which creates and recycles rock and which drives the position of continents and ocean currents, the chemistry and composition of Earth, climate and weather, the geography and topography of Earth, and the composition of the atmosphere. Each of these mentioned factors influences life's existence, abundance, and distribution on Earth (Ward and Brownlee 2000).

2.7.1 Human Evolution

For the past 3.8 billion years, life on Earth has been evolving. Understanding the basic questions of the evolution of life on Earth, including humans, is one of the most scientifically studied fields of science (Larsen 2004; Shubin 2009). Much has been learned through careful study, thought, and technological advances over more than 200 years. This gained knowledge has opened our eyes to how we became who we are, how all life on Earth has changed, and why it has changed over the past several billion years (Eldredge 2000; Gould 2002).

To understand how *Homo sapiens* evolved from early lineages of hominins, the members of the group that includes modern humans and our closest extinct relatives and ancestors, science has carefully studied ancient bones and stone tools, examined our genes, and evaluated the geology and changing environments that assisted in shaping our ancestors' world that guided their evolutionary path (Bohme et al. 2019; Smithsonian 2021).

The evolutionary transition to what we shall call more human-like characteristics occurred at what is believed to be approximately 6 million years ago (Beard 2004; Pattison 2020). This occurred when our earliest known ancestor, named *Sahelanthropus*, began the slow transition. However, it would still take more than 5 million years for *Homo sapiens* to appear. A plethora of different species lived, evolved, and went extinct along the way. Many of these early human-like versions intermingled and may have interbred at some points (Smithsonian 2021). During this period of evolutionary development, human ancestors were constantly changing. Body morphology changed, brains changed, and the ability to think changed, as evidenced by the tools and developing technology (Pettitt 2022; Smithsonian 2021).

The cumulative assessment of the lines of evidence increasingly suggests that the origin of the species *Homo sapiens* originated in Africa but not necessarily in a single time or place (Tattersall 2009). Instead, based on current evidence, diverse groups of human ancestors lived in habitable regions around Africa, evolving physically and culturally in relative isolation until changes in climate occurred, which influenced behavior in ways that intermittently brought them together and mixed and exchanged everything from genes to tool technology. Eventually, this behavior influenced the genetic makeup of modern humans (Smithsonian 2021). Genetic makeup assists in examining migration behavior, movements, and evolution of *Homo sapiens*. The oldest current DNA recovered of an early human ancestor comes from Spain's Atapuerca Mountains where teeth and bones were recovered from 28 different individuals. From these remains, a partial genome was reconstructed and the evidence revealed that the remains were from the oldest known Neanderthals and dated to 430,000 years ago (Smithsonian 2021). Although our own DNA has traces of DNA from Neanderthals and another hominin species termed **Denisovans** suggests a common ancestor, it is not apparent who that was (Smithsonian 2021).

Stone tools have been used along our lineage for approximately 3.3 million years. Approximately 1.75 million years ago, a new suite of tools, namely hand axes and cutting tools, came into use and were used until about 400,000 years ago when thrusting spears appeared and were used to hunt large prey. At around 300,000 years ago, a leap in technology took place whereby finely crafted tools

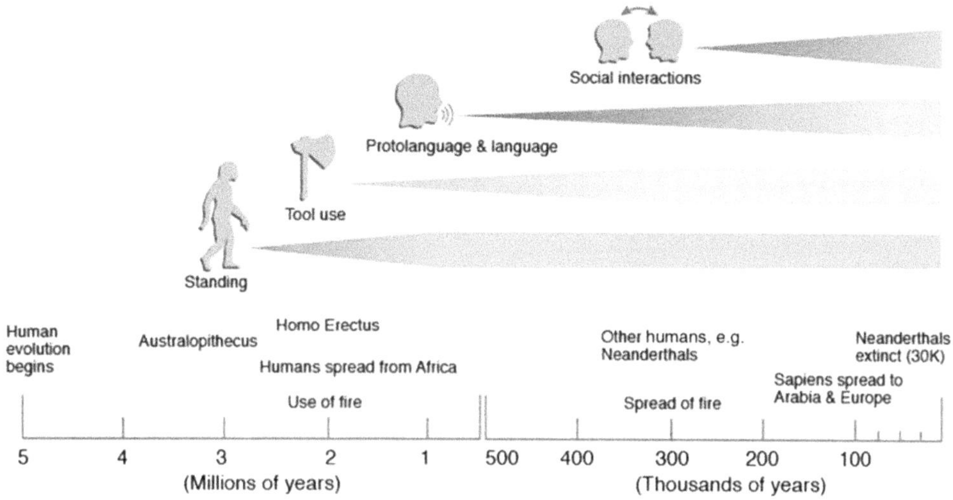

FIGURE 2.30 Summary of aspects of human evolution. (From Smithsonian. 2021. *Understanding the Essential Timeline of Homo sapiens Evolution.* www.smithsonian.org/essential-timeline-of-homo-sapens-evolution. (Accessed May 31, 2024).)

appeared which were flaked points that were attached to spears, which greatly improved hunting success (Pettitt 2022; Smithsonian 2021). In addition, at around this same time, the innovation of cooking meat using fire was discovered. Cooking meat improves diet and nutrition, which likely is one factor that influenced increased brain size (Bohme et al. 2019).

Many genetic analyses trace our beginning to Africa. However, the evidence also indicates that hominids had a propensity to wander. This is evidenced by fossil remains of hominids from Israel, Greece, and even as far away as China (Smithsonian 2021).

Several advantageous periods at around 50,000 to 60,000 years ago caused by variations in climate gave early humans the opportunity to leave Africa and migrate to the Arabian Peninsula and the Middle East. While other earlier hominids still living at the time may have interbred with early humans, their lineage appears to have faded out or had been overwhelmed by the later migration (Smithsonian 2021; Stringer 2012; Fagan 2011).

For most of our history, *Homo sapiens* have not been the only humans. We coexisted with other hominin species, and as our genes make clear, we interbred with other hominin species until as recently as 15,000 years ago (Smithsonian 2021). Figure 2.30 presents a current summary of hominid evolution.

What is the reason why other human species suffered extinction and we as *Homo sapiens* survived? Neanderthals actually had slightly larger brains, so having a larger brain is not likely the reason. A current theory is that *Homo sapiens* developed complex communication skills and other hominin species did not. The development of complex communication skills made it possible for a small or large group of *Homo sapiens* to cooperate as one toward a common goal. Complex communication also permitted *Homo sapiens* to develop storytelling capabilities that can be used to unite individuals toward achieving a common goal. An example is building cities. Figure 2.31 is a photograph of the late Greek and early Roman civilizations in Split, Croatia, where the exchange of information was a key factor in our social and urban development.

Throughout our history, the transfer of information has been key in developing civilization and achieving goals that could not be achieved by an individual. This is accomplished by storytelling. Storytelling permitted information to be handed down through generations which can be used for motivation and kinship beyond the immediate family (Harari 2024), and to portray this concept in the simplest way, "two brains are better than one."

FIGURE 2.31 Photograph showing late Greek and early Roman civilization dating back to as far as 200 BCE in Split, Croatia. (Photograph by Jennie Macaluso, with permission.)

Storytelling is the first information revolution. Others include the written word, the published word, the library, the television, the computer, the internet, and the development of artificial intelligence. Throughout our history, these vehicles of exchange and storage of information have been used to bond groups in various ways, including branding, nationalism, misinformation, disinformation, propaganda, slavery, teaching, war, and justifying destruction.

2.7.2 Human Civilization

According to Stephens et al. (2019), humans began leaving a lasting impact on Earth's surface starting 10,000 to 8,000 years ago. This time period marks the transition from hunter-gatherers to farming and pastoralists. Growing food allowed the world's human population to grow from about 10 million in the Later Pleistocene Epoch to more than 8 billion today. However, this came at a cost. The varied diet of foragers was replaced with a much more limited diet of domesticated plants and animals. This transition tended to favor a reduced nutritional quality. Now, much of the world's population relies on three foods, which include rice, corn, and wheat. Another important change in the human diet was the introduction of dairy around 5,000 years ago in northern Europe (Larsen 2023).

As humans expanded the use of agriculture, they began creating communities which changed social behavior. This laid the foundation for kinship, community organization, and conflict (Harari 2018; Larsen 2023). Currently, the amount of land that humans have developed is staggering (see Figure 2.32).

In addition, as cities grew and populations increased, some of the locations where humans settled were within geologic hazards. The outcome of locating a city within a geologic hazard was that when the hazard was triggered, loss of human life could be significant. Figure 2.33 shows Mount Vesuvius near the settlement of Pompeii where more than 3,000 residents died when the volcano suddenly erupted in 79 AD. Figure 2.34 shows the mummified remains of a person from Pompeii who died as a result of the eruption. Figure 2.34 also shows the speed at which the pyroclastic flow

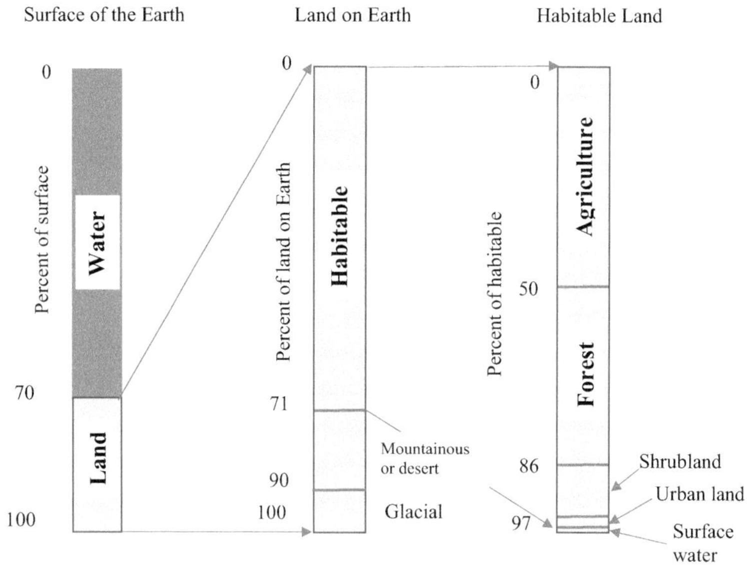

FIGURE 2.32 Breakdown of land and land use on Earth.(From United Nations. 2024. *Human Development Report*. https://hdr.undp.org. (Accessed March 9, 2024).)

FIGURE 2.33 Photograph showing Pompeii in the shadow of Mount Vesuvius. Italy, where more than 3,000 residents died in 79 AD from a volcanic eruption of Mount Vesuvius located in the background. (Photograph by Daniel T. Rogers.)

overtook Pompeii because the person shown in the photograph died while sleeping in bed along with many other people.

The modern Earth narrative, the scientific story of the 4.54-billion-year natural history of Earth generally does not include significant discussion of humans because our species has been in existence for such a short time. Scientific research centers on the natural history of Earth, geologic time, and biological evolution. Discussing Earth in this fashion is important as we begin the third millennium of our current calendar. It provides a clear viewpoint and understanding of the increasing impact of human population growth and associated stresses placed on the Earth system.

Humans have significantly impacted the air, water, land, and biosphere of the entire planet in such a short period of time that Earth has simply not been able to adapt. In fact, many scientists

FIGURE 2.34 Mummified remains of a resident of Pompeii killed while sleeping from the eruption of Mount Vesuvius in 79 AD. (Photograph by Daniel T. Rogers.)

now believe we have moved the needle of geologic time to a new geologic time period termed **Anthropocene**, meaning the age of humans. Many scientists believe that there is widespread consensus for this new designation. The idea that human activity has become the dominant force on Earth during recent recorded history is well established. What is debated, across many scientific disciplines, is not whether the Anthropocene exists, but at what point did our actions as a species become a climate-driving force equal to major geologic processes. An additional debated point is that it occurred in such a short period of time, will it truly be a new geologic period or will humans simply go extinct and these impacts will be recorded as no more than a short-term event in Earth history (Stephens et al. 2019)?

For thousands of years, human civilization has been busy changing the environment for their own benefit and with little or no regard for the natural environment. Historically, the focus was on the exploitation of the natural environment rather than protecting the natural environment. This disregard for the natural environment even extended to the United States where tens of thousands of examples of polluting land from anthropogenic sources was a common and even accepted practice.

Now that climate change has caught our attention, we are noticing and experiencing several of the detrimental realities that include higher temperatures, severe drought, flooding, and stronger slowly moving storm systems. This has raised our collective awareness of the fact that we can impact our environment on a global scale, and it does not include just the atmosphere.

2.7.3 Future Human Development

Technological innovations in medicine and public health have improved our survival to adulthood and permitted more humans to successfully breed. As survival improves, the influence of natural selection decreases and no longer is a major factor that shapes the human population. In the wild, natural selection is the most powerful evolutionary force, but other factors may take over when technology grants a second chance to those who would have died in the wild. However, evolution still occurs in other ways including mutations and factors that are not influenced by technology. Over the last 3,000 years, various genes that affect metabolic function such as lactose digestion and insulin signaling have evolved (Warmflash and Lents 2017).

Another important factor to discuss that is a result of human technology is the concept of artificial selection. The term artificial selection implies a process affecting evolution the same way as natural selection, but due to human activity as opposed to natural processes. An example is the evolution of dogs from wolves. This transition occurred over several thousand years. During this transition, humans began to consciously breed dogs for certain desired traits or for personal enjoyment (Warmflash and Lents 2017).

Many believe we are at a crossroads in our brief history on Earth. What we choose to do in the near future to address how humans have impacted the Earth will dictate how we and many other species survive in the future. Humans have a choice to continue along their current path or to choose a different path. Hopefully, if a different path is chosen, then it will include learning how we can exist in nature instead of destroying nature and changing the path of evolution.

2.8 SUMMARY AND CONCLUSION

It is important for each one of us to realize that we are but one of about 4,500 species of mammals that exist on Earth and that we are but one species in an estimated 30 to perhaps more than 100 million species that form Earth's biosphere. We are totally dependent on the biosphere for the food we eat, the air we breathe, and all the other requirements to maintain life. We find ourselves on Earth after billions of years of Earth's existence by chance, not by design. Our species has become so successful and with such speed at modifying our environment that many other life forms have been overcome and dominated by human technological and cultural innovations.

It is equally important for each of us to be aware of geologic hazards because humans can do little, if anything in some instances, to stop them. Therefore, we must be diligent to either avoid them or prepare to survive them through mitigation methods when they do occur. During the course of this book, we shall examine the relationship between geology and humans through actions that can benefit both in a sustainable manner.

REFERENCES

American Geological Institute (AGI). 1962. *Dictionary of Geological Terms*. Dolphin Books. New York, NY. 788pp.

Anderton, R. 1985. *Clastic Facies Models and Facies Analysis*. Vol. 18. Geological Society of London, Special Publication. London, England.

Beard, C. 2004. *The Hunt for the Dawn Monkey*. University of California Press. London, England. 348pp.

Bennett, M. R., and Doyle, P. 1997. *Environmental Geology; Geology and the Human Environment*. John Wiley & Sons. New York, NY. 640pp.

Black, R. 2022. *The Last Days of the Dinosaurs: An Asteroid, Extinction, and the Beginning of Our World*. St. Martin's Press. New York, NY. 287pp.

Bohme, M., Braun, R., and Breier, F. 2019. *Ancient Bones: Unearthing the Astonishing New Story of How We Became Human*. Greystone Books. Vancouver, Canada. 337pp.

Brusatte, S. 2018. *The Rise and Fall of the Dinosaurs: A New History of a Lost World*. Harper Collins Publishers. New York, NY. 408pp.

Canyonlands Natural History Association. 2024. *Geologic Cross Section of Canyonlands National Park*. www.cnha.org. (Accessed March 7, 2024).

Christopherson, R. W. 2008. *Geosystems; An Introduction of Physical Geography* 7th Edition. Prentice Hall. Upper Saddle River, NJ. 528pp.

Dana, J. D. 1985. In Hurlbut, C. S., and Klein C. eds. *Manual of Mineralogy*. 7th Edition. John Wiley & Sons. New York, NY. 834pp.

Eldredge, N. 2000. *The Pattern of Evolution*. W.H. Freeman & Company. New York, NY. 219pp.

Fagan, B. 2011. *Cro-Magnon: How the Ice Age Gave Birth to the First Modern Humans*. Bloomsbury Press. New York, NY. 295pp.

Flint, F. F., and Skinner, B. J. 1974. *Physical Geology*. John Wiley & Sons. New York, NY. 426pp.

Gould, S. J. 2002. *The Structure of Evolutionary Theory*. Harvard University Press. Cambridge, MA. 1433pp.

Harari, Y. N. 2018. *Sapiens; A Brief History of Humankind*. Harper Collins Publishers. New York, NY. 443pp.

Harari, Y. N. 2024. *Nexus: A Brief History of Information Networks from the Stone Age to AI*. Harper Collins Publishers. New York, NY. 492pp.

Hongren, Z. 2004. Geology of Italy. *Journal of International Geoscience*. Vol. 26, no. 3.

Hutton, J. 1788. Theory of the Earth; or an Investigation of the Laws Observable in the Composition, Dissolution, and Restoration of the Land Upon the Globe. *Transactions of the Royal Society of Edinburgh*. Vol. 1, Part 2, pp. 209–304.

Italian Geological Survey (IGS). 2024. *Volcanic History of Mount Etna*. www.italianjournalofgeosciences.it/297/article1115. (Accessed October 19, 2024).

Kious, W. J., and Tilling, R. I. 1996. *The Dynamic Earth: The Story of Plate Tectonics*. United States Geological Survey. United States Government Printing Office. Washington, DC.

Krumbein, W. C. and Sloss, L. L. 1963. *Stratigraphy and Sedimentation*. 2nd Edition. W. H. Freeman. San Francisco, CA.

Larsen, C. S. 2023. The past 12,000 Years of Behavior, Adaptation, Population, and Evolution Shaped Who We Are Today. *Proceedings of the National Academy of Sciences*. Vol. 120, no. 4, pp. 120–124.

Larsen, L. J. 2004. *Evolution: The Remarkable History of a Scientific Theory*. Random House. New York, NY. 347pp.

Lawrence Livermore National Laboratory. 2024. *Where on Earth Did the Water Come From*. www.llnl.gov/article/48436/where-did-water-come-from. (Accessed March 4, 2024).

National Atmospheric and Space Administration (NASA). 2024a. *Earth Fact Sheet*. www.science.nasa.gov/earth/facts. (Accessed March 7, 2024).

National Atmospheric and Space Administration (NASA). 2024b. *60 Years and Counting – Mysteries of the Universe*. (Accessed March 3, 2024). www.nasa.gov/specials/60counting/overview.html

National Geographic Society. 2024. *Erosion*. https://NGS.org/education.nationalgeographic/resource/erosion. (Accessed March 8, 2024).

National Oceanic and Atmospheric Administration (NOAA). 2024a. *Introduction to the Atmosphere*. www.noaa.gov/jetstream/atmisphere. (Accessed March 3, 2024).

National Oceanic and Atmospheric Administration (NOAA). 2024b. *Layers of the Atmosphere*. www.noaa.gov/jetstream/atmosphere/layers-of-atmosphere. (Accessed March 3, 2024).

Pattison, K. 2020. *Fossil Men: The Quest for the Oldest Skeleton and the Origins of Humankind*. Harper Collins Publishers. New York, NY. 534pp.

Pettijohn, F. J. 1975. *Sedimentary Rocks*. 3rd Edition. Harper & Row. New York, NY. 525pp.

Pettitt, P. 2022. *Homo sapiens Rediscovered: The Scientific Revolution Rewriting Our Origins*. Thames and Hudson. New York, NY. 304pp.

Rogers, D. T. 2020. *Urban Watersheds: Geology, Contamination, Environmental Regulations, and Sustainability*. 2nd Edition. CRC Press. Boca Raton, FL, 610pp.

Rogers, D. T. 2023. *Environmental Compliance Handbook: Sustainability and Future Environmental Regulations*. CRC Press. Boca Raton, FL. 548pp.

Rogers, M. H., and Rogers, D. T. 2022. Incorporating Climate Change Modeling to Parcel Level Sustainability Efforts Globally. *International Journal of Sustainable Development*. Vol. 11, no. 4.

Seyfert, C. K., and Sirkin, L. A. 1973. *Earth History and Plate Tectonics; An Introduction to Historical Geology*. Harper & Row. New York, NY.

Shubin, N. 2009. *Know Your Inner Fish: A Journey into the 3.5-Billion-Year History of the Human Body*. Vintage Books. New York, NY. 237pp.

Smithsonian. 2021. *Understanding the Essential Timeline of Homo sapiens Evolution*. humanorigins.si.edu/evidnce/human-evolution-interactive-timeline. (Accessed May 31, 2024).

Stanley, S. M. 1999. *Earth System History*. W. H. Freeman. New York, NY. 650pp.

Stephens, L., Fuller, D., Boivin, N., and Ellus. E., 2019. Archaeological Assessment Reveals Earth's Early Transformation through Land Use. *Science*. Vol. 365, no. 6456, pp. 897–902.

Stringer, C. 2012. *Lone Survivors: How We Became the Only Humans on Earth*. St. Martin's Griffin. New York, NY. 520pp.

Tattersall, I. 2009. *The Fossil Trail: How We Know What We Think We Know About Human Evolution*. 2nd Edition. Oxford University Press, New York, NY. 327pp.

United Nations. 2024. *Human Development Report*. https://hdr.undp.org. (Accessed March 9, 2024).

United States Geological Survey (USGS). 2006. *The Geologic Story of the Great Plains*. Geologic Survey Bulletin 1493. USGC. Washington, DC. 79pp. https://hdr.undp.org

United States Geological Survey (USGS). 2022. *Volume of Material Erupted from Volcanoes*. www.usgs.gov/media/images/comparison-erutption-sizes-using-volume-magma-erupted. (Accessed October 21, 2024).

United States Geological Survey (USGS). 2023. *Igneous Rocks*. www.usgs.gov/factswhat-are-igneaous-rocks. (Accessed September 26, 2024).

United States Geological Survey (USGS). 2024a. *Age of the Earth*. https://pubs.usgs.gov/gip/geotime/age.html. (Accessed March 3, 2024).

United States Geological Survey (USGS). 2024b. *Inside the Earth*. www.usgs.gov/gip/dynamic/inside.html. (Accessed March 3, 2024).

United States Geological Survey (USGS). 2024c. *Geologic Map of North America*. www.usgs.gov/media/images/geologic-map-of-north-america.. (Accessed May 3, 2024).

United States Geological Survey (USGS). 2024d. *Precambrian Time – The Early History of Earth*. www.usgs.gov/publications/precambrian-time-story-early-earth. (Accessed March 4, 2024).

United States Geological Survey (USGS). 2024e. *The Paleozoic*. www.usgs.gov.youth-and-education-in-science/paleozoic. (Accessed March 4, 2024).

United States Geological Survey (USGS). 2024f. *The Mesozoic*. www.usgs.gov.youth-and-education-in-science/mesozoic. (Accessed March 4, 2024).

United States Geological Survey (USGS). 2024g. *The Cenozoic*. www.usgs.gov.youth-and-education-in-science/cenozoic. (Accessed March 5, 2024).

United States Geological Survey. 2024h. *World Map Showing Plate Boundaries and Active Volcanoes*. http://vulcan.wr.usgs.gov/Glossary/PlateTectonic/Maps/map_plate_tectonics_world.html. (Accessed March 5, 2024).

United States Geological Survey (USGS). 2024i. *Geological Survey Bulletin 1393*. The Geologic Story of Arches National Park. www.usgs.gov/publications/geologic-story-arches-national-park. (Accessed March 7. 2024).

United States Geological Survey (USGS). 2024j. *What Is a Fault and What are the Different Types*. www.usgs.gov/faqs/what-is-a-fault-and-what-are-different-types. (Accessed March 5, 2024).

United States Geological Survey (USGS). 2024k. *Diagrams Showing Fault Types*. www.usgs.gov/tools/interactive-us-fault-map. (Accessed March 5, 2024).

United States Geological Survey (USGS). 2024l. *Volcanoes*. www.usgs.gov/volcanoes. (Accessed March, 2024).

United States Geological Survey (USGS). 2024m. *Erosion*. www.usgs.gov/mission-areas/water-resources/science/science-topis/erosion-and-sedimentation. (Accessed March 8, 2024).

United States Geological Survey (USGS). 2024n. *Erosion through Chemical Dissolution*. https://usgs.gov/publications/statistical-compliation-napap-chemical-erosion-oberservations. (Accessed March 8, 2024).

United States Geological Survey (USGS). 2024o. *Caves and Their Geology and Chemistry*. www.usgs.gov/centers/astrogeology-science-center/scince/caves. (Accessed March 8, 2024).

United States National Park Service. 2024. *Geologic Time Scale*. www.nps.gov/subjects/geology/time-scale.htm. (Accessed March 4, 2024).

Utah Geologic Survey. 2024. *What Is a Formation*. https://geology.utah.gov/map-pub/survey-notes/glad-you-asked/what-is-a-formation. (Accessed March 6, 2024).

Ward, P. D., and Brownlee, D. 2000. *Rare Earth: Why Complex Life is Uncommon in the Universe*. Springer-Verlag, New York, NY. 333pp.

Warmflash, D., and Lents, N. H. 2017. *Future of Human Evolution*. National Center for Biotechnology Information and National Institute of Health. www.visionlearning.com/en/library/biology/2/future-of-human-evolution. (Accessed June 1, 2024).

Wicander, R., and Monroe, J. 2007. *Historical Geology*. Thompson Brooks/Cole. 434pp.

3 General Geology of Urban Regions

3.1 INTRODUCTION

The study of near-surface geology, especially in urban areas, will become increasingly important in the future. In part, this is because the world's human population continues to increase and the population in urban areas is increasing at a rate much greater than the human population as a whole. More than 3 billion people currently live in urban areas. By the year 2030, the human population in urban areas is projected to be nearly 5 billion or 60% of the 8.6 billion people projected to exist on the planet. The current level of population has placed enormous pressure on our environment and the projected increases are likely to cause additional stress. Therefore, a thorough understanding of the subsurface geologic environment in urban areas is critical to understand and protect the cities and places we live.

3.2 GEOLOGY KNOWLEDGE GAP IN URBAN AREAS

Environmental geologic challenges faced in the typical urban setting can in part be attributed to the poor understanding of their subsurface geology. Lack of attention to this area of inquiry by the scientific community is one factor contributing to this knowledge gap. Geologists generally have not spent very much effort or time studying the geology of urban areas because they (Rogers 2020):

- have only become important in the last 50 years with the onset of significant environmental regulations
- are covered by buildings and roads and other human-constructed surfaces
- are difficult to access because many desirable study locations are on private property
- have been disturbed and modified due to development
- are typically located on unconsolidated sediments of Quaternary age and were historically considered overburdened and not as scientifically valuable as the bedrock located beneath the unconsolidated deposits.
- are very complex and time-consuming to study
- involve issues that may take decades to be noticed by the general community as compared with other more sudden types of geologic hazards such as earthquakes, tsunamis, and volcanic eruptions.

Most urban areas of the world are not located directly upon bedrock. Urban landscapes tend to be located on continental and marine sedimentary depositional environments including alluvial, beach, eolian, fluvial, glacial, lacustrine, and volcanic-type deposits. Many other cities are also located on sedimentary deposits of marine origin. Table 3.1 lists 29 urban areas in the United States and

DOI: 10.1201/9781003463153-4

TABLE 3.1
Major Urban Centers in the United States and Their Sedimentary Environments

	Geologic Environment								
Urban Area	**Alluvial**	**Beach**	**Eolian**	**Fluvial**	**Glacial**	**Lacustrine**	**Marine**	**Volcanic**	**Bedrock**
Atlanta[1]	X			X		X			
Baltimore[2]	X			X			X		
Boston[3]	X			X	X		X		X
Chicago[4]		X		X	X	X			
Cincinnati[5]	X			X	X				X
Cleveland[5]		X		X	X	X			
Dallas[6,7]	X		X	X		X			X
Detroit[8,9]		X		X	X	X			
Denver[10]	X		X	X	X				X
Houston[6,7]	X			X			X		
Indianapolis[30]				X	X	X			
Kansas City[11,12]	X			X	X				
Las Vegas[13]	X		X	X		X			X
Los Angeles[14]	X	X	X	X		X	X		X
Miami[15]	X			X			X		
Minneapolis[16]	X			X	X	X			X
New Orleans[17]		X		X		X	X		
New York[18]	X	X		X	X	X	X		X
Philadelphia[19,20]		X		X		X	X		X
Phoenix[21,22]	X		X	X		X			X
Pittsburgh[19]	X			X	X	X			X
Portland[23]	X			X	X	X	X	X	X
San Antonio[6,7]	X		X	X		X			X
Saint Louis[24]	X	X		X	X				X
Salt Lake City[25]	X	X		X	X	X			X
San Francisco[26]	X	X		X			X		X
Seattle[28]	X	X		X	X	X	X	X	X
Tampa[15,27]	X	X		X		X	X		X
Washington, DC[29]	X			X		X			X

Sources: King and Beikman (1974) and others.

[1]Abrams 1984; [2]USGS 2009a; [3]Kaye 1976; [4]Alder 1902; [5]Coogan 1996; [6]USGS 2009b; [7]Spearing 1991; [8]Rogers1996; [9]Rogers 1997; [10]Halka 1980; [11]Aber 1984; [12]Gentile 1984; [13] Page et al. 2005; [14]Bilodeau et al. 2007; [15]Florida Geological Survey 1994; [16]Ojakangas and Matsch 1982; [17]Snowden 1980; [18]USGS 2003; [19]Barnes and Sevon 1996; [20]Goodwin 1964; [21]Allen 2003; [22]Halka 1983; [23]Trimble 1963; [24]Missouri Department of Natural Resource 2004; [25]Stokes 1989; [26]Stoffer and Gordon 2001; [27]USGS 2005, [28]Stewart 1992; [29]USGS 2007; [30]Indiana Geological Survey 2024.

identifies the dominant geology beneath each area by type of depositional environment. As you can see, all of the sedimentary depositional environments are represented. Table 3.2 lists 25 other major urban areas of the world and identifies the dominant geology beneath each urban area.

As most urban regions of the United States and the world are dominated by recent (generally Quaternary in age) unconsolidated sedimentary deposits, we shall focus efforts in this chapter to describe sedimentary processes in greater detail. Those subject areas of sedimentary processes are those that are listed in Tables 3.1 and 3.2 and include alluvial, beach, eolian, fluvial, glacial, lacustrine, marine, and volcanic.

TABLE 3.2
Major Urban Centers of the World and Their Sedimentary Environments

	Geologic Environment								
Urban Area	**Alluvial**	**Beach**	**Eolian**	**Fluvial**	**Glacial**	**Lacustrine**	**Marine**	**Volcanic**	**Bedrock**
Beijing[1]	X			X		X			
Berlin[2]	X			X		X			
Buenos Aires[3]	X			X		X	X		
Cairo[4]	X	X	X	X		X			X
Cape Town[5]	X	X		X			X		X
Jakarta[24]	X	X		X		X	X	X	
Johannesburg[6]	X			X		X			
London[7]	X		X	X		X	X		X
Madrid[8]	X		X	X		X			
Mexico City[9]	X		X	X		X		X	X
Monterrey[23]	X			X		X			X
Montreal[22]		X		X	X		X		X
Moscow[10]	X			X		X			
New Dehli[11]	X			X		X			
Paris[12]	X			X		X	X		
Rio de Janeiro[13]	X	X		X		X	X		X
Rome[14]	X			X		X			X
Santiago[20]	X			X					
Sao Paulo[15]	X			X			X		
Shanghai[1]	X			X		X	X		
Singapore[16]	X	X		X		X	X		
Sydney[17]	X	X		X		X	X		X
Tokyo[18]		X		X		X	X		X
Toronto[19]	X			X		X			X
Vancouver[20]	X			X	X	X	X		

Sources: [1]Chinese Academy of Geological Sciences 2024; [2]University of Berlin 2024; [3]Pereyra and Rimoldi 2003; [4]University College of London 2024; [5]University of Cape Town 2024; [6]University of Johannesburg 2024; [7]British Geological Survey 2024a; [8]Gutierrez et al. 2014; [9]Jose et al., 2019; [10]Russian Academy of Sciences 2024; [11]Wadia 1953; [12]French Geological Survey 2024; [13]Brazil Geological Survey 2024a; [14]Geological Survey of Italy 2024; [15]Brazil Geological Survey 2024b; [16]British Geological Survey 2024b; [17]Australian Museum 2024; [18]Geological Survey of Japan 2024; [19]Geological Survey of Canada 2024; [20]Sepulveda et al. 2006; [21]British Columbia Geological Survey 2020; [22]Boyer et al. 1985: [23]Montalvo-Arrieta et al. 2010; [24]Darman and Sidi 2000.

3.3 SEDIMENTARY PROCESSES

Sedimentary deposits make up only 5% of the Earth's crust but account for approximately 75% of the rocks exposed at the earth's land surface (Flint and Skinner 1974; Pettijohn 1975). The study of sedimentary deposits involves surficial processes that include the accumulation, consolidation, deformation, origin, age, content, and stratigraphy of sediments. Sedimentary deposits result from a sequence of processes characterized by their removal from one location, some form of transport (e.g., down-gradient by gravity, water, or in the atmosphere by wind), and then final deposition in a sink. In sinks, sediments may not be at peace. Some may be subjected to heat and pressure and become metamorphic rocks; others may move slowly with the ocean floor and become part of a subducting tectonic plate and end up as molten magma; and many will co-evolve with the regional and local biota to create unique ecosystems such as beaches, river deltas, and dunes. The vast majority of sedimentary rocks were created in a marine environment as vast shallow seas transgressed and regressed across the continental land mass.

3.3.1 Clastic Sedimentary Processes

There are three broad categories of removal processes involved with the formation of clastic sedimentary rocks: weathering, mass-wasting, and erosion. The first process involved in the formation of a sedimentary clastic deposit is the weathering (breakdown) of bedrock that provides the sediment or clastic material necessary to form a sedimentary deposit. Bedrock is defined as the continuous solid rock of the continental crust (Flint and Skinner 1974). Once bedrock begins to erode, it forms a material known as regolith – or weathered rock devoid of organic material.

3.3.1.1 Weathering

Weathering is a destructive process by which rocks and minerals are broken down through exposure to atmospheric agents such as air, wind, water, and ice (Christopherson 2008). Weathering processes fall into two general categories: physical and chemical. Physical weathering involves the fragmentation of larger rock into smaller pieces by a mechanical process, without changing the chemical composition of the rock. Examples of physical weathering include: (1) abrasion, where rock is dislodged from the impact of grains carried by wind, water, or ice; (2) fragmentation, from the downslope movement of rock in landslides and rockfalls; (3) frost wedging, caused by wetting and freezing cycles; and (4) thermal expansion, from the contraction of rock by heating and cooling.

Chemical weathering is the breakdown of rocks and minerals through reactions between the rocks and minerals and atmospheric constituents including water, oxygen, and carbon dioxide. This process alters the chemical composition of the rocks involved. The most common types of reactions include: (1) solution, where molecules and elements in rocks and minerals dissolve directly into water; (2) oxidation and hydration, consisting of reactions between oxygen, water, and iron-bearing minerals that assist in weakening and breaking down rocks; and (3) hydrolysis, a complex weathering reaction forming clays. A more detailed treatment of chemical weathering is presented later in this chapter within the context of sediment composition and karst topography.

3.3.1.2 Mass-Wasting

Once weathering occurs and a portion of the bedrock is separated from itself and becomes regolith, it becomes prey for other forces that move the freed materials downslope under the influence of gravity. The several types of movement acting on regolith to transport the material downslope are collectively termed **mass-wasting**. Mass-wasting is the movement of rock or soil downslope by gravity without the aid of moving water, glaciers, or wind (Flint and Skinner 1974). On any slope, mass-wasting processes depend on the interplay between gravity, slope angle, moisture content, cohesion, and friction. When the gravitational force acting on a slope exceeds its resisting force, slope failure or mass-wasting occurs. The slope materials' strength and cohesion and the amount of internal friction between the materials contribute to maintaining slope stability and are collectively known as the slope's shear strength. The steepest angle that a cohesionless slope can maintain without losing its stability is known as its angle of repose (Pudasaini and Hutter 2007).

Mass-wasting can affect large bodies of regolith or just one small rock particle. In addition, mass-wasting is not confined to continental regions; it also takes place in marine environments where submarine landslides occur on steep slopes and material subsequently spreads out onto the sea floor. In areas of low relief, mass-wasting may still take place but at a very slow rate. By contrast, in mountainous areas, mass-wasting can be very fast, as in a landslide.

Factors affecting the potential or rate of mass-wasting include (1) a change in slope angle; (2) weakening of material through weathering; (3) increased water content; (4) changes in vegetation cover; and (5) overloading (Wicander 2005).

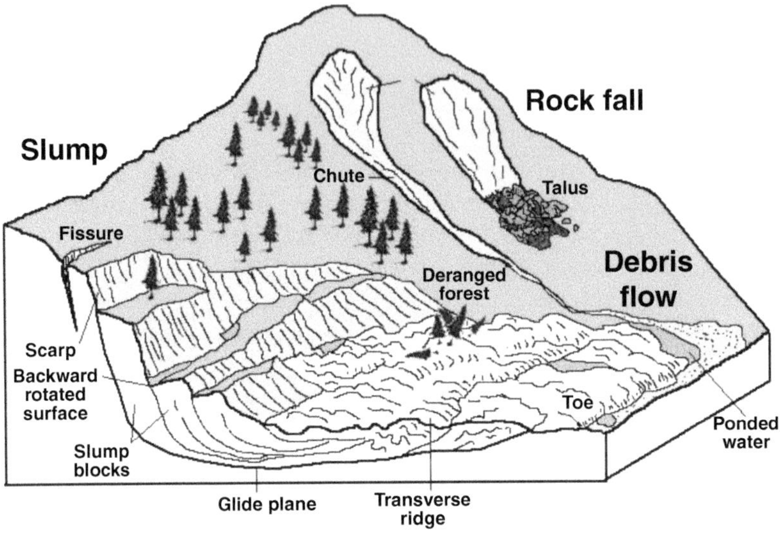

FIGURE 3.1 Examples of mass-wasting. (From United States Geological Survey. 2024a. *USGS Glossary*. http://3dparks.wr.usgs.gov/haywardfault/images/slump.jpg. (Accessed March 10, 2024).)

FIGURE 3.2 Example of a rock fall. (Photograph by Daniel T. Rogers.)

Anthropogenic sources increase the rate of mass-wasting commonly through the building of roads requiring the construction of road cuts through uneven terrain. This practice creates situations where the cut-back slope is constructed beyond the angle of repose. Overloading commonly occurs from anthropogenic sources such as heavy loading on roadways and building structures. In addition, irrigation may significantly increase the rate of mass-wasting in urban areas of varied terrain.

Types of mass-wasting include:

- *Rockfall or debris fall:* The rapid descent of a rock mass, vertically from a cliff or by leaps down a very steep slope (see Figure 3.1 and 3.2).
- *Rockslide or debris slide:* The rapid, sliding descent of a rock mass down a slope (see Figure 3.1).
- *Slump:* The downward slipping of a coherent body of rock or regolith along a curved surface or rupture. The original surface of the slumped mass and any flat-lying planes within it become rotated as the mass slides downward. See Figures 3.1 and 3.3.

FIGURE 3.3 Example of slumping. (Photograph by Daniel T. Rogers.)

- *Debris flow:* The rapid downslope plastic flow of a mass of regolith or debris. Commonly forms as an apron-like or tongue-like area at its terminus with a very irregular surface (see Figure 3.1).
- *Mudflow:* The rapid downslope movement of a very plastic and almost fluid mass of regolith or debris. Very similar to a debris flow but with a higher water content (see Figure 3.1).
- *Soil creep:* Slow movement of slope. Soil creep is the most common type of mass-wasting and is always present on any slope where there is soil. This process tends to go unnoticed because the process is slow. Some of the effects of soil creep are shown in Figure 3.1.

Gravity is not the sole force in operation during mass-wasting. Other transport mediums such as wind, flowing water, ice, or molten lava may be involved. Although flowing water is excluded from the process by definition, water nevertheless plays an important role in mass-wasting. Flowing water can erode the base of a hillslope and weaken the mid-slope region, leading to an increase in the effects of gravity. In a process known as piping, water migrating near the soil surface from seepage fronts of groundwater may subsequently produce slope instability and increase the "lubrication" of a pre-existing plane of weakness. More frequently, however, water entering the soil simply increases the pore pressure, reducing grain-to-grain contact of the soil particles and resulting in slope failures such as debris flows.

3.3.1.3 Stream Transport

After the downslope transport of material, streams carry a large portion of these solids away. The larger and heavier rocks become part of the channel bottom, while a small fraction of the solid material reacts chemically with water and becomes a dissolved load that takes the same journey as the stream. Most of the downslope material, however, becomes suspended load and eventually ends up in the continental shelves as a sedimentary deposit.

3.3.1.3.1 Erosion

Before the transport and deposition of clastic sedimentary material occur, the material must be eroded. Erosion is defined as the complex group of related processes by which rock is broken down physically and chemically and its products removed (Flint and Skinner 1974). The rate of erosion depends on many factors, including:

- Climatic factors
  - Amount and intensity of precipitation
  - Average temperature
  - Temperature range

- Wind speed and changes in wind direction
- Storm intensity and frequency
- Geologic factors
  - Sediment composition
  - Sediment size and shape
  - Sediment porosity and permeability
  - Slope of surface
  - Orientation of sediment surface (tilted, faulted, or folded)
- Biological factors
  - Vegetative cover type and amount
  - Animal habitation behaviors
  - Land use

Estimates of the natural erosion rate in North America range between 1 and 3 centimeters every 1,000 years (Flint and Skinner 1974). This estimate, however, does not include anthropogenic sources of erosion that can double its rate (United States Soil Conservation Service 1992). In addition, the rate of erosion is not uniform. Some areas may experience significant amounts of erosion over a relatively short period of time, whereas in other areas, erosion may not be apparent. Gravity is responsible for the erosion associated with mass-wasting, while precipitation initiates a sequence of splash erosion (where a soil particle is dislodged), leading to sheet erosion from overland flow, and in extreme cases, rills (small indentations) or gullies may form. Strong downslope winds on unvegetated slopes can increase the potential for rills and gully formation; this occurs in southern California during the Santa Ana wind season.

In urban environments, the creation of an impervious surface during development activities increases the volume and velocity of surface runoff, and erosion is generally accelerated. Other types of anthropogenic processes that enhance or accelerate erosion include:

- Removal of vegetative cover through logging
- Overgrazing of livestock
- Changing native vegetation
- Overwatering of vegetation
- Slash and burn agriculture and forestry practices, especially on steep slopes
- Cultivation or farming
- Removal of leaf litter, small shrubs, and grasses

On coastlines, wind, waves, and storms promote the most significant erosion. Wind produces waves and storms produce high wind and big waves. Large storm waves have been known to produce 2,000 pounds of pressure per square foot. The pure energy of waves and the chemical content of the water combine to erode sediment and rock along a coastline (Bird 1985). Other factors such as rising sea levels, tides, tidal currents, and drainage patterns from land naturally influence coastal erosion.

Waves erode shorelines in three main ways:

- *Hydraulic action:* Waves strike a steep slope and dislodge material
- *Abrasion:* Solid materials such as sand or rocks grind together causing them to become rounded and smaller in size.
- *Corrosion or dissolution:* Seawater reacts with the sediments along the shoreline and slowly dissolves portions of the solid material

Erosion of sand along a coastline is much easier for the waves to erode than rock. Along some beaches, there is an annual cycle where sand is removed from a beach one season, only to be returned

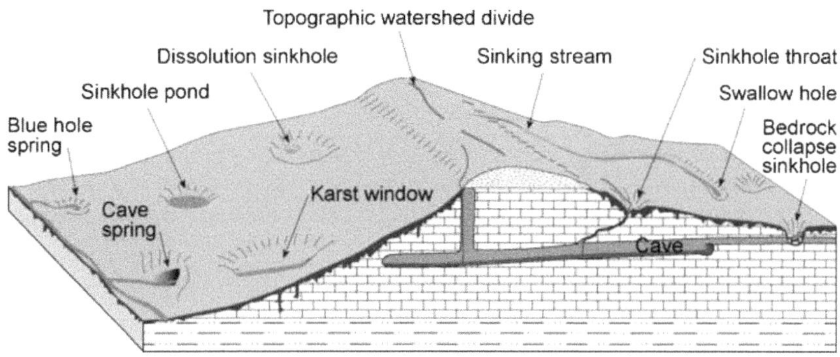

FIGURE 3.4 Common features of karst topography. (From Kentucky Geological Survey. 2024. *Karst Land in Kentucky*. www.uky.edu/KGS/water/general/karst/index.htm. (Accessed March 10, 2024).)

by waves during the successive season. In addition, a steady or seasonal wind direction at an oblique angle toward the shoreline produces waves and crashes along the beach at an angle. The release of energy produces a longshore current**.** Longshore currents are responsible for the erosion and migration of beach sediments along the beachfront from the action of waves. When a longshore current is present, sediments in the surf zone move parallel to the beach in the direction of the wind. Seasonal winds drive the longshore current and transport large quantities of sand along a beach.

In the United States, coastal erosion is significant as approximately 15% of the land area in the United States is coastal, but 40% of the population lives near the coastline. This distribution of population places many of our urban environments under significantly more stress than other less urbanized locations.

Within coastal zones, some anthropogenic activities influence – and in some cases – make coastal erosion more severe (National Oceanic and Atmospheric Administration (NOAA) 2024):

- Dredging channels may change the bathymetry offshore, leading to an increase in the energy of waves striking the shoreline
- Mining beach sand
- Construction of sea walls and spits
- Rising sea levels caused by global warming

3.3.1.3.2 Groundwater Erosion

A significant type of erosion often overlooked results from the movement of groundwater. Groundwater erosion frequently occurs in areas with significant deposits of non-clastic chemical sedimentary rocks such as limestone, dolostone, and evaporite deposits.

Groundwater erosion can eventually lead to a type of landform called **karst topography**. Karst topography is a landscape created by groundwater slowly dissolving sedimentary rock such as limestone (USGS 2024b). The resulting landforms from this type of erosion include sinkholes and caves. A sinkhole is a rounded depression of the ground and typically forms when a cave in the subsurface collapses. A cave is a natural feature beneath the surface formed by groundwater dissolving away the rock. At many locations throughout the United States, extensive cave systems have formed. Figure 3.4 shows the most common features of a karst area.

3.3.2 Non-Clastic Chemical Sedimentary Processes

Non-clastic sedimentary processes involve the precipitation of minerals from water by inorganic or biological processes. Limestone, dolostone, chert, and evaporites are the most common sedimentary rocks formed. Proper identification of sedimentary deposits from chemical processes requires the identification of the primary mineral.

As noted by Deer et al. (1966), common minerals of chemical sedimentary rocks include the following:

- *Limestone:* Composed of calcium carbonate ($CaCO_3$) and predominantly forms in a marine environment from the accumulation of shells, shell fragments, or direct precipitation from water (usually seawater).
- *Dolostone:* Composed of magnesium–calcium carbonate ($CaMg[CO_3]_2$) and associated with a marine-type environment.
- *Chert:* Composed of silica (SiO_2) and may form from microscopic organisms that secrete silica shells around their bodies or as a secondary precipitant
- *Halite:* A mineral composed of sodium chloride (NaCl), and is an evaporite deposit
- *Gypsum:* A sulfate mineral ($CaSO_4 \cdot 2H_2O$), and is an evaporite deposit

As noted by Deer et al. (1966), common minerals of chemical sedimentary rocks include the following:

- *Limestone:* Composed of calcium carbonate ($CaCO_3$) and predominantly forms in a marine environment from the accumulation of shells, shell fragments, or direct precipitation from water (usually seawater).
- *Dolostone:* Composed of magnesium-calcium carbonate ($CaMg[CO_3]_2$) and associated with a marine-type environment.
- *Chert:* Composed of silica (SiO_2) and may form from microscopic organisms that secrete silica shells around their bodies or as a secondary precipitant
- *Halite:* A mineral composed of sodium chloride (NaCl), and is an evaporite deposit
- *Gypsum:* A sulfate mineral ($CaSO_4 \cdot 2H_2O$), and is an evaporite deposit

3.3.2.1 Limestone and Dolostone

Limestone and dolostone are the dominant chemical and biochemical sedimentary deposits that form in marine environments. Limestone deposits usually form in reef-type environments, and they rarely form in terrestrial environments (karst is a landscape formed by groundwater solution – it is not a deposit). In general, significant limestone deposition requires a marine environment with three characteristics: clear water, warm water, and shallow water (Bathurst 1975; Wilson 1975)

Deposition of limestone sediments now occurs within the equatorial belt of warm ocean currents between latitudes of 30 degrees north and south, but deposits as far as 40 degrees north and south have been observed (Wilson 1975). As noted above, warm tropical waters are only one of three criteria for limestone deposition – clear water is also required, particularly for limestone reefs. As a result, significant limestone production does not occur in environments located near significant sources of clastic sediments such as a river delta (e.g., the Mississippi) because significant clastic sediment influx tends to choke limestone sedimentation. In addition, severe storms such as hurricanes may interrupt limestone sediment formation, or in some cases destroy the entire production of limestone sediment in local and regional areas. Large storms and hurricanes have much higher wave action and potentially stronger water currents. The waves from these events erode and destroy portions of near-surface reefs and also transport significant amounts of suspended sediments. As the water turbidity increases from these suspended sediments, interruption of limestone formation may occur. With respect to limestone formation in shallow water, the controlling factors are the availability of sunlight and warmer water located near the surface. Limestone will not form in the deeper parts of the ocean because $CaCO_3$ dissolves in cold water.

Throughout geologic time, limestone production has been significant. Plate tectonics coupled with the rise and fall of sea levels throughout the Paleozoic and Mesozoic have resulted in thick sequences of limestone deposits throughout the world, including a significant portion in the United States as a result of the transgressing seas. In fact, more than 10% of bedrock in the United States is composed of limestone, with significant deposits located within the eastern United States (United States Geological Survey (USGS) 2003). Formation of dolostone typically occurs as a slow replacement process and generally follows limestone deposition (Wilson 1975). About half of the calcium in the $CaCO_3$ molecule is replaced with magnesium forming dolostone ($CaMg[CO_3]_2$). Because of the addition of magnesium in its molecular structure, dolostone is more resistant to chemical erosion than limestone.

3.3.2.2 Evaporites

An evaporite forms as a result of the evaporation of water (Blatt and Tracey 1996). As the list at the beginning of the section shows, common evaporite deposits include the minerals halite and gypsum. During the evaporation of water, the concentration of constituents dissolved in the water becomes concentrated until the time when the minerals begin to precipitate. Complete evaporation leaves behind mineral deposits such as halite and gypsum. Evaporite deposits generally form when water accumulates in closed basins in arid and semi-arid environments.

3.3.3 Formation of Sediments

Once bedrock is weathered to produce sediment, it is transported by the agents of erosion, particularly water, and also by wind and ice. During transport, the sediment will be reduced in size, a process known as sorting and the grains will become progressively more rounded. Minerals such as feldspar and micas will weather clay minerals and may be separated from the rest of the sediment at the margins of lakes and oceans. Once sediment is deposited it may be composed of tightly compacted individual mineral grains of varying size, composition, thickness, and geometric shapes. Sediments form in three basic sedimentary environments:

1. *Continental or terrestrial environments:* All those occurring on the land surface.
2. *Marine:* Deposited by or within the ocean.
3. *Transitional environmental:* The boundary between continental or terrestrial and marine environments.

Common examples of the geological processes forming sediments are listed in Table 3.3.

3.3.4 Rock Formation: Diagenesis and Lithification

After sediments are formed, the processes of diagenesis turn the sediment into rock. Diagenesis describes all the physical, chemical, or biological changes sediment undergoes after it has been deposited.

The types of diagenesis a sedimentary deposit may experience include the following (Pettijohn 1975):

- Compaction
- Deformation
- Dissolution
- Cementation
- Authigenesis
- Replacement

TABLE 3.3
Common Sedimentary Geological Processes That Form Unconsolidated Sediments

Location	Type of Geological Process
Continental	• Fluvial – deposited by a river such as a sand bar or floodplain deposits • Lacustrine – deposited by a lake • Glacial – deposited by a glacier • Alluvial fan – deposited by water flowing from a steep slope onto a level plain • Certain volcanic-related deposits – deposited by mud and debris flows caused by a volcanic eruption • Ash – deposited from a volcanic eruption • Eolian – deposited by the wind • Dune – deposited by the wind • Landslides – generally forms under the force of gravity • Rockslides and rockfalls – generally form on steep unvegetated slopes in mountainous regions • Soil creep – forms on slopes and moves very slowly • Mudflows – deposited by water typically during or shortly after a severe rain or flood event; may be associated with a sloped surface
Marine	• Reef – typically non-clastic chemical sedimentary deposits such as limestone • Layers of fine-grained sediments deposited from settling • Submarine landslides and slumping
Transitional	• Delta – deposited by a river discharging to a lake or ocean • Beach – deposited along the shoreline

Sources: Pettijohn, F. J. 1975. *Sedimentary Rocks*. 3rd Edition. Harper & Row. New York, NY and Krumbein, W. C., and Sloss, L. L. 1963, *Stratigraphy and Sedimentation*. 2nd Edition. W. H. Freeman. San Francisco, CA.

- Recrystallization
- Hydration
- Bacterial action
- Concretion development

The most important diagenetic processes acting upon a sedimentary deposit are compaction and cementation (lithification). Compaction occurs over time within a sedimentary deposit where individual grains are re-arranged to form a more tightly packed sediment. The process of compaction forces out water that may be trapped in pore spaces between mineral grains. Compaction is a function of burial depth and is the first step to a sediment's journey of becoming a rock. **Cementation** involves the precipitation of minerals at grain-to-grain contacts and within the deposit's pore spaces. Only after cementation has occurred can a sedimentary deposit be considered a rock. Although compaction may have significantly reduced the amount of pore space within a sedimentary deposit, some pore space usually remains. Fluids containing minerals migrate through the remaining pore spaces and precipitate cementing materials within the remaining pore spaces and finalize the conversion of sediment to rock. The most common cements in sedimentary deposits are calcium carbonate ($CaCO_3$), silica (SiO_2), and iron oxides (Flint and Skinner 1974).

3.4 IDENTIFICATION AND CLASSIFICATION OF SEDIMENTARY DEPOSITS

To this point, sedimentary rocks have been subjected to significant stress. They have been torn from their original home, broken apart and perhaps changed chemically, and then put into nature's equivalent of the trash compactor and cement mixer. We now have them right where we want them – in an observable form near the Earth's surface – so we can proceed with their identification and classification. The knowledge gained from these procedures allows the investigator in urban areas to identify sedimentary deposits of potential commercial value, including the sand and gravel used in construction and consumer products and the glass and clay used in the manufacture of bricks. The most important outcome, however, from these procedures is the critical knowledge gained for understanding how contaminants behave and migrate in these geological environments.

3.4.1 Identification and Classification Methods

The successful identification and classification of sedimentary deposits require an evaluation of several features including the degree of sorting, grain shape, grain size, composition, thickness, and other characteristics.

3.4.1.1 Sorting

Sorting is the spatial arrangement of particles during their transport and deposition by size. Sedimentary deposits range from very well sorted to very poorly sorted. The degree of sorting in a sedimentary deposit or layer is controlled by the method of transport and deposition. Sedimentary layers formed from wind transport are generally considered well sorted because the wind can only transport fine-grained materials. Sedimentary deposits formed from the action of glaciers are generally poorly sorted because glaciers are not selective – they move everything. Figure 3.5 shows gradational examples of individual grains that range from well rounded to angular. Figure 3.6 shows an example of a poorly sorted material.

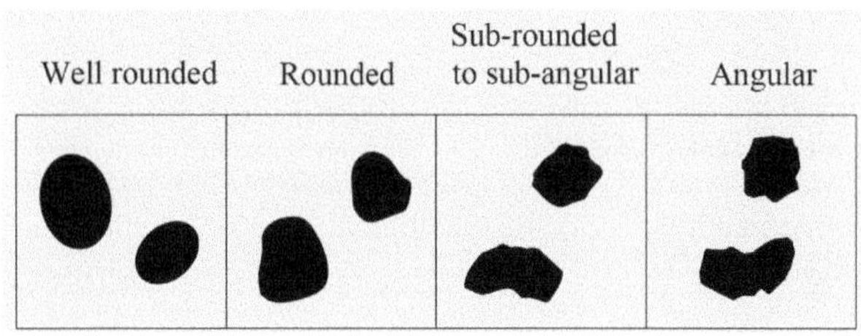

FIGURE 3.5 Examples of well-sorted, well-rounded, and angular sedimentary materials.

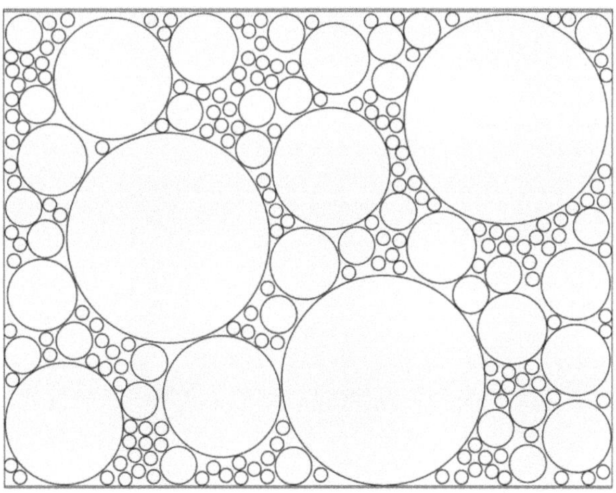

FIGURE 3.6 Example of a poorly sorted sedimentary material.

Individual grains or particles can be sorted by grain size; where the largest grains tend to settle first, and by composition; and where those grains with the highest specific gravity minerals tend to settle first.

3.4.1.2 Grain Shape

Individual grains also vary in shape. The shapes of individual grains are characterized by their degree of roundness; the degree of edge and corner removal; and their sphericity, the degree to which the grain approaches the shape of a sphere. Roundness and sphericity vary significantly and depend upon several factors, including composition, the method of transport, and the distance traveled. In general, roundness and sphericity increase the farther the material is transported before it is deposited. Knowing (1) the shape of the grains and (2) that the largest grains and those with the highest specific gravity settle out first during their deposition enables geologists to make inferences about these deposits. For example, when examining a sedimentary deposit, it may be possible to determine their travel times and distance, potential energy of the transport system, and method of transport.

3.4.1.3 Grain Size

Individual grains are also classified according to size and given specific names depending on the grain size. For instance, a grain that has a size between 1/16 and 2 mm is classified as a sand grain, and a grain having a size between 4 and 64 mm is termed a **pebble**. The Wentworth (1922) scale of grain sizes is the common method for classifying grain size and the ranges are listed in Table 3.4.

TABLE 3.4
Grain size Classification

Wentworth Size Class	Millimeters	Inches
Boulder	256 or greater	10.08 or greater
Cobble	64 to 256	2.52 to 10.08
Pebble	4 to 64	0.16 to 2.52
Very coarse gravel	32 to 64	1.26 to 2.52
Coarse gravel	16 to 32	0.63 to 1.26
Medium gravel	8 to 16	0.31 to 0.63
Fine gravel	4 to 8	0.16 to 0.31
Granule (very fine gravel)	2 to 3	0.08 to 0.16
Very coarse sand	1 to 2	0.04 to 0.08
Coarse sand	0.5 to 1	0.02 to 0.04
Medium sand	0.25 to 0.5	0.01 to 0.02
Fine sand	0.125 to 0.25	0.005 to 0.01
Very fine sand	0.0625 to 0.125	0.002 to 0.005
Silt	0.004 to 0.0625	0.0002 to 0.002
Coarse silt	0.031 to 0.0625	0.001 to 0.005
Medium silt	0.0156 to 0.031	0.0005 to 0.001
Fine silt	0.0078 to 0.0156	0.0003 to 0.0005
Very fine silt	0.0039 to 0.0078	0.0002 to 0.0003
Clay	Less than 00039	Less than 0.0002

Source: New Jersey Department of Environmental Protection (NJDEP). 2005. *Field Sampling Procedures Manual*. Trenton, NJ.

3.4.1.4 Composition

The grains in a sedimentary deposit tend to be resistant to mechanical or chemical weathering. The composition of sand deposits is highly variable and depends upon the source of the material. One of the most resistant minerals common in sedimentary deposits is quartz, a dominant mineral of beach, river, and dune deposits. The quartz content in modern-day sand deposits from major rivers and beaches in the United States ranges between 67% and 99%, with the second most common mineral being feldspar. By contrast, in tropical regions a dominant component in sand deposits is limestone. Sand deposits comprise approximately 25% of all clastic sedimentary deposits (Pettijohn 1975). There are four basic types of sand deposits Pettijohn et al. 1987; Boggs 2000):

1. Quartz arenites consist almost entirely of quartz grains and are usually well-sorted and well-rounded. Quartz arenite deposits result from extensive weathering, so they consist of only the most resistant minerals.
2. Arkoses are sand deposits with a composition of 25% feldspar or more. The grains tend to be poorly rounded and less well sorted than quartz arenites. Arkoses originate from rapidly eroding granitic and metamorphic rock where physical weathering is dominant.
3. Lithic sand deposits contain many fragments derived from fine-grained materials, mostly shales, volcanic rocks, and fine-grained metamorphic rocks.
4. Greywacke is a heterogeneous mixture of rock fragments and angular grains of quartz and feldspar; the sand grains become surrounded by a fine-grained clay matrix.

Composition is also used to help evaluate the degree of sorting in many sedimentary deposits. Clay minerals are the predominant minerals constituting clay deposits, followed by quartz, feldspar, carbonates, and iron oxide minerals, respectively (Pettijohn 1975; Hillier 2003). Clay and silt deposits tend to be more erodible than quartz because they contain higher amounts of organic material in addition to their mineral components of micas and feldspars. Types of clay minerals include kaolinite, montmorillonite–smectite, illite, and chlorite.

Clay minerals typically form over long periods of time from the gradual chemical weathering of rocks, usually silicate-bearing, by low concentrations of carbonic acid and other diluted solvents. The formation of carbonic acid involves the reaction between water, such as rain, and carbon dioxide in the atmosphere. As rain falls through the atmosphere, rainwater dissolves small quantities of carbon dioxide, so when the rain reaches the ground it is a weak carbonic acid. As water migrates through the soils and sediments, the strength of the acid may increase significantly through the addition of carbon dioxide created by the bacterial decay of vegetation. Once carbonic acid forms, it ionizes to form hydrogen and bicarbonate ions. Hydrogen ions are extremely effective in decomposing minerals through a process of hydrolysis. Hydrolysis is a chemical process in which a molecule is cleaved into two parts by the addition of a molecule of water (Greenwood and Earnshaw 1997). Clay deposits may be formed in place but usually are eroded and transported from their original location to generally low-energy depositional environments such as large lakes and marine deposits.

3.4.1.5 Thickness

Thicknesses of unconsolidated sedimentary deposits can vary enormously over short distances. Just a few feet of separation can result in significant variation. The thickness of unconsolidated sediments beneath many urban regions ranges from merely a few feet to more than 1,000 feet (Wright 1989). The deposition of these materials forms a complex interconnected web that may drastically change both vertically and horizontally. It is not uncommon to observe changes in sediment type and origin of formation in only a few inches in the vertical or horizontal directions.

3.4.2 Bedded or Stratified Depositional Features

Depositional features provide clues about the life cycle of sedimentary deposits (Wicander and Monroe 2007), and provide valuable information about the environment where the deposits formed. Some of the more commonly observed features include the following:

- *Bedding or stratification:* The layering of sediments as they are formed and deposited (Wicander and Monroe 2007). The layering of sediments is the result of vertical differences in composition, texture, and grain fabric. The most widespread deposition structure of sedimentary deposits is their bedding or stratification. Bedding layers may be very thin, on the order of less than a millimeter, to very thick, on the order of several feet. One variety of bedding layers is termed **laminae,** which are defined as very thinly bedded layers on the order of a millimeter (Potter et al. 1980). Careful inspection of stratified sediments must be conducted because compaction may enhance and distort bedding irregularities of hydraulic origin. There are several different types of stratification, including (Potter et al. 1980; Pettijohn et al. 1987):
  - Even parallel
  - Even discontinuous parallel
  - Wavy parallel
  - Wavy non-parallel
  - Discontinuous wavy parallel
  - Discontinuous wavy non-parallel
  - Discontinuous curved parallel
  - Curved non-parallel
  - Structureless
  - Mottled
  - Brecciated
  - Algal
  - Graded
  - Cross
- *Parallel horizontal bedding* is characterized by distinct beds or laminae, with either sharp or gradational boundaries that are continuous and parallel. This type of stratification is the most common in fine-grained sediments such as silts and clays. *Discontinuous, but even parallel stratification* is closely related and represents a stratification of slightly less uniformity.
- *Wavy stratification* may be parallel, non-parallel, or discontinuous and is predominantly formed by small ripples of silt and sand. When these particles are interbedded with finer-grained material they are called **flaser bedding**. Wavy bedding of all types is an indication of higher energy and implies a small but discontinuous micro-relief on the bottom where the sediment accumulates. *Discontinuous wavy, non-parallel bedding* is a less orderly arrangement and probably represents a higher level of current flow. *Curved parallel or curved non-parallel bedding* is locally present in many fine-grained sediments such as clays and probably indicates lateral accretion in channel scours.
- *Structureless bedding* represents rapid sedimentation with little or no change or variation in the source material. Figure 3.7 shows many types of stratification in fine-grained sediments.
- *Cross-bedding* are sets of bedded sediments at various, on-lapping, and inclined angles. Cross-bedding indicates changing directions and perhaps energy in the direction of origin of the sediments being deposited. In sedimentary deposits, cross-bedding typically indicates the deposit in question was affected by the forces of wind or changing water currents. Cross-bedding occurs most often within coarser-grained sediments such as sand. Figure 3.8 is a diagram showing the concept of cross-bedding.

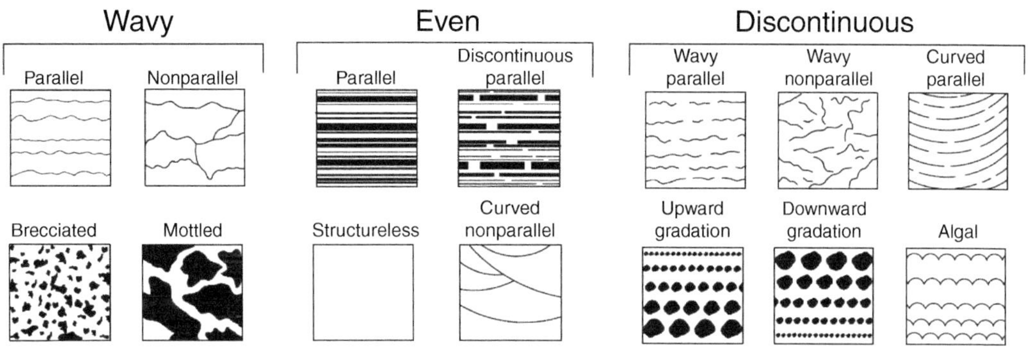

FIGURE 3.7 Stratification types in fine-grained sediments.

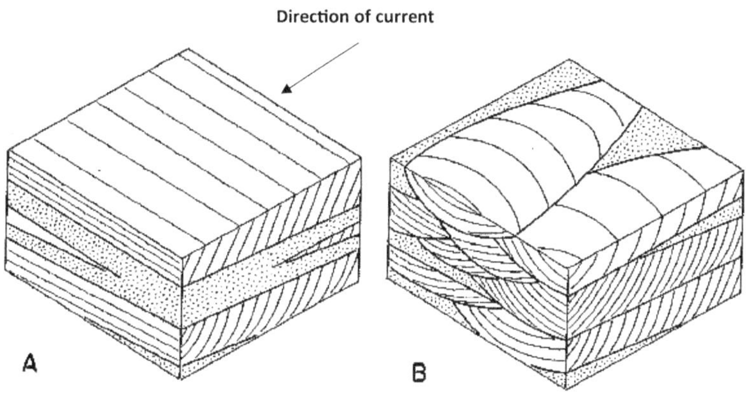

FIGURE 3.8 Planar tabular (A) and trough or festoon (B) cross-bedding.

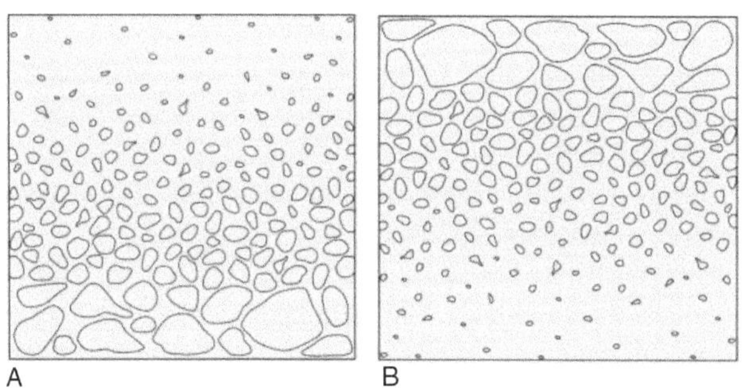

FIGURE 3.9 (A) Diagram showing an example of a deposit from a transgressive sequence. (B) Diagram showing an example of a deposit from a regressive sequence.

- Transgressive and regressive sequences leave behind distinct and different forms of deposits (see Figure 3.9A and 3.9B). A progressively finer-grained vertical sequence such as a clay deposit over a sand deposit with gradations in between is generally indicative of a transgression (increase in water depth). Transgressions can be caused by the melting of polar ice caps, displacement of ocean water by undersea volcanism, or by localized sinking or subsidence of the land in coastal areas. Continental deposits, such as lacustrine deposits, show a similar type of deposition of layers which indicates a rising water level.

A progressively coarser-grained vertical sequence, for example, sand over clay, is generally indicative of regression (a decrease in water depth). Regression can be caused by a buildup of ice in the polar ice caps, or localized uplift of the land in coastal areas. Continental deposits show a similar type of depositional layering which indicates a lowering of lake levels.

3.4.3 Other Depositional Features

Other depositional features may include remnants or identification marks useful for identifying the environment and method of creating the deposit. In addition to bedded and stratified features, these other depositional features provide significant clues to geologists investigating urban sites and help them distinguish whether the deposits being examined originated from a natural process or formed by anthropogenic activity.

- *Ripple marks:* It formed within wind-blown dune deposits or along beach deposits generally in shallow water. Ripple marks are essentially very small dune-like deposits whose long axes are perpendicular to the air or water current that formed them. Ripple marks commonly form either as asymmetrical ripple marks or as oscillation ripple marks. Asymmetrical ripple marks form in shallow water or as dune deposits where there is a dominant direction of water current or wind. Oscillation ripple marks typically form in shallow water within the surf zone where there are competing and opposite directions of water flow.
- *Mudcracks:* These are also commonly termed **desiccation cracks**. These features are formed when a fine-grained deposit – often clay-sized – undergoes drying and then is buried, preserving the feature.
- *Raindrop prints:* These are small circular impressions formed from raindrops hitting the soft surface of exposed sediment.
- *Fossils:* Observing fossils within a sedimentary deposit offers geologists many insights as to how the deposit formed, the age of the deposit, and many other factors. When placed in a stratigraphic sequence with deposits from both above and below, fossils are used to reconstruct the history of the region. For instance, observing fossils or remnants of fish indicates the deposit formed underwater, and observing terrestrial fossils, such as a mammoth or mastodon, indicates the land surface was nearby.
- *Bioturbation and tracks:* Bioturbation is the physical rearrangement of the soil profile by soil organisms. The resulting features are also described as trace fossils and can range in scale from worm burrows to animal tracks.
- *Varves:* These form in a variety of marine and lacustrine depositional environments from seasonal variations in clastic, biological, and chemical layering of sediments. Varves are a special variety of stratification. The classic varve deposit appears as alternative light and dark layering of sediment in a glacial lake. The light layers usually are composed of silt and fine-grained sand and are higher energy deposits because they travel farther, typically originating as glacial meltwater. These light layers are also typically coarser-grained compared with the darker clay layers and are deposited in the summer months when there is little or no ice cover. In lakes, the layering pattern is a function of their seasonal freezing and thawing and is also influenced by the seasonal fluctuations in sediment loading, which is heavier in the ice-free months.

3.5 SEDIMENTARY DEPOSITIONAL ENVIRONMENTS

Sedimentary depositional environments represent a combination of physical, chemical, and biological processes associated with the deposition of a specific type of sediment. Grain size, shape,

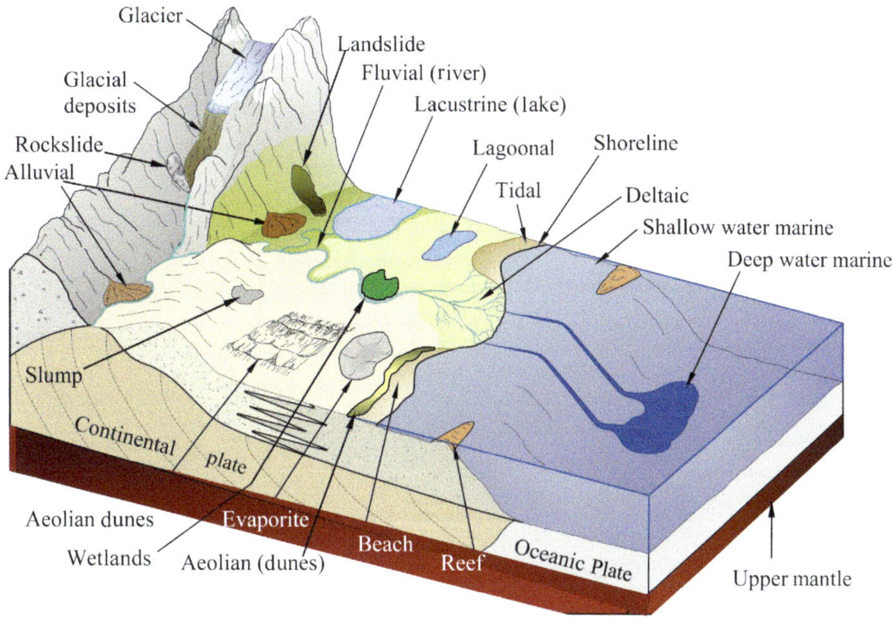

FIGURE 3.10 Sedimentary depositional environments (not to scale).

sorting, composition, thickness, anthropogenic disturbance, natural disturbance, unconformities, and other features present clues as to how, where, and when both clastic and chemical sedimentary deposits formed. As sedimentary deposits form at or very near the surface of the ground, they also communicate to geologists what the climatic conditions were at the time of deposition (Wicander and Monroe 2007). This demonstrates the interdisciplinary nature of geology as a significant portion of the formation and preservation of sedimentary deposits is atmospheric and weather-related. Figure 3.10 shows several sedimentary depositional environments.

An example of a depositional structure containing direct evidence about the environment of its deposition is an extensive cross-bedded sand deposit, indicating there were either dunes or a beach nearby. Fish fossils embedded in this sand would also indicate the sand was deposited in shallow water. This pattern may have significant implications because it could signal the historical climatic shifts and changes, and perhaps present clues related to atmospheric-related hazards (e.g., floods) affecting the site and its surrounding area. Table 3.5 lists the dominant types of clastic sedimentary deposits and corresponding depositional environments in which they typically form.

3.5.1 Alluvial

Alluvial deposits are typically formed by fast-flowing stream water. They range in size from fine silt to boulders and also include debris flows and rock slides from adjacent mountain slopes. One example of an alluvial deposit is an alluvial fan. An alluvial fan is a flat to gently sloping mass of sediment and loose rock material shaped like a fan. These deposits are created at the base of a mountain slope where fast-moving steams coming out of the mountains meet the relatively flat surfaces of a basin floor or broad valley. At this junction, the stream gradient is significantly decreased, which removes much of the energy available for sediment transport by the flowing water (Bloom 1978). Alluvial fans are common in the western United States.

TABLE 3.5
General Depositional Environments of Select Clastic Sedimentary Deposits

Location	Depositional Environment	Sedimentary Process	Type of Deposit
Continental	Alluvial Fan	Stream deposits from fast-flowing water. Also includes debris flows and rock slides from adjacent mountain slopes	Conglomerate and coarse sands that are poorly to moderately sorted
	Rivers (fluvial)	Deposition from fast- and slow-moving water current on channel beds and bars. Low-velocity deposition on floodplains during overbank flow generally during floods.	Sand, moderately to well-sorted and potentially cross-bedded in higher-energy-type deposits, to clay and silt in floodplain deposits
	Deserts (eolian or aeolian)	Sediment transport by wind (eolian or sometimes referred to as aeolian). Leads to formation and migration of dune deposits.	Sand, very well-sorted and rounded, quartz-rich
	Glaciers	Wide-range of deposits from unsorted till in moraines to well sorted, coarse- to fine-grained deposits near unique glacial-type deposits termed **kames** and **kettles**	Fine-grained conglomerate-type deposits; poorly sorted to moderately well-sorted kame deposits
	Lakes (lacustrine)	Deposition of fine-grained sediments in deep water to coarser-grained in beach deposits. Also, evaporation of lake water causing chemical precipitation, especially in arid environments.	Finely bedded silt and clay deposits indicating deeper water formation to coarser-grained sand indicating beach deposits
	Wetland	Mainly composed of decayed vegetative material and very fine-grained sediments	Buried peat; when buried and exposed to heat and pressure may develop into coal
Transitional	Deltas	Deposition of fast to slow-moving water at the mouth of a river or stream entering a large body of water. Re-working of sediments by wave action. Deposition of organic debris in swamps	Sand moderately- to well-sorted to silt and clay. May be fossiliferous and may also contain layers of peat
	Beach	Re-working of sediment layers by waves. May also include formation of dunes.	Sand fine- to coarse-grained, moderately- to well-sorted, quarts rich
	Tidal flats	High- to low-velocity flow in channels and may also have sheet flow. Periodic wetting and drying.	Sand to clay deposits, fine-grained and well sorted, may be organic and fossiliferous, mudcracks and trace fossils
Marine	Reefs and lagoons	High to low energy flow through submarine locations and deposition of increasingly finer material with waning flow. May exhibit submarine disturbance indicative of sliding and debris flows	Conglomerate, sand, to clay. Moderately well-sorted, typically exhibits graded bedding (coarser to finer grained upward sequence). Also unsorted sliding or debris flow deposits.
	Deeper water	Abyssal plain type deposits which typically exhibit very fine non-descript dark muddy-type deposits	Very fine-grained clay

Source: Wicander, R., and Monroe, J. 2007. *Historical Geology*. Thompson Brooks/Cole.

3.5.2 Fluvial

Fluvial deposits are sediments deposited by a river or stream. Deposition results from fast- and slow-moving water currents on channel beds and bars. Low-velocity deposition occurs on floodplains during floods. Here, the deposits are typically fine-grained silts and clays and may extend over large areas. In fact, according to Walker and Coleman (1987), the Atlantic and Gulf coastal plains in large part owe their origins to fluvial deposition over the last 65 million or more years by the numerous rivers that drain off the Appalachian Mountains. Fluvial sediments occur in four main types (Leopold et al. 1992):

1. Channel-floor sediments deposited on the bottom of the river or stream; sizes range from fine sand to gravel depending on flow velocity.
2. Bar sediments deposited along the river and stream banks, especially along the inside portions of river or stream channel bends (Figure 3.11). Deposits are typically sand and gravel with some silt.
3. Channel-top and bar-top sediments deposited on top of the bar sediments are formed in the shallow water regions in channels and abandoned channels. These types of deposits are generally sand and silt-sized deposits.
4. Floodplain sediments are deposited when the water level rises above the confines of the river or stream bank and overflows into the surrounding area. These deposits tend to be finer-grained and are typically composed of sand, silts, and clays.

3.5.3 Lacustrine

Lacustrine sediments are lake deposits characterized by the presence of a range of fine-grained sediments in deep water to coarser-grained sediments in shallow water near the shoreline. Large lakes, such as the Great Lakes, may have well-defined and significant beach deposits composed of sand and gravel. In some arid regions, such as the southwestern portion of the United States, evaporation of lake water leaves behind evaporite deposits. Fine-grained lacustrine deposits tend to be thinly layered or stratified. Figure 3.12 is an example of a thinly layered lacustrine deposit.

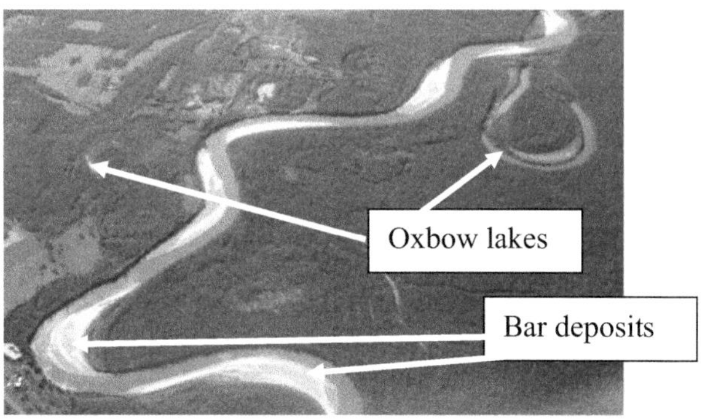

FIGURE 3.11 Example of fluvial bar sediment formation and oxbow lakes in a fully mature stream. (Photograph by Daniel T. Rogers.)

FIGURE 3.12 Fine-grained lacustrine deposit. (Photograph by Daniel T. Rogers.)

3.5.4 Eolian

Eolian (sometimes referred to as aeolian) deposits result from the wind, and in the United States occur primarily in arid or semi-arid climates such as the southwest because these areas often have a large supply of unconsolidated sediments. Many other regions, however, also exhibit eolian deposition generated during the last 2 million years (USGS 2024c). Some of these locations include parts of Nebraska, Kansas, Texas, South Dakota, Colorado, Oklahoma, Illinois, Iowa, sections of the east coast, southwest, and many other areas (Gustavson et al. 1991; Gibbard and Kolfschoten 2004).

3.5.5 Wetlands

A wetland is an area whose soil is saturated with water permanently or seasonally but is saturated long enough to support aquatic plants. Wetlands also include areas partially or totally covered by shallow pools of water. Examples of wetlands are swamps, marshes, and bogs. Deposits in wetlands are primarily composed of decaying vegetative matter. Sedimentary deposits derived from wetlands are typically peat deposits and some very fine-grained sediments such as clay (Mitsch 2009). Extensive wetland deposits are located in the Midwest and portions of the southeastern United States.

3.5.6 Glacial

Glacial sediments are deposits associated directly or indirectly with glaciers (Benn and Evans 1999). Although most of us do not live near active glaciers, the northern half of North America experienced several episodes of glacial advance and retreat during the last 2 million years (Imbrie and Imbrie 1979; Wright 1989). In 1894, Chamberlin was the first to attempt to draw a map of North America depicting the extent of glaciation during the Pleistocene (Figure 3.13).

The following discussion refers to Figure 3.14 and is based on the description of glacial deposits found in Benn and Evans (1999) and Bennett and Glasser (1996):

- Moraines – These are formed by the deposition of material from a glacier and are exposed after the glacier has receded. These features usually appear as glacial till. Glacial till is material directly deposited from glacial ice. Glacial till usually appears as linear mounds of unsorted material, or as a very poorly sorted mixture of rock, gravel, and boulders within a matrix of finer-grained silt- and clay-sized material. There are three main types of moraines:
    - *Lateral moraine:* formed on the sides of glaciers

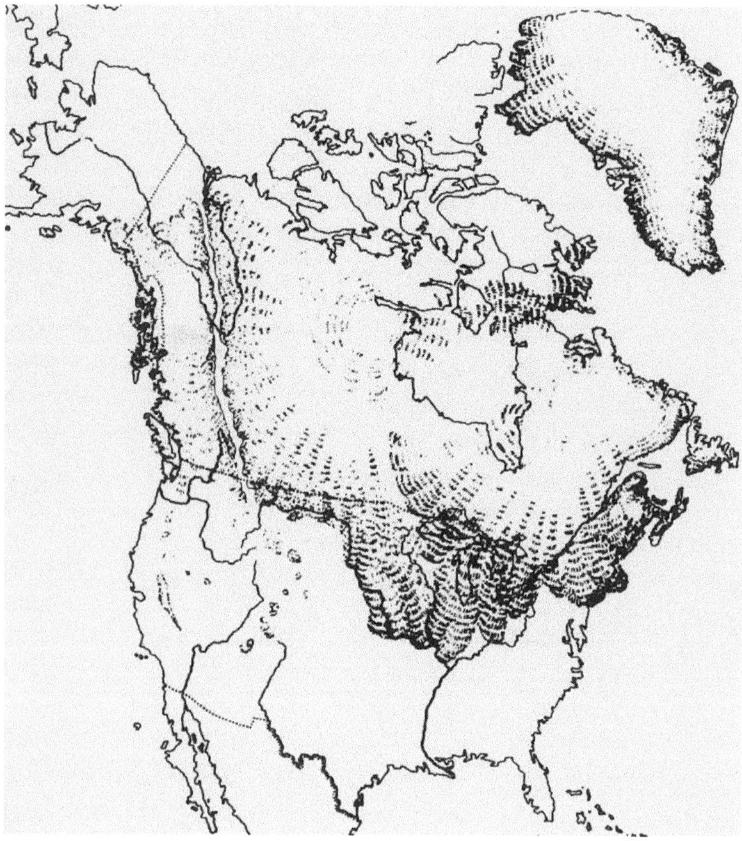

FIGURE 3.13 Extent of Pleistocene Age glaciers. (From Chamberlin, T. C., and Salisbury, R. D. 1909. *A College Textbook of Geology*. The Lakeside Press, R.R. Donnelley & Sons Company. Chicago, IL.)

  - *End moraine:* formed at the end of a glacier or at locations where the glacier had been stagnant during a retreating or regressive phase
  - *Ground moraine:* formed under the ice between the lateral moraine. Deposits of the ground moraine variety tend to be finer-grained than lateral or end moraines. Ground moraines may also be termed **lodgment tills**.
- *Kames:* Ice contact features shaped by glacial meltwater; these form conical-shaped masses of coarse-grained sand and gravel with little or no fine-grained material.
- *Kettles:* These are depressions formed when portions of ice from the glacial front become partially or wholly buried with glacial outwash.
- *Eskers:* These are formed beneath glaciers in ice-walled tunnels of outwash; they typically deposit coarse-grained gravel deposits in an irregular pattern.
- *Outwash:* Outwash are glacial sediments deposited by the action of meltwater transporting glacial till away from the glacier front. Glacial meltwater streams from a glacial front may form a type of pattern called **braided streams**. Braided streams consist of a maze of interconnected channels with high sediment loads.
- *Drumlins:* These are elongated hills composed of glacial till; they are much steeper on one side than the other. The steeper slope indicates the direction of ice movement.
- *Glacial lakes:* Many glacial lakes form between the ice front and end moraines of a retreating glacier. The end moraines form an effective barrier for glacial meltwater to form. Deposits in glacial lakes tend to be finer-grained and may contain varves.

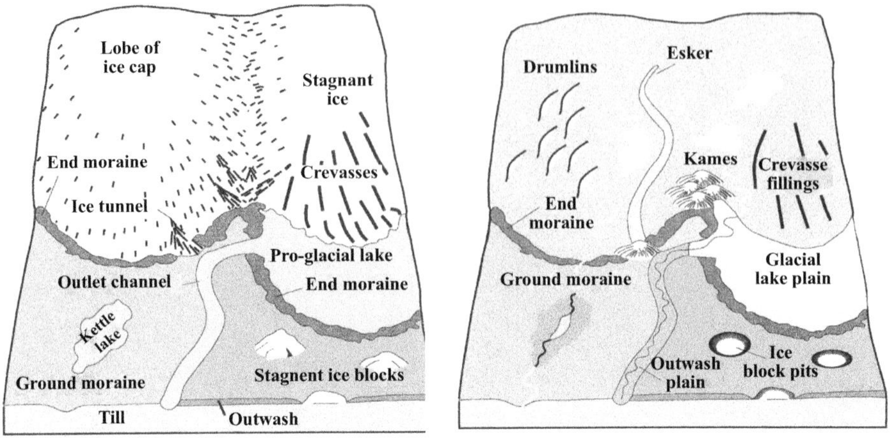

FIGURE 3.14 Types of glacial deposits as they are formed in the glacial sedimentary environment. (Modified from Farrand, W. R. 1988. *The Glacial Lakes around Michigan*. Michigan Department of Natural Resources, Geological Survey Division Bulletin Number 4. Lansing, MI.)

- *Glacial erratic:* A piece of rock that differs from the size and type of native rock to the area in which it is located.
- *Glacial grooves:* Sets of parallel furrows that have been ground out of rock surfaces by moving glaciers.

Glacial lacustrine lakes or pro-glacial lakes as shown in Figure 3.14 were very common throughout North America during the Late Pleistocene (Benn and Evans 1998). Many urban areas such as Chicago, Detroit, Cleveland, and Buffalo are located along or near the shoreline of these glacial lakes (Farrand 1988; Rogers 1996; Benn and Evans 1998). Other notable large Pleistocene glacial lakes include: Lake Bonneville which occupied the area where Salt Lake City in Utah is located (Green and Currey 1988; Morrison 1991); Lake Missoula and others that were located in Montana, Idaho, and eastern Washington (Alt 2001); and Lake Agassiz in the northern United States and southern Canada (Imbrie and Imbrie 1986). Many of these Pleistocene glacial lakes were as large as the current Great Lakes and formed or have influenced many of the landscapes of urban areas in the northern United States.

Most glaciers in the world are melting and are retreating at increasing rates due to the effects of climate change. Figure 3.15 is an example of a high-latitude glacier and ice field in Canada. Figure 3.16 is an example of a retreating mountain glacier on Mount Rainier in Washington. Figure 3.17 is an example of a glacial erratic in Yosemite National Park in California.

An example of a rock glacier is shown in Figure 3.18. A rock glacier is a mass of rock, much like a talus slope, but also has an ice component and moves downslope much like a glacier under the force of gravity.

3.5.7 Non-Clastic Chemical Depositional Environments

As discussed earlier, chemical sedimentary deposits form as minerals precipitated from water through organic or inorganic processes. Identification of sedimentary deposits from chemical processes requires the identification of the primary mineral.

Common evaporite deposits include halite and gypsum. All water bodies on the surface and in aquifers contain dissolved constituents including salts. For an evaporite deposit to form, the water

FIGURE 3.15 Athabasca Glacier and Columbia Ice Field, Alberta, Canada. (Photograph by Daniel T. Rogers.)

FIGURE 3.16 Glacial retreat of the Nisqually Glacier on Mount Rainier, Washington. (Photograph by Daniel T. Rogers, August 2019.)

must evaporate into the atmosphere, causing the minerals dissolved in the water to precipitate. Evaporites generally form when water accumulates in closed basins in arid and semi-arid environments. Enclosed basins in arid environments are favored because they provide high evaporation rates due to low humidity and lack an outlet for the water to drain, which essentially traps the water. Examples of extensive evaporite deposits include areas in and around the Great Salt Lake of Utah and parts of Nevada.

Many limestone deposits form from the accumulation of shells or precipitants from microorganisms. The deposition of limestone strata is thus a byproduct and indicator of biological activity in the geological record (Blatt and Tracey 1996; Taylor and Wilson 2003). Over time these deposits build and sometimes become hundreds of feet thick. A common environment where limestone forms is a reef. Currently, some areas in the Gulf of Mexico, portions of Indonesia, and the Great Barrier Reef located off the northeast coast of Australia are the most active centers of limestone production (Wilson 1975).

FIGURE 3.17 Glacial erratics in Yosemite National Park, California. (Photograph by Daniel T. Rogers.)

FIGURE 3.18 Rock Glacier, Jasper National Park, Alberta, Canada. (Photograph by Daniel T. Rogers.)

3.6 DISTURBANCE

Disturbance of sedimentary deposits occurs constantly from the onset of sediment formation. There are several anthropogenic and natural types of disturbance, and when conducting studies in an urban environment these should be described and evaluated.

3.6.1 Anthropogenic Disturbance

One common concern for any investigator evaluating shallow unconsolidated deposits is whether any disturbance has occurred since deposition. This concern is ratcheted up in urban and developed areas as anthropogenic disturbance often includes either excavation or filling. Often this type of activity can be readily observed – but this is not always the case. Typically, to evaluate the existence of anthropogenic disturbance in a specific area, geologists rely on a few clues for assistance. Some of these clues include the following:

- Evidence of historical development
- Evidence of landfilling
- Evidence of grading or land surface disturbance
- Non-naturally occurring debris located within sediments
- Evidence of excavation
- Lack of original deposition structures within the deposits in question
- Inconsistency or evidence of stressed or absent vegetative cover
- Topography inconsistent with the surrounding area or region

Figure 3.19 is a photograph consisting of landfill material. In this case, several metal drums were excavated indicating anthropogenic disturbance.

FIGURE 3.19 Example of anthropogenically deposited material. (Photograph by Daniel T. Rogers.)

3.6.2 Naturally Occurring Disturbance

Disturbance can also occur naturally, although at times it may be difficult to determine whether the disturbance is natural or anthropogenic. For instance, spoils from dredging historically were spread over the land surface in low areas to enable the development of these areas. Dredging a river bottom and spreading the material over the land surface is accomplished by pumping a combination of river water and bottom sediment to the desired area of disposal. Using water as the transport agent for the sediment may create some difficulty in identifying the source of the disturbance because the water and this mixture of sediments and other materials can leave depositional structures similar to natural deposits. Therefore, special care should be undertaken when evaluating sediment layers in urban settings.

Examples of naturally occurring types of disturbance include the following (Pettijohn 1975):

- Diagenesis, typically involving compaction and lithification
- Erosional removal of a portion or an entire deposit (causing an unconformity)
- Bioturbation involving the disturbance of very near-surface soil layers typically by worms and other macroinvertebrates
- Vegetative disturbance from trees with large and deep root systems
- Differential compaction or subsidence; these events can obscure, bend, or offset original depositional deposits and sequences and make interpretation of deposition challenging
- Redistribution of sand deposits by wind along a beach, forming dune deposits often difficult to recognize because many dune deposits do not contain easily recognizable depositional structures
- Offsets in bedding planes or sediment layers from faulting or other tectonic activity. The results of tectonic offsets may make the matching of depositional layers impossible depending on the distance of the offset.
- Landslides along steep slopes or unvegetated hill slopes resulting from fire or flooding.
- Chemical disturbance from the migration of water from the surface through the sedimentary layers; leaving behind a chemical precipitant or dissolving a portion of the original constituents of the sediments. Given the correct circumstances, this type of disturbance may lead to a karst topography.

The upper portion of Figure 3.20 shows an example of a disturbance likely originating from a naturally occurring source because the line separating the disturbed portion from the non-disturbed

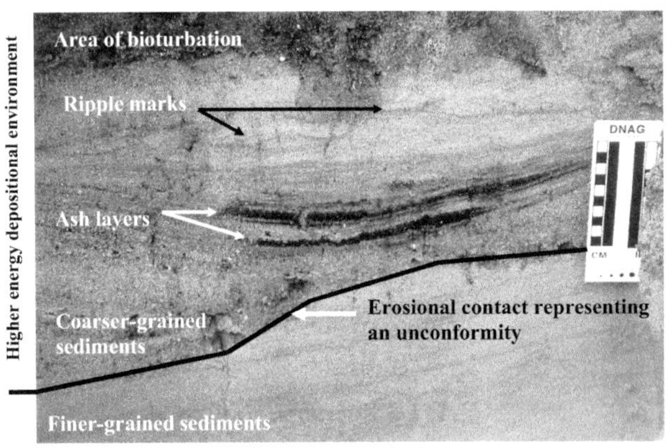

FIGURE 3.20 Late Pleistocene unconsolidated sedimentary deposit in the Rouge River watershed in southeastern Michigan composed of sand (lighter color) and ash (black). (Photograph by Daniel T. Rogers.)

portion is not sharp, as would be characteristic of an anthropogenic source. Instead, the boundary is transitional and gradual in most areas, suggesting the disturbance was probably a combination of bioturbation or differential subsidence.

3.7 SUMMARY AND CONCLUSION

The geology of any urban area literally forms the foundation of where we live. Moreover, the geology of any urban region shapes the manner in which we live: our lifestyles; food production; natural resource availability; and economic activity, and ultimately determines our standard of living and our sustainability. We have explored the different types of geological environments where urban developments are located and discovered that most urban areas exist in geological settings of sedimentary origin. We have also discovered why the geology of urban areas is very complex and changes rapidly. In addition, we learned every urban area of the United States, and for that matter, the world has a unique set of geologic deposits formed from more than two – and in some cases up to eight or more – distinct geologic processes.

We have explored a representative sample of the many types of geological environments of urban areas throughout the United States and the world. What is abundantly apparent from this assessment is that many urban areas are located in geological settings dominated by sedimentary deposits located near water. There exists a complex, dynamic, co-dependent, and sometimes delicate relationship between the presence of water and geology. When urban areas are placed into the mix, these complexities and delicate relationships may be thrown out of balance. Although not always intentional, the resulting human impacts may alter our lives or the environment negatively, and in some cases catastrophically. What determines our sustainability is only in small part determined by urban location; most critical are the specific actions we undertake to understand and protect the urban environment.

Finally, we have briefly introduced the most important factor determined by geology at any location, and especially in urban areas: the presence and availability of water, which is the most precious resource required for living organisms, and their continued sustainability. Nace (1969), a noted global hydrologist once stated, "the story of the growth of civilization and science could be written largely in terms of human concern with water." The next chapter explores water in the urban environment from a scientific perspective and encompasses a wide range of processes from its basic molecular structure and chemical behavior to its actions on the surface, in the atmosphere, and below the ground.

REFERENCES

Aber, J. S. 1984. History of Kansas Glacial Geology. *Journal of Earth Science History*. Vol. 3, pp. 134–142.

Abrams, C. F. 1984. *Geology of Greater Atlanta*. Georgia Geological Survey Bulletin 96. Georgia Geological Survey. Atlanta, GA.

Alder, W. C. 1902. *Geology of Chicago: The 1902 Chicago Folio*. Geologic Atlas of the United States Number 81. United States Geological Survey. Washington, DC.

Allen, R. 2003. *The Rocks of the Valley of the Sun*. Phoenix, AZ. Gemland, Los Angeles CA. www.Gemland.com. (Accessed November 4, 2019).

Alt, D. 2001. *Glacial Lake Missoula and Its Humongous Flood*. Mountain Press Publishing Company. Missoula, MT.

Australian Museum. 2024. *Geology of Sydney Harbour. Geology of Sydney Harbour*. The Australian Museum. https://australianMuseum/lear/animals/wildlife-sydney/geology-of-sydney-harbour (Accessed March 9, 2024).

Barnes, J. H., and Sevon, W. D. 1996. *The Geological Story of Pennsylvania*. Educational Series Book No. 4. Pennsylvania Geological Survey. Harrisburg, PA.

Bathurst, R. G. C. 1975. *Carbonate Sediments and Their Diagenesis*. Development in Sedimentology Book 12. Elsevier Scientific. New York, NY.

Benn, D. I., and Evans D. J. 1998. *Glaciers and Glaciation*. John Wiley & Sons. New York, NY.

Benn, D. L. and Evans, D. J. 1999. *Glaciers and Glaciation*. John Wiley & Sons. New York, New York. 496p.

Imbrie, J., and Imbrie, K. P. 1986. *Ice Ages: Solving the Mystery*. Harvard University Press. Cambridge, MA. 296p.

Bennett, M. R., and Glasser, N. F. 1996. *Glacier Geology: Ice Sheets and Landforms*. John Wiley & Sons. New York, NY.

Bilodeau, W. C. Bilodeau, S. W., Gath E. M., Oborne, M., and Proctor, R. J. 2007. Geology of Los Angeles. *Environmental and Engineering Geoscience Journal*. Vol. 13, pp. 99–160.

Bird, E. C. 1985. *Coastal Changes: A Global Review*. John Wiley & Sons. New York, NY.

Blatt, H., and Tracey, R. J. 1996. *Petrology: Igneous, Sedimentary, and Metamorphic*. 2nd Edition. W. H. Freeman and Company. New York, NY.

Bloom, A. L. 1978. *Geomorphology*. Waveland Press, Inc. Long Grove, IL.

Boggs, J. R. 2000. *Principles of Sedimentology and Stratigraphy*. 3rd Edition. Merril Publishing Company. Toronto, Canada.

Boyer, L., Bensoussan, A., Durand, M., and Grice, R. H. 1985. Geology of Montreal, Province of Canada. *Journal of Environmental and Engineering Geoscience*. Vol. 22, no. 4, pp. 329–94.

Brazil Geological Survey. 2024a. *Geology of Rio de Janeiro*. www.sgb.gov.br/geology-of-rio-de-janeiro. (Accessed March 9, 2024).

Brazil Geological Survey. 2024b. *Geology of Sao Paulo*. www.sgb.gov.br/geology-of-sau-paulo. (Accessed March 9, 2024).

British Columbia Geological Survey. 2020. *Surficial Geologic Map of the Vancouver Region*. www2.gov.bc.ca/gov/content/industry/mineral-exploration/british-columbia-geology. (Accessed June 9, 2024).

British Geological Survey. 2024a. *Geology of London*. www.inhs.ork.uk/index.php/articles-british/249-geology-of-london. (Accessed March 9, 2024).

British Geological Survey. 2024b. *Geology of Singapore*. www.bgs.ac.uk/geology-projects/3d-geology-of-singapore. (Accessed March 9, 2024).

Chamberlin, T. C., and Salisbury, R. D. 1909. *A College Textbook of Geology*. The Lakeside Press, R.R. Donnelley & Sons Company. Chicago, IL.

Chinese Academy of Geological Sciences. 2024. *Overview of the Geology of China*. Institute of Geology-Chinese Academy of Geological Sciences. cags.ac.cn. (Accessed March 9, 2024).

Christopherson, R. W. 2008. *Geosystems: An Introduction to Physical Geography*. 7th Edition. Prentice Hall. Upper Saddle River, NJ.

Coogan, A. H. 1996. *Ohio's Surface Rocks and Sediments*. Bulletin 70. Ohio Geological Survey Bulletin 70. Ohio Geological Survey. Columbus, OH.

Darman, H., and Sidi, F. H. 2000. *An Outline of the Geology of Indonesia*. Indonesian Geologist Association. Jakarta, Indonesia. pp. 1–11.

Deer, W., Howie, R. A., and Zussman, J. 1966. *An Introduction to Rock Forming Minerals*. Longman Group Limited. Essex, England.

Farrand W. R. 1988. *The Glacial Lakes around Michigan*. Geological Survey Division Bulletin Number 4. Michigan Department of Natural Resources. Lansing, MI.

Flint, F. F., and Skinner, B. J. 1974. *Physical Geology*. John Wiley & Sons. New York, NY. 426pp.

Florida Geological Survey. 1994. *Florida's Geological History and Geological Resources*. Florida Geological Survey. Tallahassee, FL.

French Geological Survey. 2024. Geology of Paris. https://brgm.fr/en/news/press-release/geologic-map-of-paris-general-public. (Accessed March 9, 2024).

Gentile, R. J. 1984. Geology of the Circumferential Highway System at Kansas City, Missouri. In: *National Association of Geology Teachers – Central Section Guidebook*. NAGT. Kansas City, MO.

Geological Survey of Canada. 2024. *Geology of Toronto, Ontario. Bringing Southern Ontario's Geology to Virtual Life*. canada.ca. (Accessed March 9, 2024).

Geological Survey of Italy. 2024. *Geology of Rome*. www.isprambiente.gov.it/en/publications/technical-periodicals/descriptive-memories-of-the-geologi-map-of-rome-from-the-historical-center-to-the-outskirts. (Accessed March 9, 2024).

Geological Survey of Japan. 2024. *Geology of Tokyo*. Geology of Japan | Geological Survey of Japan, AIST | 産総研 地質調査総合センター / Geological Survey of Japan, AIST. gsj.jp. (Accessed March 9, 2024).

Gibbard, P., and van Kolfschoten, T. 2004. The Pleistocene and Holocene Epochs. In: Gradstein, F. M., Ogg, J. G., and Smith, A. G. (eds.), *A Geologic Time Scale.* Cambridge University Press. UK.

Goodwin, B. K. 1964. *Guidebook to the Geology of the Philadelphia Area.* Philadelphia Geological Society. Philadelphia, PA.

Green, S. A., and Currey, D. R. 1988. The Stansbury Shoreline and Other Transgressive Deposits of the Lake Bonneville Cycle. In: *In the Footsteps of G. K. Gilbert – Lake Bonneville and Neotectonics of the Eastern Basin and Range Province, Guidebook for Field Trip Number Twelve.* Geological Society of America. Boulder, CO.

Greenwood, N. N., and Earnshaw, A. 1997. *Chemistry of the Elements.* 2nd Edition. Oxford Press. Butterworth-Heinemann, UK.

Gustavson, T. C., Baumgardner, R. W., Caran, S. C., Holliday, V. T., Mehnert, H. H., O'Neill, J. M., and Reeves, C. C. 1991. Quaternary Geology of the Southern Great Plains and an Adjacent Segment of the Rolling Hills. In: Morrison, R. B., editor. *The Decade of North American Geology. Vol. K-2. Quaternary Nonglacial Geology: Conterminous United States.* The Geological Society of America. Boulder, CO. pp. 477–501.

Gutierrez, F., Gutierrex, M., and Martin-Serrano, A. 2014. *The Geology and Geomorphology of Spain. Landscapes and Landforms of Spain.* Springer Nature. Madrid, Spain. 343pp.

Halka, C. 1980. *Roadside Geology of Colorado.* Mountain Press Publishing Company. Denver, CO.

Halka, C. 1983. *Roadside Geology of Arizona.* Mountain Press Publishing Company. Denver, CO.

Hillier, S. 2003. Clay Mineralogy. In: Middleton, G. V., Church, M. J., Coniglio, M., Hardie, L. A., and Longstaffe, F. J., editors. *Encyclopedia of Sediments and Sedimentary Rocks.* Kluwer Academic Publishers. Dordrecht, The Netherlands.

Imbrie, J., and Imbrie, K. P. 1979. *Ice Ages; Solving the Mystery.* Harvard University Press. Cambridge, MA.

Indiana Geological Survey. 2019. *Indiana Geology.* Indiana Geological Survey. Bloomington, Indiana. https://legacy.igws.indiana.edu/surficial. (Accessed March 9, 2024).

Jose, L. A., Layer, P. W., Macias, J. L., Morales-Casique, E. and Farcia-Paloma, A. 2019. *Geology and Stratigraphy of the Mexico Basin (Mexico City).* Taylor & Frances Online. www.tanfonline.com/doi/full/10.1080/1744567.2019.1593251 (Accessed March 9, 2024).

Kaye, C. A. 1976. *The Geology and Early History of the Boston Area of Massachusetts: A Bicentennial Approach.* USGS Special Report 1476. USGS. Washington, DC.

Kentucky Geological Survey. 2024. *Karst Land in Kentucky.* www.uky.edu/KGS/water/general/karst/index.htm (Accessed March 10, 2024).

King, P. B., and Beikman, H. M. 1974. *Geologic Map of the United States; Excluding Alaska and Hawaii.* Vol. 1. United States Geological Survey, Governmental Printing Office. Washington, DC. 2,500,000. 2 sheets.

Krumbein, W. C., and Sloss, L. L. 1963. *Stratigraphy and Sedimentation.* 2nd Edition. W. H. Freeman. San Francisco, CA.

Leopold, L. B., Wolman, M. G., and Miller, J. P. 1992. *Fluvial Processes in Geomorphology.* The Dover. New York, NY.

Missouri Department of Natural Resources. 2004. *Geologic Map of Missouri.* Missouri Department of Natural Resources, Geological Survey Division. Rolla, MO.

Mitsch, W. J., Gosselink, J.G., Anderson, C. J., and Zhang, L. 2009. *Wetlands Ecosystems.* John Wiley & Sons. New York, NY.

Montalvo-Arrieta, J. C., Chavez-Cabello, F., Velasco-Tapia., G., and Navarro, D. L. 2010. *Causes and Effects of Landslides in the Monterrey Metropolitan Area, NE Mexico.* Nova Science Publications. New York, NY. 32pp.

Morrison, R. B. 1991. Quaternary Stratigraphic, Hydrologic, and Climatic History of the Great Basin, with Emphasis on Lake Lohontan, Bonneville, and Tecopa. In: Morrison, R. B., editor. *Quaternary Nonglacial Geology of Conterminous United States. The Geology of North America.* Geological Association of America. Boulder, CO. pp. 283–320.

Nace, R. L., 1969. *Water and Man – A World View.* UNESCO. Paris, France.

National Oceanic and Atmospheric Administration (NOAA). 2024. *Our Changing Planet.* https://cpo.noaa/climate-tool/our-changing-planet-the-U-S-global-change-research-program-for-fiscal-year-2021. (Accessed March 10, 2024).

New Jersey Department of Environmental Protection (NJDEP). 2005. *Field Sampling Procedures Manual.* Trenton, NJ.

Ojakangas, R. W., and Matsch, C. L. 1982. *Minnesota's Geology*. University of Minnesota Press. Minneapolis, MN.

Page, W. R., Lundstrom, S. C., Harris, A. G., Langenheim, V. E., Workman, J. B., Mahan, S. A., Paces, J. B. Dixon, G. L. Rowley, P. D., Burchfiel, B. C., Bell, J. W., and Smith, E. I. 2005. *Geologic and Geophysical Map of the Las Vegas 30′ × 60′ Quadrangle, Clark and Nye Counties, Nevada, and Inyo County, California*. United States Geological Survey Pamphlet to Accompany Scientific Investigations Map 2814. USGS. Washington, DC.

Pereyra, F. X., and Rimoldi, H. 2003. Geological and Environmental Aspects of the Development of Megacities: The Case of Buenos Aires Metropolitan Area (AMBA), Argentina. *Bulletin of Engineering Geology and the Environmental*. Vol. 62, pp. 341–351.

Pettijohn, F. J. 1975. *Sedimentary Rocks*. 3rd Edition. Harper & Row. New York, NY.

Pettijohn, F. J., Potter, P. E., and Siever, R., 1987. *Sand and Sandstone*. 2nd Edition. Springer-Verlag. New York, NY.

Potter, P. E., Maynard, J. B., and Pryor, W. A. 1980. *Sedimentology of Shale*. Springer-Verlag. New York, NY.

Pudasaini, S. P., and Hutter K. 2007. *Avalanche Dynamics: Dynamics of Rapid Flows of Dense Granular Avalanches*. Springer-Verlag. New York, NY.

Rogers, D. T. 1996. *Environmental Geology of Metropolitan Detroit*. Clayton Environmental Consultants. Novi, MI.

Rogers, D. T. 1997. Surface Geologic Map of the Rouge River Watershed, Southeastern Michigan. Wayne County, Michigan. Vol. 1, 63,360. 2 sheets.

Rogers, D. T. 2020. *Urban Watersheds: Geology, Contamination, Environmental Regulations, and Sustainability*. 2nd Edition. CRC Press. Boca Raton, FL 610pp.

Russian Academy of Sciences. 2024. *Specific Features of Geological Risk Assessment in Moscow, Russia*. https://karpinskyinstitute.ru/en. (Accessed March 9, 2024).

Sepulveda, S. A., Rebolledo, S., Lara, M., and Padilla, C. 2006. *Landslide Hazards in Santiago, Chile: An Overview*. The Geological Society of London. London, England. IAEG no 105, pp. 1–8.

Snowden, J. O. 1980. *Geology of Greater New Orleans*. New Orleans Geological Society. New Orleans, LA.

Spearing, D. 1991. *Roadside Geology of Texas*. Mountain Press Publishing Company. Denver, CO.

Stewart, M. 1992. *Karst Characteristics of the USF Geology Park*. Department of Geology, University of Southern Florida. Tampa, FL.

Stoffer, P. W., and Gordon, L. 2001. *Geology and Natural History of the San Francisco Bay Area – A Field Trip Guidebook*. USGS Bulletin 2188. USGS. Washington, DC.

Stokes, W. L. 1989. *Geology of Utah*. Utah Museum of Natural History Occasional Paper No. 6. Utah Geologic Survey, Salt Lake City, UT.

Taylor, P. D., and Wilson, M. A. 2003. Paleoecology and Evolution of Marine Hard Substrate Communities. *Earth-Science Reviews*. Vol. 62, pp. 1–103.

Trimble, D. E. 1963. *Geology of Portland, Oregon and Adjacent Areas*. USGS Bulletin 1119. USGS. Washington, DC.

United States Geological Survey (USGS). 2003. *Geology of the New York City Region*. http://3dparks.wr.usgs.gov/nyc/index.html. (Accessed March 15, 2024).

United States Geological Survey (USGS). 2005. *The New Geologic Map and Geodatabase of Seattle*. University of Washington. Seattle, WA. https://pubsusgs/of/2005/1252. (Accessed March 15, 2024).

United States Geological Survey (USGS). 2007. *Geology of the National Capital Region – A Field Trip Guidebook*. USGS Circular 1264. USGS. Washington, DC.

United States Geological Survey (USGS). 2009a. *Geology of the Mid-Atlantic Urban Corridor (GOMAC)*. USGS Circular 1148. USGS. Washington, DC.

United States Geological Survey (USGS). 2009b. *Geology of Texas*. http://mrdata.usgs.gov/sgmc/tx.html. (Accessed March 17, 2024).

United States Geological Survey (USGS). 2024a. *USGS Glossary*. www.usgs.gov/scientific-collections/glossary-terms. (Accessed March 10, 2024).

United States Geological Survey (USGS). 2024b. *Karst and the United States Geological Survey*. www.usgs.gov/mission-ares/water-resources/science/karst-aquifers. (Accessed March 10, 2024).

United States Geological Survey (USGS). 2024c. *Eolian History of the United States*. www.usgs.gov/gip/deserts/eolian. (Accessed March 10, 2024).

United States Soil Conservation Service. 1992. *Soil Erosion by Water*. United States Department of Agriculture. Washington, DC.

University College of London. 2024. *Geology of Egypt*. www.ucl.as.uk/earth-sciences. (Accessed March 9, 2024).

University of Berlin. 2024. *Geology of Berlin*. Institute of Geological Sciences. www.fu-berlin.de. (Accessed March 9, 2024).

University of Cape Town. 2024. *Geology of Cape Town*. https://science.uct.ac.za/department-geological-sciences/geology-of-caprtown-0. (Accessed March 9, 2024).

University of Johannesburg. 2024. *Geology of South Africa*. https://pure.uj.za/en/organisations/geology. (Accessed March 9, 2024).

Wadia, D. N. 1953. *Geology of India*. Archeological Survey of India. New Delhi, India. 592pp.

Walker, H. J., and Coleman, J. M. 1987. Atlantic and Gulf Coast Province. In: Graf, W. L., editor. *Geomorphic Systems of North America. Vol. 2. Decade of North American Geology*. Geological Society of America. Boulder, CO. pp. 51–110.

Wentworth, C. K. 1922. A Scale of Grade and Class Terms for Clastic Sediments. *Journal of Geology*. Vol. 30, pp. 377–392.

Wicander, R. (2005). *The Changing Earth: Exploring Geology and Evolution*. Thompson Brooks/Cole. New York, NY.

Wicander, R., and Monroe, J. 2007. *Historical Geology*. Thompson Brooks/Cole

Wilson, J. L. 1975. *Carbonate Facies in Geologic History*. Springer-Verlag. Heidelberg, Germany.

Wright, H. E. 1989. The Quaternary. In: Bally, A. W., and Palmer, A. R., editors. *The Decade of North American Geology. Vol. A. The Geology of North America – An Overview*. The Geological Society of America. Boulder, CO. pp. 523–536.

4 Water and the Hydrogeology of Urban Areas

4.1 INTRODUCTION

Water is required to sustain life on Earth. Water plays a central role in maintaining our survival as a species, preservation of the natural environment, and therefore achieving a sustainable planet (United Nations 2003). From an additional sustainability point of view, water also greatly assists in the recycling of Earth by breaking down rocks and transporting them to the oceans to create new rocks. Yet, there are immense challenges facing humans with respect to securing water for their basic needs and long-term quality of life. Although almost three-fourths of Earth's surface is covered with water, most of this water is not potable, and more than 95% of freshwater is either frozen, underground, or in a gaseous phase.

Uniform distribution of solar energy from the sun is not possible on Earth, mainly because the Earth is a sphere, is tilted on its axis, has a non-uniform arrangement of land and water, and has differences in surface elevation. Surpluses and deficits of energy arise at different locations and create uneven atmospheric pressures and densities. In a thermodynamic system, these inequalities try to even themselves out, so fluids move in a quest to achieve overall energy equilibrium. Air (via wind) transports part of the solar energy it has absorbed and moves it from zones of higher pressure to zones of lower pressure. Water moves the excess energy it has absorbed from the equatorial region in a general poleward direction via ocean currents. All of this movement results in the Earth having a peculiar and unpredictable precipitation pattern where some areas receive precipitation almost daily while others may not get any for years.

Precipitation from the atmosphere recharges the underground water reservoirs and plays a central role in shaping the geology of urban areas through weathering, sediment transport and deposition, chemical precipitation and dissolution, and erosion. The flow of water at the surface is controlled hierarchically by a system of topographically bounded spatial units called watersheds. Small watersheds consisting of small streams feed larger watersheds with larger streams; for example, the Scioto River watershed in Ohio delivers its flows into the Ohio River, which then flows into the larger Mississippi River. Some of the precipitation is intercepted by vegetation, some pools on the surface, and a good part of it moves downward through the soil where it recharges groundwater.

This chapter opens with a brief discussion of basic water chemistry, and the geography of Earth's water, and examines how water moves cyclically between the atmosphere and the oceans. Understanding basic water chemistry and refreshing ourselves about the global water cycle provide the knowledge base for the more detailed material later in the chapter and later sections of this book. Key objectives of this chapter include an investigation of the multi-faceted nature of surface water and groundwater as discrete entities, while also considering their interactions. The chapter concludes with a discussion of the anthropogenic effects on surface water and groundwater.

DOI: 10.1201/9781003463153-5

4.2 WATER BASICS

This section begins with the basic chemistry and properties of water. The focus then turns to water's abundance and distribution on Earth, with special attention given to the work water accomplishes as it moves through its global cycle. We conclude the chapter with a discussion of potable water.

4.2.1 Basic Water Chemistry

A water molecule is composed of two hydrogen atoms bonded with one atom of oxygen. This bonding ratio gives water a chemical formula of H_2O (Hill et al. 1993). The hydrogen atoms are attached to each side of the oxygen atom, and this configuration results in a **polarity** or directional charge. The water molecule is positively charged at the hydrogen end and negatively charged on the opposite side where the oxygen atom is located. As opposite electrical charges attract, water molecules have an affinity for each other, and in its liquid state water forms a linear configuration (Hill et al. 1993).

The angular arrangement of the water molecule shown in Figure 4.1 results from the properties of its chemical bonds. Hydrogen atoms are attached to the oxygen atom in a **covalent bond**, with each black dot representing an electron. Note the circled pairs of electrons on the two sides opposite the hydrogen atoms. The pairing of electrons creates a slightly greater negative charge, which acts to push the two hydrogen atoms closer. This push results in a slightly bent tetrahedral molecule with an angle of approximately 107.5 degrees. The polarity of a water molecule is significant because it permits water to:

- Stick together with other water molecules, as in a drop of water.
- Interact with other polar molecules – this is why other substances are able to dissolve in water. Water is known as a universal solvent because it dissolves more substances than any other liquid.
- Interact with electricity

Some important chemical properties of water are given below:

- It freezes at 0°C (32°F).
- It boils at 100°C (212°F) at mean sea level.
- Liquid water is denser than ice
- It has a high **specific heat capacity**
- It has a very high surface tension.
- In is vapor form, it is considered a greenhouse gas

Water is unique because it is the only substance on Earth found in all three states (liquid, solid, and gas) within our planet's temperature ranges. Liquid water is essential to life, as it comprises approximately 60% of the human body by weight and 70% of the human brain. Some organisms are 90% liquid water by weight.

Without water, humans simply would not exist. The ability of water to dissolve so many different substances allows cells to use the nutrients, minerals, and chemicals in biological processes. In natural systems, wherever water goes – the air, the ground, streams and lakes, biota, or through our bodies – it takes valuable chemicals, minerals, and nutrients picked up along the way. This transport ability also means water can carry substances harmful to humans and the environment. If these contaminants are present at a sufficient concentration and the exposure is long enough, potentially harmful effects can occur. The damage can be immediate and obvious, as when oil is washed up on a beach and kills waterfowl – or slowly developing and silent – a scenario represented all too often by the benzene that originated from a leaking gasoline tank and carried by groundwater into a drinking water well.

4.2.2 Water Cycle

Water is constantly on the move through the four spheres of the geosphere: the atmosphere, biosphere, hydrosphere and lithosphere. This movement of water is cyclical and is called the hydrologic cycle or water cycle. In this cycle, water changes phase many times and exists as a liquid (surface waters and groundwater), solid (snow and ice), or gas (water vapor). The global movement of water is initiated by solar energy, which evaporates surface water into the atmosphere. Much of this water vapor condenses and falls as some form of precipitation on a distant land surface where it evaporates; flows back into the oceans through rivers and streams; is taken up by vegetation and slowly released into the atmosphere as evapotranspiration; or infiltrates into the ground. Groundwater also migrates back to the oceans (Jones 1997; Alley et al. 1999). Figure 4.1 depicts the water cycle on Earth.

As with all physical systems, the water cycle performs some work. These three main chores are accomplished: (1) evaporation desalinates large quantities of ocean water and creates freshwater that is potentially potable; (2) nutrients are transported – a process called the biogeochemical cycles – which moves carbon, phosphorus, and other nutrients between their reservoirs. For example, the carbon cycle transports enormous quantities of carbon between the atmosphere and the oceans by precipitation, with much of this carbon held within the carbonic acid (H_2CO_3) formed in the atmosphere by the combination of CO_2 and water vapor; and (3) landscapes are altered by flowing water (erosion and sedimentation) and water's phase changes between ice and liquid (ice wedging and

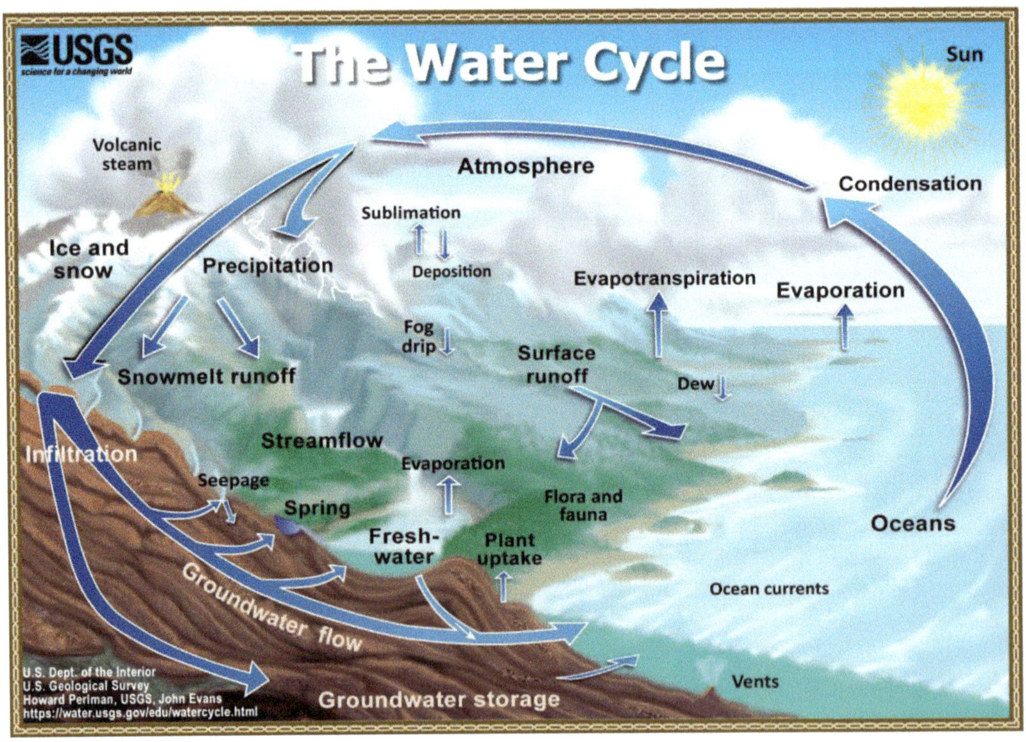

FIGURE 4.1 The water cycle. (From United States Geological Survey. 2024a. *Summary of the Water Cycle.* water.usgs.gov/edu/watercycle.html. (Accessed March 10, 2024).)

frost heaving). Water, with its higher viscosity and weight per unit volume than air, is the key agent of landscape change on Earth.

4.2.3 Distribution of Water on Earth

The vast majority of water on the Earth is saline and resides in the oceans. Although the water in the oceans is important for life on Earth, it is the freshwater contained in ice and snow, rivers and streams, lakes, swamps, and groundwater that provides people with the water they need to live. The oceans contain 97% of water on Earth and are saline and only 3% is freshwater. Of that 3% of freshwater, the majority is ice and groundwater, leaving just 0.3% of Earth's total water in lakes, rivers, streams, and swamps. Table 4.1 presents a more detailed accounting of Earth's water reserves, specifically where the water resides, the estimated quantities within these reservoirs, and the amounts of freshwater as a percentage of total water (Shiklomanov 1993; Jones 1997).

The major lessons learned from the characteristics of Earth's water are: (1) most of our water is not potable and the accessible portion of potable water is small; and (2) contamination of a significant amount of our potable water resources can and does create regional shortages, despite the existence of an adequate supply of fresh water per capita on Earth.

4.2.4 Potable Water Use

The term **potable water** means water used for any human purpose (Fuhrer et al. 1999), and includes uses other than drinking water. In the United States, potable water use includes the following (modified from Fuhrer at al. 1999):

- Drinking and bottling (including beverages such as soft drinks)
- Washing

TABLE 4.1
Earth's Water by Source, Volume, and Type

Water Source	Water Volume (Cubic Miles)	Percent Freshwater	Percent Total Water
Oceans, Seas, and Bays	321,000,000	–	96.5
Ice Caps, Glaciers, and Permanent Snow	7,773,000	68.7	1.74
Groundwater	5,614,000	–	1.7
• Fresh	2,526,000	30.1	0.76
• Saline	3,088,000	–	0.94
Soil Moisture	3,959	0.05	0.001
Ground Ice and Permafrost	71,970	0.86	0.022
Lakes	42,320	–	0.013
• Fresh	21,830	0.26	0.007
• Saline	20,490	–	0.006
Atmosphere	3,095	0.04	0.001
Swamps	2,752	0.03	0.0008
Rivers	509	0.006	0.0002
Biological Water	269	0.003	0.0001
Total	**332,500,000**	**–**	**100**

Source: Shiklomanov, I. 1993. World Fresh Water Resources. In: Gleick, P. T., editor. *Water in Crisis: A Guide to the World's Fresh Water Resources*. Oxford University Press. New York, NY, and Jones, J. A. 1997. *Global Hydrology: Processes, Resources, and Environmental Management*. Longman Publishers. London, England.

- Recreation
- Heating and cooling
- Fire fighting
- Irrigation
- Industrial use
- Commercial use
- Livestock
- Mining
- Electricity production
- Wastewater treatment

In 2000, people in the United States used an average of 407.5 billion gallons of water per day. Of this total amount, 79% or 323 billion gallons was from a surface water source, and 21% or 84.5 billion gallons originated from groundwater (United States Geological Survey (USGS) 2004). Over the past 50 years, the usage of water for electricity production increased by more than 500%, and since 1965 has been in the category of highest water use. Water used for irrigation purposes has also risen 50% between 1950 and 2000 – another significant increase.

From 1970 to 2000, the use of fresh surface water in the United States has remained relatively constant at a usage rate of approximately 275 billion gallons of water per day. During this same period, the use of fresh groundwater has increased from approximately 65 billion to 85 billion gallons of water per day. In many states, groundwater is the key source of potable water for irrigation; Nebraska, Texas, and California are examples. Some states, such as California, have designated specific areas for **groundwater recharge**. Here, during the short rainy season surface water seeps into the ground and is stored until it is needed for human use during dryer periods.

Over 90% of the people in Idaho, New Mexico, Mississippi, and Florida use groundwater for drinking water purposes. By contrast, the populations of Colorado, Rhode Island, and Kentucky obtain less than 30% of their drinking water from groundwater (USGS 2004).

The data on water use indicate humans require and use enormous quantities of water every day to provide energy, food, and to sustain their lives. However, the geography of water often dictates specific usage patterns. Many regions of the United States lack abundant fresh water supplies, and they are required to use water more prudently because of its relative scarcity compared to other areas. The arid southwest and the humid region of the Great Lakes demonstrate this contrast (Meyer 1989). Because of this geo-climatic factor and other considerations (one being the water law doctrines existing in eastern and western states), water is not equally protected throughout the United States, and the methods of conservation and recycling of water vary widely depending on location.

We now turn our attention to watersheds – the organizers of fresh water supplies. The focus here will be on urban watersheds and the specific implications their surface waters and groundwater have for regional geology, contaminant sources and sinks, and the long-term sustainability of the landscape.

4.3 SURFACE WATER IN URBAN AREAS

The importance of surface water in urban areas is significant. This water acts as a source of potable water, and supports agriculture, commerce, and transportation. From an environmental perspective, surface water is responsible for the erosion and deposition of a significant portion of the sediments lying beneath urban areas, and when surface water is in the form of a stream, it carries suspended and dissolved materials into other bodies of water.

Nature organizes streamflow by adopting the most energy-efficient method of transport within a physical framework consisting largely of the local geology and climate. In areas of tough fractured rock, the streams follow the fractures, whereas in sedimentary environments consisting of softer rock, a dendritic or tree-like pattern of drainage develops. If it rains more, then a denser stream network evolves. The primary objective of a stream is to transport the water it receives and the materials it carries without delay so that the next time a precipitation event occurs the transport can continue. The organizational structure that nature came up with to perform these tasks is called a watershed.

4.3.1 Watersheds

A **watershed** is defined as an extent of land where water as rain or snowmelt drains downhill into a body of water such as a river, lake, swamp, estuary, reservoir, wetland, bay, sea, or ocean (Winter et al. 1998). In some cases, the drainage areas for water at the surface and below within a watershed may differ. When the areas are different, the groundwater watershed is usually larger than the surface watershed (De Barry 2004). Watersheds act as a conduit for the transport of water and materials at the surface and below the surface (groundwater). Figure 4.2 shows the functional regions of a surface watershed and their geofluvial characteristics, i.e., flow volumes and erosion/deposition balance.

The basic moniker used for watersheds varies. Other terms describing a watershed include catchment, catchment area, catchment basin, drainage area, drainage basin, river basin, or water basin. For consistency and to avoid the use of a thesaurus, we use the term watershed throughout the book. The entire land surface of the Earth can be subdivided into watersheds. Some of the more notable watersheds in North America include those drained by the rivers listed in Table 4.2 (Kammerer 1990).

Whether water drains off the land into successively larger watersheds or drains from a watershed like the Mississippi, it ultimately discharges into an ocean. The Atlantic Ocean receives surface water from 47% of the land surface of the world, and the Pacific and Indian Oceans each drain 13% of the land surface. The remaining surface drainage either ends up in the Arctic Ocean or does not reach any ocean. Non-ocean-bound surface water drains into enclosed basins, such as the Great Basin located in the western portion of the United States (Kammerer 1990). Table 4.3 lists the major urban areas of the United States and nearby surface water features.

Of the 29 urban centers listed in Table 4.3, 12 are located near the ocean, and 8 more are located near bodies of water deep and wide enough to support large-scale shipping. The 12 centers located near an ocean include Baltimore, Boston, Houston, Los Angeles, Miami, New Orleans, New York,

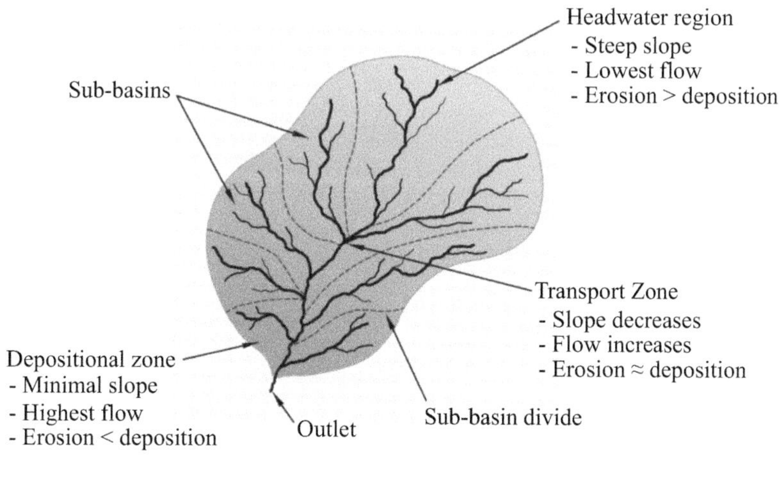

FIGURE 4.2 Basic watershed and transport functions.

TABLE 4.2
Some Major Rivers of the United States

River	Length (Miles)	Drainage Area (Square Miles)	Discharge (Cubic Feet/Second)	Major Tributaries
Missouri	2,500	529,000	76,200	Yellowstone River Platte River
Mississippi	2,339	1,150,000	593,000	Illinois River Ohio River Missouri River Arkansas River
Rio Grande	1,885	336,000	10,000	Pecos River
Arkansas	1,460	161,000	41,000	Canadian River Cimarron River
Colorado	1,450	246,000	15,000	Green River
Columbia	1,152	258,000	265,000	Snake River
Ohio	975	203,000	281,000	Allegheny River Monongahela River
St. Lawrence	760	396,000	348,000	Great Lakes Ottawa River

TABLE 4.3
Major Urban Areas of the United States and Associated Surface Water Features

Urban Area	Surface Water Features
Atlanta	Chattahoochee River
Baltimore	Atlantic Ocean
Boston	Atlantic Ocean
Chicago	Lake Michigan and Chicago River
Cincinnati	Ohio River
Cleveland	Lake Erie
Dallas	Trinity River and White Rock River
Detroit	Detroit River
Denver	South Platte River
Houston	Gulf of Mexico
Indianapolis	White River
Kansas City	Missouri River
Las Vegas	Colorado River
Los Angeles	Pacific Ocean
Miami	Atlantic Ocean
Minneapolis	Mississippi River
New Orleans	Mississippi River and Gulf of Mexico
New York	Atlantic Ocean and Hudson River
Philadelphia	Atlantic Ocean and Delaware River
Phoenix	Gila River and Salt River
Pittsburgh	Ohio River
Portland	Pacific Ocean, Columbia River and Willamette River
Saint Louis	Mississippi River and Missouri River

TABLE 4.3 (Continued)
Major Urban Areas of the United States and Associated Surface Water Features

Urban Area	Surface Water Features
Salt Lake City	Great Salt Lake and Jordan River
San Antonio	Salado River, San Antonio River, and Olmos River
San Francisco	Pacific Ocean
Seattle	Pacific Ocean
Tampa	Gulf of Mexico
Washington, DC	Potomac River

Philadelphia, Portland, San Francisco, Seattle, and Tampa. The eight cities located near bodies of water large enough to support large-scale shipping include Chicago, Cincinnati, Cleveland, Detroit, Kansas City, Minneapolis, Pittsburgh, and Saint Louis. The remaining seven urban locations are located near water bodies that are used for energy sources (hydroelectric), recreational functions, and sources of drinking water. These seven remaining urban areas include Atlanta, Dallas, Denver, Indianapolis, Las Vegas, Phoenix, Salt Lake City, and San Antonio.

Now let us take one step further by examining the major cities of the world by population. Table 4.4 lists the major urban cities of the world. What is noticeable is that each is near a major body of water and each is located on unconsolidated deposits, usually of fluvial, lacustrine, or glacial origin.

4.3.2 Surface Water Drainage Development and Patterns

Climate and geology were mentioned earlier as important factors influencing the development of surface water flow patterns and drainage density. A more comprehensive list of factors involved in these processes includes the following (Leopold et al. 1992):

- Climate
- Vegetation
- Topography
- Rainfall
- Geology

Rivers and streams go through an evolutionary process as they age, and follow the general sequence depicted in Figure 4.3(United States National Park Service 2024).

A young stream is characterized by a V-shaped valley, rapids and waterfalls, and no floodplain as depicted in Zone 1 of Figure 4.3. Over time, a young stream deepens its valley as depicted in Zone 2 of Figure 4.3. As the stream matures, it begins to develop a floodplain and sand and gravel bars. The valley deepening slows as depicted in Zone 3 of A fully mature stream exhibits a very wide floodplain with pronounced meanders and cut-off meanders called **ox-bow lakes** as depicted in the right portion of Zone 3 of Figure 4.3. Young streams can also be characterized as predominantly eroding streams and fully mature streams as depositional streams (Leopold et al. 1992).

The drainage development pattern of a watershed may yield some clues about its subsurface geology. For instance, a dendritic drainage pattern with a higher drainage density in the headwater region than in the depositional zone may indicate higher drainage capacity and finer-grained

TABLE 4.4
Water Bodies and Near-Surface Geology Near the Major Cities of the World

City	Location	Estimated Metropolitan Population (Millions)	Dominant Surficial Geology	Water Body
Tokyo	Japan	37.8	Unconsolidated – fluvial	Pacific Ocean
Shanghai	China	34.8	Unconsolidated – fluvial	Pacific Ocean
Jakarta	Indonesia	31.7	Unconsolidated – fluvial	Pacific Ocean
Delhi	India	26.4	Unconsolidated – fluvial	Yamuna River
Seoul	Korea	25.5	Unconsolidated – fluvial, lacustrine	Han River
Beijing	China	24.9	Unconsolidated – fluvial, lacustrine	Yongding River
New York City	USA	23.8	Unconsolidated – glacial, alluvial	Atlantic Ocean
Mexico City	Mexico	21.6	Unconsolidated – lacustrine	Lerma River
Sao Paulo	Brazil	21.2	Unconsolidated – fluvial, alluvial	Atlantic Ocean
Cairo	Egypt	20.5	Unconsolidated – fluvial, eolian	Nile River
Los Angeles	United States	18.7	Unconsolidated – fluvial, alluvial	Pacific Ocean
Moscow	Russia	16.9	Unconsolidated – fluvial, glacial	Moskva River
Istanbul	Turkey	15.2	Unconsolidated – fluvial, eolian, alluvial	Turkish Straits
London	England	14.2	Unconsolidated – fluvial	Thames River
Buenos Aires	Argentina	13.1	Unconsolidated – fluvial, alluvial	Atlantic Ocean
Paris	France	12.6	Unconsolidated – fluvial	Seine River
Rio de Janeiro	Brazil	12.3	Unconsolidated – fluvial, alluvial	Atlantic Ocean
Chicago	USA	9.8	Unconsolidated – fluvial, glacial, lacustrine	Lake Michigan
Johannesburg	South Africa	9.6	Unconsolidated – fluvial, lacustrine	Jukskei River
Riyadh	Saudi Arabia	7.7	Unconsolidated – eolian, alluvial	Wadi Hanifa River
Santiago	Chile	6.7	Unconsolidated – fluvial, alluvial, lacustrine	Pacific Ocean
Berlin	Germany	6.1	Unconsolidated – fluvial	Spree River
Toronto	Canada	5.9	Unconsolidated – fluvial, glacial, lacustrine	Lake Ontario
Sydney	Australia	5.0	Unconsolidated – fluvial, alluvial	Pacific Ocean

Source: United Nations. 2024. *World Population Prospects*. United Nations Department of Economic and Social Affairs. Population Division. https://esa.un.org/unpd/wpp/data. (Accessed March 10, 2024).

sediments. In the headwater region, silty clay underlies the streams, whereas in the depositional zone, the sediments are porous with a high infiltration capacity. As the clay sediments in the headwater region are not very permeable and restrict infiltration, most of the precipitation drains as surface water. The result is a denser stream drainage pattern required to drain the same amount of land. This same drainage pattern can also be formed by a steeper slope gradient in the headwater region than in the depositional zone.

There are other processes at work that determine a watershed's drainage density, so we must consider other factors in addition to soil infiltration capacity. For example, groundwater at the surface shapes the Earth's topography through a process known as seepage erosion. In combination with overland flow, seepage erosion contributes to the initiation and growth of channel networks (Dunne 1990; Abrams et al. 2009). Many of the United States' urban watersheds of glacial-lacustrine origin have higher drainage densities in sandy soils, and this pattern could be related to the initiation of channels by groundwater. The exact causes of channel density have not been resolved (Abrams et al. 2009), but the regional geology provides a good starting point for gaining important information about the surface streams. As this discussion reveals, there is a close relationship between surface water and groundwater. At many locations, surface water recharges groundwater and in turn,

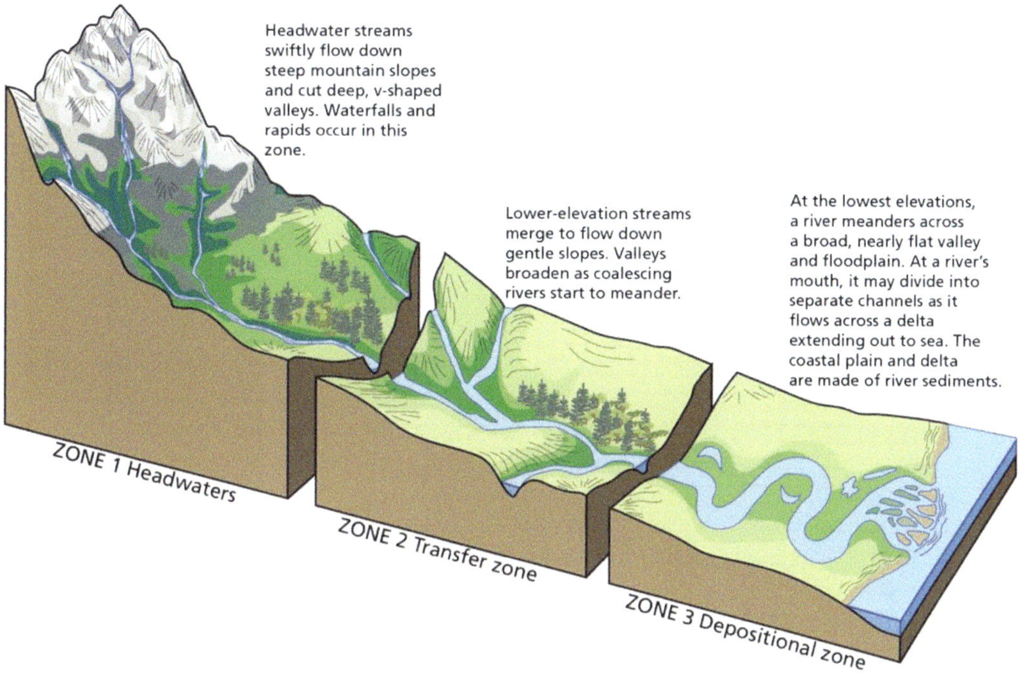

FIGURE 4.3 Stream evolution process. (National Park Service. 2024. *River Systems and Fluvial Landforms.* www.nps.gov/subjects/geology/fluvial-landforms.htm. (Accessed March 10, 2024).)

groundwater discharges to surface water, but this is only one part of the bigger picture. We now turn our attention to groundwater.

4.4 GROUNDWATER

Groundwater is defined as any water beneath the surface of the ground (Heath 1983), and is often described in terms of its occurrence in an aquifer. An **aquifer** is a mappable geologic unit(s) composed of water-saturated porous media capable of storing and transmitting significant quantities of water under ordinary conditions (Freeze and Cherry 1979). For example, the unconsolidated geologic material composing many aquifers consists of sands and gravels, and these formations are deemed aquifers because they meet the requirements just defined. Unconsolidated geologic material composed of silt and clay can also make the grade for aquifer status, but sometimes these finer-grained materials do not transmit water quickly enough to fulfill the definition.

Early civilizations such as the Persians and Romans recognized the importance of groundwater as a source of drinking water and as a supplement to precipitation for crop irrigation. The presence of water in the subsurface is influenced by fundamental geologic processes, including the subsurface rock type and the character of the tectonic activity influencing the region (e.g., folding or faulting). Today, the distribution and abundance of groundwater is a primary concern facing the growing human population on Earth, especially those within arid and semi-arid watersheds. Although the use of groundwater dates back thousands of years, it has only begun to be understood in the last couple of centuries – and major gaps in our knowledge still persist today. Scientists now understand the role of groundwater in the hydrologic cycle but still lack detailed information about groundwater in urban areas. This information includes groundwater's geographic distribution and migration pathways, anthropogenic influence, contaminant behavior and transport, and its connection with and influence on surface water.

At the watershed scale, more knowledge is required to understand the geology of groundwater and non-groundwater-bearing strata, as well as its quality, flow, residence time, age, and anthropogenic contributions and impacts to this resource. In addition, we need a deeper understanding of the fundamental hydrologic and hydrogeologic processes where groundwater plays a critical role, such as infiltration, surface runoff, and the recharge of surface waters. The many reasons underlying the lack of understanding concerning groundwater can be organized into a few broad categories:

1. Hydrogeological
   a. An absence of readily available groundwater in a particular area. There is a tendency to study only things we can see.
   b. An abundance of groundwater in a particular area. In this case, groundwater may be taken for granted and accepted without critical inquiry.
   c. An in-depth knowledge and understanding of geological and hydrological principles is required. These skills are limited to a relatively small segment of the population.
2. Geographical
   a. Other readily available and exploitable sources of water such as a large river or lake are nearby. As a result, groundwater is ignored.
3. Financial
   a. Groundwater is expensive to investigate, study, and develop as a water source. Money woes can limit curiosity.

Groundwater has only been studied outside of academic circles when something negative occurs. This includes events where contaminated groundwater is consumed from an impacted aquifer, or when the water supply runs out or is significantly reduced. This is particularly disturbing because approximately 50% of the population in the United States obtains its drinking water from groundwater, and more than 40% of irrigation water comes from groundwater.

The importance of groundwater cannot be overstated, because groundwater accounts for over 95% of all the fresh water on the planet – excluding the hard-to-reach water stored in ice caps and glaciers. The estimated total world reserve of groundwater is estimated to be approximately 2.5 million cubic miles (USGS 2004), which is about 100 times more than all the fresh water in lakes, streams, rivers, and swamps on Earth. To envision this amount of water is difficult, but suppose all of the groundwater in the world was spread out on the surface of the United States, the water depth would be close to a half mile. Let's go global. If all the groundwater was placed on the land surface of the Earth it would be 150 feet deep.

The generally poor understanding of groundwater is compounded in urban areas because the near-surface unconsolidated sediments here are also poorly understood. This is somewhat alarming because a significant portion of the human population currently resides in urban areas and this population is projected to increase rapidly in the future (United Nations 2003 and USGS 2024c).

Groundwater plays an important part in the water cycle on the planet: a cycle involving a complex set of interactions between water and the surface, subsurface, and atmosphere. The next section builds the foundation for understanding groundwater within this context.

4.5 FUNDAMENTAL CONCEPTS OF GROUNDWATER

Describing the occurrence, distribution, behavior, and migration of groundwater beneath the surface is best accomplished from the perspective of geographic scale. At the macro- or regional scale, groundwater exists in well-defined geologic formations (aquifers), and at the micro-scale, groundwater assumes many forms, including any water unavailable for flow such as the water surrounding minerals grains due to surface tension.

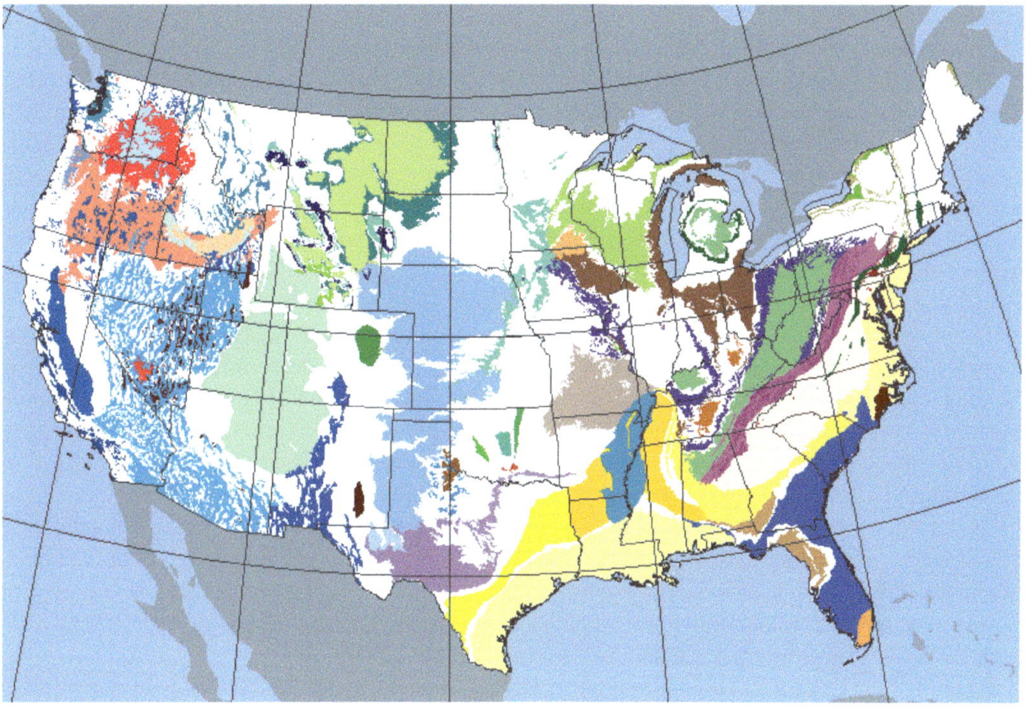

FIGURE 4.4 Distribution of major U.S. aquifers. (From United States Geological Survey. 2024d. *Principle Aquifers of the 48 Conterminous United States*. http://water.usgs.gov/GIS/browse/aquifers_us.jpg. (Accessed March 10, 2024).)

The term "significant quantities" used to describe water in an aquifer is a relative term and depends on whether the water is intended for human exploitation. The outcome is largely dependent upon location and geography. For instance, an aquifer in a desert region may not yield as much water as an aquifer located elsewhere but it still may be considered significant due to its geographical location. The majority of aquifers in the world considered high-yielding are composed of sand and gravel deposits; for example, the high-yielding Ogallala aquifer stretching from the Texas panhandle to North Dakota in the Great Plains region is sand and gravel. Figure 4.4 shows the distribution of aquifers in the conterminous United States. Most of these aquifers are composed of sedimentary rock and underlie a majority of the country's major urban areas.

4.5.1 Aquifer Types and Basic Structure

There are two types of aquifers, confined and unconfined, based on presence or absence of a water table. A water table or phreatic surface is the upper extent of the zone of saturation in direct contact with atmospheric air pressure through void spaces in the overlying geologic material. An **unconfined aquifer**, therefore, is defined as a saturated geologic material where the surface of the saturated layer is equal to atmospheric pressure (Heath 1983). The depth to the water table surface can range from just a few feet or less in humid regions to more than 1,000 feet in dry desert regions. Unconfined aquifers are usually the first saturated waters encountered beneath the surface. If this water is of limited areal extent, it is commonly known as a perched aquifer or perched water depending on the amount and extent of saturated materials.

Perched water or perched aquifers are commonly encountered in urban environments. Many building foundations, sewers, and roads are backfilled with materials such as sand and gravel with higher **permeability** – capable of being penetrated by water – than the indigenous soils often composed of clay. In these cases, water eventually saturates the more permeable materials surrounding the foundation and perched water develops. In a pure geologic sense, this underground configuration of water does not fit the definition of an aquifer because it is not a mappable geologic unit, does not occur in a naturally occurring formation, and exists because of anthropogenic activities. **Confined aquifers** are permeable geologic formations or strata bounded above and below by relatively impervious geologic material – clay layers or other similar materials – and contain water at a pressure greater than atmospheric pressure (Heath 1983). The low permeability layers or strata separate the higher permeability aquifer material from direct contact with atmospheric pressure and impede the upward and downward movement of water.

These low-permeable strata consist of aquitards and aquicludes and are commonly referred to as confining units (Heath 1983). An **aquitard** is a low-permeability geologic formation not capable of yielding significant quantities of water. Aquitards are often composed of clay or other fine-grained materials. An **aquiclude** is an extremely low-permeability geologic formation that in practice is not permeable. As most unconsolidated formations and geologic units exhibiting low permeability are composed of clay, and clay has a high water storage potential of up to 40% (Freeze and Cherry 1979), an additional distinction between aquitards and aquicludes arises. Specifically, an aquitard is considered a geologic unit with low permeability and high storage potential, whereas an aquiclude is a geologic unit characterized by low permeability and low storage potential (Payne et al. 2008). However, as all-natural materials have some natural permeability, the trend today is to move away from the somewhat arbitrary terms aquitard and aquiclude (USGS 1984).

As shown in Figure 4.5, the confining layer above the confined aquifer creates a column of water with very limited exposure to the ground surface. This overburden of water creates additional pressure within these formations, so any well installed within a confined aquifer will have a water level higher than the elevation where the water was encountered. If the water level in the well rises but does not reach the surface, the well is termed an **artesian well**. When the pressure in a confined aquifer is sufficient to allow the water level to reach or exceed the ground surface, the well is called a **flowing artesian well** (Driscoll 1986). Figure 4.5 shows a schematic of an unconfined and a confined aquifer and also shows artesian and flowing artesian wells. The elevation water that can reach in wells drilled into a confined aquifer is called the piezometric surface, and is indicated by the dashed line in Figure 4.5.

Many types of aquifers can exist in certain geological environments where thick sequences of variable unconsolidated deposits are present. In these cases, the term commonly applied is an **aquifer system**.

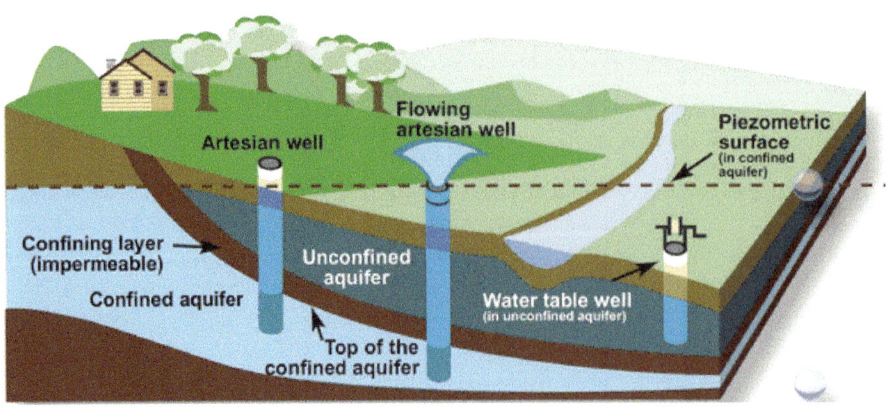

FIGURE 4.5 Example of an unconfined and confined aquifer and artesian and flowing artesian wells. (From Environment Canada. 2024. *Aquifers and Wells*. www.ec.gc.ca/water/images/nature/gedwtr/a5f3e.html. (Accessed March 10, 2024).)

In areas of karst topography, groundwater flows through bedrock and initially concentrates between bedding layers, fractures, or faults within the sedimentary deposits. Rock types prone to the development of karst topography contain a high amount of calcite ($CaCO_3$), such as limestone. Through time, chemical weathering along some of the groundwater migration pathways begins to dissolve away the rock, and enlarges the pathway. If this enlargement continues for enough duration, a cavity or void space called a **cavern** is formed. As the cavern grows with continued groundwater migration, it may connect with other caverns and lead to increased groundwater flow as void space increases. If water levels drop due to natural groundwater fluctuation or increased rates of groundwater flow, the caverns fill with air instead of water and become a cave. A **cave** is a natural cavity beneath the surface created by groundwater dissolving rock (Burger and Dubertret 1984; Ford and Williams 2007).

Karst formations are cavernous and have high rates of permeability. At many locations with well-developed karst topography, surface streams disappear and flow underground through networks of caves only to re-appear at the surface sometimes several miles away. Flow rates under these circumstances do not fit into the range of usual groundwater velocities; they approach those of surface streams. Overall groundwater flow conditions in karst topography become difficult to evaluate because not only do the flow rates vary widely but also flow directions can be unpredictable. These flow properties of karst-resident groundwater are a function of the fracture and dissolution patterns of the host rock (Ford and Williams 2007). Figure 4.6 is a diagram of karst topography.

Evaluating an aquifer in karst topography requires mapping the fracture pattern of the host rock. The mapping process begins with tracer testing. A tracer die is placed at an upgradient location, and the monitoring devices or observation points are set up at several downgradient locations. The elapsed time between the upgradient introduction of the die and the first observation of the die at downgradient

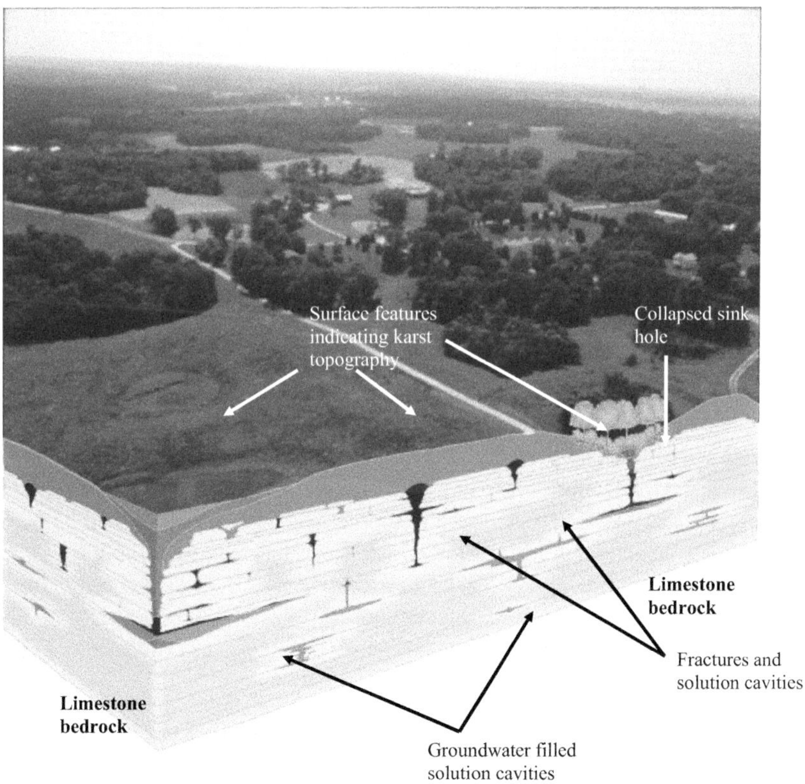

FIGURE 4.6 Karst topography. (From Illinois State Geological Survey. 2024. *Geobit 7 Karst*. www.isgs.edu/maps-data-pub/publications/geobits7.shtml. (Accessed March 10, 2024).)

locations allows for an estimation of the rate of groundwater flow, and general groundwater flow direction, and provides information on the fracture pattern of the aquifer (Ford and Williams 2007).

4.5.2 Heterogeneity and Homogeneity

Heterogeneity and homogeneity are measures of variability used to classify geological diversity. In geological terms, **heterogeneity** is a term applied to highly variable or poorly sorted geologic materials, whereas **homogeneity** refers to geologic formations without much variability and are characterized as well-sorted (Heath 1983). Aquifers are typically composed of coarse-grained sediments derived from processes characterized as high energy. The types of geological processes leading to the formation of aquifers in unconsolidated sedimentary deposits are fluvial, glacial, lacustrine, beach, or eolian (USGS 2024b). These sediments consist of layered sands and gravels and may also be cross-bedded. As a rule, no two aquifers – or for that matter – no two locations within an aquifer are exactly alike. As aquifers are a member of a larger set of geologic units (e.g., glacial deposits), some of those principles used to evaluate the heterogeneity and homogeneity of the larger set also apply to aquifers:

- The size and shape of any geologic unit are inherently different when evaluated from a three-dimensional perspective. Confined and unconfined aquifers range in thickness and aerial distribution, and composition of matrix materials (e.g., sand, gravel, fractured limestone).
- Each geologic unit has a unique arrangement and network of interconnecting features created by the depositional method. Karst aquifers are demonstrably different in this respect than sand and gravel aquifers.
- Comparison of heterogeneity and homogeneity is possible at several different geographical scales. The mega-scale is a geologically dependent formational scale and is regional in extent. An example here is the heterogeneity of large river deposits compared to a glacial moraine deposit. The macro-scale refers to the variability within the formation itself. In this case, you could compare the variability of two sand bars deposited by the same river. The micro-scale refers to the variability and arrangement of depositional layers and individual grains within a specific deposit.

When an aquifer has high variability, it is called heterogeneous. An aquifer of low variability may be labeled as homogeneous or relatively homogeneous. The probability of encountering a truly homogeneous aquifer is effectively zero, as all aquifers demonstrate some degree of variability, with most displaying a high degree of variability.

The terms **isotropic** and **anisotropic** are used in describing the degree of variability related to a given direction for a specified geologic feature. For instance, a geologic formation may exhibit a more homogeneous texture within a specific bedding layer trending in a certain direction. This phenomenon then qualifies as a layer displaying either more of an isotropic tendency or anisotropic tendency in a given direction from a defined location. Figure 4.7 has examples showing the different combinations of homogeneous, heterogeneous, isotropic, and anisotropic concepts (Heath 1983).

The concept of heterogeneity describing the variability within an aquifer is significant and requires a sound scientific understanding. Heterogeneity has a profound effect on the migration and fate of contaminants present in groundwater, and influences how those contaminants are removed, destroyed, or chemically altered through specially designed remedial technologies (Payne et al. 2008).

4.5.3 Hydraulic Conductivity

Hydraulic conductivity (K) is a parameter used to measure the relative ease with which water flows through an aquifer or other geologic formations such as an aquitard. **Hydraulic conductivity** (K) is defined as the ability of saturated geologic media to conduct water under an induced hydraulic

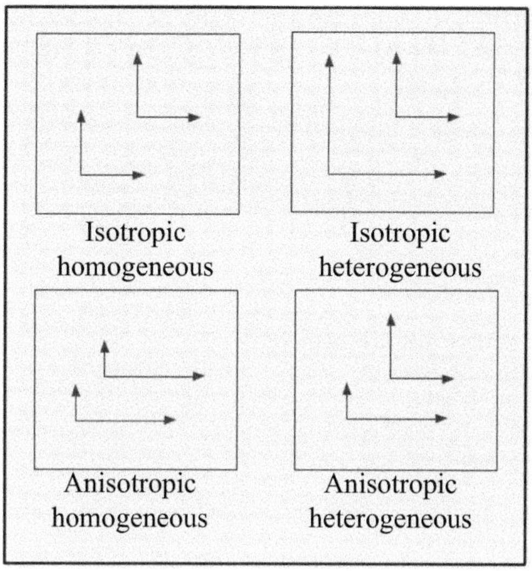

FIGURE 4.7 Visualization of variability concepts.

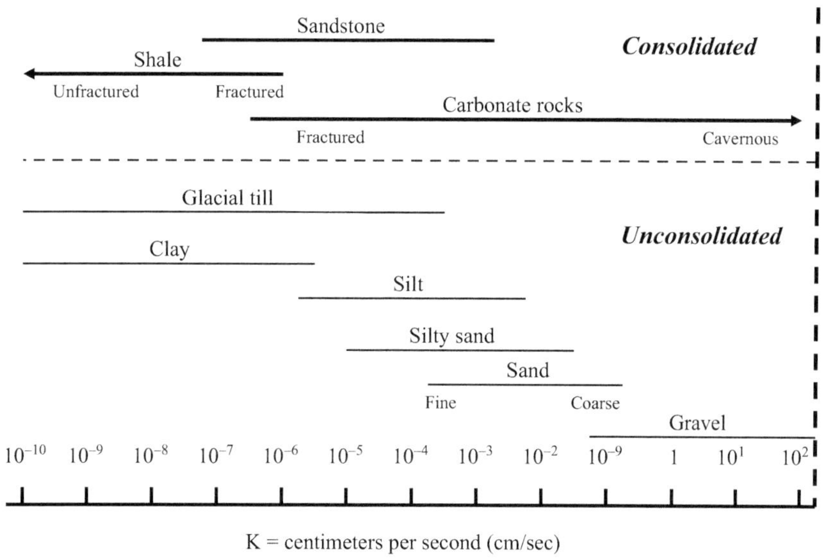

FIGURE 4.8 Ranges of hydraulic conductivities for different geologic materials. (Modified from Heath, R. C. 1983. *Basic Ground-Water Hydrology*. United States Geological Survey, Water Supply Paper 2220. United States Government Printing Office. Alexandria, VA.)

or pressure gradient (Payne et al. 2008). Geologic formations composed of sands and gravels typically have a high hydraulic conductivity or K value (e.g., 10^{-2} cm/sec to 10^{+1} cm/sec). Geologic formations composed of clay which typically might be considered an aquitard or even an aquiclude frequently have very low hydraulic conductivities (e.g., 10^{-6} cm/sec to 10^{-8} cm/sec). Figure 4.8 lists the average range of hydraulic conductivities for different geologic materials (Heath 1983).

In 1856 a researcher from France named Henry Darcy conducted an experiment to evaluate the flow of water through sand beds. His experiments resulted in the development of Darcy's Law. **Darcy's Law** predicts that the flow of water (Q (cm^3/sec)) through a sand bed is *proportional* to its hydraulic conductivity (K), area (A (cm^2)) and the water level difference between each end of the sand bed (dH (cm)), and *inversely proportional* to the thickness (L) of the sand bed.

Equation 4.1 presents Darcy's Law:

$$Q = K \times A \times \frac{dH}{dL} \tag{4.1}$$

where K is the hydraulic conductivity (cm/sec)
A is the area (cm^2)
dH is the change in height of the bed of sand
dL is the length of the bed of sand

4.5.4 Groundwater Movement

Groundwater is in constant motion, although groundwater flow rates are usually much slower than those of surface streams. Unlike surface streams, groundwater is pulled downward by gravity through an intricate network of passageways between the pore spaces in the soil and sediment.

This area is called the **vadose zone** (Heath 1983). The vadose zone (also called the zone of aeration or unsaturated zone) is the subsurface area where water is present, but air still occupies part of the overall pore space. Figure 4.9 shows the location of the vadose zone relative to the water table and saturated zone.

A thin layer of water – called **hygroscopic water** – will always be attracted to the surface of mineral grains. This attraction is a result of unsatisfied ionic charges on the surface of the mineral grains combined with the polar nature of water molecules. This is termed the force of **molecular attraction**.

The thickness of the vadose zone varies by location. In desert locations, the vadose zone may be several hundred feet to more than a thousand feet thick; in humid regions where there is significant precipitation, the thickness may be just a few feet or even less. There are also seasonal variations in the thickness of the vadose zone. During prolonged periods of drought, the vadose zone may become much thicker as opposed to periods of prolonged rainfall when the thickness of the vadose zone may shrink significantly. At most locations, there are seasonal variations in the thickness of the vadose zone.

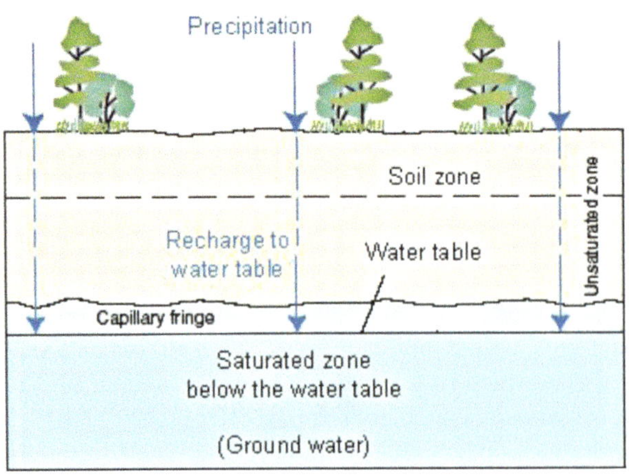

FIGURE 4.9 Diagram of the vadose zone. (From United States Geological Survey. 2024a. *Summary of the Water Cycle*. http://ga/water.usgs.gov/edu/watercyclesummary.html (Accessed March 10, 2024).)

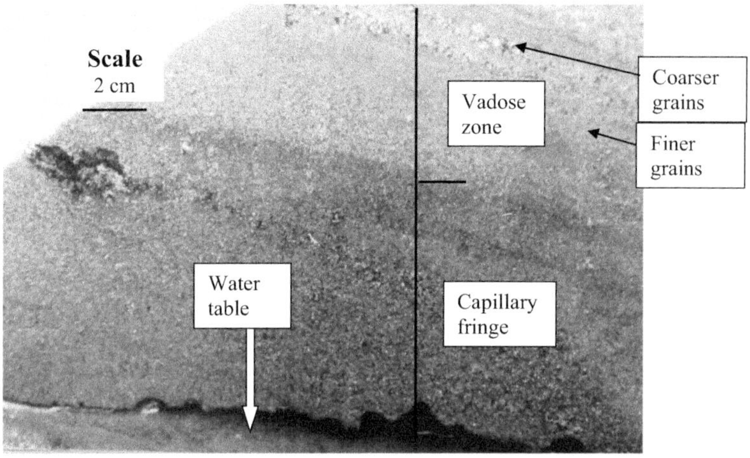

FIGURE 4.10 Photograph showing vadose zone, capillary fringe, and water table surface. (Photograph by Daniel T. Rogers.)

As surface water continues its journey downward through the unsaturated zone, it encounters the capillary fringe. The **capillary fringe** is the subsurface layer where groundwater seeps up from the water table by capillary action to fill pore spaces (USGS 2024b). At the base of the capillary fringe, the pore spaces are filled with water due to tension saturation. The thickness of the capillary fringe is dependent upon the balance between the soil/sediment's adhesion binding capability and surface tension, and the gravitational force of the lifted water mass. What factors prevail here? In most cases, the thickness of the capillary fringe is less in course-grained materials than in finer-grained materials (Payne et al. 2008). Figure 4.10 shows the vertical arrangement of the subsurface zones just described. Figure 4.9 also demonstrates that gradation in grain size can occur over very short distances.

Once water reaches the water table, it migrates downward through the available interconnected pore spaces under the force of gravity and from zones of higher pressure to areas of lower pressure. Groundwater can also move upward from areas of higher pressure to areas of lower pressure. Imagine a U–shaped tube partially filled with water. If pressure is exerted on one side of the tube, the water on the other side rises proportionally in response to the pressure exerted on the other side. This is commonly also referred to as upwelling.

Similar variations in pressure occur beneath the surface of the Earth. Pressure is higher beneath hills and lower in valleys (Heath 1983). Elevated land contains more mass than valleys and therefore exerts more force per unit area – the definition of pressure. The lower pressure below valleys creates a pressure gradient toward the surface, allowing deeper groundwater to flow upward and reach surface water.

At the micro-scale within the saturated zone, groundwater does not travel in a straight line. Figure 4.11 shows a potential pathway a particle of water may take when moving within the saturated zone.

4.5.5 Primary Porosity

To reiterate, groundwater travels through the subsurface through the interconnected pores between the individual mineral grains. **Porosity** is the percentage of the volume of the sediment that is open space (pore space), which is depicted in Figure 4.12.

Well-sorted and well-rounded sediments have a higher degree of porosity compared to poorly sorted sediments because there are no fine-grained particles to fill the pore space between the

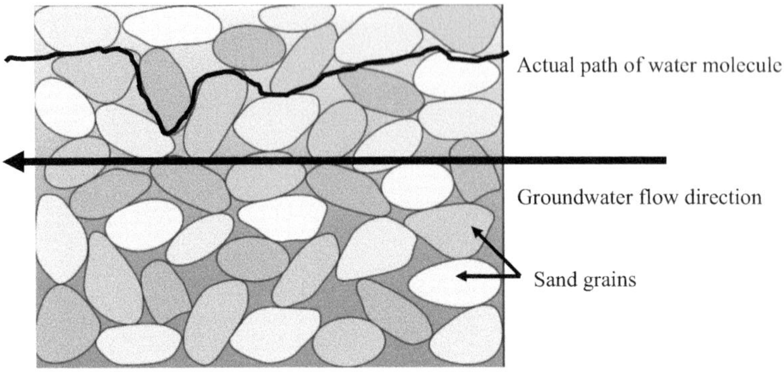

FIGURE 4.11 Groundwater flow on the micro-scale.

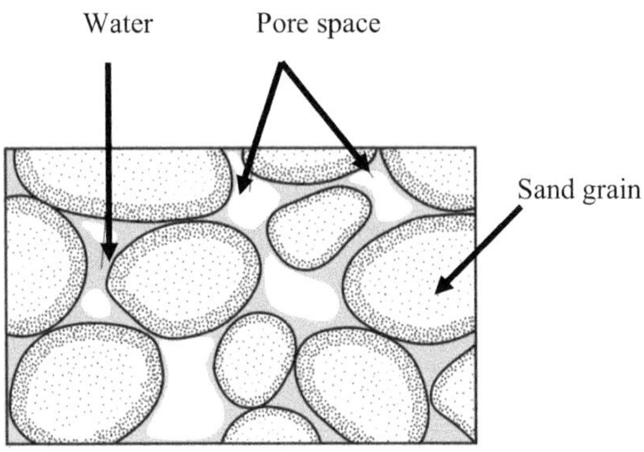

FIGURE 4.12 Example of pore space in sediments.

well-sorted and rounded grains. Figure 4.16 depicts well-sorted materials and also qualitatively demonstrates the difference between a well-sorted, well-rounded, large-grained material compared to a smaller-grained, less well-rounded, well-sorted material.

Porosity (Ǿ) is defined by Equation 4.2:

$$\phi = \frac{V_v}{V_T} \tag{4.2}$$

where V_v is the volume of void space and V_T is the total bulk volume of the material including the solid and void components.

Porosity is composed of the interstitial space between soil materials; the water here consists of the hygroscopic water bound tightly to the material surface and the other water capable of draining freely under the influence of gravity. Water capable of draining freely under the influence of gravity is termed **specific yield (S_y)** or **effective porosity (n_e)**; and it represents the approximate measures of the interconnected pore space groundwater flows through within geologic strata or layers. The remaining water is termed **specific retention (S_r). Total porosity** is the sum of effective porosity and specific retention and is commonly expressed as Equation 4.3 (Heath 1983):

$$N = S_y + S_r \tag{4.3}$$

FIGURE 4.13 Presence of groundwater due to an unconformity. (Photograph by Daniel T. Rogers.)

where N = total porosity
S_y = specific yield
S_r = specific retention

As stated earlier, a poorly sorted sediment is not as porous as a well-sorted sediment because the fine-grained particles occupy spaces between the larger grains. An additional type of primary porosity, not often discussed and more difficult to evaluate, occurs between non-conformable contacts or unconformities. As discussed in Chapter 2, an unconformity represents a buried erosional surface separating two different rock masses or deposits. This surface represents a plane or surface of discontinuity often exhibiting a different and highly localized porosity. As a result, this surface is often saturated with water, especially if the geologic units above and below are low-permeable formations. These types of contacts do not contain large quantities of groundwater, but they often contain enough water to enable their flows to be measured.

An example of this type of porosity is presented in Figure 4.13. Figure 4.13 is a photograph of an excavation through a fine-grained lacustrine deposit composed of silty clay overlying a ground moraine or lodgment till deposit of glacial origin. The unconformable contact between the two different and distinct geologic units is represented at the floor of the excavation. Groundwater is not encountered in the upper lacustrine deposit nor is it encountered within the ground moraine deposit. Instead, groundwater is encountered at the unconformable contact between the two geologic units.

This type of porosity is significant because it represents an apparent anomaly when evaluating the porosity and permeability of the two geologic units individually. By themselves, the geologic units may not exhibit significant porosity, but the contact between the two units creates enough porosity for groundwater to be present with enough perhaps to permit flow.

4.5.6 Secondary Porosity

Secondary porosity refers to the porosities that form within a geological unit after the material has been deposited (Freeze and Cherry 1979). Types of secondary porosity include the following:

- Fractures caused by differential compaction
- Fractures caused by freeze–thaw cycles
- Fractures caused by vegetative root fragments
- Bioturbation
- Vugs (small to medium-sized cavities inside rocks), fissures, or solution cavities caused by dissolution from migrating waters or other chemicals that dissolve or precipitate minerals

Freeze and Cherry (1979) note the extent of freeze–thaw within some soils in colder climates can reach up to 30 feet beneath the surface. This may be particularly significant when examining the integrity of an aquiclude at shallow depths. Any of the above-listed types of secondary porosity can also influence the integrity of an aquiclude from shallow to sometimes significant depths beneath the ground surface.

Secondary porosity in the form of vertical fractures and root fragments combined with the porosity observed at unconformable contacts may assist in explaining why groundwater is observed in some locations. Geologic units previously characterized as impermeable or not having enough groundwater to permit flow may indeed yield measurable water. The awareness of this possibility may help when evaluating anthropogenic influences on the geology and hydrogeology of urban areas for groundwater and surface water protection purposes (Howard and Gerber 1997).

4.5.7 Water Table and Hydraulic Gradient

The water level in a well penetrating an unconfined aquifer undergoing horizontal flow will be equal to the level of the water table. When these water levels are joined across multiple wells in two-dimensional space, a water table plane or **potentiometric surface** is defined (Freeze and Cherry 1979). This constantly changing surface is a measure of the upper physical boundary of the water table. Fluids flow between the differences in pressure or elevations existing at distinct locations. For example, groundwater flows from areas of high pressure to areas of lower pressure; or in hydrologic terms, groundwater flows from areas of higher potential to areas of lower potential. A *pressure* differential is called a pressure gradient, whereas the *elevation* difference between wells is called a **hydraulic gradient**. The hydraulic gradient between the water levels in Well B and Well A, as depicted in Figure 4.14, is often referred to as hydraulic head. To calculate the hydraulic head

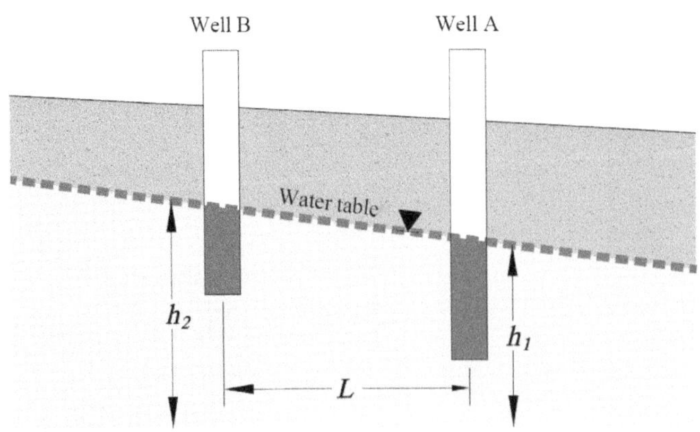

FIGURE 4.14 Hydraulic gradient.

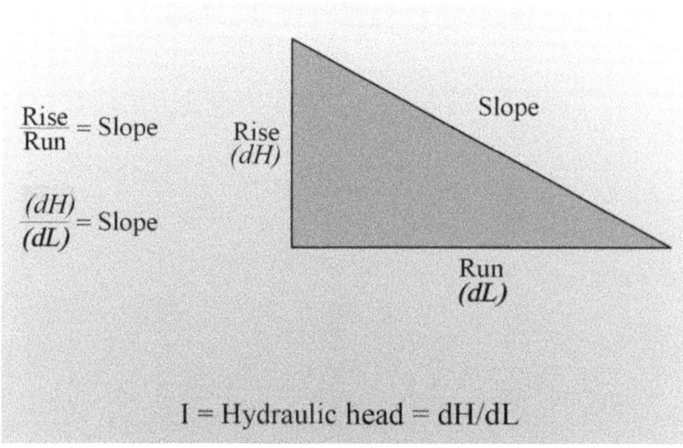

FIGURE 4.15 Calculating the hydraulic head.

between two wells, take the ratio of the rise or height difference between the two wells (*dH*) to the run or horizontal distance between the two wells (*dL*). This ratio is shown in Figure 4.15.

4.5.8 Determining the Direction of Groundwater Flow

Calculating the direction of groundwater flow requires at least three data points that accurately measure the level of groundwater under atmospheric pressure. The vertical accuracy required is usually within 0.01 of a foot. Having three wells is the minimum number necessary to determine the direction of groundwater flow, and if only three wells are used their location, pattern, or placement should be in a triangular pattern. This information is needed to determine the general direction of groundwater flow between the three wells:

- Relative geographic position of the wells
- Distance between the wells
- Elevation of water in each well

Calculating the direction of groundwater flow involves the following steps (Heath 1983):

1. Identify the well with the intermediate water level
2. Interpolate the location of the intermediate water level well along a straight line between the wells having the highest and lowest relative water levels
3. Create a line segment of equal potential to the intermediate well by drawing a straight line from the well with the intermediate water level to its water equivalent elevation plotted along the straight line between the wells having the highest and lowest relative water elevations.
4. Draw a perpendicular line to the water-level contour just plotted with the well of highest or lowest water elevation. This line parallels the general direction of groundwater flow.
5. Calculate the difference between the water level of the well and the intermediate well-level contour by using the distance between the well and the contour to reveal the hydraulic gradient.

An example of the methodology explained above is shown in Figure 4.16.

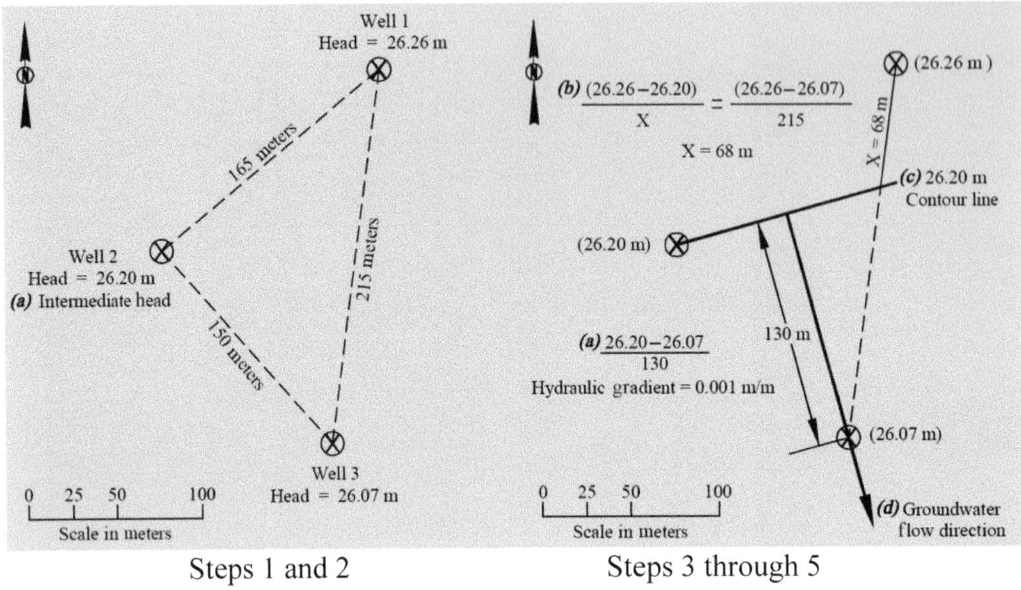

FIGURE 4.16 Determining the direction of groundwater flow. (Modified from Heath, R. C. 1983. *Basic Ground-Water Hydrology*. United States Geological Survey, Water Supply Paper 2220. United States Government Printing Office. Alexandria, VA.)

4.5.9 Estimating the Rate of Groundwater Flow

To calculate the rate or velocity of groundwater flow requires site-specific knowledge or estimation of the following parameters:

- Hydraulic conductivity (K) of the saturated geologic strata
- Effective porosity (n_e) of the saturated geologic strata
- Hydraulic gradient (dH/dL) or slope of the groundwater surface

Once the above parameters are measured or estimated, the rate or velocity of groundwater flow can be calculated using the stream economy equation and Darcy's Law. The stream economy equation is as follows:

$$Q = Av \tag{4.4}$$

where Q = discharge (volume of water per unit time (cm^3/sec)
A = cross-sectional area (at right angle to the water flow direction) cm^2
v = velocity (cm/sec)

$$Q = K \times A \times \frac{dH}{dL} \tag{4.5}$$

where K is the hydraulic conductivity (cm/sec)
A is the area (cm^2)
dH/dL is the hydraulic gradient

Step 1 of Equation 4.6 given below shows that groundwater velocity can be computed by substituting Av (which equals Q) in Equation 4.4 for the Q in Equation 4.5. After canceling the area (A) on both sides of the equation, the velocity (v) is equal to hydraulic conductivity (K) multiplied by the hydraulic gradient (step 2). Groundwater, however, does not migrate in open space; it travels through aquifer materials impeding its velocity. To account for this, the effective porosity of the saturated geologic strata (specific yield) is required to accurately estimate the velocity of groundwater flow (step 3).

$$\text{Step 1: } \mathrm{Av} = \mathrm{k} \times \mathrm{A} \times \frac{dH}{dL} \rightarrow \text{Step 2: } \mathrm{v} = \mathrm{k} \times \frac{dH}{dL} \rightarrow \text{Step 3: } \mathrm{v} = \frac{\mathrm{k}}{n_e} \times \frac{dH}{dL} \qquad (4.6)$$

To illustrate the difference in the velocity of groundwater flow through two different materials, aquifers composed of sand and clay with the same hydraulic gradients are compared:

Sand aquifer example:

$$K = 0.1 \text{ cm/sec}$$

$$dH/dL = 1 \text{ cm}/1{,}000 \text{ cm}$$

$$n_e = 0.22$$

$$v = \frac{K}{n_e} \times \frac{dH}{dL}$$

$$v = 0.1 \text{ cm/sec} \times 1/0.22 \times 1/1{,}000 = 2.2^{-5} \text{ cm/sec}$$

Clayey aquifer example:

$$K = 0.000001 \text{ cm/sec}$$

$$dH/dL = 1 \text{ cm}/1{,}000 \text{ cm}$$

$$n_e = 0.20$$

$$v = \frac{K}{n_e} \times \frac{dH}{dL}$$

$$v = 0.000001 \text{ cm/sec} \times 1/0.20 \times 1/1{,}000 = 5^{-9} \text{ cm/sec}$$

Given the same hydraulic gradient, the velocity of groundwater flow in the sand aquifer is approximately 10,000 times greater than the velocity of groundwater flow in the clay aquifer.

The residence time of water as groundwater can range from days to thousands of years and is a function of several variables. The dominant influence is depth, manifested by the strong correlation between increasing groundwater depth and residence time. Other factors influencing residence time include the following:

- The composition of the aquifer material
- Effective porosity

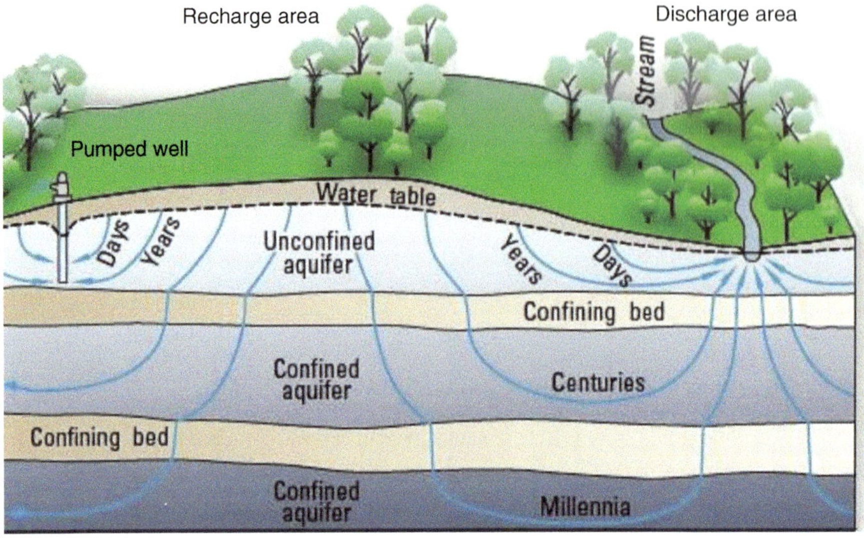

FIGURE 4.17 Range in groundwater residence time. (From Focazio, M. J., Reilly, T. E., Rupert, M. G., and Helsel, D. R. 2001. *Assessing Ground-Water Vulnerability to Contamination: Providing Scientifically Defensible Information for Decision Makers*. United States Geological Survey Circular 1224. USGS. Denver, CO).)

TABLE 4.5
Estimated Depth and Residence Time of the World's Water Supply

Parameter	Equivalent Depth (meters)	Approximate Residence Time
Oceans and Seas	2,500	4,000 years
Lakes and Reservoirs	0.250	10 years
Swamps	0.007	1–10 years
River Channels	0.003	2 weeks
Soil Moisture	0.130	2 weeks to 1 year
Groundwater	120	2 weeks to 10,000 years
Ice caps and Glaciers	60	10–10,000 years
Atmospheric water	0.025	10 days
Biospheric water	0.001	1 week

Source: Environment Canada. 2024. *Aquifers and Wells*. www.ec.gc.ca/water/images/nature/gedwtr/a5f3e.htm (Accessed March 10, 2024).

- Hydraulic gradient
- Distance to point of discharge

The span of residence times for groundwater is depicted in Figure 4.17, and the residence times for water in other environments are listed in Table 4.5.

4.5.10 Groundwater Withdrawal

Withdrawal of groundwater is typically accomplished by pumping a well installed within the aquifer. As groundwater is pumped from a well, the water level in the well begins to decline as

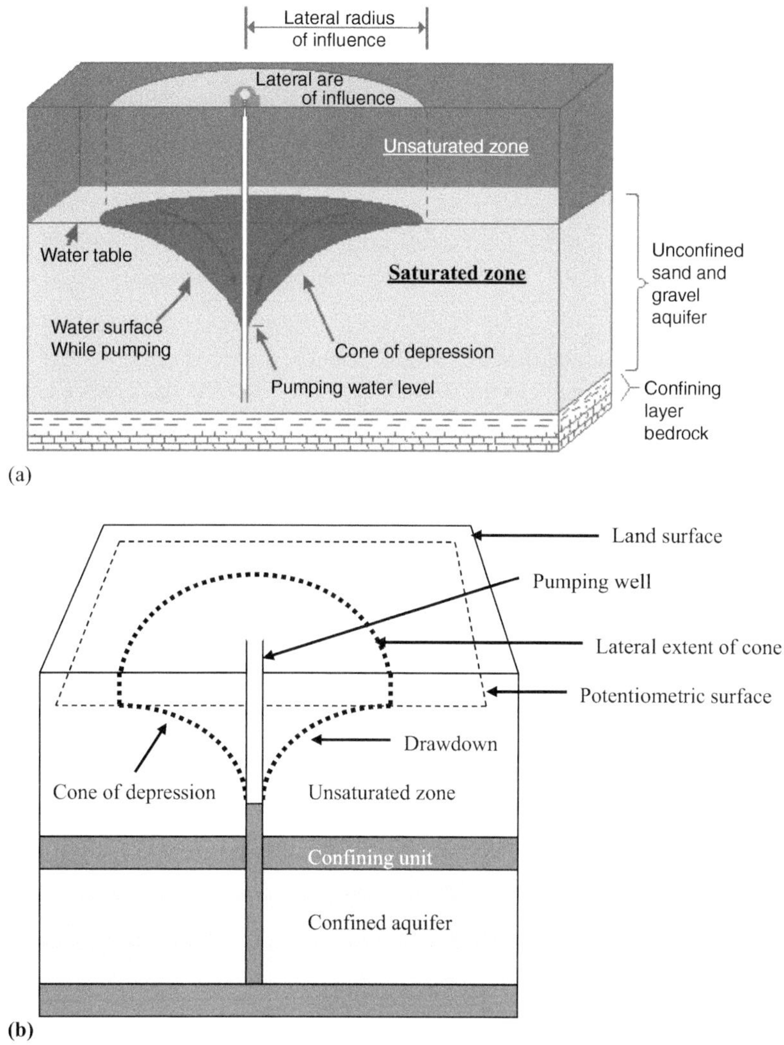

FIGURE 4.18 A, Cone of depression from a pumping well in an unconfined aquifer. B, Cone of depression from a pumping well in a confined aquifer. (Modified from Heath, R. C. 1983. *Basic Ground-Water Hydrology*. United States Geological Survey, Water Supply Paper 2220. United States Government Printing Office. Alexandria, VA.)

water is removed from storage in the well. A hydraulic gradient is created as the water level falls below the level of the surrounding aquifer. Water then begins to migrate from the aquifer into the well. The water level will continue to decline within the well and the flow rate of water into the well will increase until the inflow rate is equal to the withdrawal rate. This process occurs only when the pumping rate does not exceed the capacity of the aquifer formation to transmit groundwater to the well. In this case, the diameter of the well may be too small, the pump should be placed lower in the well, or the withdrawal rate should be decreased. Water from the aquifer converges on the well from all directions and the hydraulic gradient gets steeper near the well. For these reasons, the three-dimensional shape of water withdrawal from a well is termed a **cone of depression**. The cones of depression shown in Figure 4.18 result from a pumping well withdrawing water from unconfined and confined aquifers. Transmissivity (T) is a measure of the amount of water that can be transmitted horizontally through a unit width by the full saturated thickness of the aquifer under a hydraulic gradient of 1. Taking into account the definition of hydraulic conductivity, transmissivity (T) is actually equal to the hydraulic conductivity (K) multiplied by the thickness of the aquifer (b) (Equation 4.7) (Heath 1983):

$$T = Kb \tag{4.7}$$

where T = transmissivity
K = hydraulic conductivity
b = aquifer thickness

The aquifer parameters of transmissivity (T) and storage coefficient (S) are variables that can dictate the shape of the cone of depression. The storage coefficient is the volume of water that a permeable unit will absorb or expel from storage per unit surface area per unit change in head. For a confined aquifer, the expansion of the cone of depression in response to depressurizing (pumping) is very small, but the compression pressures can be significant. This permits the cone of depression to expand and deepen rapidly when pumped, so a lower storage coefficient (S) value will create a deeper and wider cone than a higher storage coefficient. An aquifer with a low transmissivity will develop a deep and narrow cone of depression and an aquifer with a high transmissivity will develop a shallow and wide cone of depression (Heath 1983).

4.6 SURFACE WATER–GROUNDWATER INTERACTION

Surface water and groundwater are fundamentally connected, and their interaction occurs everywhere in all watersheds. It is difficult to separate the two because they "feed" off of each other and are interconnected through the water cycle.

4.6.1 Stream Interaction with Groundwater

Groundwater accounts for most, and often all, of baseflow in rivers and streams. **Baseflow** is defined as groundwater seepage into a stream channel (Freeze and Cherry 1979). This influx of groundwater is very important to streams and rivers because it provides flow during periods of low precipitation or drought, and helps sustain aquatic life (Winter et al. 1998). On hillslopes, many streams originate as groundwater as shown in Figure 4.19.

When groundwater flows toward a surface stream and discharges into it, the stream is termed a **gaining stream**. When a stream loses water to groundwater, the stream is termed a **losing stream**.

Another stream interaction with groundwater is termed disconnected, and it occurs when the surface water stream and groundwater are separated by the unsaturated or vadose zone. This usually occurs when the sediments within the unsaturated zone beneath the stream have low hydraulic conductivity and impair significant communication between the two. Another cause of this interaction is

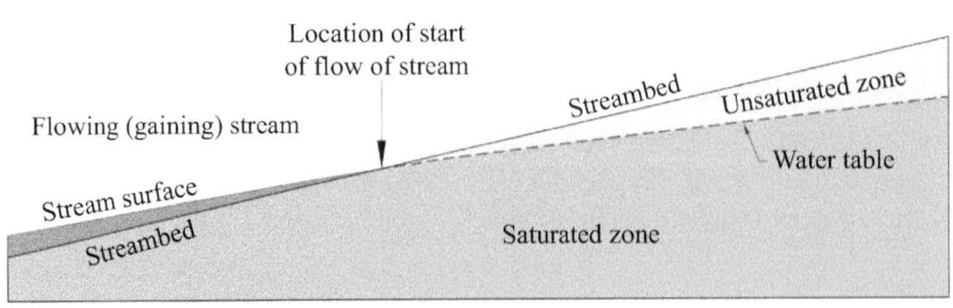

FIGURE 4.19 Beginning of stream flow from groundwater.

a lowering of the water table due to drought or the local geology. As streams and rivers flow through varied terrain and geology, it is possible one stream may exhibit the three types of connections (Alley et al. 1999).

At any specific location, a stream is either gaining, losing, or disconnected depending on geographic, geologic, hydrologic, and anthropogenic factors.

During one study of the Rouge River watershed in southeastern Michigan, the direction of groundwater flow was evaluated at several hundred sites within the watershed. Groundwater flow direction was established using a minimum of 6 and a maximum of 80 monitoring wells per site, and each monitoring well was surveyed by a licensed surveyor. In a detailed study of the lower branch of the Rouge River, Murray and Zhou (2000) installed 54 groundwater monitoring wells along the river along with 18 well points installed into the river's bed sediment.

The results of these studies conclusively show that groundwater flow is discharging through the bed sediment into the river and in a downgradient direction, consistent with the direction of surface water flow. This is significant for an urban watershed like the Rouge for these reasons: (1) the direct connection between groundwater and surface water means that contaminants released within the watershed may impact shallow groundwater; and (2) this contamination could also reach surface water and the lower Great Lakes (Rogers 1997; Rogers and Murray 1997; Murray and Zhou 2000).

4.6.2 Lake Interaction with Groundwater

Lakes interact with groundwater in three basic ways. Some lakes receive groundwater through the entire bed of the lake, some lakes have water loss through the entire bed of the lake, and perhaps most commonly, lakes receive groundwater at some locations of the bed and have water loss to groundwater at other locations. These three scenarios are depicted in Figure 3.41 (Winter et al. 1998). Scenario A represents a lake receiving groundwater. Scenario B represents water loss from a lake to groundwater and scenario C represents both water loss and water gain from groundwater to a lake.

In natural lakes not controlled by dams, the water level does not change as rapidly as the water level in streams and rivers, where it may fluctuate widely in response to large rain events. Evaporation affects the water levels of lakes more than streams and rivers because: (1) lakes have larger surface areas, and (2) there is less shading from the vegetation around a lake than from the dense riparian vegetation often present along a stream or riverbank (Winter et al. 1998).

4.6.3 Wetland Interaction with Groundwater

Wetlands prevent rapid drainage of water from the land surface, and, similar to lakes and streams, wetlands can receive groundwater inflow, recharge groundwater, or do both. The wetlands situated in topographic depressions of the land surface exhibit interactions similar to lakes and streams. Figure 4.28a depicts a wetland receiving groundwater inflow and Figure 4.28b shows a wetland discharging water to groundwater (Winter et al. 1998).

The United States has just over 100 million acres of wetlands remaining (Dahl and Allord 1994). The remaining wetland areas in the United States are concentrated along the southeast coast, the Mississippi River Valley, and the formerly glaciated regions of the upper Midwest.

4.6.4 Ocean Water, Groundwater, and Surface Water Interaction

Coastal regions may be more complex than any other region because surface water and groundwater interact with saline water at the ocean surface and saline groundwater. In addition, other forces such as ocean currents, storms, and tidal fluctuations may significantly influence these interactions. Therefore, understanding the geology of coastal regions is the first step toward understanding this

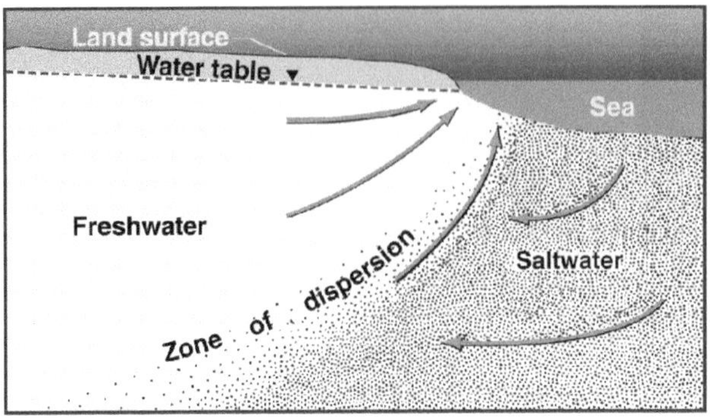

FIGURE 4.20 Generalized diagram showing groundwater, surface water, ocean water interaction.

complex system (Barlow 2003). Figure 4.20 shows a generalized view of surface water, groundwater, and ocean water interaction (Heath 1983).

Under normal conditions, the seaward movement of freshwater prevents saline water from encroaching coastal aquifers. The **zone of dispersion or transition zone** is a diffuse boundary where freshwater and saline water mix (Barlow 2003).

One method used to estimate the depth to the zone of dispersion is called the Ghyben–Herzberg relation named after the two scientists who developed it independently in the late 1800s (Barlow 2003). The thicknesses of freshwater and seawater are expressed by Equation 4.8:

$$z = \frac{\rho_f}{\rho_s - \rho_f} h \tag{4.8}$$

where ρ_f = density of freshwater
ρ_s = density of seawater
h = thickness of the freshwater zone above sea level
z = thickness of the freshwater zone below sea level

Freshwater has a density of 1.000 g/cm^3 at 20°C, whereas seawater has a density of approximately 1.025 g/cm^3. Although the density differential between seawater and freshwater is small, Equation 4.8 demonstrates that 40 feet of freshwater exists below sea level for every 1 foot of freshwater above sea level.

4.7 ANTHROPOGENIC INFLUENCES ON SURFACE WATER AND GROUNDWATER

Urbanization has a profound influence on the hydrogeology of an urban area because urbanization creates an anthropogenic infrastructure. This infrastructure is intended to mimic the natural systems it replaces but ends up changing many or all of the characteristics of the surface water and groundwater in developed areas (Vuono and Hallenbeck 1995; Zaporozec and Eaton 1996; Kibel 1998). For example, surface water drainage patterns are changed, runoff velocity increases, groundwater levels are raised and lowered, and groundwater recharge areas can expand or disappear (Burn et al. 2007; Garcia-Fresca 2007; Howard et al. 2007; Mohrlok et al. 2007). While many of us are aware of anthropogenic influences on surface water, it is much more difficult to recognize, evaluate, and monitor anthropogenic influences on groundwater because they are not readily observable.

4.7.1 Anthropogenic Surface Water Influences

A more detailed overview of the anthropogenic influences on surface water includes the following (Kaufman and Marsh 1997):

- Changing drainage patterns, most notably the creation of ephemeral drainage networks (e.g., driveways and streets) in the headwater zones of streams
- Increasing runoff volume and velocity
- "First flush" pollution, which occurs when road oil and other auto related liquids are washed off the pavement during intense thunderstorms
- Increased erosion and sediment loading
- Wetland destruction
- Dam construction for hydroelectric power and reservoirs
- Rising groundwater elevations in areas near dams and reservoirs
- Contaminated groundwater discharging to surface water in urban areas
- Using surface water for irrigation
- Discharge of industrial and sanitary wastewater to surface water
- Diversion of polluted stormwater into surface water bodies
- Combined sewer overflows
- Infiltration/inflow problems within the sewer and water supply pipe networks of older urban areas
- Decreased baseflow resulting from the construction of impervious surfaces such as roads and buildings
- Pruning (removal) of streams due to new development
- Direct and indirect discharges of contaminants to surface water

Some of the key forces behind drainage pattern modification include the following:

- Flood control through (1) construction of levees, (2) deepening of streams and rivers through dredging, (3) construction of stormwater retention ponds, and (4) straightening of meanders and bends in streams and rivers, and the removal of bankside vegetation – processes collectively termed **channelization**
- Canal construction
- Road construction
- Landscaping
- Urban sprawl; the practice of creating an urban landscape equivalent to mono-crop agriculture. Mixed residential, retail, and commercial land use within neighborhoods is replaced by regional malls and large box retail stores with massive parking lots.

One of the more notable anthropogenic changes to surface water flow was the reversal of the flow of the Chicago River in 1900 after the completion of the Chicago Sanitary and Ship Canal. This flow modification made the transport of goods and shipping between the Great Lakes and the Mississippi River possible. Reversing the flow of the river also provided a clean source of water and changed the local nickname of the river, which had been known as the "Stinking River" (Friends of the Chicago River 2024).

Withdrawing groundwater close to a surface water body provides another example of the human influences on surface water flow. In this scenario, pumping large quantities of groundwater near a surface water source may result in some depletion of water from the surface water source caused by the pumping action (Alley et al. 1999). In one case known to the authors, contaminated surface water was drawn into drinking water wells as a result of over-pumping. This recently happened in a rural community in the town of Franklin, Michigan, located in the northwestern part of the Rouge

River watershed. Drinking water wells located adjacent to the Rouge River were drawing river water into the wells which was consumed by residents for several months before the situation was detected and rectified. The consumption of contaminated drinking water led to a variety of serious intestinal illnesses of several Franklin residents.

4.7.1.1 Destruction of Wetlands

Wetlands are nature's sponge and filter. They perform the following functions: adsorbing excess water; cleaning and biologically degrading harmful chemicals such as oil and other petroleum compounds; assisting in keeping lakes and rivers clean and free of excess sediment; slowing the erosion process; decreasing flooding; and are important areas of groundwater recharge.

In the early 1600s, the United States had an estimated 221 million acres of wetlands, but by the mid-1980s their extent had been reduced to approximately 103 million acres – a 54% reduction (Dahl and Allord 1994). Seven states including Indiana, Illinois, Missouri, Kentucky, Iowa, California, and Ohio have lost more than 80% of their wetlands. Although much of wetland loss is attributed to agricultural land conversion, other anthropogenic causes include the following:

- Construction of dikes and levees along rivers to divert or minimize flooding
- Logging
- Mining
- Road construction
- Building construction
- Urban expansion
- Non-native plant invasion
- Overgrazing

As recently as the 1970s, destruction of wetlands in the United States was encouraged. It was not until the passage of the Emergency Wetland Resources Act of 1986 that wetlands became protected as a vital ecologic and hydrological resource. The States with the highest loss of wetlands include Ohio, Indiana, Illinois, Iowa, Missouri, and California (Dahl and Allord 1994).

Approximately 150 species of birds including the bald eagle, great blue heron, and belted kingfisher, and more than 200 species of fish depend on wetlands for their survival. Mammals such as the muskrat, white-tailed deer, and numerous species of reptiles, amphibians, and invertebrates also rely on wetlands. Wetlands exceed all other land types in wildlife productivity and diversity because they are ecotones – a region where two ecosystems overlap and exhibit a synergistic (non-linear) effect with respect to habitat quality. In the case of wetlands, the overlapping ecosystems are an aquatic lowland area and the adjacent drier upland terrain.

4.7.1.2 Dam Construction

Dams are constructed for hydroelectric power generation, flood control, improvement of navigation, and water supply, to provide a water source for irrigation and recreation (see Figure 4.30). Dams cause the water table to rise and may also increase head pressures in confined aquifers if conditions are favorable. Two major concerns about dams are whether the rise in groundwater levels will weaken the surrounding geologic material, and if this rise will impact the natural groundwater flow patterns. These concerns were realized during the Teton Dam collapse on June 5, 1976, when a leak appeared in the earthen dam as it was being filled. The leak first appeared approximately 130 feet below the crest of the dam. The dam collapsed a couple of hours later (Randle et al. 2000). Eleven people lost their lives in the collapse of the Teton Dam. The collapse also caused more than 30 landslides, killed more than 13,000 cattle, caused more than $300 million in property damage, and flooded portions of Rexburg, Wilford, Sugar City, Salem, and Hibbard, Idaho (Randle et al. 2000).

4.7.2 Anthropogenic Groundwater Influences

Documentation of anthropogenic influences on groundwater in urban areas is a recent phenomenon (Barber 1997). The late arrival of this technique occurred because the connection between anthropogenic processes at the surface and shallow groundwater in urban areas is complex and difficult to investigate – or was simply not thought to be important.

Some anthropogenic influences on groundwater include the following (Sharp 1997; Howard 1997):

- Excessive groundwater withdrawal resulting in saltwater intrusion along coastal areas
- Significant decreases in groundwater surface elevations in arid regions or any region where a consistently high volume of groundwater withdrawal occurs
- Decreasing natural groundwater recharge resulting from urban development and the construction of impervious surfaces.
- Decreasing residence time of water in shallow aquifers in urban areas through construction and dewatering efforts.
- Decreasing the residence time of water in deeper aquifers through excessive withdrawal.
- Increasing anthropogenic groundwater recharge resulting from leaking sewers, septic tanks, and injection wells.
- Ground surface subsidence resulting from excessive groundwater withdrawal – an event frequently evident in urban regions where groundwater withdrawal is excessive.
- Flooding in urban areas due to the increase in impervious surfaces as a byproduct of urban development.
- Excessive sewer leakage resulting in groundwater quality degradation and potential flooding.
- Increased potential for land subsidence and landslides in areas near dams and reservoirs due to rising groundwater levels.
- Direct and indirect discharges of contaminants to groundwater through accidental releases, leaking sewers, septic systems, injection wells, as well as residential, commercial, and industrial discharges.

4.7.2.1 Saltwater Intrusion

Fresh groundwater in the United States is surrounded by saline groundwater both vertically and laterally. This is especially evident along the coastlines as noted in Section 4.6.4, although much of the interior of the country is underlain by deep saline aquifers (Alley et al. 1999). This juxtaposition of freshwater and saltwater creates the potential for saltwater intrusion and rendering the freshwater non-potable. This scenario can occur from installing and pumping a water well placed too close to a saltwater source.

4.7.2.2 Changes in Groundwater Flow

A minimum of three monitoring wells are necessary to estimate the direction of groundwater flow. Because urban areas are complex, using a minimum of three monitoring wells is usually not sufficient. The flow patterns of groundwater in urban areas vary considerably due to anthropogenic influence, especially if the water table is shallow (less than 20 feet beneath the surface), so using more than three monitoring wells is necessary. Although no precise number of monitoring wells necessary to adequately characterize the groundwater flow pattern beneath a specific location has been established, it should be noted that groundwater flow is often complex and that a minimum of three monitoring wells may establish a general flow directly, though it may not tell the whole story.

The following considerations can help with the evaluation:

- Size and shape of a site: Large or irregularly shaped sites require more monitoring wells to evaluate groundwater flow patterns.

- Surface water features: If a stream is located along the boundary or dissects the site.
- Potential presence of multiple aquifers or perched aquifers: The presence of multiple aquifers may have different groundwater flow directions
- Seasonal fluctuations in surface water discharge and precipitation patterns
- Anthropogenic influences including the following:
  - Utility corridors or sewers: Utility corridors or sewers potentially influence groundwater flow in two basic ways: (1) leaking water supplies act to recharge groundwater, and (2) storm sewers and other utility corridors act as preferential groundwater flow pathways when the water table intersects the utility corridor.
  - Extraction or pumping wells
  - Deep building foundations
  - Land disturbance
  - Artificial recharge areas
  - Pipelines
  - Wastewater discharge points

4.7.2.3 Artificial Recharge

Artificial groundwater recharge is defined as the practice of increasing the amount of water by artificial means that enters an aquifer, which has been performed in the United States for over 50 years (Todd 1959). This definition has been expanded to include direct recharge of groundwater for potable sources through spreading basins, pits, and injection through wells. Not included is seepage of sewerage or deep well injection of wastes (Aiken and Kuniansky 2002).

Many states practice artificial recharge techniques to replenish groundwater reserves for potable purposes when other sources of potable are either not available or too expensive. Florida, New York, and California are examples (Aiken and Kuniansky 2002), and other states in the high plains region of the United States including Kansas, Nebraska, Colorado, South Dakota, and North Dakota are also using artificial recharge techniques.

The artificial recharge technique is most commonly employed in a **spreading basin.** Here, water is spread over a large surface area composed of permeable sediments such as sand and gravel and permitted to seep into the ground and migrate to the aquifer below.

There are several advantages to artificial recharge of groundwater:

- It limits loss of water through evaporation.
- It can lower the cost that dam construction or the construction of water towers.
- It can store very large quantities of water.
- It mitigates the impact of subsidence.
- It permits the **conjunctive use** of groundwater and surface water. During wet periods, groundwater is stored for later use during dry weather and reduces the stress placed on surface water reservoirs for supply.

The disadvantages of artificial recharge include the following:

- Ineffective recovery leading to water loss
- Storing water beneath urban areas may lead to degrading groundwater quality
- May increase dissolution if the chemistry of infiltrating water is significantly different from native groundwater chemistry
- Changes in natural groundwater flow paths and directions.

Some countries, especially those in more arid regions, are evaluating and experimenting with techniques of artificial recharge using stormwater and reclaimed water (Dillon et al. 2007; Howard et al. 2007; Limaye 2007). These techniques may become commonplace in the future as our climate changes and population increases generate a greater need for freshwater. Urban areas, where large populations use large quantities of water, can improve groundwater recharge through an understanding of the hydrogeological environment. This understanding can help avoid aquifer contamination and groundwater mounding (Howard et al. 2007).

4.7.2.4 Land Subsidence

Land subsidence is defined as the gradual settling or sudden sinking of the Earth's surface caused by the movement of subsurface materials (Galloway et al. 2001). Subsidence is a global issue. In the United States, more than 17,000 square miles of land surface in 45 states has experienced some form of land subsidence. This is equivalent to a land area the size of Vermont and New Hampshire combined.

The principal causes of subsidence include the following (National Research Council 1991):

- Aquifer system compactions
- Drainage of organic soils
- Underground mining
- Hydrocompaction
- Natural compaction
- Sinkholes
- Thawing permafrost

More than 80% of the identified subsidence in the United States is a result of exploitation of groundwater resources (Galloway et al. 2001). In the future, land subsidence may become exacerbated by the increasing rates of withdrawal and exploitation of groundwater. Regions especially prone to subsidence are the southwest portion of the United States; areas overlain by more soluble geological deposits such as limestone, gypsum, or halite; and areas where groundwater withdrawal is significant or a necessity because no dependable surface water source is available.

4.8 SUMMARY AND CONCLUSION

Geology and water work together to shape our urban environment. Human impact on our environment has also been significant, and within every urban area, there exists a unique set of environmental concerns from geologic, hydrogeologic, and anthropocentric perspectives. This will become clearer when we move on to explore the presence, migration, and persistence of contaminants in urban areas.

Water is a very good solvent, and it has the ability to dissolve and erode substances and materials at the surface and below. Water often transports these substances and materials over great distances. An appropriate quote heard in geological circles states, "Water destroys everything that the Earth creates and puts it in the oceans (reference source unknown)."

In the next chapter, we will combine what we have learned thus far and examine natural hazards and risks posed by different geology and hydrogeologic phenomena.

REFERENCES

Abrams, D. M., Lobkovsky, A. E., Petroff, A. P., Straub, K. M., McElroy, B., Mohrig, D. C., Kudrolli, A., and Rothman, D. H. 2009. Growth Laws for Channel Networks Incised by Groundwater Flow. *Nature Geoscience*. Vol. 2, pp. 193–196. Published online DOI: 10.1038/NGEO432

Aiken, G. R., and Kuniansky, E. L. 2002. *United States Geological Survey Artificial Recharge Workshop Proceedings, Sacramento, California*. United States Geological Survey Open-File Report 02-89. USGS. Washington, DC.

Alley, W. M., Reilly, T. E., and Franke, O. E. 1999. *Sustainability of Groundwater Resources*. United States Geological Survey Circular 1186. USGS. Denver, CO.

Barber, C. 1997. Perspectives on Innovations and Solutions for Groundwater Management. In: Chilton, J., et al., editors. *Groundwater in Urban Environment*. Balkema Publishers. The Netherlands. pp. 3–10.

Barlow, P. M. 2003. *Ground Water in Freshwater – Saltwater Environments on the Atlantic Coast*. United States Environmental Protection Agency (USEPA) Circular 1262. USEPA. Washington, DC.

Burger, A. C., and Dubertret, L. 1984. *Hydrogeology of Karstic Terrains: Case Studies*. International Association of Hydrogeologists, International Contributions to Hydrogeology. London, UK. Vol. 1.

Burn, S., Eiswirth, M., Correll, R., Cronin, A., DeSilva, D., Diaper, C., Dillon, P., Mohrlok, U., Morris, B., Rueedi, J., Wolf, L., Vizintin, G., and Vott, U. 2007. Urban Infrastructure and Its Impact on Groundwater Contamination. In: Howard, K. W. F., editor. *Urban Groundwater – Meeting the Challenge*. Taylor & Francis. London, England.

Dahl, T. E., and Allord, G. J. 1994. *Technical Aspects of Wetlands; History of Wetlands in the Conterminous United States*. United States Geological Survey Water Supply Paper 2425. USGS. Washington, DC.

De Barry, P. A. 2004. *Watersheds: Processes, Assessments, and Management*. John Wiley & Sons. New York, NY.

Dillon, P., Toze, S., and Pavelic, P. 2007. Conjunctive Use of Urban Surface Water and Groundwater for Improved Urban Water Security. In: Howard, K. W. F., editor. *Urban Groundwater – Meeting the Challenge*. Taylor & Francis. London, England

Driscoll, F. G. 1986. *Groundwater and Wells*. 2nd Edition. Johnson Filtration Systems. St. Paul, MN.

Dunne, T. 1990. The Role of Subsurface Water in Earth-Surface Processes and Landforms. In: Higgins, C. G., and Coates, D. R., editors. *Groundwater Geomorphology*. Vol. 252. Geological Society of America Special Papers, Geological Society of America. Boulder, CO. pp. 1–28.

Environment Canada. 2024. *Aquifers and Wells*. www.canada.ca/en/environment-climate-change-services/water-overview/sources/groundwaterhtml. (Accessed March 10, 2024).

Focazio, M. J., Reilly, T. E., Rupert, M. G., and Helsel, D. R. 2001. *Assessing Ground-Water Vulnerability to Contamination: Providing Scientifically Defensible Information for Decision Makers*. United States Geological Survey Circular 1224. USGS. Denver, CO.

Ford, D. C., and Williams, P. *Karst Hydrogeology and Geomorphology*. 2nd Edition. John Wiley & Sons. New York, NY.

Freeze, R. A., and Cherry J. A. 1979. *Groundwater*. Prentice-Hall Inc. Englewood Cliffs, NJ.

Friends of the Chicago River. 2024. *Brief History of the Chicago River*. www.chicagoriver.org/blog/2019/4/friends-of-the-chicago-river-the-chicago-river-history-the-amazing-journey. (Accessed March 25, 2024).

Fuhrer, G. J., Gilliom, R. J., Hamilton, P. A., Morace, J. L., Nowell, L. H., Rinella, J. F., Stoner, J. D., and Wentz, D. A. 1999. *The Quality of Our Nation's Water – Nutrients and Pesticides*. United States Geological Survey Circular 1225. USGS. Reston, VA.

Galloway, D., Jones, D. R., and Ingebritsen, S. E. 2001. *Land Subsidence in the United States*. United States Geological Survey Circular 1182. USGS. Denver, CO.

Garcia-Fresca, B. 2007. Urban-enhanced Groundwater Recharge: Review and Case Study of Austin, Texas, USA. In: Howard, K. W. F., editor. *Urban Groundwater – Meeting the Challenge*. Taylor & Francis. London, UK. pp. 3–18.

Heath, R. C. 1983. *Basic Ground-Water Hydrology*. United States Geological Survey, Water Supply Paper 2220. United States Government Printing Office. Alexandria, VA.

Hill, J. W., Feigl, D. M., and Baum, S. J. 1993. *Chemistry and Life*. Macmillan Publishing Company. New York, NY.

Howard, K. W. F. 1997. Incorporating Policies for Groundwater Protection in Urban Planning. In: Chilton, J., et al., editors. *Groundwater in Urban Environment*. Balkema Publishers. The Netherlands. pp. 31–40.

Howard, K. W. F., Di Biase, S., Thompson, J., Maier, H., and Van Egmond, J. 2007. Stormwater Infiltration Technologies for Augmenting Groundwater Recharge in Urban Areas. In: Howard, K. W. F., editor. *Urban Groundwater – Meeting the Challenge*. Taylor & Francis. London, England.

Howard, K. W. F., and Gerber, R. E. 1997. Do Tills Beneath Urban Toronto Provide Adequate Groundwater Protection? In: Chilton, J., et al., editors. *Groundwater in Urban Environment*. Balkema Publishers. The Netherlands. pp. 433–438.

Illinois State Geological Survey. 2024. *Geobit 7 Karst*. www.isgs.edu/maps-data-pub/publications/geobits7.shtml. (Accessed March 10, 2024).

Jones, J. A. 1997. *Global Hydrology: Processes, Resources, and Environmental Management.* Longman Publishers. London, England.

Kammerer, J. C. 1990. *Largest Rivers in the United States.* United States Geological Survey Open-File Report 87-242. USGS. Washington, DC.

Kaufman, M. M., and Marsh, W. M. 1997. Hydro-Ecological Implications of Edge Cities. *Landscape and Urban Planning.* Vol. 36, pp. 277–290.

Kibel, P. S. 1998. The Urban Nexus: Open Space, Brownfields, and Justice. *Boston College Environmental Affairs Law Review.* Vol. 25, no. 3, pp. 589–618.

Leopold, L. B., Wolman, M. G., and Miller, J. P. 1992. *Fluvial Processes in Geomorphology.* The Dover Publishing Company. New York, NY.

Limaye, S. D. 2007. Aspects of Urban Groundwater Management and Use in India. In: Howard, K. W. F., editor. *Urban Groundwater – Meeting the Challenge.* Taylor & Francis. London, England.

Meyer, G. 1989. Fresh Water of the North American Continent. In: Bally, A. W., and Palmer, A. R., editors. *The Decade of North American Geology. Vol. A. The Geology of North America – An Overview.* The Geological Society of America. Boulder, CO. pp. 537–554.

Mohrlok, U., Cata, C., and Bucker-Gittel, M. 2007. Impact on Urban Groundwater by Wastewater Infiltration into Soils. In: Howard, K. W. F., editor. *Urban Groundwater – Meeting the Challenge.* Taylor & Francis. London, England.

Murray, K. S., and Zhou, X. 2000. *Relationship between Groundwater, Surface Water Quality and Land Use in an Urban Watershed.* International Symposium on Hydrogeology and the Environment. Wuhan, P.R. China.

National Research Council. 1991. *Mitigating Losses from Land Subsidence in the United States.* National Academic Press. Washington, DC.

Payne, F. C., Quinnan, J. A., and Potter, S. T. 2008. *Remediation Hydraulics.* CRC Press. Boca Raton, FL.

Randle, T. L., Bountry, J. A., and Klinger, R. 2000. *Geomorphology and River Hydraulic of the Teton River Upstream of the Teton Dam, Teton River, Idaho.* United States Department of the Interior, Bureau of Reclamation. Denver, CO.

Rogers, D. T. 1997. The Influence of Groundwater on Surface Water in Michigan's Rouge River Watershed. *Proceedings of the American Water Resources Association, Conjunctive Use of Water Resources; Aquifer Storage and Recovery Conference.* Long Beach, CA. pp. 173–180.

Rogers, D. T., and Murray, K. S. 1997. Occurrence of Groundwater in Metropolitan Detroit. In: Chilton, J., et al., editors. *Groundwater in the Urban Environment. Volume 1. Problems, Processes, and Management.* Balkema Publishers. The Netherlands. pp. 155–161.

Sharp, J. M. 1997. Ground-water Supply Issues in Urban and Urbanizing Areas. In: Chilton, J., et al. editors. *Groundwater in the Urban Environment.* Balkema Publishers. The Netherlands, pp. 67–74,

Shiklomanov, I. 1993. World Fresh Water Resources. In: Gleick, P. T., editor. *Water in Crisis: A Guide to the World's Fresh Water Resources.* Oxford University Press, New York, NY.

Todd, D. K. 1959. *Annotated Bibliography on Artificial Recharge of Groundwater in the United States Through 1954.* United States Geological Survey Water Supply Paper 1477. USGS. Washington, DC.

United Nations. 2003. *International Decade for Action; Water for Life 2005–2015.* United Nations. New York, NY. https://www.unwater.org/news/water-for-life-decade-2005-2015. (Accessed March 20, 2024).

United Nations. 2024. *World Population Prospects: The 2024 Revision Population Database.* United Nations Population Division. http://esa.un.org/unpp.p2k0data.asp. (Accessed March 10, 2024).

United States Geological Survey (USGS). 1984. *National Water Summary 1983 – Hydrologic Event and Issues.* United States Geological Survey Water-Supply Paper 2250. USGS. Washington, DC.

United States Geological Survey (USGS). 2004. *Estimated Use of Water in the United States.* USGS Circular 1268. USGS. Washington, DC.

United States Geological Survey (USGS). 2024a. *Summary of the Water Cycle.* www.usgs.gov/special-topics/water-science/science/water-cycle. (Accessed March 10, 2024).

United States Geological Survey (USGS). 2024b. *Water Properties.* www.usgs.gov/special-topis/water-science-school/science/water-properties. (Accessed March 10, 2024).

United States Geological Survey (USGS). 2024c. *Science in Your Watershed.* http://water.usgs.gov/wsc/map_index.html. (Accessed March 10, 2024).

United States Geological Survey (USGS). 2024d. *Principle Aquifers of the 48 Conterminous United States.* http://water.usgs.gov/GIS/browse/aquifers_us.jpg. (Accessed March 10, 2024).

United States National Park Service. 2024. *River Systems and Fluvial Landforms*. www.nps.org/subjects/geology/fluvial-landfors. (Accessed March 10, 2024).

Vuono, M., and Hallenbeck, R. P. 1995. Redeveloping Contaminated Properties. *Journal of Risk Management*. Vol. 42, pp. 58–69.

Winter, T. C., Harvey, J. W., Franke, O. E., and Alley, W. M. 1998. *Ground Water and Surface Water: A Single Resource*. United States Geological Survey Circular 1139. USGS. Denver, CO.

Zaporozec, A., and Eaton, T. T. 1996. Ground-water Resource Inventory in Urbanized Areas. In: *Hydrology and Hydrogeology of Urban and Urbanizing Areas*. American Institute of Hydrology (AIH) Annual Meeting Proceedings. St. Paul, MN.

5 Geologic and Environmental Hazards

5.1 INTRODUCTION

Geologic, environmental, and anthropogenic-induced hazards can be thought of as three sets of hazards that interact and are linked to one another through geology and human behavior. Geologic hazards include earthquakes, landslides, tsunamis, volcanic eruptions, coastal hazards, subsidence and sinkholes, and erosion. Environmental hazards include extreme atmospheric events such as flooding, tornados, hurricanes, snowstorms, drought, severe thunderstorms, extreme heat and cold, wildfires, and others. Environmental hazards are influenced by the geology of a particular area including topography, specific location, rock types, geomorphology, drainage, vegetation, elevation, and other physical geological features. Anthropogenic-induced hazards include many aspects of human behavior that have negatively impacted humans and the natural world. Table 5.1 lists the different hazard types for each of the three categories.

In this chapter, we will address geologic and environmental hazards. In forthcoming chapters, we will address anthropogenic-induced hazards and explore how these hazards interact with the nature and the geologic environment. For purposes of this book, a **hazard** refers to a natural geologic event, environmental event, or an anthropogenic-induced event. **Risk** is a measure of the magnitude of an event and the frequency of occurrence. Risk can be applied by evaluating the probability of a negative outcome or impact from a geologic or environmental hazard, an extreme atmospheric event, or exposure to a contaminant from an anthropogenic source. **Sensitivity** is a measure of how resilient a target population or ecological sector is to the hazard. To put it in terms that perhaps can be better understood, we do not always have control over the hazard, but we can lower our vulnerability of a negative outcome as a result of understanding and quantifying the risk. Equation 5.1 presents this concept in more mathematical terms whereby a geologic hazard may exist, but for risk to be realized, and hazard must be actualized and occur within a certain magnitude. The next variable is sensitivity of anthropogenic or environmental exposure to the risk. The outcome is then a measure of the impact severity.

$$\text{Geologic/Environmental Hazard} \times \text{Magnitude and Risk of Occurrence} \times \text{Sensitivity} = \text{Impact Severity} \tag{5.1}$$

Natural geological, environmental, and anthropogenic-induced hazards and associated impacts are of key concern to life on Earth. Understanding and monitoring these hazards through scientific study to identify and quantify risk potential are critical to improve risk reduction methods so that sustainability potential for survival is greatly improved (British Geological Survey (BGS) 2024).

In this chapter, we shall discuss geologic and environmental risks separately for organizational purposes. We shall discuss each from a scientific point of view so that a deeper understanding is

DOI: 10.1201/9781003463153-6

TABLE 5.1
Types of Hazards Discussed

Geologic Hazards	Environmental Hazards	Anthropogenic-Induced Hazards
Earthquake	Hurricanes	Habitat destruction
Tsunami	Tornados	Natural resource exploitation
Volcanic eruption	Severe storms	Water cycle disruption
Landslide	Extreme precipitation	Deforestation
Coastal hazard	Flooding	Overfishing
Subsidence and sink holes	Drought	Overhunting and poaching
Erosion	Extreme heat	Invasive species
Aeolian processes (dust)	Extreme cold	Species extinction
	Wildfires	Light pollution
	Lightning	Noise pollution
	Solar radiation	Ocean pollution
	Hail	Air pollution
	High winds	Water pollution
	Severe winter weather	Land pollution
	Marine heat waves	Chemical contaminants
	High-tide flooding	Agriculture
	Derechos	Land development
		Endangered species
		Pathogens
		Erosion
		Particulate matter or dust
		Nuclear radiation
		Certain buildings
		Certain wind turbines
		Climate change

gained. We will use this more in-depth understanding to evaluate specific differences and range of different potential variations with each hazard type and will then discuss the potential severity of risk qualitatively.

5.2 GEOLOGIC HAZARDS

Geologic hazards that will be discussed in this section include earthquakes, volcanic eruptions, landslides, tsunamis, coastal hazards subsidence and sinkholes, and erosion.

5.2.1 Earthquakes

Earthquakes are sudden and often destructive. They are terrifying and literally shake us to our core. They are mysterious and powerful acts of nature that humble all who have lived through an earthquake. An earthquake is what happens when two blocks of the Earth suddenly slip past one another. The surface where they slip is called a **fault** or **fault plane**. The location below the Earth's surface where the earthquakes starts is called the **hypocenter**, and the location directly above it on the surface of the earth is called the **epicenter** (United States Geological Survey (USGS) 2024a).

Sometimes an earthquake has what is termed **foreshocks**. These are smaller earthquakes that happen in the same location before the large earthquake occurs. At this point, scientists cannot tell that an earthquake is a foreshock until the larger earthquake occurs. The largest, main earthquake is called the **mainshock**. Mainshocks always have what is termed **aftershocks** that follow the mainshock. Aftershocks, which are small than the mainshock, can continue for weeks, months, or in some instances, for years after the mainshock (USGS 2024a).

5.2.1.1 Seismic Waves

While the edges of faults are stuck together, and the rest of the block is moving, the energy that would normally cause the blocks to slide past one another is being stored up. When the force of the moving blocks finally overcomes the friction of the jagged edges of the fault and it unsticks, all the stored energy is suddenly released. The energy radiates outward from the fault in all directions in the form of **seismic waves** similar to the ripples in a pond. The seismic waves then shake the earth as they move through it, and when the waves reach the surface of the earth, they shake the ground and everything on it (USGS 2024a).

The size of an earthquake depends on the size of the fault and the amount of slip on the fault. However, this is sometimes impossible to directly measure, so scientists use what is called **seismograph** at the surface. A seismograph records the intensity of the shaking. When an earthquake occurs, the base of the seismograph shakes as well, but a hanging weight does not. The difference in position between the shaking part of the seismograph and the motionless part is what is recorded. Figure 5.1 shows a recorded earthquake from a seismograph.

Shaking of the ground occurs as a result of an earthquake. Although the physics of seismic waves caused by an earthquake is complex, ground shaking can be explained in terms of body waves, compressional, or P waves, and shear, or S waves. P and S waves are characterized as high-frequency waves. Other types of seismic waves created by earthquake are termed Rayleigh and Love, and are characterized as low-frequency waves and occur after the P and S waves (USGS 2024b). When an earthquake occurs, seismic waves are created and are termed “P” waves and “S” waves. P waves propagate through the Earth at a speed of approximately 15,000 miles/h and are the first waves to arrive at a given location and cause ground vibration. S waves arrive next, and cause side-to-side or horizontal motion (USGS 2024b).

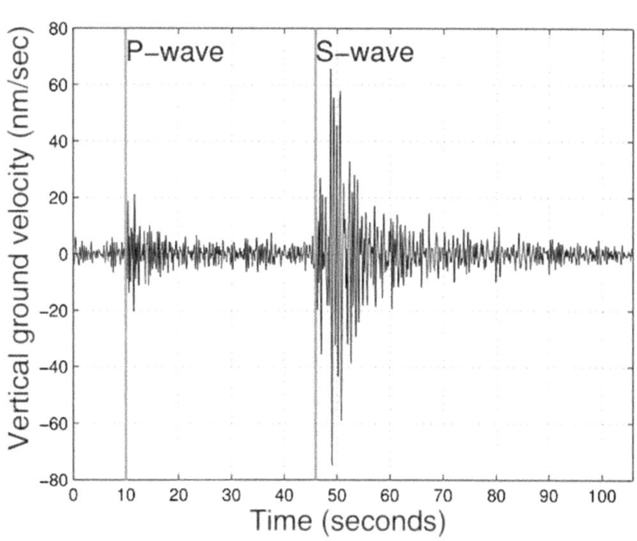

FIGURE 5.1 Earthquake recorded using a seismograph. (United States Geological Survey (USGS). 2024a. *The Effects of Earthquakes*. www.usgs.gov/programs/earthquake-hazards/science-earthquakes. (Accessed March 11, 2024).)

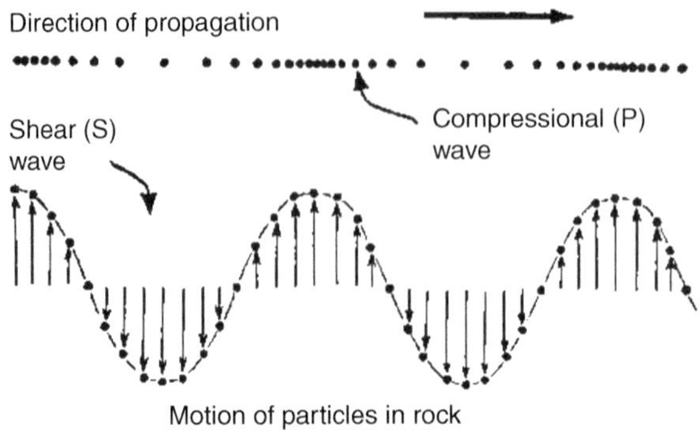

FIGURE 5.2 Diagram showing the difference between P and S waves. (United States Geological Survey (USGS). 2024a. *The Effects of Earthquakes*. www.usgs.gov/programs/earthquake-hazards/science-earthquakes. (Accessed March 11, 2024).)

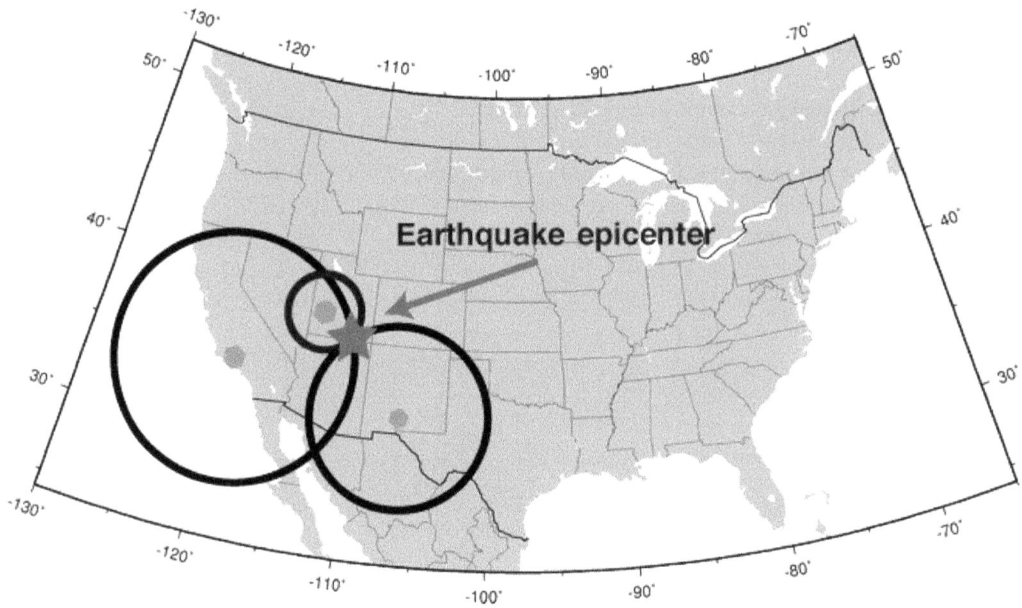

FIGURE 5.3 Locating an epicenter of an earthquake using the triangulation method. (United States Geological Survey (USGS). 2024a. *The Effects of Earthquakes*. www.usgs.gov/programs/earthquake-hazards/science-earthquakes. (Accessed March 11, 2024).)

Figure 5.1 shows each wave as recorded by the seismograph. P waves and S waves shake the ground differently as depicted in Figure 5.2. To understand how this works, let's compare P and S waves to lightning and thunder. Light travels faster than sound, so during a thunderstorm you would first see the lightning and then you will hear the thunder. If you are close to the lightning, the thunder will arrive closer to when you saw the lightning. If you are farther away from the lightning, the sound of the thunder will take a longer period of time to reach you. Therefore, P waves are like the lightning, and S waves are like the thunder.

Under close examination at the amount of time between the P and S wave on a seismogram recorded on a seismograph, scientists can tell how far away the earthquake was from that location. However, they cannot tell in what direction from the seismograph the earthquake was, only how far away it was. If one were to draw a circle on a map of the distance and was conducted at two other seismograph stations, then by a process of triangulation, the epicenter of the earthquake can be identified (see Figure 5.3).

5.2.1.2 Earthquake Magnitude, Energy Released, and Shaking Intensity

An earthquake magnitude, energy released, and shaking intensity are all related measurements of an earthquake that are often confused with one another. Their dependencies and relationships can be complicated, and even one of these concepts alone can be confusing. The severity of an earthquake can be expressed in terms of both *intensity* and *magnitude*. However, the two terms are quite different. Intensity is based on the observed effects of ground shaking on humans, buildings, and natural features. It varies from place to place within the distributed region depending on the location of the observer with respect to the earthquake epicenter (USGS 2024c). Magnitude is related to the amount of seismic energy released at the hypocenter of an earthquake. It is based on the amplitude of the earthquake waves recorded on instruments which have a common calibration. The magnitude of an earthquake is thus represented by a single, instrumently determined value (USGS 2024c). Factors that contribute to evaluating the destructive nature of an earthquake depend on many factors that include the depth of an earthquake, geologic materials in which the earthquake occurs, geologic materials in which the seismic waves travel, type of near-surface geological materials, the presence of fluids, strength of rock material, and other physical aspects (USGS 2024c).

An earthquake has a single magnitude. The shaking that an earthquake creates is dependent on many values and typically varies from place to place depending on distance, surface materials, and other factors. The magnitude of an earthquake is expressed in whole number and decimal fractions and is based on a logarithmic scale. Each whole number increase in magnitude represents a 10-fold increase in measured amplitude as measured on a seismogram. The **Richter Scale (M_L)** has historically been used since it was developed in the late 19th century. However, the **Moment Magnitude (M_0)** is used more commonly and is based on physical properties of the earthquake derived from an analysis of all the waveforms recorded from the shaking. First, the seismic moment is computed, and then it is converted to a magnitude designed to be roughly equal to the Richter Scale in the magnitude range where they overlap (see Equation 5.2) (USGS 2024c).

$$\text{Moment } (M_0) = \text{Rigidity} \times \text{Area} \times \text{Slip} \tag{5.2}$$

where rigidity is the strength of the rock along the fault
area is the area of the fault that slipped
slip is the distance the fault slipped

Another way to measure the size of an earthquake is to compute how much energy an earthquake released. The amount of energy radiated by an earthquake is a measure of the potential for damage. An earthquake releases energy at many frequencies, and to compute an accurate value, all frequencies of shaking for the entire event must be included (USGS 2024c). While a whole number increase in magnitude represents a 10-fold increase in the measured amplitude, it represents an energy release that is 32 times greater. The energy can be converted into yet another type called the **Energy Magnitude (M_e)**. However, as the Energy Magnitude and Moment Magnitude measure two different properties of an earthquake, their values are not the same. The energy release can also be approximated by converting the moment magnitude, M_w, to energy using Equation 5.3 (USGS 2024c):

$$\text{Log } E = 5.24 = 1.44\ M_w \tag{5.3}$$

where M_w is the moment magnitude

The effect of an earthquake on the Earth's surface is called the **intensity**. The intensity scale consists of a series of certain key responses such as people awakening, movement of furniture, damage to chimneys, and, finally, total destruction. Although many types of scales exist, the one used in the United States is the Modified Mercalli (MM) Intensity Scale. It was developed in 1931 by American seismologists Harry Wood and Frank Neumann. The MM Intensity Scale is composed of increasing levels of intensity that range from imperceptible shaking to catastrophic destruction. It does not have a mathematical basis but is instead an arbitrary ranking based on visual effects (USGS 2024d).

5.2.1.3 Earthquake Facts

Some earthquake facts are as follows (USGS 2024e, 2024f):

- Approximately 500,000 detectable earthquakes occur every year globally.
- Approximately 100,000 earthquakes yearly are felt by humans.
- Approximately 100 earthquakes occur globally that cause actual damage.
- Approximately 75% of the population of the United States are under a risk of experiencing an earthquake that can be felt (see Figure 5.4).
- Most earthquakes occur along the edges of the Pacific Tectonic Plate (see Figure 5.5).
- The world's largest recorded earthquake occurred in 1960 in Chili and its intensity was measured at 9.6 on the Richter Scale.
- The largest recorded earthquake in the United States occurred in 1964 in Alaska and its intensity was measured at 9.2 on the Richter Scale.
- P waves generated by an earthquake travel at a speed of approximately 15,000 miles/h. S waves travel at a slower rate.
- The country recording the largest loss of life from earthquakes is China estimated at over a million people. Other countries with significant loss of life include Haiti, Japan, Turkey, United States, Azerbaijan, Mexico, Chili, Tibet, Nepal, Indonesia, Ecuador, Russia, Sumatra, and many others.

5.2.1.4 Physical Effects of Earthquakes

The physical effects of earthquakes include the following (USGS 2024g):

- Ground shaking
- Surface faulting
- Liquefaction
- Flow failures
- Landslides
- Tsunamis

Each type of physical effect is described below.

5.2.1.4.1 Ground Shaking

Ground shaking is a term used to describe the vibration of the ground during an earthquake. As a generalization, the severity of ground shaking increases as the magnitude increases and decreases as the distance from the epicenter increases.

In land-use zoning and earthquake-resistant design, knowledge of the amplitude, frequency composition, and the time duration of ground shaking is needed. These quantities can be determined from empirical (observed) data correlating the with the magnitude and the distribution of the Modified Mercalli intensity of the earthquake, distance of the building from causative fault, and the physical properties of the soil and rock underlying the building. The subjective numerical value of the MM

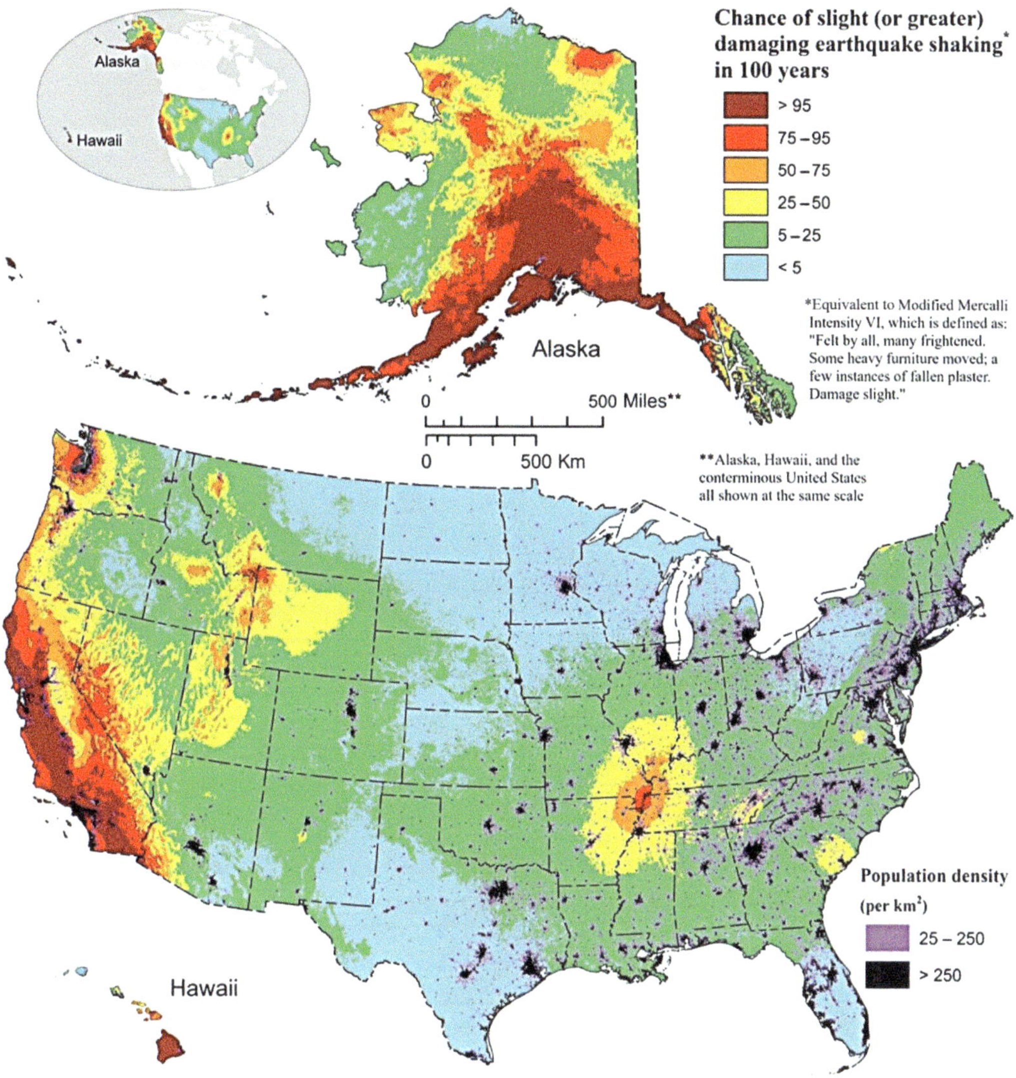

FIGURE 5.4 Earthquake risks in the United States. (United States Geological Survey (USGS). 2024f. *Earthquake Hazard Map*. www.earthquakes.usgs.gov/earthquakes/map. (Accessed March 12, 2024).)

Intensity Scale indicates the effects of ground shaking on humans, buildings, and the surface of the Earth. Injuries and loss of life from ground shaking have historically been significant with the majority of loss originating from building collapse (USGS 2024g).

5.2.1.4.2 Surface Faulting

Surface faulting is the differential movement of the two sides of the fracture at the Earth's surface whether it can be caused by a strike-slip, normal, or reverse (thrust) fault. Although displacements of these kinds can result from landslides and other shallow processes, surface faulting, as the term is used here, applies to differential movements caused by deep forces within the Earth, the slow movement of sedimentary deposits toward the Gulf of Mexico, and faulting associated with salt domes. The displacements, lengths, and widths of surface fault ruptures show a wide range. While injuries do occur from surface faulting, significant loss of life is not a high risk (USGS 2024g).

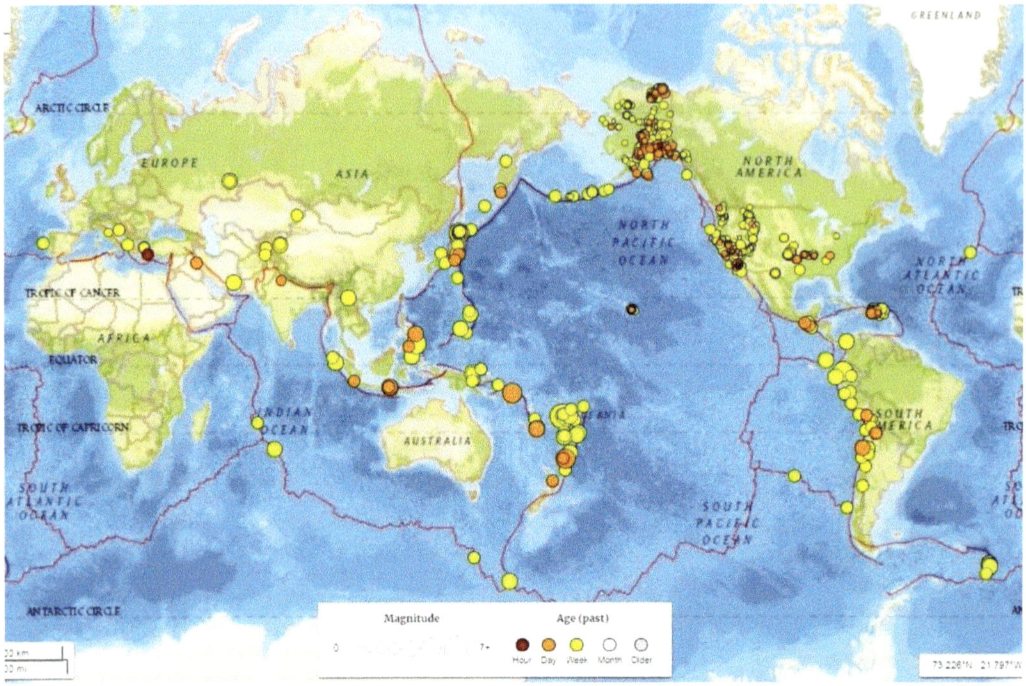

FIGURE 5.5 Earthquakes worldwide. (United States Geological Survey (USGS). 2024c. *Earthquake Magnitude, Energy Release, and Shaking Intensity*. www.usgs.gov/earthquake-hazards/earthquake-magnitude-energy-release-and-shaking-intensity. (Accessed March 12, 2024).)

5.2.1.4.3 *Liquefaction*

Liquefaction is a physical process that takes place during some earthquakes that may lead to ground failure. As a consequence of liquefaction, clay-free-soil deposits, primarily sands and silts, temporarily lose strength and behave as viscous fluids rather than solids. Liquefaction takes place when seismic waves pass through a saturated granular soil layer, distort the granular structure, and cause some of the void spaces to collapse. Disruptions to the soil generated by these collapses cause transfer of the ground-shaking load to grain-to-grain contacts in the soil layer to the pore water. This transfer of load increases pressure in the pore water, and can cause either drainage to occur or, if drainage is restricted, a sudden buildup of pore-water pressure. When the pore-water pressure rises to about the pressure caused by the weight of the column of soil, the granular soil layer behaves like a fluid rather than like a solid for a short period. In this condition, deformations can easily occur (USGS 2024g). Liquefaction is restricted to certain geologic and hydrologic environments, mainly areas where sand and silts were deposited in the last 10,000 years and where groundwater is within 30 feet of the surface. Generally, the younger and looser the sediment and the higher the piezometric surface (water table), the more susceptible a soil is to liquefaction (USGS 2024g).

5.2.1.4.4 *Flow Failures*

Flow failures, consisting of liquified soil or blocks of intact material riding on a layer of liquified soil, are the most catastrophic type of ground failure caused by liquefaction. These failures commonly move several tens of feet and, if geometric conditions permit, several tens of kilometers. Flows travel at velocities as great as many tens of kilometers per hour. Flow failures usually form in loose saturated sand or silts on slopes greater than 3 degrees (USGS 2024g).

5.2.1.4.5 Landslides

Past experience has shown that several types of landslides take place in conjunction with earthquakes. The most abundant types of earthquake-induced landslides are rock falls and slides of rock fragments that form on steep slopes. Shallow debris flows or slides forming on steep slopes and soil and rock slumps and block slides forming on moderate to steep slopes also take place, but they are less abundant. Reactivation of dormant slumps or block slides caused by earthquakes is rare.

5.2.1.4.6 Tsunamis

Tsunamis are water waves that are caused by sudden vertical movement of a large area of the sea floor during an undersea earthquake. Tsunamis are often call tidal waves, but they are not. Tsunamis are not caused by tidal action of the Moon and Sun. The height of a tsunami in the deeper portions of the ocean is typically less than a meter, but the distance between wave crests can be very long, more than 100 km. The speed at which the tsunami travels decreases as water depth decreases. In the mid-Pacific Ocean, where the water depths reach 5 km, tsunami can travel at speeds of more than 700 km/h. As tsunamis reach shallow water around an island or continental shelf, the height of the waves increases, sometimes by an order of magnitude or more, and sometimes reaching a height of 30 m. The large distance between wave crests prevents tsunamis from dissipating energy as a breaking surf. Instead, tsunamis cause water levels to rise rapidly along coast lines (USGS 2024g). Tsunamis and earthquake ground shaking differ in their destructive characteristics. Ground shaking causes destruction mainly in the vicinity of the causative fault, but tsunamis cause destruction both locally and at a very distant location from the location of tsunami formation (USGS 2024g).

5.2.2 Volcanic Eruptions

Volcanic eruptions are one of Earth's most spectacular and most dangerous natural phenomena. They are awe inspiring and yet terrifying at the same time. They remind us that the planet we call home is dynamic and always changing and that we are mere spectators. One would have a hard time finding a more spectacular display of geological power or a potent reminder of Earth's dangers than an erupting volcano. Some facts about volcanoes are presented in Figure 5.6.

When erupting, all volcanoes pose a degree of risk to people and infrastructure. However, the risks are not equal from one volcano to another. This is because each volcano is unique in eruptive style and geographic location. Assessing the relative threats posed by volcanoes identifies which volcanoes warrant the greatest risk-mitigation efforts.

What are the potential risks of volcanoes? The list is rather extensive and may include the following (USGS 2024i):

- Earthquakes
- Lave flows
- Falling ash
- Volcanic gases
- Acid rain
- Volcanic bombs
- Lahars (mud of debris flows)
- Landslides
- Pyroclastic flows
- Fumaroles
- Surface and groundwater contamination
- Explosive eruptions
- Vegetation destruction

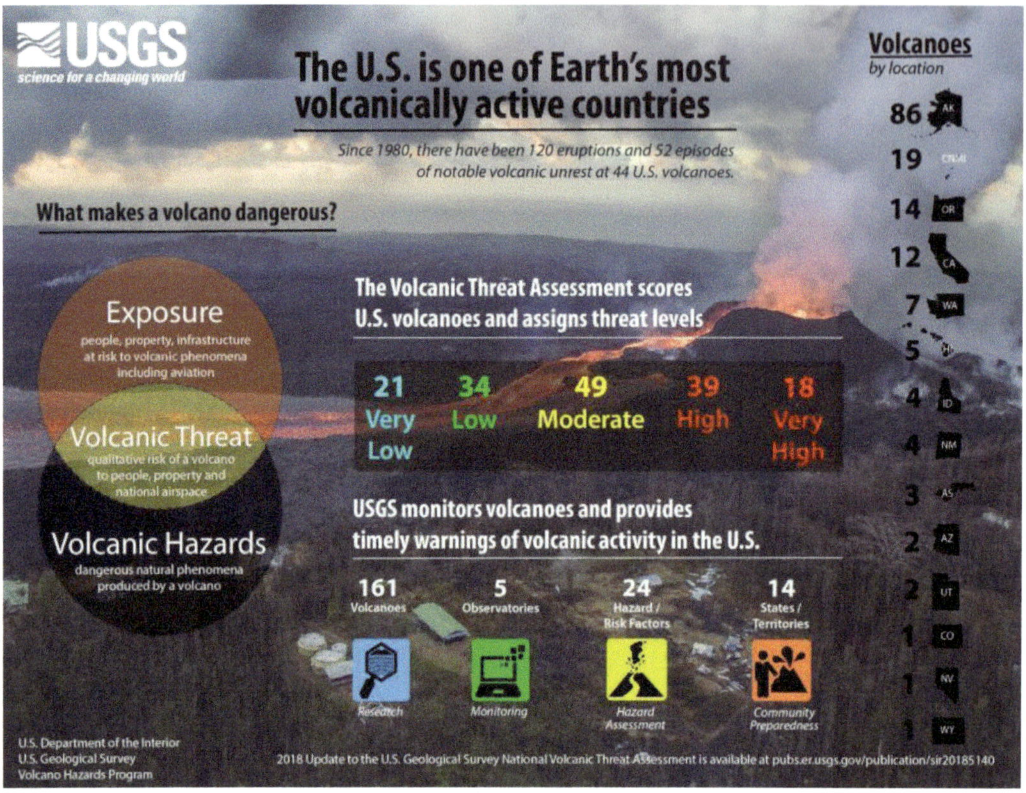

FIGURE 5.6 Facts about volcanoes in the United States. (United States Geological Survey (USGS). 2024h. *Volcanoes: Just the Facts*. www.usgs.gov/volcanoes-just-the-facts. (Accessed March 12, 2024).)

- Long-term erosion
- Loss of animal and human life

Figure 5.7 shows a cross section of a volcano showing many of the risks (USGS 2024i).

More than 50 volcanoes in the United States have erupted one or more times in the past 200 years. The most volcanically active regions are in Alaska, Hawaii, California, Oregon, and Washington. Volcanoes produce a wide variety of hazards that can injure or kill humans and destroy vast areas of land and habitat. Some of the volcanic hazards are described in the sections below (USGS 2024i).

5.2.2.1 Eruption Columns and Clouds

An explosive eruption blasts solid and molten rock fragments, termed **tephra**, and volcanic gases into the air with tremendous force. The largest rock fragments, termed **volcanic bombs**, usually fall back to the ground within 5 km of the vent. Smaller fragments (less than 1 cm), composed of volcanic glass (obsidian-like), minerals, and ash (very fine particles), rise in the air, forming huge billowing **eruption column** (see Figure 5.7). Eruption columns can grow rapidly and reach more than 20 km into the atmosphere and reach beyond the troposphere and into the lower portions of the stratosphere and form what is termed an **Eruption Column** and **Cloud**.

The volcanic ash in the cloud can pose a serious hazard to aviation. During the past 20 years, about 80 commercial aircraft have been damaged by inadvertently flying into ash clouds, and several have nearly crashed because of engine failure. Large eruption clouds can extend hundreds of miles downwind, resulting in **ash fall** over enormous areas. Ash from the Mount Saint Helens eruption of

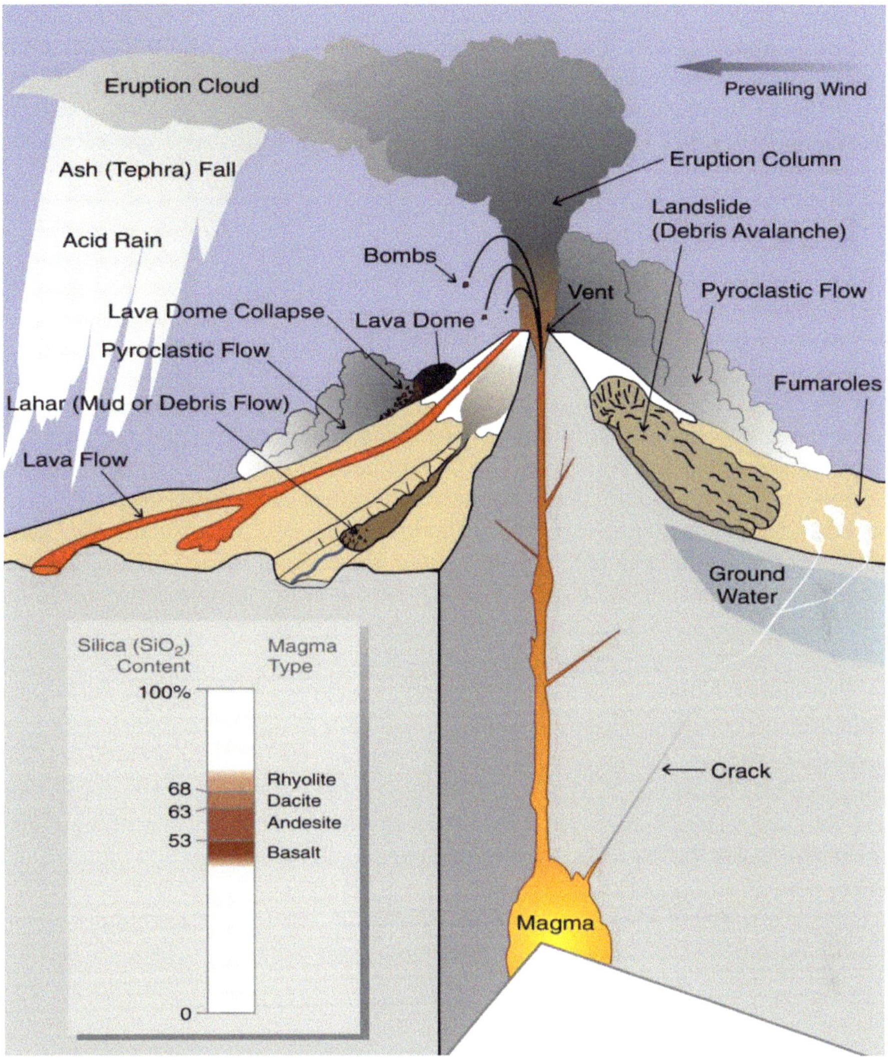

FIGURE 5.7 Volcano hazards. (United States Geological Survey (USGS). 2024i. *What Are Volcano Hazards?* www.usgs.gov/volcano-hazards. (Accessed March 12, 2024).)

May 18, 1980, fell over an area of more than 40,000 km^2. Heavy ash fall can collapse buildings, and even minor ash fall can damage crops, electronics, and machinery such as automobiles.

5.2.2.2 Volcanic Gases

Volcanic gases are emitted during eruptions. Even when a volcano is not erupting, cracks in the ground can allow gases to reach the surface through small openings termed **fumaroles**. More than 90% of all gas emitted by volcanoes is water vapor (steam), most of which is heated groundwater. Other common volcanic gases are carbon dioxide, sulfur dioxide, hydrogen sulfide, and fluorine. Sulfur dioxide gas can react with water droplets in the atmosphere and create **acid rain**, which causes corrosion and harms vegetation. Carbon dioxide is heavier than air and can be trapped in low areas in concentrations that are lethal to people and animal life. Fluorine, in high concentrations is toxic, and can be adsorbed onto volcanic ash particles that later fall to the ground. The fluorine on the particles can poison animals.

Cataclysmic eruptions, such Mount Pinatubo (Philippines), inject huge amounts of sulfur dioxide gas into the stratosphere, where it combines with water to form an aerosol (mist) of sulfuric acid. By reflecting solar radiation, such aerosols can lower the Earth's average surface temperature of extended periods of time by several degrees centigrade. These sulfuric acid aerosols also contribute to the destruction of the ozone layer by altering chlorine and nitrogen compounds in the upper atmosphere (USGS 2024i).

5.2.2.3 Volcanic Flows and Domes

Molten rock (magma) that pours or oozes onto the Earth's surface is called lava and forms **lava flows**. The higher a lava's content of **silica** (silicon dioxide, SiO_2), the less easily it flows. For example, low-silica **basalt** lava can form fast-moving (20 to 40 km/h) streams or can spread out in broad thin sheets as much as several kilometers wide. Since 1983, Kilauea Volcano on the island of Hawaii has erupted basalt lava flows that have destroyed more than 200 houses and severed the nearby coastal highway. In contrast, flows of higher-silica content, termed **andesite** and **dacite** lava, tend to be thick and sluggish, traveling only a short distance from a vent. Dacite and **rhyolite** lavas often squeeze out of a vent to form irregular mounds called lava domes. Between 1980 and 1986, a dacite **lava dome** at Mount Saint Helens grew to approximately 300 m high and 320 m wide (USGS 2024i). Figure 5.8 is a photograph of the north side of Mount Saint Helens showing the dacite lave dome in the center of the crater in August 2018. As you can see in the photo, vegetation is coming back to the area, but the destructive nature of the eruption 38 years later is still very visible.

5.2.2.4 Pyroclastic Flows

High-speed avalanches of hot ash, rock fragments, and gas can move down the sides of a volcano during explosive eruptions or when the steep side of a growing lava dome collapses and breaks apart. The pyroclastic flows can be as hot as 1,000°C and move at speeds of 150 to 250 km/h. Such flows tend to follow valleys and are capable of knocking down and burning everything in their paths. Low-density pyroclastic flows, called **pyroclastic surges**, can easily overflow ridges hundreds of meters high.

The climactic eruption of Mount Saint Helens on May 18, 1980, generated a series of explosions that formed a huge pyroclastic surge. The lateral blast that occurred destroyed an area exceeding

FIGURE 5.8 Mount Saint Helens in August 2018. (Photograph by Daniel T. Rogers.)

350 km^2. Trees with a trunk diameter were mowed down like blades of grass as far at 25 km from the volcano.

5.2.2.5 Volcano Landslides

A **landslide** or **debris avalanche** is a rapid downhill movement of rocky material, snow, and/or ice. Volcano landslides range in size from small movements of loose debris on the surface a volcano to massive collapses of the entire summit or sides of a volcano, as was the case with Mount Saint Helens (see Figure 5.7). Steep volcanoes are susceptible to landslides because they are built up partly of layers of loose volcanic rock fragments. Some rocks on volcanoes have also been altered to soft, slippery clay minerals by circulating hot acidic groundwater. Landslides on volcano slopes are triggered when eruptions, heavy rainfall, or large earthquakes cause these materials to break free and move downhill. At least five large landslides have swept down the slopes of Mount Rainier, which is approximately 100 km north of Mount Saint Helens in Washington State during the last 6,000 years. The largest volcano landslide in historical time occurred at the start of the May 18, 1980 eruption of Mount Saint Helens (see Figure 5.8). Figure 5.9 is a photograph of Mount Rainier showing a volcano landslide that occurred within the last 6,000 years.

5.2.2.6 Lahars

Mudflows or **debris flows** composed mostly of volcanic materials on the flanks of a volcano are call **lahars**. These flows of mud, rock, and water can rush down valleys and stream channels at speeds of 30 to 40 km/h and can travel more than 75 km from a volcano. Some lahars contain so much rock debris (perhaps as high as 60% to 90%) that they look like fast-moving rivers of wet concrete. Close to their source, these flows are powerful enough to rip up and carry large trees, houses, and huge house-sized boulders kilometers downstream. Farther downstream they entomb everything in their path in mud.

Historically, lahars have been one of the deadliest volcano hazards. They can occur both during an eruption and when a volcano is quiet. The water that creates lahars can come from melting snow and ice (especially water from a glacier melted by a pyroclastic flow or surge), intense rainfall, or the break-out of a summit crater lake. Large lahars are a potential hazard to many communities downstream from glacier-clad volcanoes such as Mount Rainier. Many of the volcanoes in the pacific northwest of the United States have established evacuation routes in case of any volcano hazard (see Figure 5.10).

FIGURE 5.9 Mount Rainier showing historical landslide. (Photograph by Daniel T. Rogers.)

FIGURE 5.10 Road sign identifying volcano evacuation route. (Photograph by Daniel T. Rogers.)

5.2.3 Landslides

Landslide is a general term used to describe the downslope movement of soil, rock, and organic materials under the effects of gravity, as well as the landform that results from such movement. Figure 5.11 shows a basic landslide and components.

5.2.3.1 Landslide Classification

Landslides can be classified into different types on the basis of the type of movement and the type of material involved. In brief, material in a landslide mass is either rock or soil, or both. The type of movement describes the actual internal mechanic of how the landslide mass is displaced and includes fall, topple, slide, or flow. Thus, landslides are described using two terms that refer respectively to material movement (rock, soil, or both) and actual internal mechanics of how the landslide mass was displaced. A rotational slide is a type of landslide on which the rupture is curved upward (spoon-shaped). A translational landslide is a landslide on which the material moves out, or down and outward, along a relatively planar surface with little rotational movement or backward tilting. Figure 5.12 shows a series of diagrams presenting the different types (USGS 2024j).

5.2.3.2 Landslide Causes

The following is a list of landslide causes or triggers of landslides from slopes that are not fully stable (USGS 2024j).

- *Slope saturation:* Slope saturation by water is a primary cause of landslides. Saturation can occur in the form of intense rainfall, snowmelt, changes in groundwater levels, and surface water level changes along coastlines, earth dams, and in the banks of lakes, reservoirs, canals,

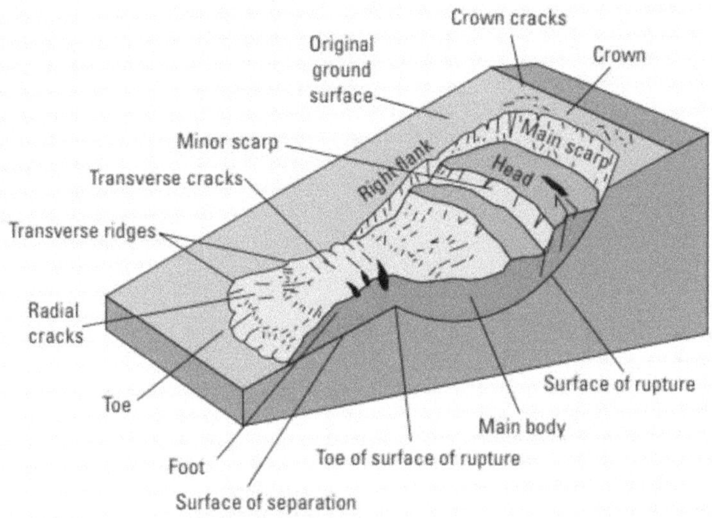

FIGURE 5.11 Basic components of a landslide. (United States Geological Survey (USGS). 2024j. *The Landslide Handbook – A Guide to Understanding Landslides.* www.usgs.gov/handbook2081.pdf. (Accessed March 13, 2024).)

and rivers (USGS 2024j). Landslides and flooding are closely associated because both are related to precipitation, runoff, and the saturation of ground by water.

- *Seismic activity:* Many mountainous areas that are vulnerable to landslides have also experienced at least moderate rates of earthquake activity. Earthquakes in steep landslide-prone areas greatly increase the likelihood that landslides will occur.
- *Volcanic activity:* Landslides due to volcanic activity represent some of the most devastating types of failures. Volcanic lava can melt snow rapidly, which can form a deluge of rock, soil, ash, and water that accelerates rapidly on the steep slopes of volcanoes, devastating anything in its path.
- *Human activities:* Expanded development onto new land creating towns and cities is the primary cause of how human contribute to the occurrence of landslides. Disturbing vegetation, oversteepening slopes, and undercutting the bottom of slopes are examples of anthropogenic-induced causes of landslides. Figure 5.13 is a good example of building a road on a steep slope whereby constructing the road undercut a slope and the result was a landslide.

5.2.3.3 Landslide Effects

Fast-moving landslides such as a debris flow are the most destructive type of landslide to structures. This is because they typically give little to no warning when they occur and move too quickly for any mitigation measures to be initiated. One of the greatest potential consequences of landslides is to the transportation industry and includes roads and rail lines (USGS 2024j).

Landslides also have effects on the natural environment that include the following (USGS 2024j):

- Changing the morphology of the Earth's surface
- Damaging or destroying vegetation
- Destroying habitat

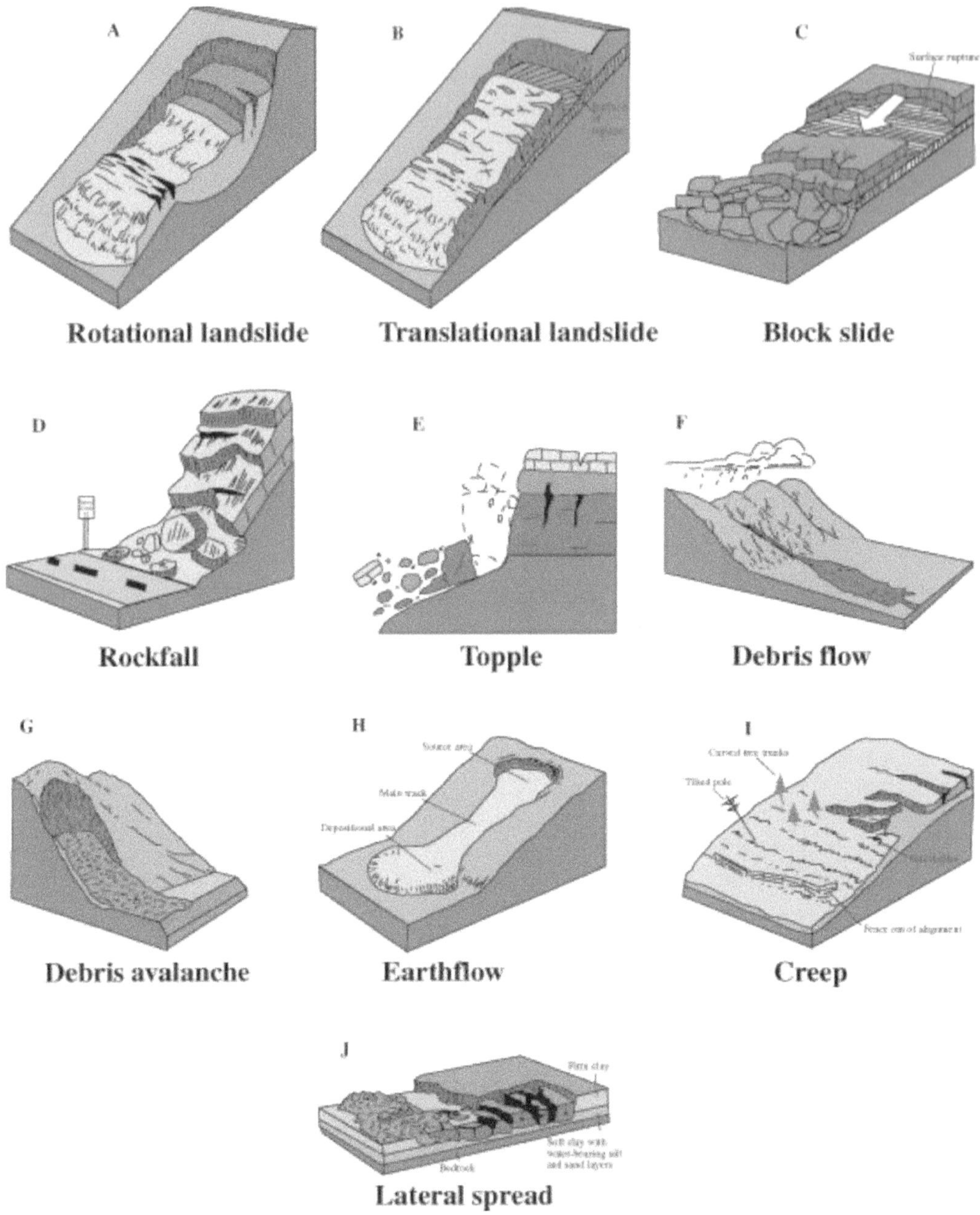

FIGURE 5.12 Types of landslides. (United States Geological Survey (USGS). 2024j. *The Landslide Handbook – A Guide to Understanding Landslides*. www.usgs.gov/handbook2081.pdf. (Accessed March 13, 2024).)

Examples of landslides that occur in the natural environment include the following:

- *Submarine landslide:* Submarine landslide is a general term used to describe the downslope mass movement of geologic material from shallower to deeper regions of the ocean. These commonly occur along continent's slope.

FIGURE 5.13 Photograph of a translational landslide. (United States Geological Survey (USGS). 2024j. *The Landslide Handbook – A Guide to Understanding Landslides*. www.usgs.gov/handbook2081.pdf. (Accessed March 13, 2024).)

- *Coastal cliff retreat:* Coastal cliff retreat is a common effect of landslides and occurs as rock and soil falls, slides, topples, flows, and avalanches.
- *Landslide dams:* Landslide dams can naturally occur when a large landslide blocks the flow of a river and causing a lake to form behind the dam.

5.2.4 Coastal Hazards

While coastal change is natural, hazards arise when these changes put humans at risk. Some of the abrupt hazards include earthquakes, landslides, and tsunamis. Other hazards where perhaps there is a short time to prepare or evacuate include hurricanes and extreme storm events (USGS 2024k). Figure 5.14 is a diagram that shows the different possibilities in response to most coastal hazards.

As we have discussed in this chapter and previous chapters, costal hazards commonly include the following (USGS 2024k):

- Excessive erosion during storm events
- Cliff retreat
- Landslides
- Tsunamis
- Earthquakes
- Sea-level rise due to storms or hurricanes

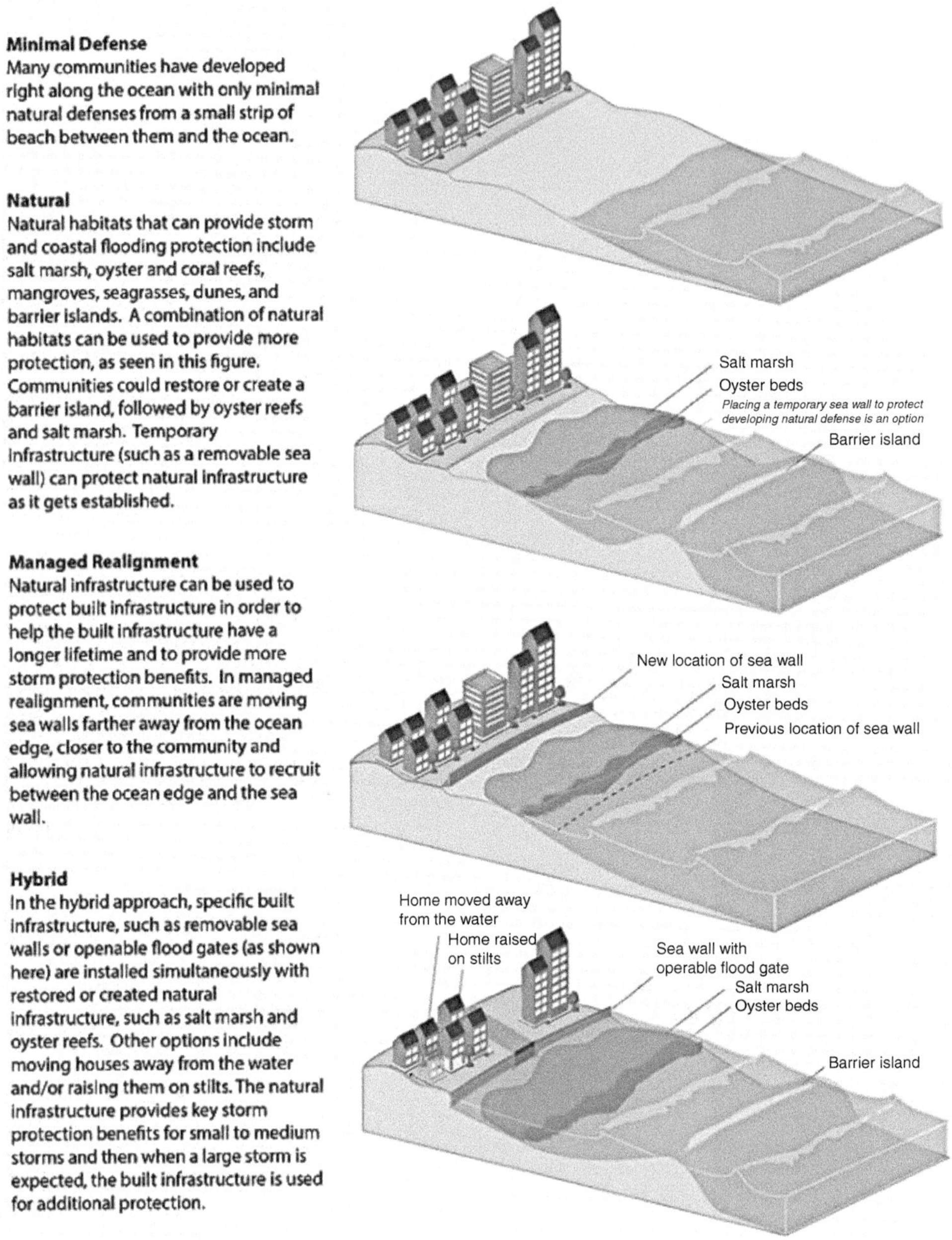

FIGURE 5.14 Potential responses to minimize coastal hazards. (United States Geological Survey (USGS). 2024k. *Coastal Hazards*. www.usgs.gov/coastal-hazards. (Accessed March 13, 2024).)

The above list is generally referred to as visible hazards. Coastal hazards also have many invisible hazards that include toxins, pathogens, saltwater intrusion, and wastewater discharge that can threaten coastal ecosystems, wildlife, and human health if the exposure is high enough and for an extended period of time. These contaminants can enter coastal waterways through seepage, spills, runoff, or through groundwater migration. In addition, these contaminants may also enter the coastal environment as a result of storm damage or other natural event (USGS 2024k).

5.2.5 Subsidence and Sinkholes

Sinkholes are just one of many forms of ground collapse or subsidence. **Land subsidence** is a gradual settling or sudden sinking of the Earth's surface. The principal causes of land subsidence are aquifer-system compaction, drainage of organic soils, underground mining, hydrocompaction, natural compaction, sinkholes, and thawing permafrost. Land subsidence can affect areas that are thousands of kilometers in size (USGS 2024l).

A **sinkhole** is a depression in the ground that has no natural external surface drainage. Basically, this means that when it rains, all of the water stays inside the sinkhole and typically drains into the subsurface slowly. Sinkholes are most common in what is termed **karst terrain**. These are regions where the type of rock below the land surface has the potential to dissolve by groundwater migrating through the rock. Soluble rocks include salt beds and domes, gypsum, limestone, and other rocks high in carbonate (USGS 2024l).

More than 80% of known land subsidence in the United States is a consequence of groundwater use. Subsidence is often an overlooked environmental issue of land and water-use practices. Increasing land development threatens to exacerbate existing land-subsidence problems and initiate new issues.

5.2.6 Erosion

As we learned in Chapter 2, erosion is a geological process whereby earth materials are worn away and transported by natural forces such as wind or water. Types of erosion include physical and chemical. The types of physical erosion include the following (USGS 2024m):

- Water erosion
- Ocean erosion
- Wind erosion
- Ice erosion
- Thermal erosion
- Mass-wasting
- Solar erosion
- Eroding animals

Erosion induced by anthropogenic activities in many ways has changed landscapes over large areas and has inflicted enormous harm on the environment. According to Kemp et al. (2020) and Owens (2020), erosion from anthropogenic activities is the primary agent of geomorphic change in Earth and the rates of change have far outpaced natural geologic processes. Anthropogenic activities that increase the natural rate of erosion include the following:

- Climate change
- Land development
- Deforestation
- Agriculture
- Habitat destruction
- Natural resource exploitation

Erosion from anthropogenic sources has impacted large ecosystems and has degraded soil health on a worldwide scale. In addition, erosion induced by agricultural has choked river systems and has negative effects on the oceans from increased sedimentation and transport of contaminants (Thaler et al. 2022).

5.2.7 Aeolian Processes and Dust

Dust originates from several types of natural sources. Dust may be heavy enough to see but light enough to be carried by the wind. Dust is generally defined as microscopic particles of material. Dust can be made up of pollen, bacteria, smoke, ash, salt crystals, and small bits of soil or rock, including sand.

Significant sources of dust are from desert regions such as the Sahara Desert in northern Africa. In fact, it is not uncommon for dust originating from the Sahara Desert to reach North and South America. Another significant source of dust can be volcanic eruptions which can emit millions of metric tons of dust high into the atmosphere (USGS 2023).

Significant accumulation of dust in the atmosphere can influence climate. In fact, a strong correlation exists between aeolian activity and the decadal droughts related to the Pacific Decadal Oscillation in the southern Pacific Ocean (USGS 2023). Aeolian processes are defined as a type of geomorphic process that involves the erosion, transport, and deposition of particles caused by wind at or near the surface of the Earth. An example of an aeolian deposit are dunes composed of sand in desert regions.

Dust can be considered a natural risk to human health and the environment and can also be considered, in some instances, as a pollutant from an anthropogenic source such as a wildfire that was started by human activity or source. Another example is the Dust Bowl in the 1930s where large dust storms occurred during drought conditions in the central United States that were exacerbated by poor agricultural practices (USGS 2023). Therefore, dust can be considered a geologic hazard and an anthropogenic-induced hazard, and it depends on the circumstances and source of the dust.

Inhalation of dust can cause a range of health problems, including the following (United States Environmental Protection Agency (USEPA) 2025a):

- Allergic reactions
- Asthma
- Eye irritation
- Sneezing
- Coughing
- Hay fever
- Chronic breathing and lung problems
- Heart disease
- Lung infections
- Reduced life span
- Death

5.2.8 Summary of Geologic Hazards

Geologic hazards occur as a result of natural processes within and on the Earth. Table 5.2 lists the deadliest earthquakes and volcanic eruptions that have occurred within recent human history. Table 5.3 lists the deadliest tsunamis and landslides that have occurred in recent human history. Note that geologic hazards are concentrated near tectonically active locations where plate boundaries collide.

In the next section, we discuss environmental hazards. The occurrence of environmental hazards, just as with geologic hazards, is connected to geology through tectonic activity and the position of continents, topography, specific location, proximity to the ocean, mountain ranges, and climate.

TABLE 5.2
Deadliest Earthquakes and Volcanic Eruptions

Deadliest Geologic Hazards			
Geologic Hazard	**Location**	**Year/Magnitude, If Applicable**	**Estimated Deaths**
Earthquakes[1]	Shaanxi, China	1556/8.0	830,000
	Port-Au-Prince, Haiti	2010/7.0	316,000
	Antakya, Turkey	115/7.5	260,000
	Antakya, Turkey	525/7.0	250,000
	Tangshan, China	1976/7.5	242,769
	Gyzndzha, Azerbaijan	1139/Unknown	230,000
	Sumatra, Indonesia	2004/9.1	227,899
	Gansu, China	1920/8.3	200,000
	Dvin, Armenia	893/Unknown	150,000
	Tokyo, Japan	1923/7.9	142,807
Volcanic eruptions[2]	Tambora, Indonesia	1815	92,000
	Krakatau, Indonesia	1883	36,000
	Mont Pelee, Martinique	1902	30,000
	Nevado del Ruiz, Columbia	1985	25,000
	Unzendake, Japan	1792	15,000
	Lakagigar, Iceland	1783	9,350
	Mount Vesuvius, Italy	79	3,360

Source: [1]National Geophysical Data Center. 2024. *Significant Earthquake Database*. https://ourworldindata.org/the-worlds-deadliest-earthquakes. (Accessed May 7, 2024). [2]United States Geological Survey (USGS). 2024n. *Deadliest Volcanic Eruptions*. www.usgs.gov/deadliest-volcanic-eruptions. (Accessed May 7, 2024).

5.3 ENVIRONMENTAL HAZARDS

Geology through tectonic activity determines growth of mountain ranges, volcanic activity, and position of the continents. These actions have profound effects on global ocean and atmospheric circulation patterns, and therefore greatly influence weather and climate patterns. In general terms, the influences of plate tectonics include the following (National Oceanic and Atmospheric Administration (NOAA) 2025):

- A decrease in coastal areas and amount of dry land increases when the continents come together.
- More ice sheets form and sea level falls, and temperatures decrease when continents move toward the poles.
- Mountain ranges affect atmospheric circulation patterns globally and create orographic precipitation patterns regionally.
- Changes in the rate of tectonic activity can influence volcanic activity, which affects the amount of solar radiation at the Earth's surface.
- Sea-level rise increases when tectonic activity increases seafloor spreading.
- Increased formation of limestone and coal locks carbon in rocks and limits the amount of atmospheric heating by limiting the amount of greenhouse gases emitted into the atmosphere.

TABLE 5.3
Deadliest Tsunamis and Landslides

Deadliest Geologic Hazards				
Geologic Hazard	**Location**	**Year**	**Wave Height (m)**	**Estimated Deaths**
Tsunamis[1]	Sumatra, Indonesia	2004	50	230,000
	North Pacific Coast, Japan	2011	10	18,000
	Lisbon, Portugal	1755	30	60,000
	Krakatau, Indonesia	1883	37	40,000
	Enshunada Sea, Japan	1498	Unknown	31,000
	Nankaido, Japan	1707	25	30,000
	Sanrika, Japan	1896	38	22,000
	Northern Chile	1868	21	25,000
	Ryuku Islands, Japan	1771	15	12,000
	Ise Bay, Japan	1586	6	8,000
Landslides[2]	Haiyuan, China	1920		200,000
	Vargus, Venezuela	1999		30,000
	Khait, Tajikistan	1949		28,000
	Armero, Columbia	1985		23,000
	Yungay, Peru	1970		22,000
	Diexi, China	1933		9,300
	Northern India	2013		6,054
	Huaraz, Peru	1941		5,000
	Nevado Huascaran, Peru	1962		4,500
	Vajont, Italy	1963		2,500
	Kelud, Indonesia	1919		200
	Papua New Guinea	2024		2,000

Source: [1]United States Geological Survey. 2024o. *Deadliest Tsunamis*. www.usgs.gov/deadliest-tsunamis. (Accessed May 7, 2024). [2]United States Geological Survey (USGS). 2024p. *Deadliest Landslides*. www.usgs.gov/dealiestest-landslides. (Accessed May 7, 2024).

Environmental hazards as addressed in this book generally refer to extreme atmospheric events and include the following:

- Floods and erosion
- Wildfires
- Hurricanes
- Tornadoes
- Extreme precipitation events
- Extreme heat
- Extreme cold
- Drought
- Severe thunderstorms
- Derecho
- Severe winter weather
- Lightning
- Hail
- High winds

- Solar radiation
- High-tide floods
- Marine heat wave

The following sections will explain each environmental risk and the geological connection.

5.3.1 Floods and Erosion

We have discussed erosion in this chapter and previous chapters and now we will discuss erosion as it relates to flooding. Floodwaters typically carry materials from eroded stream and river banks. The sediments are carried within the floodwaters until the energy or the flowing stream or river decreases sufficiently below the energy required to keep the sediment suspended in the water column. At this point, the once suspended sediment settles out and can clog riverbeds and streams, smother aquatic organisms, and destroy habitats (USGS 2024m).

Some of the risks posed by floods include the following (National Geographic Society 2024a):

- *Harming wildlife:* Floods can harm many different types of wildlife through drowning, disease proliferation, and habitat destruction.
- *Sedimentation and erosion:* Floodways often alter the landscape and cause harm as discussed above.
- *Contamination:* Floodwaters often carry contamination such as pesticides, herbicides, industrial chemical, petroleum and petroleum compounds, oil, debris, and sewerage.
- *Spread disease:* Floods are the leading cause of weather-related infectious disease outbreaks.
- *Nutrient removal:* While floods can return nutrients, they also have the possibility to remove them through erosion and thereby exposing root systems to the atmosphere.
- *Recharge groundwater:* While floods can recharge groundwater, they can also contaminate the groundwater from contaminants dissolved within the floodwaters.

Floods can be so significant that they put human health at risk. As an example, it 1931, floodwater overwhelmed the banks of the Yangtze River in China. The flood was responsible for killing hundreds of thousands of people. It was one of the worst floods in recorded history (National Geographic 2025).

5.3.2 Wildfires

Extreme heat can lead to more frequent, severe, and prolonged heat waves and droughts, and can make wildfires much worse (National Geographic Society 2024b). Wildfires are made worse because air temperatures are higher and dryer which makes the soil and vegetation drier. Figure 5.15 is a photograph of a wildfire near Boulder, Colorado. It is unknown how the fire started but it spread quickly due to high winds that were reported to near 150 km/h at times. At the time of the fire, the region was in a drought (United States National Forest Service 2024). Many wildfires occur naturally but some are intentional or accidental. An example of an accidentally caused wildfire has been known to occur during episodes of high winds that damage electrical transmission lines. Naturally caused wildfires generally occur from lightning strikes. Meteorologists are not yet able to forecast wildfire outbreaks, but there are three conditions that must be present in order for a wildfire to burn that include fuel, oxygen, and a heat or ignition source (National Geographic Society 2024c). According to the National Geographic Society (2024c), four out of five wildfires are started by people. Since 1975 the percentage of wildfires caused by lightning have increased from 2% to 5%.

FIGURE 5.15 Photograph showing recent wildfire near Boulder, Colorado. (United States National Forest Service. 2024. *Wildfires Information*. www.NFS.gov/wildfire-information. (Accessed March 13, 2024).)

5.3.3 Atmospheric Events

According to the National Atmospheric and Space Administration (NASA 2025) and the NOAA (2024a), extreme atmospheric events include many aspects of weather and climate on Earth. We have introduced and have already discussed some of these phenomena and will introduce the rest in this section. However, as many of you know, climate change is altering our climate on Earth in numerous ways. We will discuss the effects of climate change in the next section of this book, but here we will also briefly introduce extreme atmospheric events and will discuss many in greater detail in the next section. According to NASA (2024a) and NOAA (2024a), extreme atmospheric events include the following:

- Hurricanes
- Tornadoes
- Extreme precipitation events
- Extreme heat
- Extreme cold
- Drought
- Severe thunderstorms
- Derecho
- Severe winter weather
- Lightning

- Hail
- High winds
- Solar radiation
- High-tide floods
- Marine heat wave

According to the Intergovernmental Panel on Climate Change (IPCC) Sixth Assessment report (IPCC 2021), the anthropogenically caused rise in greenhouse gases has increased the frequency and intensity of extreme atmospheric events on Earth. Research indicates that risks of extreme atmospheric events will increase as global temperatures continue to increase. An interesting and somewhat disturbing fact concerning climate change is the impact of increasing global atmospheric temperatures. According to the physics of the atmosphere, an increase in temperature of a degree Celsius means that the atmosphere can hold more moisture. In fact, this translates into the atmosphere being able to hold 7% more water vapor per degree Celsius increase in temperature. Therefore, if the atmosphere has the ability to hold more water vapor, it can also precipitate more. An additional impact of climate change in the temperature difference between the tropical areas and the arctic regions is not as great with a warmer climate. The result is that the steering currents of air in the upper regions of the atmosphere tend to slow. Also, the result is that storms systems move more slowly over the land surface and that precipitation has the potential to increase over a fixed point. The result of these two factors strongly suggests that increased moisture content with slower moving storm systems increases precipitation probabilities (American Geophysical Union 2024).

An additional factor that influences the severity, duration, and intensity of extreme atmospheric events is the local and regional geology of any location on Earth. A few examples to consider include the following:

- Mountainous versus flat terrain
- Rock type
- Sediment type
- Local surface water bodies, streams, and rivers
- Presence of groundwater
- Presence of identified geologic hazards
- Location of cities and the built environment
- Many other influences

When discussing extreme atmospheric events, it is important to consider the damage that they cause to the natural environment and on the loss of life and property damage. For example, in just a 3-year period from 2020 through 2022, the total estimated cost of extreme atmospheric events in the United States alone that together include wildfires, tornadoes, hurricanes, severe storms, and flooding exceeded $2.6 trillion dollars. The number of events exceeding $1 billion dollars was 60 (NOAA 2024b). According to NOAA (2024c, 2024d), the total cost of extreme atmospheric for the year 2023 will increase compared to the previous 3-year average. Figure 5.16 is a summary of the extreme atmospheric events for 2023 through November in the United Sates (NOAA 2024d).

As we discuss extreme atmospheric events in the following sections, keep in mind how and why these events occur and the influence of geology on these outcomes.

5.3.3.1 Hurricanes

Hurricanes are one of nature's most powerful storms. They produce strong winds, storm surge flooding, and heavy rainfall that can lead to inland flooding, tornadoes, and currents (NOAA 2024b). A **hurricane** is a low-pressure system with organized thunderstorm activity and it forms

FIGURE 5.16 Satellite image of Hurricane Humberto off the southeastern coast of the United States in 2019. (National Oceanic and Atmospheric Administration (NOAA). 2024c. *Hurricane Formation.* www.noaa.gov/education/hurricane-formation. (Accessed March 14, 2024).)

over tropical or subtropical waters and gains energy from the warm ocean water. Circulation of a low-pressure system in the Northern Hemisphere is counterclockwise (see Figure 5.16).

In order for a hurricane to form, the following conditions must be satisfied (NOAA 2024c):

- Ocean water temperatures greater than 27°C
- Ocean depths should exceed 50 m
- An unstable atmosphere driven by differences in temperature, where temperature decreases with height
- Moist air near the mid-level of the atmosphere
- Little change in wind speed or direction with increasing height (known as wind shear)

As the low pressure begins to deepen (or atmospheric pressure decreases), the storm becomes stronger. At a sustained wind speed of approximately 55 km/h, the storm is termed a **tropical storm**. When wind speeds increase to approximately 118 km/h (74 miles/h), the storm then becomes a hurricane as a Category 1. There are four categories that describe hurricane intensity, which are determined by a measure of its sustain windspeed and is termed the **Saffir–Simpson Hurricane Wind Scale**, and are as follows (NOAA 2024c):

- *Tropical depression:* sustained wind speed less than 53 km/h (39 miles/h)
- *Tropical storm:* sustained wind speed from 53 to 118 km/h (39 to 74 miles/h)
- *Category 1 Hurricane:* sustained wind speed from 120 to 154 km/h (75 to 95 miles/h)
- *Category 2 Hurricane:* sustained wind speed from 155 to 180 km/h (96 to 110 miles/h)
- *Category 3 Hurricane:* sustained wind speed from 181 to 209 km/h (111 to 129 miles/h)
- *Category 4 Hurricane:* sustained wind speed of 210 to 253 km/h (129 to 156 miles/h)
- *Category 5 Hurricane:* sustained wind speed of 254 km/h or greater (156 miles/h or greater)

According to the United Sates Federal Emergency Management Agency (FEMA), the estimated total loss of life from just hurricanes in the United States from 1900 to 2004 totaled 7,836. Worldwide fatalities from hurricanes from multiple sources were estimated at greater than 400,000 (FEMA 2024). Refer to Figure 5.18 for a breakdown.

5.3.3.2 Tornadoes

A **tornado** is a narrow, violently rotating column of air that extends from a thunderstorm to the ground surface. By definition, if the rotating column does not contact the ground surface, it is called a **funnel cloud**, not a tornado. If a tornado forms over water, it is called a **water spout**. If a water spout travels from water to land and comes into contact with the surface of the ground, it is referred to as a tornado (NOAA 2024e). Because wind is invisible, it is sometimes difficult to see a tornado unless it forms a condensation funnel made up of water droplets, dust, and debris. Tornadoes can be among the most violent extreme atmospheric events (NOAA 2024e). Figure 5.17 shows a photograph of a tornado. Figure 5.18 shows a diagram of a thunderstorm that can produce a tornado.

Approximately 1200 tornados form in the United States per year, with most occurring in the Great Plains from Texas to North Dakota and the mid-west and southern states (NOAA 2024e). Tornados can form any time of year but are much more frequent in spring and early summer. In addition, tornados favor a time period usually between 4 and 9 pm to form when the atmospheric conditions are more unstable and favor tornado formation (NOAA 2024e).

FIGURE 5.17 Photograph of a tornado showing a condensation funnel of water droplets. (National Oceanic and Atmospheric Administration (NOAA). 2024e. *Tornado Basics*. www.noaa.gov/education/tornadobasics. (Accessed March 14, 2024).)

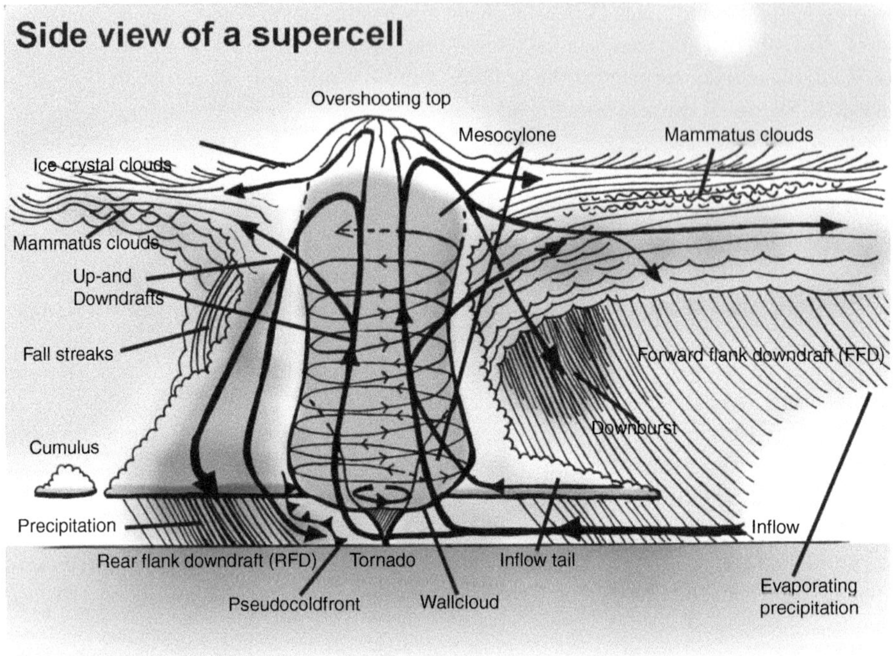

FIGURE 5.18 Cross section of a supercell or mesocyclone. (National Oceanic and Atmospheric Administration (NOAA). 2024a. *Extreme Atmospheric Events*. www.noaa.gov/extreme-atmospheric-events. (Accessed March 14, 2024).)

TABLE 5.4
Enhanced Fujita Tornado Strength Scale

EF-Scale	Class	Wind Speed MPH km/h	Description
EF-0	Weak	65 to 85 105 to 137	Gale
EF-1	Weak	86 to 110 138 to 177	Moderate
EF-2	Strong	111 to 135 178 to 217	Significant
EF-3	Strong	136 to 165 218 to 266	Severe
EF-4	Volent	166 to 200 267 to 322	Devastating
EF-5	Violent	>200 >322	Incredible

Source: United States National Weather Service. 2024a. *Enhanced Fujita Scale*. www.weather.gov/tae/ef_scale. (Accessed March 14, 2024).

To determine the strength of a tornado, meteorologists and other experts examine the damage caused by a tornado. From an evaluation of wind damage, the wind speed of a tornado can be estimated. Currently, scientists use the **Enhanced Fujita Scale** to estimate the strength of a tornado (see Table 5.4).

When assigning a wind speed rating to a tornado, investigators use 28 damage indicators such as building type, structures, and trees. For each damage indicator, there are 8 degrees of damage ranging from the beginning of visible damage to complete destruction (NOAA 2024e).

The science of tornado formation is not yet completely understood, and much research remains. What is known is that the most destructive tornadoes form from what is termed **supercell**

thunderstorm. A supercell thunderstorm is a rotating thunderstorm with well-defined Doppler radar-confirmed circulation and is called a **mesocyclone**. To form a mesocyclone, one of the keys factors necessary is strong wind shear, or a change in wind speed and direction with increased height in a thunderstorm. Mesocyclones do not always produce a tornado. In addition to producing a tornado, mesocyclones can also produce large hail, severe straight-line winds in excess of 100 km/h, frequent cloud to ground lightning, and heavy rain (see Figure 5.18) (NOAA 2024e).

Recent research suggests that tornado formation is dictated mainly at the storm scale, in and around the mesocyclone. Recent theories and results from current research projects suggest that when a mesocyclone forms, tornado development is related to the temperature difference across the edge of the downdraft air wrapping around the mesocyclone (NOAA 2024e).

According to NOAA, the 10-year average for fatalities from tornadoes is 71 people per year. The total estimate cost of property damage from tornadoes over the last 10 years exceeded $1 billion dollars per year (NOAA 2024f).

5.3.3.3 Severe Thunderstorms

A **thunderstorm** is defined as a rain shower during which you hear thunder (NOAA 2024a). As thunder originates from lightning, all thunderstorms have lightning. Severe thunderstorms can be an extreme weather event. Severe thunderstorms can produce strong straight-line winds exceeding 100 km/h, large hail, heavy precipitation that can lead to flooding, and frequent cloud to ground lightning.

Thunderstorms are usually formed through a process called **convection**, which is defined as upward movement of air that transports whatever is in the air along with the upward movement, especially any moisture in the air (NOAA 2024a). A thunderstorm is classified as severe when it contains one or more of the following, hail (with a diameter of 1.25 cm), wind gusts exceeding 75 km/h, or a tornado (NOAA 2024a).

Three basic ingredients are required for a thunderstorm to form, which include moisture, rising air, and a lifting mechanism to lift the air. In summer months, the sun typically heats the ground surface and surrounding air and as warm air rises, this typically begins the process (NOAA 2024a). As the air rises, it transfers heat from the surface of the earth to the upper levels of the atmosphere. The water vapor begins to cool, releases heat, condenses, and then forms a cloud. The cloud eventually grows upward into areas where the temperature is below freezing (NOAA 2024a).

Worldwide, there are approximately 16 million thunderstorms per year. At any given moment, there are about 2,000 thunderstorms in progress. There are about 100,000 thunderstorms each year in the United States and about 10% of them reach severe classification (NOAA 2024a).

In the United States, thunderstorms are most common during the spring and summer and favor the southern states. Many hazardous weather events are associated with thunderstorms. Under favorable conditions, rainfall from thunderstorms can cause flooding and even flash flooding, which kills more people each year than hurricanes, tornadoes, and lightning combined. Lightning is also responsible for starting about 5% of all wildfires in the United States. Some thunderstorm have strong winds and even produce derechos (which we cover in the next section), which can exceed 180 km/h and can knock down trees and mobile homes (NOAA 2024a).

Thunderstorms have three stages in their life cycle: the developing stage, the mature stage, and the dissipating stage. The developing stage of a thunderstorm is marked by a cumulus cloud type that is pushed upward by the rising air (updraft). The cumulus cloud soon looks like a tower as the updraft continues to develop. There is little to no rain in this stage but there could be some lightning.

The thunderstorm enters the mature stage when the updraft continues to feed the storm, but precipitation begins to fall out of the storm creating a downdraft and rain-cooled air spreads out along the ground forming a gust front. The mature stage is the most likely time for hail, heavy rain, frequent lightning, strong winds, and tornadoes. Eventually, a large amount of precipitation is produced and the updraft is overcome by the downdraft which starts the dissipating stage (NOAA 2024a).

5.3.3.4 Lightning

Lightning is not only spectacular, but also dangerous. Worldwide approximately 2,000 people are killed every year because of lightning. Many more survive lightning strikes but may suffer from a variety of symptoms including memory loss, dizziness, weakness, numbness, and many other ailments (National Geographic Society 2024d).

Lightning is essentially a giant spark of electricity in the atmosphere that occurs between clouds, the air, or the ground. In the early stages of development, air behaves as an insulator between the positive and negative charges in a cloud and between the cloud and the ground surface. When the opposite charges build up enough, the insulating capacity of the air breaks down and there is a rapid discharge of electricity, which is what we know as lightning. The flash of lightning temporarily equalizes the charged regions in the atmosphere until the opposite charges begin again to build up. Lightning is most often associated with thunderstorms but may also occur during and within a volcanic eruption, extremely intense wildfires, nuclear detonations, heavy snowstorms, and large hurricanes (NOAA 2024g).

Lightning causes thunder. Energy from the lightning itself heats the air briefly to perhaps as much as 50,000°F. This causes the air to explode outward. The huge pressure in the initial outward shock wave decreases rapidly with increasing distance and with approximately 10 m or so becomes small enough to be perceived as the sound of thunder. In some circumstances, thunder can be heard as far away as 40 km (NOAA 2024g).

5.3.3.5 High Winds

In general terms, high winds are caused by atmospheric pressure variations. While any area can experience high winds, some areas may be more favored to experience high winds. The main cause of wind is the uneven heating of the surface of the Earth from the sun. This uneven heating causes pressure differences which then creates wind. Wind is a process in which the atmosphere attempts to reach equilibrium. Wind is the great equalizer of the atmosphere, transporting heat, moisture, pollutants, and dust. In some instances, wind can transport items around the globe. Landforms, processes, and impacts of wind are called **aeolian landforms**, which is a geological process that forms deposits such as dunes (sand deposits) and loess (silt deposits) (National Geographic Society 2024e).

In simple terms, warm air in equatorial regions rises higher into the atmosphere and migrates toward the poles creating a low-pressure system. At the same time, cooler, denser air moves off the surface of the Earth toward the equator to replace the heated air. This is a high-pressure system. Winds generally blow from higher pressure to lower pressure. This occurs because of the circular nature and rotation of the Earth and is known as the Coriolis Effect. The Coriolis Effect makes wind systems twist counterclockwise in the Northern Hemisphere and clockwise in the Southern Hemisphere (NOAA 2024h).

From an environmental hazard point of view, high winds that are generated that may cause harm to humans and the environment are winds that usually occur at speeds in excess of 80 km/h. Environmental hazards that have wind components that can therefore cause harm include hurricanes, tornadoes, thunderstorms, blizzards, and derechos.

Other instances where high winds are created that can cause harm have a geological component such as certain landforms including mountain ranges whereby wind is funneled through canyons or between mountains. This is called a **katabatic** or **catabatic wind** or a **drainage wind** that carries high-density air from higher elevations down a slope under the force of gravity. In the United States, the Chinook winds and Santa Ana winds are produced through this phenomenon. Parts of Europe, specifically the Alps, Greenland, Antarctica, and other parts of the world, also have experienced these types of winds (NOAA 2024h). Still more geologically induced wind hazards can be created by volcanic eruptions, large and sudden landslide, and pyroclastic flows.

5.3.3.6 Derecho

A derecho is a widespread, long-lived wind storm. Derechos are associated with bands of rapidly moving showers or thunderstorms. A derecho has a common Doppler radar signature known as a bow echo. Bow echoes typically arise when a storm's rain-cooled outflow winds are strong and move preferentially in one direction. A derecho may be associated with a single bow echo or with multiple bows. Bow echoes, in turn, may consist of an individual storm, or may be composed of a series of adjacent storms, such as a squall line, that together take the shape of one large bow echo (NOAA 2024i).

According to the National Weather Service (NWS), derecho winds must exceed 90 km/h (57 miles/ h), and travel at least 650 km (400 miles) and have a width of 100 km (60 miles). Some derecho winds have exceeded 160 km/h (100 miles/h). The strong derecho wind speed occurred in Wisconsin on July 4, 1977 and reached a measured wind speed of 185 km/h (115 miles/h) (NOAA 2024i).

Some of the most intense derechos occur in summer where they form on the fringes of heat waves. The relationship between heat waves and derechos is more than statistical. It turns out that the meteorological conditions favorable for large-scale heat waves also are conducive for derechos. The most common area for derecho development is in the summer months in the upper Mississippi valley and upper Great Lakes to the Ohio Valley and northeastern United States (NOAA 2024i).

There are two types of derechos, serial and progressive. A serial derecho is most common during the spring and the fall months. These are produced by multiple bow echoes embedded in an extensive squall line (typically many hundreds of kilometers). This type of derecho is typically associated with a strong, migrating low-pressure system with strong to low- to mid-tropospheric flow. A progressive derecho is associated with a relatively short line of thunderstorms, (approximately 100 to 200 km) that at times may take the shape of a single bow echo (NOAA 2024i). Figure 5.19 shows a typical bow echo of a large derecho that originated in Minnesota that occurred on July 15, 2024, in the upper Midwest. This particular derecho was responsible for 10 active tornadoes on the ground at the same time in the vicinity of Chicago, Illinois (NWS 2024). Figure 5.20 shows a shelf cloud created by a derecho in northern Illinois in 2021.

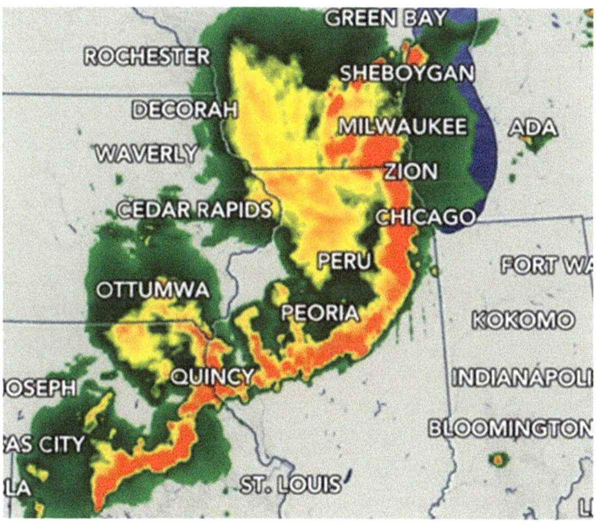

FIGURE 5.19 Radar image of a derecho in the upper Midwest on July 15, 2024. (National Weather Service (NWS). *Radar Image July 15, 2024, Chicago, Illinois*. www.nws.gove/radar/july15/chicago/illinois. (Accessed July 16, 2024).)

FIGURE 5.20 A shelf cloud on the leading edge of a derecho-producing thunderstorm in 2021 in Hampshire, Illinois. (National Oceanic and Atmospheric Administration (NOAA). 2024i. *About Derechos*. www.spc.noaa.gov/about-derechos. (Accessed March 15, 2024).)

5.3.3.7 Snowstorms and Blizzards

Snowstorms and blizzards are rated on a scale called the Northeast Snowfall Impact Scale (NESIS) The NESIS scale has five categories that include Extreme, Crippling, Major, Significant, and Notable. The NESIS scale differs from other meteorological indices, in that it uses population information in addition to meteorological data (NOAA 2024j). The United States National Weather Service (2024b) defines a blizzard as a storm with large amounts of snow or blowing snow, winds greater than 50 km/h, and visibility of less than 0.4 km and lasts for more than 3 hours. In order for a winter storm to develop, three attributes are necessary and include cold air, lift in the atmosphere, and moisture to create snow (NOAA 2024j). Benefits of snowstorms and even blizzards include providing water to streams and river during warm periods through snowpack melting. California and many other western states benefit greatly from the storage of water in mountain snowpack (NOAA 2024k).

5.3.3.8 Extreme Cold

Extreme cold weather occurs every year in the continental United States and affects millions of people. The arctic air can be dangerous or deadly. People exposed to extreme cold are susceptible to frostbite and hypothermia. Areas most prone to frostbite are uncovered skin and the extremities, such as hand and feet. Hypothermia occurs when the body loses heat faster than the body can produce heat (United States National Weather Service 2024c).

Exposure to cold at temperatures below 0°C can lead to hypothermia from prolonged exposure if you are not properly clothed (NOAA 2024h). Cold weather can also affect crops. In late spring or early fall, cold air outbreaks can damage or destroy farm produce, as well as residential plants and flowers (United States National Weather Service 2024c).

5.3.3.9 Extreme Heat

In the United States. extreme heat wave frequency has increased from an average of 2 per year in 1961 to slightly more than 6 per year in 2021. Heat wave duration and intensity have also increased during the same period. A persistent period of unusually hot days, defined as a temperature exceeding the 85th percentile in average temperature, is referred to as an extreme heat event (USEPA 2024c). Unusually hot days and heat wave events are a natural part of the day-to-day variation in weather. However, as Earth's climate warms, hotter than usual days and nights become more frequent, become even hotter, and last longer (USEPA 2024c). Increasing in extreme heat waves can lead to more heat-related illnesses and deaths, especially if people and communities do not take steps to adapt. Extreme heat is also associated with drought conditions and affects whole ecosystems and all life forms (USEPA 2024c).

5.3.3.10 Hail

Hail is associated with high, vertical cumulonimbus clouds that are typically associated with thunderstorms. Ice particles within a cumulonimbus cloud develop from supercooled water (see Figure 5.18). The particles fall toward the bottom of the cloud under the force of gravity but sometimes are forced back up by powerful updrafts within the cloud. This process may occur more than once and usually results in increasing the size of hail pellets until the mass is greater than the force of updrafts and then the hail falls to the surface under the force of gravity National Geographic Society 2024e). Hail does significant damage to homes, automobiles, and crops. One of the most devastating hailstorms in the United States occurred near Phoenix, Arizona, in October 2010. Hail as large as 7.6 cm in diameter caused an estimated $2.8 billion in damage to property, namely homes and automobiles. The largest hailstone ever recovered in the United States was produced by a thunderstorm in South Dakota in June 2010 and was 20 cm in diameter and weighed 0.9 kg (2 pounds) (National Geography Society 2024e).

5.3.3.11 Marine Heat Wave

Marine heat waves are generally considered periods of persistent warm ocean currents defined as any time the ocean temperature exceeds the 90th percentile for a specific length of time (NOAA 2024l). Marine heat waves can have a profound negative effect on marine life including fish and coral reefs. Marine heat waves have been associated with climate change (NOAA 2024l).

5.3.3.12 Solar Radiation

Solar radiation is energy that comes from the sun. It is produced by nuclear reactions within the sun and releases large amounts of energy. The energy produced by the sun travels to the Earth through electromagnetic waves, which include X-rays, ultraviolet and infrared radiation, radio emissions, and visible light. In addition, the sun also emits radiation from solar flares which are large eruptions of energy coming off the sun and contain energy forms such as heat, magnetic energy, and ionizing radiation. The ionizing radiation released during solar flares includes X-rays and gamma rays. Other forms of energy include coronal mass ejections and proton storms. The Earth's atmosphere and electromagnetic field protect life on Earth from the harmful effects of radiation from the sun (USEPA 2024a). Radiation from solar activity is a natural part of our environment and does not threaten human health on a day-to-day basis. Ultraviolet rays from the sun pose a much higher risk to human health. Ultraviolet rays cause sunburn and have been linked to skin cancer in humans (USEPA 2025b). The ozone layer in Earth's upper atmosphere filters much of the ultraviolet radiation from the sun and is beneficial to all life on Earth.

However, as we will discuss in Chapter 8, anthropogenic activity from the use of some aerosols has eroded some of the ozone layer. Depletion of the ozone layer means that there is less protection from certain ultraviolet wavelengths of solar radiation from the sun. Studies have linked depletion of the ozone layer to certain types of skin cancer. Depletion of the ozone layer has also been linked to causing negative effects on plants, marine ecosystems, biogeochemical cycles, and solid earthen materials (USEPA 2024b).

5.3.3.13 Drought

A drought is a complicated phenomenon and can be difficult to define because drought can mean different things in different regions. A simple definition of drought is a period of time of unusually persistent dry weather that continues long enough to cause damage to crops, water supplies, and native vegetation and animal life (NASA 2024b). Droughts can be exacerbated by strong winds. Droughts can significantly increase the risk of wildfires by affecting soil moisture and groundwater. Since 1900, more than 11 million people have died and more than 2 billion people have been directly affected by drought (NASA 2024b). Most droughts occur when the regular weather patterns are interrupted, causing a disruption in the water cycle. This disruption can significantly impact the amounts of precipitation that a region receives. Scientists have discovered a link between climate change and drought (NASA 2024b).

5.3.3.14 Extreme Precipitation

In the last few decades, a larger percentage of precipitation has occurred from intense single-day events. The prevalence of extreme single-day precipitation events had remained rather stable from 1910 to 1980 but has risen substantially since then. Over the entire period from 1910 to 2020, the portion of the United States experiencing extreme single-day precipitation events increased at a rate of nearly one half a percentage point per decade (USEPA 2024d). As an example, within a 2-week period in 2022, there were five 1-in-1,000-year precipitation events within the United States, which occurred in St. Louis, eastern Kentucky, southeastern Illinois, Death Valley, and in Dallas (USEPA 2024d)

Climate change can affect the intensity and frequency of precipitation. Warmer oceans increase the amount of water that evaporates into the air. When more moisture-laden air moves over land or converges into a storm system, it can produce more intense precipitation (USEPA 2024d). For every degree Celsius that Earth's atmospheric temperature increases, the amount of water vapor in the atmosphere can increase 7% according to the laws of thermodynamics (NASA 2024c). Specifically, this is described as the Clausius–Clapeyron relation and is described in their equation published with Lord Kelvin in the 19th century. The fundamental description involves chemical thermodynamics, which specifies the temperature dependence of pressure, most importantly vapor pressure of a liquid such as water (Skliris et al. 2016).

An additional factor affecting the amount of precipitation is the decrease in the temperature difference from high-latitude compared to mid- and low-latitude locations. The decrease in temperature difference caused by climate change tends to slow the migration of storms systems as they traverse the globe. Hence, storms with increased water vapor that are also slower moving can greatly increase precipitation in a self-amplifying feedback loop (NASA 2024c; USEPA 2024d).

5.3.3.15 High-tide Flooding

Recent high-tide flooding has occurred at multiple locations within the United States. Coastal communities along the eastern and west coast of the United States reported record high-tide flooding in 2023. The trend is expected to continue due to rising sea level caused by climate change. High-tide flooding occurs when the tide is approximately 0.5 m higher than the normal (NOAA 2024m). In 2023, high-tide flooding was documented six times along the Atlantic Coast and was documented four times along the Pacific Coast. By the year 2050, projections for high-tide flooding events are expected to increase to 45 to 75 high-tide flooding days (NOAA 2024m).

5.3.4 Summary of Environmental Hazards

Environmental hazards are becoming more frequent and more intense as the human population continues to increase. The World Health Organization (WHO) has estimated that 3.6 billion, or nearly

TABLE 5.5
Deadliest Hurricanes Worldwide and in the United States

Environmental Hazard	Location	Year	Estimated Deaths
Hurricanes Worldwide[1]	Bangladesh	1970	300,000
	China	1931	300,000
	Myanmar	2008	140,000
	Bangladesh	1991	138,373
	Bangladesh	1911	120,000
	Bangladesh	1917	70,000
	China	1922	60,000
	Bangladesh	1962	51,000
	China	1912	50,000
	Bangladesh	1919	40,000
Hurricanes United States[2]	Galveston, Texas	1900	12,000
	Puerto Rico	2017	2,950
	Okeechobee, Floride	1928	2,500
	Caminada, Louisiana	1893	2,000
	Savannah, Georgia	1893	2,000
	New Orleans, Louisiana	2005	1,200
	Georgia and South Carolina	1881	700
	Louisiana and Texas	1957	416
	South Florida	1935	408
	Louisiana Coast	1856	400

Source: [1]National Oceanic and Atmospheric Administration (NOAA). 2024n. *Hurricanes in History*. https://nhc.noaa.gov/hurricanes-in-history. (Accessed May 26, 2024). [2]National Oceanic and Atmospheric Administration (NOAA). 2024o. *National Hurricane Center. Hurricanes in the United States*. www.nhc.noaa.gov/hurricanes-in-the-United-States. (Accessed May 26, 2024).

half the current human population, live within the boundaries of one or more environmental hazards (WHO 2024).

According to the WHO, the major driver is a combination of increasing human population and increased effects of climate change. Table 5.5 lists the deadliest hurricanes that have occurred in recent human history both worldwide and in the United States. Table 5.6 lists the deadliest tornadoes in the United States. Table 5.7 lists the deadliest floods worldwide and in the United States. Table 5.8 lists the largest wildfires in the United States. Table 5.9 lists the deadliest heat waves in the United States. Table 5.10 lists the worst winter storms in the United States.

5.4 CONNECTIONS BETWEEN GEOLOGY AND ENVIRONMENTAL HAZARDS

There are many connections between geology and environmental risks. The largest geological influence on global climate is plate tectonics. Some influences are obvious, and others are subtle but may be just as significant. The more obvious connections that go beyond plate tectonics include types and severity of erosion and mass-wasting, areas of deposition, topography, geography, and location. Additional connections include regional and local climate patterns, precipitation, ocean currents, heat distribution on Earth, water cycle, and amount of solar radiation.

TABLE 5.6
Deadliest Tornadoes in the United States

Environmental Hazard	Location	Year	Estimated Deaths
Tornadoes	Tri-State, Missouri, Illinois, Indiana	1925	695
	Natchez, Mississippi	1840	317
	St. Louis, Missouri	1896	255
	Tupelo, Mississippi	1936	216
	Gainesville, Georgia	1936	203
	Joplin, Missouri	2011	160
	Amite, Louisiana	1908	143
	Woodward, Oklahoma	1947	117
	New Richmond, Wisconsin	1899	116
	Flint, Michigan	1953	116

Source: National Oceanic and Atmospheric Administration (NOAA). 2024p. *Storm Prediction Center. Worst Tornadoes*. https://nhc.noaa.gov/tornadoes-in-history. (Accessed May 26, 2024).

TABLE 5.7
Deadliest Floods Worldwide and in the United States

Environmental Hazard	Location	Year	Estimated Deaths
Floods Worldwide[1]	Yangtze River, China	1931	3,700,000
	Yellow River China	1887	2,000,000
	Yellow River, China	1938	800,000
	Banqiao Dam Failure, China	1975	230,000
	Yangtze River, China	1935	145,000
	St. Felix, Netherlands	1530	100,000
	Red River, North Vietnam	1971	100,000
	Jiangsu-Anhui, China	1911	100,000
	St. Lucia, Netherlands	1287	50,000
	Eastern Guatemala	1949	40,000
Floods United States[2]	Johnstown, Pennsylvania	1889	2.209
	Statewide Ohio Flood	1913	467
	St. Francis Dam, California	1928	400
	Ohio River Flood	1937	385
	Oregon Hellner Flash Flood	1903	324
	Brazos River, Texas	1899	284
	Black Hills, South Dakota	1972	238
	Superstorm Sandy, East Coast	2012	233
	Thompson Canyon, Colorado	1976	144
	Buffalo Creek, West Virginia	1972	125

Source: [1]National Oceanic and Atmospheric Administration (NOAA). 2024q. *Floods Worldwide*. https://noaa.gov/floods-in-world-hisotry. (Accessed May 26, 2024). [2]National Oceanic and Atmospheric Administration (NOAA). 2024r. *Floods in the United States*. www.noaa/worst-floods-in-the-united-states. (Accessed May 26, 2024).

TABLE 5.8
Largest Wildfires in the United States

Environmental Hazard	Location	Year	Area Burned (Hectares)	Estimated Deaths
Wildfires in the United States	Yarnell, Arizona	2013	3,398	19
	Mendocino. California	2018	185,761	1
	Mt. Shasta, California	2021	389,754	1
	Peshtigo, Wisconsin	1871	404,600	1,500
	Northern California	2020	417,809	1
	Smokehouse Creek, Texas	2024	436,153	33
	Taylor Complex, Alaska	2004	487,762	None
	Michigan	1871	1,011,508	250
	Pacific Northwest	1910	1,213,800	87
	Miramichi, Maine	1825	1,213,800	160

Source: Western Fire Chiefs Association. 2024. *Largest Wildfires in United Sates History*. www.wfca.com/wildfire-articleas/biggest-wildfires-in-us-history. (Accessed May 26, 2024).

TABLE 5.9
Deadliest Heat Waves in the United States

Environmental Hazard	Location	Year	Estimated Deaths
Heat Wave	Eastern United States	1988	10,000
	Eastern United States	1901	9,500
	Central United States	1936	5,000
	Eastern United States	1980	1,700
	Eastern United States	1896	1,500
	Northeast United States	1972	900
	Eastern United States	1954	300
	United States	2006	220
	New York	1911	146
	Southern United States	2000	140

Source: National Oceanic and Atmospheric Administration (NOAA). 2024s. *Worst Heat Waves in the United States*. https://nhc.noaa.gov/heat-waves-in-history. (Accessed May 26, 2024).

Climate influences the types and behavior and distribution of risks and severity of impacts as a result of those risks throughout the world. Therefore, it is only fitting that we include an overview of climate of the United States. This will provide some context as to the unique attributes each area has when the connection between the geology of a given area is combined with climatic influences of a given location. All climates are the product of many geological factors, including latitude, elevation, topography, distance from the ocean, and location on a continent. No climate is uniform. Small variations, called **microclimates**, exist in every climate region. Microclimates are largely influenced by local and regional topographic features such as lakes, vegetation, and cities. In large urban areas, for example, streets and buildings absorb heat from the sun, raising the average temperature of the city higher than average temperatures of more open areas nearby that

TABLE 5.10
Worst Winter Storms in the United States

Environmental Hazard	Location	Year	Estimated Snowfall (cm)	Estimated Deaths
Winter Storms in the United States	Northeast United States	1888	127	400
	Georgia to Maine	1889	86	Unknown
	East Coast	1993	75	310
	Great Lakes	1913	70	250
	Northern Great Plains	1888	50	235
	Appalachian States	1950	50	353
	East Coast	1922	90	98
	Midwest	1940	50	145
	Midwest	1975	50	12
	Mid-Atlantic States	2010	82	10

Source: National Oceanic and Atmospheric Administration (NOAA). 2024t. *Worst Winter Storms*. www.noaa.gov/worst-winter-storms. (Accessed May 26, 2024).

are vegetated. Large bodies of water, such as the Great Lakes in the United States and Canada, can also have microclimates. Cities on the southern side of Lake Ontario, Lake Erie, Lake Superior, Lake Huron, and Lake Michigan are cloudier and receive much more snow than cities on the northern shore. This *Lake Effect* is a result of cold winds blowing across the warmer lake water (National Geographic Society 2024d).

Examining the climate map presented in Figure 5.21, one can immediately observe that the climate patterns of the eastern portion of the United States are stratified except for the Appalachian region. On the other hand, the western portion of the United States is mountainous, which significantly influences climate. This in turn influences most environmental risks in either lowering the risk or increasing the risk. For instance, elevation favors cooling weather and winter storms coming off the Pacific Ocean precipitate snow at increasing amount as elevation is gained through the orographic effect. More snow means more summer runoff and increases the opportunity for erosion to occur.

5.5 SUMMARY AND CONCLUSION

Natural geological and environmental hazards and associated impacts are of key concern to lives and livelihoods across the Earth. Understanding and monitoring geological and environmental processes and using this information to identify and quantify risk potential are critical to improve risk reduction methods so that sustainability potential for survival is greatly improved (BGS 2024).

In this chapter, we have learned that geological risks vary with respect to distribution across the globe as well as vary in intensity. This holds true for environmental risks as well. We have also learned of the connections between geology and environmental risks are intertwined and in some cases the lines between each are blurred. In the next chapter, we will add a layer of complexity and introduce how anthropogenic activities exacerbate many of these risks.

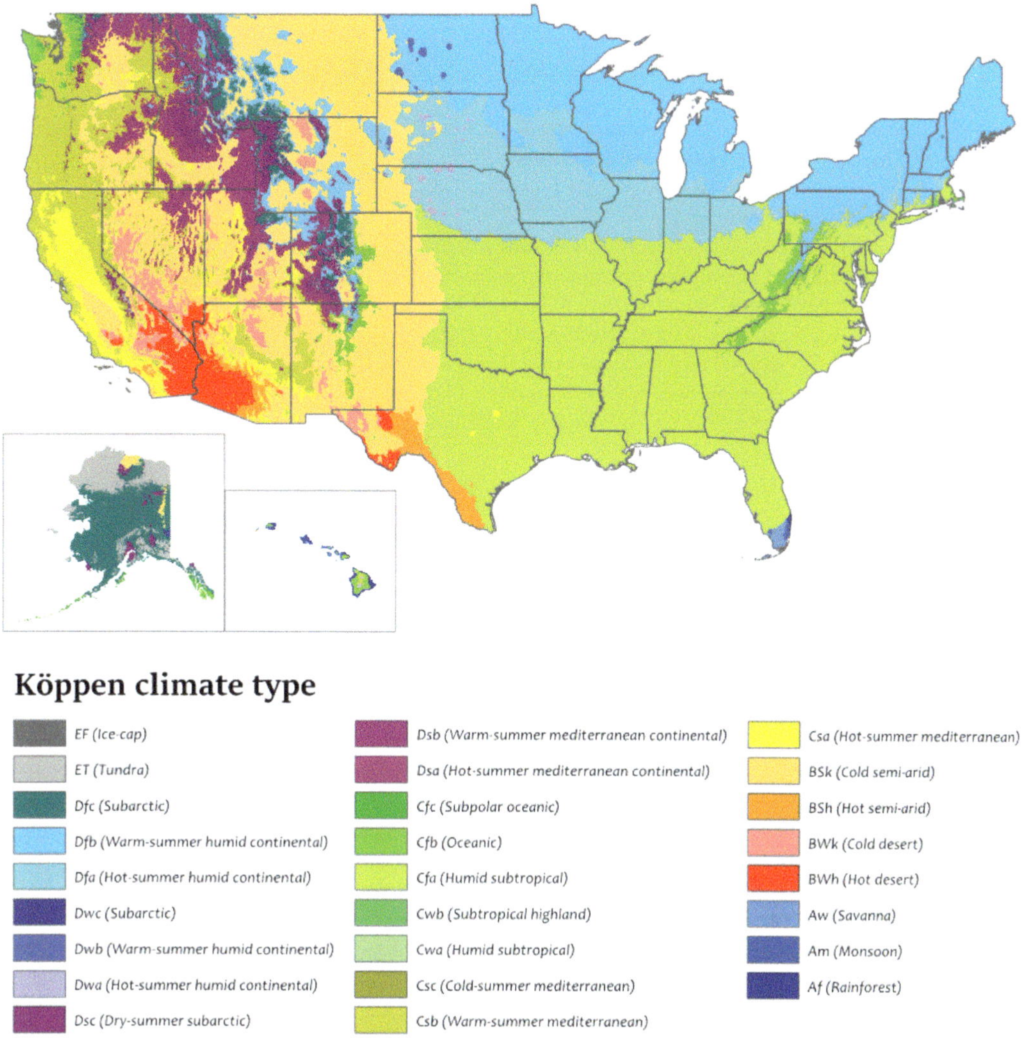

FIGURE 5.21 Climate map of the United States. (From National Oceanic and Atmospheric Administration (NOAA). 2024j. *Koppen-Geiger Climate Changes*. https://sos.noaa.gov/datasets/koppen-geiger-climate-changes-1901-2100/. (Accessed March 27, 2024).)

REFERENCES

American Geophysical Union. 2024. *Atmospheric Moisture Contents Explained*. www.agupubs.onlinelibrary.wiley.com. (Accessed March 14, 2024).

British Geological Survey (BGS). 2024. *Global Geological Risk*. www.bgs.ac.uk/geology-projects/global-geological-risk. (Accessed March 11, 2024).

Kemp, D. B., Sadler, P. M., and Vanacker, V. 2020. The Human Impact on North American Erosion, Sediment Transfer, and Storage in a Geologic Context. *Journal of Nature Communications*. Vol. 11, no. 6012, pp. 1–8.

National Atmospheric and Space Administration (NASA). 2024a. *Global Climate Change – Mitigation and Adaptation*. https://science.nasa.gov/climate-change/adaptation-mitigation. (Accessed March 14, 2024).

National Atmospheric and Space Administration (NASA). 2024b. *What Is Drought and What Are the Causes*. https://gpm.nasa.gov/resources/faq/what-drought-and-what-causes-it. (Accessed June 10, 2024).

National Atmospheric and Space Administration (NASA). 2024c. *Steamy Relationships: How Atmospheric Water Vapor Amplifies Earth's Greenhouse Effect.* www.science.nasa.gov/earth/climate-change/steamy-relationshipd-how-atmospheric-water-vapor-amplifies-earths-greenhouse-effect. (Accessed June 10, 2024).

National Atmospheric and Space Administration (NASA). 2025. *Extreme Weather and Climate Change.* https://science.nasa.gov/climate-change/exteme-weather. (Accessed February 2, 2025).

National Geographic. 2025. *The Many Effect of Flooding.* https://educationnationalgeographic.org/resource/many-effects-flooding/4th-grade

National Geographic Society. 2024a. *Extreme Environmental Events.* https://education.nationalgeographic.org/resource/influence-climate-change-extreme-enviromental-events. (Accessed March 13, 2024).

National Geographic Society. 2024b. *The Many Effects of Flooding.* https://education.nationalgeographic.org/resource/many-effects-flooding. (Accessed March 13, 2024).

National Geographic Society. 2024c. *Wildfires: How They Form, and Why They're SO Dangerous.* https://education.nationalgeographic.org/resource/wildfires. (Accessed March 13, 2024).

National Geographic Society. 2024d. *Lightning.* www.nationalgeographic.com/lightning. (Accessed June 10, 2024).

National Geographic Society. 2024e. *Hail.* www.education.nationalgeographicsociety.org/hail. (Accessed June 2, 2024).

National Geophysical Data Center. 2024. *Significant Earthquake Database.* https://ourworldindata.org/the-worlds-deadliest-earthquakes. (Accessed May 7, 2024)

National Oceanic and Atmospheric Administration (NOAA). 2024a. *Extreme Atmospheric Events.* www.noaa.gov/topics-tags/extremeevents. (Accessed March 14, 2024).

National Oceanic and Atmospheric Administration (NOAA). 2024b. *Hurricanes.* www.noaa.gov/education/reource-collections/weather-atmosphere/hurricanes. (Accessed March 14, 2024).

National Oceanic and Atmospheric Administration (NOAA). 2024c. *Hurricane Formation.* www.noaa.gov/education/reource-collections/weather-atmosphere/hurricane-formation. (Accessed March 14, 2024).

National Oceanic and Atmospheric Administration (NOAA). 2024d. *The Cost of Extreme Atmospheric Events.* www.ncei.noaa.gov/access/billions/events.pdf. (Accessed March 14, 2024).

National Oceanic and Atmospheric Administration (NOAA). 2024e. *Tornado Basics.* www.noaa.gov/education/resource-collections/weather-atmosphere/tornadoes. (Accessed March 14, 2024).

National Oceanic and Atmospheric Administration (NOAA). 2024f. *Tornado Damages and Fatalities.* www.spc.noaa.gov/climo/torn/fatalmap.php. (Accessed March 14, 2024).

National Oceanic and Atmospheric Administration (NOAA). 2024g. *Lightning.* www.nssl.noaa.gov/education/svrw101/lightning. (Accessed June 10, 2024).

National Oceanic and Atmospheric Administration (NOAA). 2024h. *Wind and Wind Hazards.* https://waether.gov/safety/wind-hazards. (Accessed June 10, 2024).

National Oceanic and Atmospheric Administration (NOAA). 2024i. *About Derechos.* www.spc.noaa.gov/abtderechos/derechofacts.htm. (Accessed March 15, 2024).

National Oceanic and Atmospheric Administration (NOAA). 2024j. *Regional Snowfall Index (RSI) and The Northeast Snowfall Impact Scale (NESIS).* www.ncei.noaa.gov/access/monitoring.rsi. (Accessed March 14, 2024).

National Oceanic and Atmospheric Administration (NOAA). 2024k. *Winter Weather. National Severe Storms Laboratory.* www.nssi.noaa.gov/education/svrwx/101/winter. (Accessed March 14, 2024).

National Oceanic and Atmospheric Administration (NOAA). 2024l. *Marine Heat Waves. Physical Sciences Laboratory.* https://psi.noaa.gov/marine-heatwaves. (Accessed June 10, 2024).

National Oceanic and Atmospheric Administration (NOAA). 2024m. *High-Tide Flooding in the United States.* https://tidesandcurrents.noaa.gov/high-tide-flooding. (Accessed June 10, 2024).

National Oceanic and Atmospheric Administration (NOAA). 2024n. *Hurricanes in History.* https://oceanservice.noaa.gov/news/historical-hussucances. (Accessed May 26, 2024).

National Oceanic and Atmospheric Administration (NOAA). 2024o. *National Hurricane Center. Hurricanes in the United States.* www.ncei.noaa.gov/pub/data/images/US_hrrican_strikes/continental_US_hurrican_strikes_1950-2022. (Accessed May 26, 2024).

National Oceanic and Atmospheric Administration (NOAA). 2024p. *Storm Prediction Center. Worst Tornadoes.* www.spc.noaa.gov/faq/tornado/killers'html. (Accessed May 26, 2024).

National Oceanic and Atmospheric Administration (NOAA). 2024q. *Floods Worldwide*. www.nssl.noaa.gov/research.flood. (Accessed May 26, 2024).

National Oceanic and Atmospheric Administration (NOAA). 2024r. *Floods in the United States*. www.ncdc.noaa.gov/stormevents. (Accessed May 26, 2024).

National Oceanic and Atmospheric Administration (NOAA) 2024s. *Worst Heat Waves in the United States*. https://cpo.noaa.gov/us-heat-waves-in-new-interagency-indicator. (Accessed May 26, 2024).

National Oceanic and Atmospheric Administration (NOAA). 2024t. *Worst Winter Storms*. www.nssl.noaa.gov/education/svrwx101/winter/types. (Accessed May 26, 2024).

National Oceanic and Atmospheric Administration (NOAA). 2025. *Plate Tectonics and Lava Lamps*. www.noaa.gov/pffice-education/outreach-communication/hands-on-science-activities/plate-tectonics-and-lava-lamps. (Accessed February 2, 2025).

National Weather Service (NWS). 2024. *Radar Image July 15, 2024, Chicago, Illinois*. www.nws.gov/radar/july15/chicago/illinois. (Accessed July 16, 2024).

Owens, P. N. 2020. Soil Erosion and Sediment Dynamis in the Anthropocene; *A review of Human Impacts during a Period of Rapid Global Environmental Change. Journal of Soils and Sediments*. Vol. 20, pp. 4115–4143.

Skliris, N., Zika, J., Nurer, G., Josey, S. A., and Marsh, R. 2016. *Global Water Cycle Amplifying as Less Than Clausius-Claperon Rate. Science Reports*. www.nature.com/articles/srep38752. (Accessed June 14, 2024).

Thaler, E. A., Kwang, J. S., Quirk, B. J., Quarrier, C. L., and Larsen, I. J. 2022. Rates of Historical Anthropogenic Soil Erosion in the Midwestern United States. *Journal of the American Geophysical Union*. Vol. 10, no. 3, pp. 220–242.

United Nations Intergovernmental Panel on Climate Change (IPCC). 2021. *Sixth Assessment Report*. www.ipcc.ch/assessment-report/ar6. (Accessed March 14, 2024).

United States Environmental Protection Agency (USEPA). 2024a. *Radiation from Solar Activity*. www.epa.gov/radtown/radiation-solar-activity. (Accessed June 10, 2024).

United States Environmental Protection Agency (USEPA). 2024b. *Health and Environmental Effects of Ozone Layer Depletion*. www.epa.gov/ozone-pollution-and-your-health/health-effects-ozone-general-population. (Accessed June 10, 2024).

United States Environmental Protection Agency (USEPA). 2024c. *Heat Waves as Climate Indicators*. ww.epa.gov/climate-indicators/climate-change-indicators-heat-waves. (Accessed June 10, 2024).

United States Environmental Protection Agency (USEPA). 2024d. *Extreme Precipitation Events as Climate Change Indicators*. www.epa.gov/climate-indicators/climate-change-indicators-heavy-precipitation. (Accessed June 10, 2024).

United States Environmental Protection Agency (USEPA). 2025a. *Health an Environmental Effects of Particulate Matter (PM)*. www.epa.gov/pm-pollution/ealth-and-environmnetal-effects-particulate-matter-pm. (Accessed February 2, 2025).

United States Environmental Protection Agency (USEPA). 2025b. *Health Effects of UV Radiation*. www.epa.gov/sunsafety/health-effects-UV-radiation. (Accessed February 2, 2025).

United States Federal Emergency Management Agency (FEMA). 2024. *Loss of Life from Natural Disasters*. https://fema.gov/protectiveactionsarticle/hurricanes. (Accessed March 14, 2024).

United States Geological Survey (USGS). 2023. *Sources, Compositions, and Effects of Atmospheric Dust*. www.usgs.gov/centers/geosciences-and-environmental-change-science-center/science/sources/compositions/and-effects-of-atmospheric-dus. (Accessed December 28, 2023).

United States Geological Survey (USGS). 2024a. *The Effects of Earthquakes*. www.usgs.gov/programs/earthquake-hazards/science-earthquakes. (Accessed March 11, 2024).

United States Geological Survey (USGS). 2024b. *The Science of Earthquakes*. www.usgs.gov/programs/earthquake-hazards/science-earthquakes. (Accessed March 12, 2024).

United States Geological Survey (USGS). 2024c. *Earthquake Magnitude, Energy Release, and Shaking Intensity*. www.uss.gov/programs/earthquake-hazards/earthquake-magnetidu-energy-release-and-shaking-intensity. (Accessed March 12, 2024).

United States Geological Survey (USGS). 2024d. *The Modified Mercalli Intensity Scale*. www.usgs.gov/programs/earthquake-hazards/modified-mercalli-intensity-scale. (Accessed March 4, 2024).

United States Geological Survey (USGS). 2024e. *Earthquakes in the United States*. www.usgs.gov/programs/earthquake-hazards/earthquakes. (Accessed March 12, 2024).

United States Geological Survey (USGS). 2024f. *Earthquake Hazard Map*. www.usgs.gov/programs/earthquake-hazards/maps. (Accessed March 12, 2024).

United States Geological Survey (USGS). 2024g. *Effects of Earthquakes*. www.usgs.gov/programs/earthquake-hazards/what-are-the-effects-earthquakes. (Accessed March 12, 2024).

United States Geological Survey (USGS). 2024h. *Volcanoes: Just the Facts*. www.usgs.gov/programs/VHP/about-volcanoes. (Accessed March 12, 2024).

United States Geological Survey (USGS). 2024i. *What Are Volcano Hazards?* www.usgs.gov/volcano-hazards. (Accessed March 12, 2024).

United States Geological Survey (USGS). 2024j. *The Landslide Handbook – A Guide to Understanding Landslides*. https://pubs.usgs.gov/circ/1325. (Accessed March 13, 2024).

United States Geological Survey (USGS). 2024k. *Coastal Hazards*. www.usgs.gov/tools/coastal-change-hazards-portal. (Accessed March 13, 2024).

United States Geological Survey (USGS). 2024l. *Land Subsidence*. www.usgs.gov/mission-areas/water-resources/acience/land-subsidence. (Accessed March 13, 2024).

United States Geological Survey (USGS). 2024m. *Floods and Erosion*. www.usgs.gov/centers/sawsc/science/science-topics/coastal-flooding-and-erosion. (Accessed March 12, 2024).

United States Geological Survey (USGS). 2024n. *Deadliest Volcanic Eruptions*. www.usgs.gov/faqs/which-volanic-eruptions-were-deadliest. (Accessed May 7, 2024).

United States Geological Survey. 2024o. *Deadliest Tsunamis*. www.usgs.gov/special-topics/water-science=school/science/tsunamis-and-tsunami-hazards. (Accessed May 7, 2024).

United States Geological Survey (USGS). 2024p. *Deadliest Landslides*. www.usgs.gov/programs/landslide-hazards/science/us-and-worldwide-landslide-data. (Accessed May 7, 2024).

United States National Forest Service. 2024. *Wildfire Information*. www.fs.usda/science-technology/fire/information. (Accessed March 13, 2024).

United States National Weather Service. 2024a. *Enhanced Fujita Scale*. www.weather.gov/tae/ef_scale. (Accessed March 14, 2024).

United States National Weather Service. 2024b. *Winter Storms and Blizzards*. www.weather.gov/winterstormsandblizzards. (Accessed March 14, 2024).

United States National Weather Service. 2024c. *Cold Weather Safety*. www.weather.gov/safety/cold. (Accessed March 15, 2024).

Western Fire Chiefs Association. 2024. *Largest Wildfires in United Sates History*. www.wfca.com/wildfire-articleas/biggest-wildfires-in-us-history. (Accessed May 26, 2024).

World Health Organization (WHO). 2024. *Environmental Risks to Human Health*. www.who.int/activities/environmnetal-health-impacts. (Accessed May 26, 2024).

Part II

Anthropogenic-Induced Environmental Hazards and Risks

6 Anthropogenic-Induced Hazards

6.1 INTRODUCTION

Many believe that we do not purposely treat our planet with malice and that much of the damage inflicted has been out of lack of knowledge or ignorance and that our labels for justifying our destructive behavior have an origin to avoid guilt and responsibility. Another common misconception is that humans did not believe that they could negatively impact the Earth on a global scale. There was also a belief that the Earth would filter pollution and break it down into harmless substances. Each of these historical beliefs is incorrect. If we do not learn from our past and continue our attempts to conquer and change nature, we will face the harsh reality that our actions will continue to have a significant negative impact on Earth's atmosphere and climate, water, and land, and threaten the survival of many species including our own (Table 6.1).

To manage the discussion in this chapter in an organized manner, we will begin by looking at the history of anthropogenic risks, then move on to assessing anthropogenic risks to air, water, and land, and conclude with a discussion of living organisms. But first, let us revisit the types of hazards including anthropogenic-induced hazards. Many will be discussed in this chapter and the remaining will be addressed in the next chapter where we discuss the exploitation of natural resources.

6.2 HISTORICAL PERSPECTIVE

For the last 11,000 years, humans have modified the Earth through agriculture, building cities, exploiting resources, and carrying out other related activities. Now with more than 8 billion people on the planet, we are experiencing the effects of our own actions by running out of habitable land, polluting the entire planet, and ultimately being responsible for the sixth major extinction in Earth's 4.54-billion-year history. Modifications to the environment have been conducted to suit our needs and to make our lives safer and more comfortable. To avoid feeling guilty and taking responsibility for our own actions, we have fooled ourselves and tried to justify the destruction of the natural environment by creating clever labels such as taming nature, Western expansion, progress, sustainable development, and falsely claiming that Earth was created just for us.

Although there is much we can be proud of for limiting pollutant effects through environmental regulation in the last 50 years, it clearly has not been enough and our species is now at a critical crossroad. What we choose to do in the next few years to address not just climate change, but other negative impacts to water, land, and other living species, which will have a profound impact on how we and other species survive in the future. One of the most significant aspects of our technological advances in the last few centuries, which has introduced a multitude of pollutants and has adversely impacted the entire planet, is that they have greatly outpaced nature's ability to evolve and adapt to human's destructive behavior.

DOI: 10.1201/9781003463153-8

TABLE 6.1
Types of Hazards Discussed

Geologic Hazards	Environmental Hazards	Anthropogenic-Induced Hazards
Earthquake	Hurricanes	Habitat destruction
Tsunami	Tornados	Natural resource exploitation
Volcanic eruption	Severe storms	Water cycle disruption
Landslide	Extreme precipitation	Deforestation
Coastal hazard	Flooding	Overfishing
Subsidence and sink holes	Drought	Overhunting and poaching
Erosion	Extreme heat	Invasive species
Aeolian processes (dust)	Extreme cold	Species extinction
	Wildfires	Light pollution
	Lightning	Noise pollution
	Solar radiation	Ocean pollution
	Hail	Air pollution
	High winds	Water pollution
	Severe winter weather	Land pollution
	Marine heat wave	Chemical contaminants
	High-tide flooding	Agriculture
	Derecho's	Land development
		Endangered species
		Pathogens
		Erosion
		Particulate matter or dust
		Nuclear radiation
		Certain buildings
		Certain wind turbines
		Climate change

Concern for the environment can be traced to the Roman Empire when the Roman Senate passed legislation to protect Rome's supply of water which addressed concern over freshwater and sewage disposal (Bannon 2024).

In the 14th century, England prohibited the burning of coal in London and the disposal of waste into waterways, specifically the Thames River (Smithsonian 2024a). In December 1682, in what would become the United States, William Penn ordered that one acre of forest be preserved for every five acres cleared for settlement (Pennsylvania Historical Association 2024). In addition, Benjamin Franklin widely wrote about the adverse effects of lead and dumping of waste (Environmental Education Associates 2021). However, it was not until 1899 that the United States government passed what is commonly referred to as the first environmental law in the United States, the Rivers and Harbors Act which was intended to clear the harbors of garbage so that ships would not be impeded in transporting goods (United States Environmental Protection Agency (USEPA) 2023a, 2023b).

It was not until the Industrial Revolution starting in the 1750s that environmental degradation of the air, water, and land became a significant issue. The Industrial Revolution marked a time of rapid transition from making products by hand to making products with machines. The stationary steam engine invented by James Watt led to the widespread use of steam power that used coal. New chemical manufacturing methods, iron production processes, the development of machine tools, and the rise of the factory system all came about in a relatively short time, which sparked the rise of the industrial age and resulted in the discharge of large amounts of waste emitted into the air,

FIGURE 6.1 Air emissions from industrial sources in Pennsylvania in the 1880s. (United States Research Archives. 2024. www.archives.gov. (Accessed March 22, 2024).)

into waterways, and discarded onto the land. Figure 6.1 shows air emissions in the late 1880s in Pennsylvania (United States Research Archives, 2024).

In the 19th century, the environmentalist, naturalist, and author, John Muir, began to bring attention to the mistreatment of the environment by humans. Muir was very troubled by the blatant disregard for and exploitation of the environment and its resources across the United States. John Muir was a preservationist and dedicated his life to a tenet that the environment should be maintained in its pristine form and its resources should not be harvested for human use. Later in his life, he became friends with then president Theodore Roosevelt. Roosevelt and Muir hiked and camped in what is now Yosemite National Park. This friendship brought more appreciation and urgency to political leaders and was a significant factor in environmental legislation and creating the National Parks system in the United States.

As recently as the first half of the 20th century, environmental awareness and legislation were generally lacking not only in the United States but also worldwide. There were a few international agreements that primarily focused on boundary waters, navigation, and fishing rights along shared waterways between countries. However, they ignored pollution and ecological issues. In fact, the most convenient and least costly method of waste disposal at the time was "up the stack or down the river" (Haynes 1954). One of the first international agreements, the Convention for the Protection of Valuable Birds Useful to Agriculture was signed in 1902, was between 12 European countries for the protection of valuable birds. It was not until the publishing of the Rachel Carson book, "Silent Spring" in 1962 which described the environmental effects of Dichlorodiphenyltrichloroethane (DDT) on birds that catapulted the environmental movement in the United States into the public eye.

FIGURE 6.2 Pollution: it is a crying shame. (From United States Advisory Council. 2021. (www.adcouncil.org/Our-Campaigns/The-Classics/pollution-keep-america-beautiful-iron-eyes-cody. (Accessed March 31, 2021).)

The 1960s and 1970s are generally described as the age of environmentalism in the United States. **Environmentalism** is defined as a value system that seeks to redefine humankind's relationship with nature (Tarlock 2001). The dawn of the environmental movement was fueled by the spirit of the Civil Rights Movement that occurred at the same time during our history. Leaders of the civil rights and environmental movement challenged political leaders over the dumping of chemicals without any care for the environment. This is when the United States realized that environmental degradation had significantly affected the air and water quality of many urban areas, especially the air in Los Angeles and the water in Lake Erie, which is one of the Great Lakes. Starting on Earth Day in 1971, the Ad Council and Keep America Beautiful campaign aired public awareness commercials on television depicting a Native American shedding a tear when looking out over the polluted landscape of the United States. The commercial stated "people start pollution, people can stop pollution" (see Figure 6.2). A cartoon that was created by Walt Kelly also during the environmental movement that perhaps stated anthropogenic-induced issues best, "We have met the enemy, and he is us" (Ohio State University 2024).

Two significant sites of environmental contamination were discovered about this time: the Times Beach in Missouri and the Love Canal near Buffalo New York. These two sites would become a catalyst for the enactment of the Comprehensive Environmental Response and Liability Act (CERCLA), commonly known as Superfund, and the Resource Conservation and Recovery Act (RCRA), which addressed the disposal of solid and hazardous waste. Times Beach and Love Canal both became the first Superfund Sites in the United States. Currently, there are over 1,500 Superfund Sites (USEPA 2023b). In many instances in the United States, there is a single incident that shocks us into action. There have been numerous environmental incidents that have caused enormous damage to the environment and loss of life. Some include the following (USEPA 2023b):

- Love Canal soil and groundwater contamination in 1978
- Valley of drums soil, groundwater, and surface water contamination in 1978
- Woburn Massachusetts
- Times Beach dioxin contamination in 1983
- Bhopal, India, toxic gas release in 1984

- Garbage barge of 1988
- Exxon Valdes oil spill in 1989
- Deepwater Horizon explosion and oil spill in 2010

6.3 ANTHROPOGENIC IMPACTS TO AIR

For thousands of years, human civilization has been busy changing the environment for its own benefit and with little or no regard for the natural environment. This disregard for the natural environment even extended to the United States where numerous examples of pollution were a common and even accepted practice. Now that climate change has caught our attention, we are noticing and experiencing several of the detrimental effects that include higher temperature, severe drought, flooding, and stronger but slower-moving storm systems. This has raised our collective awareness of the fact that we can impact our environment on a global scale, and it does not include just the atmosphere.

Risks to air from anthropogenic sources are significant, and present themselves as an array of different contaminants released in high volumes annually (USEPA 1998). Most contaminants are released into the atmosphere from anthropogenic sources at or near the surface, with most of the impacts occurring in the troposphere, and to a lesser degree the stratosphere. Contaminants released near the ground can mix throughout the troposphere in a few weeks, but it can take years or decades for them to reach the stratosphere (Hemond and Fechner-Levy 2000). Records of atmospheric contaminants and their effects can be traced to the 13th century when King Edward I banned the burning of kiln coal in London due to its impacts on air quality (Wilson and Sprengler 1996).

Climate change is perhaps the best current example that humans have not treated Earth in a sustainable way. However, as we will discover, climate change is just one example of how humans have impacted air. Climate change can act as the global wake-up call that drives action toward a more sustainable planet. Climate change has motivated more than 195 countries, businesses, and other organizations to come together and work toward limiting the effects of climate change.

From a historical perspective, the United States is the world's largest emitter of greenhouse gases (GHGs) (USEPA 2024a). Pressure is mounting for the world to take significant action to reduce GHG emissions so that the consequences can be limited. If not, the result could be cataclysmic.

Over the last few decades, the United States may not have had difficulty in accepting the science of climate change but did object to early efforts to curb GHG emissions. This is evidenced by the fact that the United States did not join the Kyoto Protocol more than 20 years ago. The United States did sign on to the Paris Agreement in 2015 under President Obama only to be taken back under the Trump administration and then returned under the Biden administration in January 2021. This will likely be significant because the most recent report on climate change from the Intergovernmental Panel on Climate Change (IPCC 2021), states:

> It is unequivocal that human influence has warmed the atmosphere, ocean and land. Widespread and rapid changes in the atmosphere, ocean, cryosphere and biosphere have occurred.

The above statement has significant implications because it concludes that GHG emissions from humans have negatively impacted life on Earth. We will dedicate a significant amount of time to discussing the science of climate change as it relates to GHGs. The reason why we will dedicate this time is because there is a significant amount of misinformation concerning GHGs and climate change in the United States. Regardless of the origin or how widespread the misinformation is, the importance of explaining the science, reviewing the facts, and learning what all of us need to do are important for all of us. However, this is only the beginning of the story. There are a multitude of other risks to air from anthropogenic sources and that is even not to mention the risks to water, land, and living organisms.

6.3.1 Overview of Anthropogenic Impacts to Air

According to the World Health Organization (WHO), 9 out of 10 humans breathe air containing high levels of pollutants and nearly every human is exposed to at least some level of pollutants from direct or indirect anthropogenic sources with every breath (WHO 2024a). Air pollution causes strokes, heart disease, lung cancer, chronic obstructive pulmonary diseases, and respiratory infections. Both **ambient** (outdoor air) and indoor air containing pollutants are responsible for approximately 7 million premature deaths worldwide as of 2018 (WHO 2024a). A sad fact is that with all the efforts of different organizations and governments of the world, air quality continues to decline and now significantly affects everyone in every country (WHO 2024a; IPCC 2021).

Before we begin our journey through anthropogenic risks to air, we must first consider the meaning of the term **environment**. According to the United Nations (2024a), the environment is a broad term that encompasses all living and nonliving things on Earth or some regions thereof. This definition identifies the natural world existing within the atmosphere, hydrosphere, lithosphere, and biosphere, and also includes the built or developed world (defined as the **anthrosphere**). There are a multitude of risks to air from anthropogenic sources (see Figure 6.3).

We shall introduce many in this chapter and will further discuss other contaminants and pollution in Chapter 8. For this chapter, we will discuss the following:

- GHGs
- Upper atmospheric ozone
- Volatile organic compounds (VOCs) and smog-forming chemicals
- Criteria pollutants that include carbon monoxide, lead, ground-level ozone, particulate matter, nitrogen dioxide, and sulfur dioxide

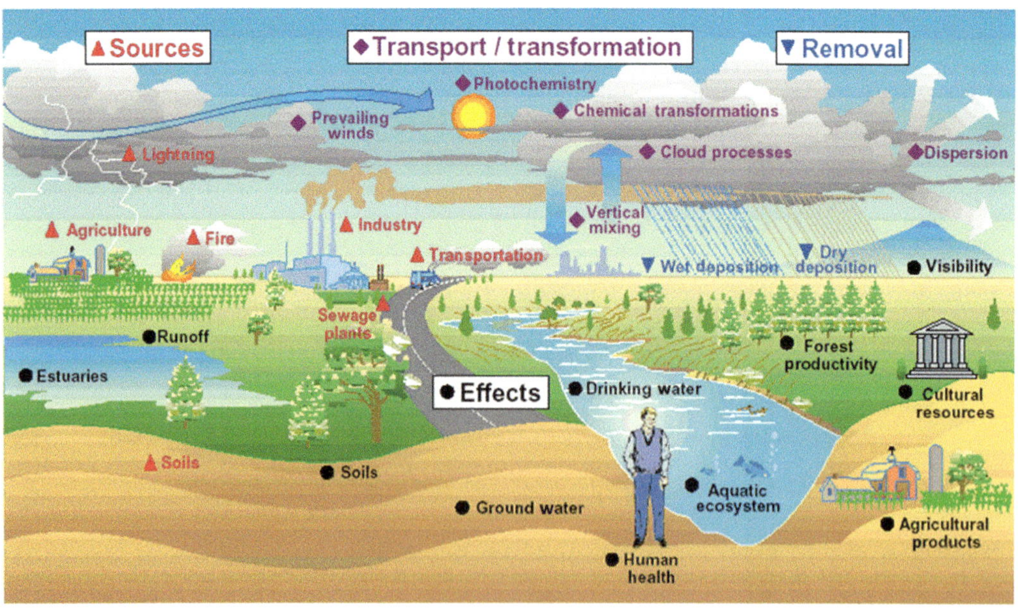

FIGURE 6.3 Sources, transport methods, and removal of air pollutants from anthropogenic sources. (From United States Environmental Protection Agency (USEPA). 1998. *Taking Toxics Out of the Air.* EPA451/K-98-001. USEPA. Washington, DC.)

The list of risks from other anthropogenic sources includes thousands of environmental pollutants and other substances or organisms. Therefore, to keep our discussion manageable, we will only discuss a few of the most significant. However, just to give you an idea of the plethora of anthropogenic risks, the following is a short list (Rogers 2020):

- VOCs
  - Dense non-aqueous phase liquids (chlorinated VOCs)
  - Light non-aqueous phase liquids (non-chlorinated VOCs)
- Polynuclear aromatic hydrocarbons (PNAs or PAHs)
- Semi-volatile organic compounds (SVOCs)
- Polychlorinated biphenyls (PCBs)
- Pesticides and herbicides
- Heavy metals (including but not limited to arsenic, barium, cadmium, copper, chromium, mercury, lead, silver, selenium, zinc, and others)
- Common fertilizers including nitrates, phosphorus, and potassium
- GHGs
- Chlorofluorocarbons (CFCs)
- Carbon dioxide
- Carbon monoxide
- Particulate matter (dust)
- Ozone
- Bacteria such as *Escherichia coli*
- Viruses
- Invasive species (the number of invasive species is over 10,000 just in the United States)
- Pharmaceuticals
  - Antibiotics
  - Steroids
  - Antibacterial chemicals
  - Hormones
  - Narcotics
  - Many other legal and illegal drugs
  - Sildenafil citrate, commonly known as Viagra
  - Benzoylmethylecgonine, commonly known as cocaine
- Cyanide
- Asbestos
- Acids
- Bases
- Radioactive compounds
- Dioxins
- Polyfluoroalkyl substances (PFAS)
- Emerging contaminants
  - Insect repellants
  - Solvents
  - Detergents
  - Plasticizers (including micro-plastics)
  - Fire retardants
  - Veterinary antibiotics
  - Lesser known pesticides
  - 1,4-Dioxane

- 1,2,3-Trichloropropane
- Perchlorates
- Others

The effects of anthropogenic impacts on air do not just end with the premature death of humans but also influence species' decline and extinction of all animal and plant life on Earth (WHO 2024a; United Nations 2024a).

6.3.2 Historical Anthropogenic Effects to Air in the United States

What moved the United States toward enacting environmental legislation to protect our air? Was there an incident that influenced the development of the Clean Air Act (CAA)? The answer is yes but it took time. The incident occurred in the fall of 1948 in Donora Pennsylvania. A yellow fog appeared on October 26 and lasted until October 31 when a rain storm dissipated the yellow fog (Smithsonian 2024b). The fog grew so bad that 20 residents died of asphyxiation and over 7,000 residents were hospitalized as a result of the severe air pollution (USEPA 2023c). Figure 6.4 is a photograph taken around noon on October 29, 1948, of the smog that became so bad that it looked like the photo was taken at night. The incident drew national attention when Walter Winchell broadcast news of the disaster on his national radio show. During the incident, the Donora Zinc works shut down its smelters to eliminate as much emissions as possible. Rain began to fall at this point as well which dissipated the smog and the Zinc works returned to its normal operations the next day (USEPA 2023c).

The incident was investigated by several governmental agencies at the time and the conclusion was that the main air pollutant that caused the health concerns was sulfur dioxide and concentrations were estimated to be at a concentration that is more than 125 times that recommended by the WHO (2024. However, this incident did not have the impact that later incidents had on developing new environmental regulations.

The Donora incident provided the scientific community with data that related to the following:

- Toxicity of certain pollutants
- Human population sensitivity variability

FIGURE 6.4 Donora Pennsylvania air pollution incident. (From United States Environmental Protection Agency (USEPA). 2023c. *Donora Smog of 1948*. EPA Alumni Association: News and Views. EPA Alumni Association: News and View. www.epaalimni.org/examples/donora/index.html (Accessed December 26, 2023).)

- Exposure data
- Expanded the knowledge and methods of toxicology

The lessons learned from this incident, in part, showed that air pollution can inflict enormous harm and even death to humans. It also helped identify the types of pollution most dangerous that would later be part of the CAA Criteria Pollutants, identified a need for environmental regulations, and demonstrated that geography and climate can be very important in influencing the behavior of contaminants and human health (USEPA 2023c).

6.3.3 Greenhouse Gases

GHGs include carbon dioxide (CO_2), methane (CH_4), nitrous oxide (N_2O), and fluorinated gases (F-gases). GHGs cause the Earth to warm up and since the start of the Industrial Revolution in the early 18th century, levels of GHGs in the Earth's atmosphere have increased (USEPA 2024a). Figure 6.5 shows this increase in GHGs from the National Oceanic and Atmospheric Administration (NOAA 2024a) global air sampling network since 1975. The combined heating influence of each GHG since 1980 is presented in Figure 6.6. It should be noted that there has been a 47% increase in heating influence from emissions since 1990.

Figure 6.7 shows the contribution of GHGs by economic sector (USEPA 2024b). Figure 6.8 shows current atmospheric CO_2 concentrations as measured on Mauno Loa in Hawaii (NOAA 2024b). As Figure 6.9 shows, the concentration of CO_2 in the atmosphere is currently at 426 parts per million. The main source of GHGs worldwide is the burning of fossil fuels for generation of

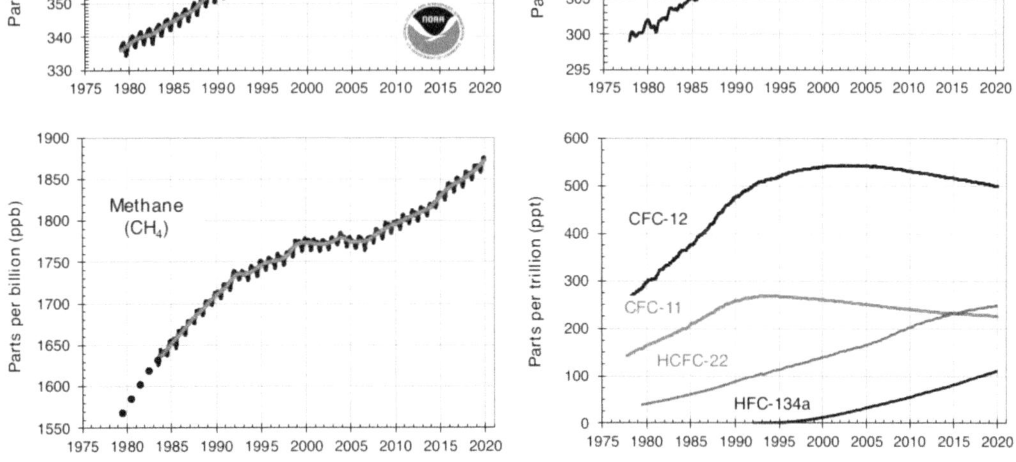

FIGURE 6.5 Global average of GHGs in Earth's lower atmosphere. (National Oceanic and Atmospheric Administration (NOAA). 2024a. *Global Monitoring Laboratory – Carbon Cycle Greenhouse Gases*. https://gml.noaa.gov/ccgg/data. (Accessed March 16, 2024).)

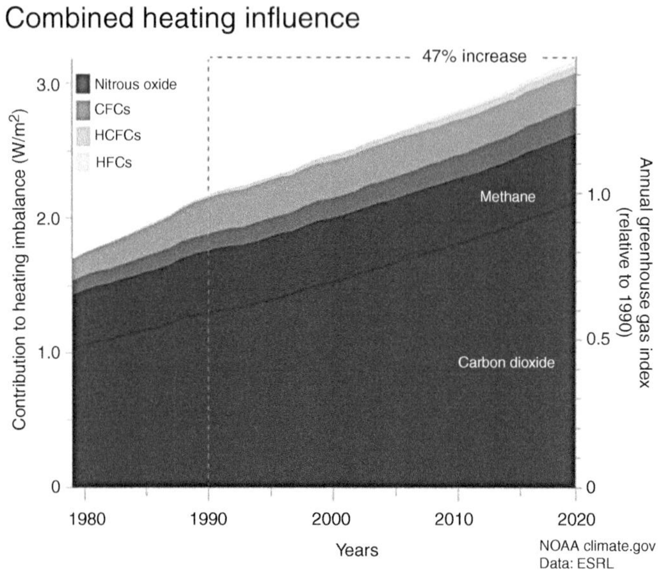

FIGURE 6.6 Combined heating influence of GHGs since 1980. (United States Environmental Protection Agency (USEPA). 2024a. *Global Greenhouse Gas Emissions Data*. www.epa.gov/ghgemissions/global-greenhouse-gas-overview. (Accessed March 15, 2024).)

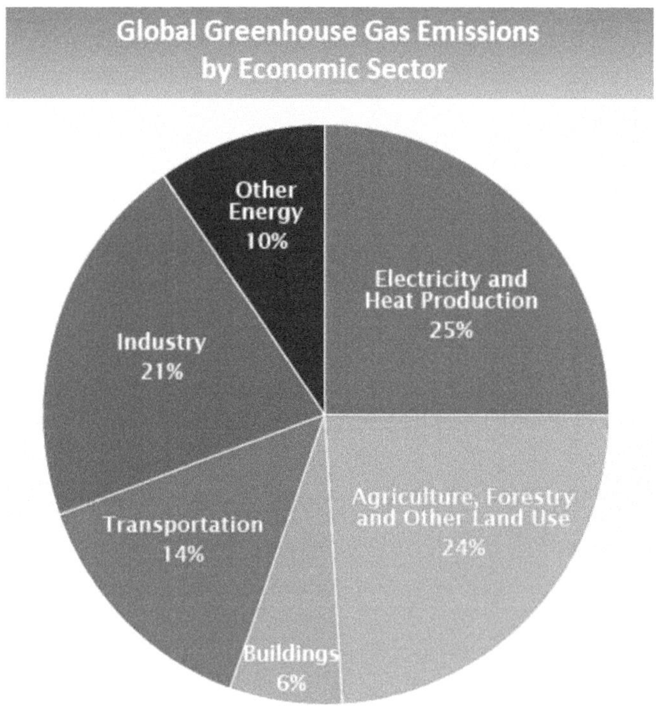

FIGURE 6.7 Source of GHG emissions in the United States by economic sector. (United States Environmental Protection Agency (USEPA). 2024b. *Sources of Greenhouse Gas Emissions*. www.epa.gov/ghgemissions/global-greenhouse-gas-overview. (Accessed March 16, 2024).)

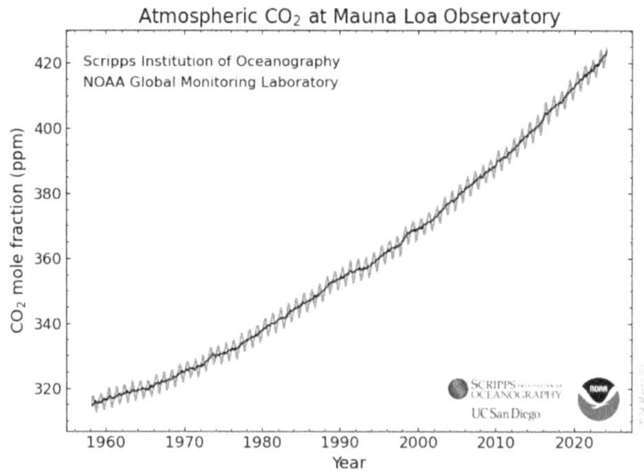

FIGURE 6.8 Atmospheric CO_2 from 1980 to 2024 at Mauna Loa, Hawaii. (National Oceanic and Atmospheric Administration (NOAA). 2024b. *Monthly Average Mauna Loa CO_2*. https://gml.noaa.gov/ccgg/trends. (Accessed March 16, 2024).)

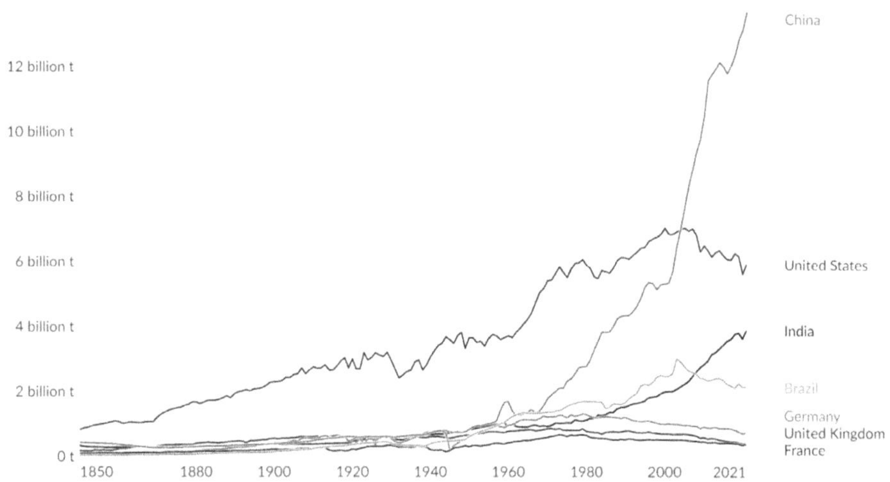

FIGURE 6.9 GHG emission by country. (United Stated Environmental Protection Agency (USEPA). 2024c. *Global Greenhouse Gas Emissions Data*. Global Greenhouse Gas Emissions Data | Greenhouse Gas (GHG) Emissions | US EPA. www.epa.gov/ghgemissions/global-greenhouse-gas-overview (Accessed March 16, 2024).)

electricity, heat, agriculture, and transportation (USEPA 2024a). In the United States, the significant sources of GHG emissions include the following:

- Transportation
- Electricity generation plants
- Industry
- Commercial and residential sources
- Agriculture

Globally, GHGs began to emit in earnest from anthropogenic sources at the beginning of the Industrial Revolution in the early 18th century when coal (a fossil fuel) became the fuel of choice.

Since around 1900, gasoline (a fossil fuel) was added and quickly became the preferred fuel, especially with the invention of the internal combustion engine and the automobile. The automobile then became affordable and with the construction of the interstate transportation system in the 1950s, GHGs from mobile sources have significantly increased. Globally since 1970, CO_2 emissions have increased by approximately 90%, with emissions from fossil fuel combustion and industrial processes contributing approximately 78% of the total GHG emissions' increase from 1970 to 2011. Agriculture, deforestation, and other land-use changes have been the second largest contributors (USEPA 2024b).

The United States is no longer the largest emitter of GHGs. China holds this dubious distinction. As of 2014, the United States was the second largest emitter of GHG with the European Union in third place. Emissions and sinks related to changes in land use are not included in these estimates. However, changes in land use can be important. Current estimates indicate that the net global GHG emissions from agriculture, forestry, and other land use exceed 8 billion metric tons of CO_2 equivalent, or about 24% of total GHG emissions. Specifically, in the United States and in Europe, changes in land use associated with human activities have the net effect of absorbing CO_2, which partially offsets the emission from deforestation and rapid urbanization in other parts of the world (USEPA 2024b). In addition, it should be noted that the United States has decreased its total GHG emissions by more than a billion tons from its peak in 2024. This is likely attributed to an increase in renewable energy sources and a decrease in using coal to generate electricity (USEPA 2024c). Figure 6.9 shows the total GHG emissions for select countries since 1850. Figure 6.10 shows the cumulative GHG emissions for select countries since 1750.

In the next section, we will examine the paleoclimate of Earth. This will provide context and perspective on the meaning and potential consequences of increasing concentrations of GHGs in the atmosphere.

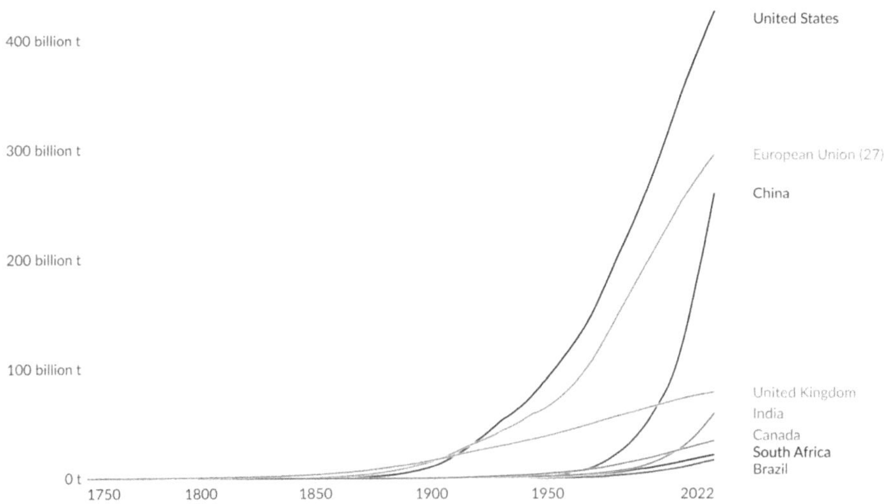

FIGURE 6.10 Cumulative GHG emissions by country. (*Source:* United Stated Environmental Protection Agency (USEPA). 2024c. *Global Greenhouse Gas Emissions Data.* Global Greenhouse Gas Emissions Data | Greenhouse Gas (GHG) Emissions | US EPA. www.epa.gov/air-emissions-inventories/national-emissions-inventory-nei (Accessed March 16, 2024).)

6.3.3.1 Defining Climate

According to NASA (2024), the differences between weather and climate are distinct and a matter of perspective. **Weather** relates to atmospheric conditions over a short period of time. **Climate** is the behavior of the atmosphere over a long period of time.

There are many components to weather, some of which include sunshine, rain, cloud cover, hail, snow, sleet, freezing rain flooding, blizzards, ice storms, thunderstorms, steady rains, excessive heat, heat waves, and excessive cold. Many consider climate as the average weather for a particular region and time period, usually over a 30-year observational interval (NASA 2024). For example, after recording rain data, lake and reservoir elevations, and satellite data, scientists can evaluate whether a particular season, such as a summer in the northern hemisphere, was drier than average. If the drier-than-average conditions persist over many summers, it would likely indicate that a change in regional climate has occurred.

Since the 1970s and 1980s, scientists noticed that our global climate could change or was changing due to increased concentrations of GHGs, which were gradually building up in the atmosphere. It was near this point in time that a growing number of scientists realized that the data supported the conclusion that the atmosphere was warming and it was because of us.

6.3.3.2 Climate Science

Paleoclimate is the study of the ancient climate of Earth (USGS 2024a). The study of paleoclimate is important because if we understand what Earth's climate was like and why, in the geologic past, then we can use that information to assist our understanding of what Earth's climate will be in future.

6.3.3.2.1 Paleoclimate Archives and Proxies

As modern records do not outline most of Earth's climatic past, scientists must gather data preserved in nature from a variety of sources referred to as proxy records. Some of these proxy records are from sources that include the following (USGS 2024a):

- Dendrochronology, the study of tree rings, is a source of information that can provide data on temperature and precipitation.
- Lake varves, which are the layers of sediment at the bottom of a lake
- Fossils including diatoms and foraminifera have provided data on what the water chemistry and temperature were like in the geologic past.
- Ice cores provide valuable information on atmospheric chemistry and climatic conditions in the geologic past from a whole host of sources including pollen and analysis of trapped air bubbles for oxygen isotope analysis. The trapped air has proven tremendously valuable as a source of direct measurement of the composition of the atmosphere at the time the ice was formed.
- Ocean floor sediment cores.
- Scierochronology, which examines corals and coral reefs.
- Speleothems, which primarily examines stalagmites deposits in caves.
- Studying the chemistry and age of rocks, stratigraphy, and petrology (which is the study of the origin, small-scale structure, and composition of rocks) provides information on continental drift, solar energy, GHGs in the atmosphere, variations in Earth's orbit, and a host of other valuable information on Earth's past climate.

Taken together, this information has assisted scientists in recreating the climate of Earth for the past 2 billion years. This information from the science of paleoclimatology, which has taken nearly 100 years of careful study and has involved thousands of scientists, has taught us that the climate of Earth is in a perpetual state of change. This research has revealed that there have been several ice ages in the geologic past as well as several periods of global warming in Earth's history.

FIGURE 6.11 Ice core showing dark and lighter bands and an ash layer that is 21,000 years old. (National Oceanic and Atmospheric Administration (NOAA). 2024d. *Climate at the Core*. https://climate.gov/news-features/climate-tech/climate-cores. (Accessed March 17, 2024).)

Additionally, scientists have determined that there can be both short- and long-term trends in climate. Scientists have become aware that Earth is a very dynamic place and that past events might assist us better understand not just the past but help us prepare for future climate changes (National Geographic Society 2024a).

Ice cores recovered from the polar regions of the Earth contain the most comprehensive, direct record of the Earth's high-latitude climate for the past 800,000 years. In addition to providing an accurate proxy temperature record, which is gathered from analysis of the stable isotope ratios preserved in the ice, and the only direct observational record of net accumulation, ice cores also provide the only direct record of atmospheric composition through gases trapped in the ice, total gas content in the ice, and isotopic ratios of those gases trapped in the ice (National Snow and Ice Data Center (NSIDC) 2024; NOAA 2024a; NASA 2024; USGS 2024b). Additional advantages of ice cores are that they often contain light and dark bands that alternate like tree rings in the ice core representing summer and winter layers of ice (see Figure 6.11). Lastly, ice cores sometimes have ash layers which can be dated and can be used as an additional line of evidence of age and are also an important age confirmation. Figure 6.11 also shows such an ash layer that was dated at 21,000 years ago.

6.3.3.2.2 *Earth's Paleoclimate*

Over the past million years, Earth's climate has alternated between cold periods (glacials) and shorter relatively warmer periods (interglacials) in a sawtooth pattern. Based in part on paleoclimate data, we know that Earth's orbital cycles affect the seasonal and geographic distribution of solar energy along with changing concentrations of heat-trapping GHGs like CH_4 and CO_2. Levels of CO_2 primarily influence the mean surface temperature of the planet (see Figure 6.12). Figure 6.12

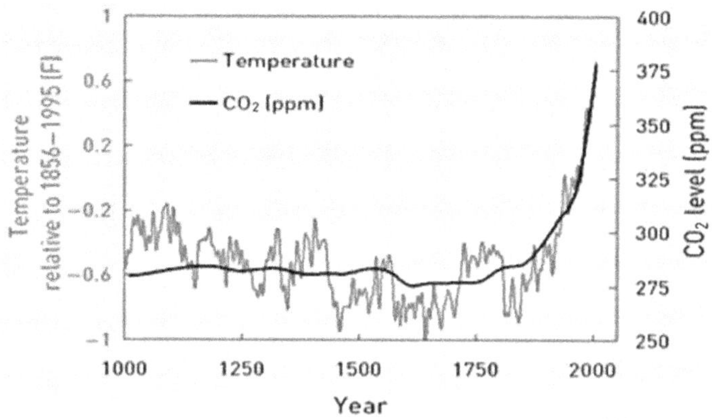

FIGURE 6.12 Temperature and CO_2 for the last 1,000 years. (United Stated Environmental Protection Agency (USEPA). 2024c. *Global Greenhouse Gas Emissions Data*. Global Greenhouse Gas Emissions Data | Greenhouse Gas (GHG) Emissions | US EPA. https://epa.gov/climate-indicators/climate-change-indicators-atmospheric-concentrations-greenhouse-gases (Accessed March 16, 2024).).

examines temperature and CO_2 concentration in the atmosphere for the past 1,000 years and shows that temperature and CO_2 concentrations have been increasing since 1750. This coincides with the beginning of the Industrial Revolution.

Using ice core data for the past 800,000 years, scientists have been able to determine atmospheric temperatures and CO_2 concentrations. The CO_2 results are shown in Figure 6.13. Examining Figure 6.13 clearly shows a correlation between CO_2 and temperature. The higher the CO_2 concentration, the more global temperature increases. A potentially disturbing fact about climate over the past 800,000 years is that the highest CO_2 concentration was approximately 300 ppm. However, the current CO_2 concentration is 422 ppm. That is more than 30% higher than at any time in the past 800,000 years.

Figure 6.14 examines Earth's climate for the past 500 million years or for the majority of the Phanerozoic Era.

During the past 500 million years of Earth's history, there have been three episodes when the Earth cooled sufficiently to produce active glaciers on a wide scale. The first was approximately 440 million years ago and did not last long (labeled as circled number 1 in Figure 6.14). The second was from approximately 370 to 270 million years ago (labeled in Figure 6.14 as a circled number 2). The period did undergo some fluctuations, but the overall length of a cooler period lasted for nearly 100 million years. The third period (labeled in Figure 6.14 as a circled number 5) experienced active glaciations that began at the start of the Pleistocene Era approximately 1.8 million years ago. The circled numbers 3 and 4 in Figure 6.14 refer to warm periods in Earth's history.

6.3.3.4 Greenhouse Gas Legislative Action

The above sections clearly establish the connection between GHG emissions and a warming atmosphere. The USEPA has been aware of this connection at least as far back in time as 1991. In 1991, USEPA stated in its report to Congress that the nature of human existence on Earth has changed dramatically during the past century. The world population has increased threefold. However, this was in 1991, now in 2024 (33 years later) the human population is estimated at over 8 billion people as opposed to just over 5 billion in 1991, which is a significant increase (United States Census Bureau 2024). USEPA (1991) further stated that industrial production has increased by a factor of 50 and the world economy has increased by a factor of 20. In all, USEPA concluded that human activity is changing the features of the Earth and the chemistry of the atmosphere.

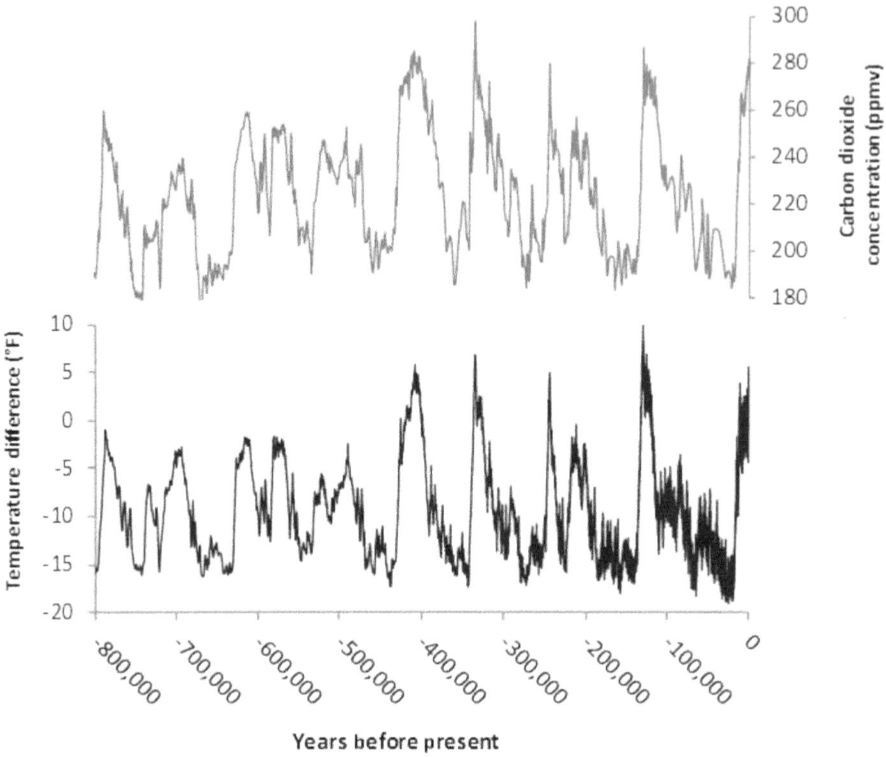

FIGURE 6.13 Atmospheric temperature and CO_2 for the last 800,000 years. (United States Geological Survey (USGS). 2024c. *Important Paleoenvironmental Perspectives on Modern Climate Change.* www.usgs.gov/programs/ecosystems-land-change-science-program/science. (Accessed March 18, 2024).)

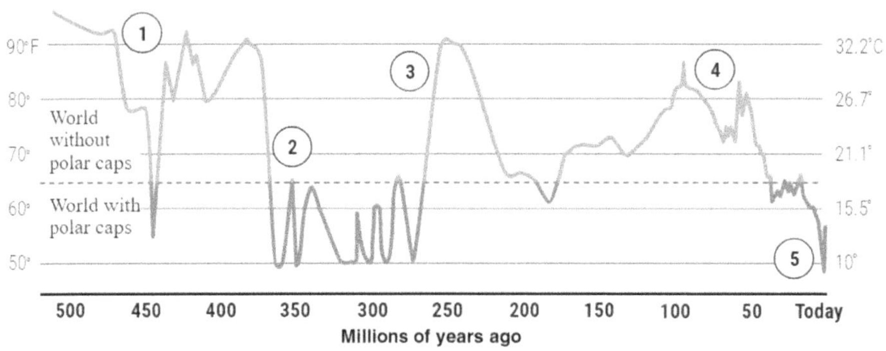

FIGURE 6.14 Atmospheric temperature for the last 500 million years. (Smithsonian. 2024c. *A 500 Million Year Survey of Earth's Climate*. www.science.org/content/article/500-million-year-history-of.earth. (Accessed March 24, 2024).)

Currently, reductions of GHS emissions are voluntary and Congress has yet to pass legislation to cut GHG emissions. USEPA (1991) stated that to have real change in reducing the effects of global climate change from GHG emissions that it must be through international cooperation. Thus far, efforts to reduce GHG emissions under an enforceable international treaty have failed, in part, due to language that limits GHG emissions in developed countries such as the United States but does not limit GHG missions in developing nations, such as China and India (USEPA 2024c).

Climate change regulatory history in the United States is brief compared to the European Union. At first, climate change underwent scientific study to evaluate if there was a basis in the fact that human activities can and have affected climate. The path leading up to regulating GHGs in the United States began with the growing concern within the scientific community since the 1970s but did not gain enough traction until 1999 when the International Center for Technology Assessment (ICTA) and several other organizations filed a petition for rulemaking to the USEPA requesting that the agency regulate GHG emission from new automobiles under the CAA. USEPA then argued that GHG emissions met the definition of "air pollutant" under the CAA, that GHG emissions contributed to the pollution that is reasonably anticipated to endanger public health and welfare, and that it was technically feasible to reduce GHG emissions. It was not until a decision by the Supreme Court on April 2, 2007, that found that the USEPA had the authority to regulate GHG emissions from new motor vehicles under the CAA (National Association of Clean Air Agencies 2013; USEPA 2024d).

In April 2009, USEPA published a finding that GHG emissions endangered public health and welfare and that GHG emissions from new motor vehicles and new motor vehicle engines were contributing to air pollution that endangers public health and welfare. In December 2009, USEPA published the final finding that has come to be known as "The Endangerment Finding." The Endangerment Finding had consequences not only for mobile sources of GHGs but also for stationary sources. In 2011, USEPA declared that GHGs were subject to regulation under the CAA for stationary sources for the Prevention of Significant Deterioration (PSD) and Title V permitting (USEPA 2024d).

6.3.3.5 Summary of Greenhouse Gases

Climate change due to GHG emissions from anthropogenic sources is now a reality. The time has come to adapt to the changing climate (Rogers and Rogers 2022a, 2022b). Adapting to climate change is disproportionate. According to USEPA (2021), climate change will most severely affect those who are least able to prepare for, and recover from, climate change impacts such as heat waves, drought, poor air quality, flooding, sea-level rise, ocean warming, reduced food production, increased health effects, species extinction, increased poverty, and other impacts. USEPA's analysis indicates that racial and ethnic minority communities are particularly vulnerable to the greatest impacts of climate change. A wide range of adaptations are suggested and include the following (Rogers and Rogers 2022b):

- Shift to renewable energy sources
- Reduce food loss and waste
- Consume less meat
- Strengthen building foundations
- Use less water
- Plant native species instead of grass
- Plant a vegetable garden
- Halt deforestation
- Embrace electric vehicles, public transportation, and other forms of transportation
- Embrace low-carbon urban planning techniques
- Move inland and do not build nearest the ocean
- Move from flood-prone areas
- Address poverty
- Invest in climate justice
- Align global climate change policy

There are a multitude of other actions that can be undertaken to reduce GHG emissions and prepare for climate change realities. Actions may be too late to restrict climate change from the United Nations' goal of not to exceed 1.5°C But it is not a lost cause yet. It is on all of us to support immediate action to the most needed (Concern USA 2022).

6.3.4 Other Anthropogenic Impacts to Air

GHGs are not the only anthropogenic impacts that have impacted air Other impacts that are also very significant include upper atmospheric ozone, smog, acid rain, and criteria pollutants. Each is further discussed in the following sections.

6.3.4.1 Upper Atmospheric Ozone

Ozone (O_3) is a gas present in Earth's upper atmosphere and at ground level. The ozone occurring at ground level is considered an air pollutant by USEPA (2024e), whereas the ozone occurring in the upper atmosphere is not considered a pollutant, because it is beneficial to life on Earth by acting as a filter of ultraviolet (UV) radiation produced by the sun (USEPA 2024f). Ozone from the ground level to a height of approximately six miles in the troposphere is the main constituent of urban smog (USEPA 2024f).

Until the mid-1990s, ozone in the stratosphere was gradually being destroyed by anthropogenically produced chemicals referred to as ozone-depleting substances (ODS), including (USEPA 2024f) the following:

- CFCs
- Hydrochlorofluorocarbons (HCFCs)
- Halons
- Methyl bromide
- Carbon tetrachloride
- Methyl chloroform

These substances are commonly used in coolants, foaming agents, fire extinguishers, solvents, pesticides, and aerosol propellants. Once released into the environment they tend to degrade very slowly (USEPA 2024f). It should also be noted that hydrofluorocarbons (HFCs) have been substituted widely for CFCs. While this substitution has slowed the harmful effects brought on by free chlorine in the upper atmosphere, HFCs are potent GHGs. Ground-level ozone is not emitted directly into the air but is created by chemical reactions between oxides of nitrogen (NO_x) and VOCs in the presence of sunlight. Emissions from automobile exhaust, gasoline vapors, chemical solvents, electrical generating facilities, and certain factories are some of the major sources that emit compounds leading to the generation of ozone (USEPA 2024f). Table 6.2 shows the percent contributions of these major sources.

Ground-level ozone is of particular concern in urban regions of the United States. During the summer, strong sunlight and hot weather produce the conditions necessary for producing harmful levels of ozone. Over-exposure to ground-level ozone can result in difficulty in breathing and other respiratory effects (USEPA 2024e), especially for the elderly, very young, and those with existing respiratory ailments.

6.3.4.2 Volatile Organic Compounds and Smog Forming Chemicals

VOCs are organic compounds that generally volatilize or evaporate readily under normal atmospheric pressure and temperatures. They usually have a high vapor pressure, low-to-medium solubility, and low molecular weight. As a result of these chemical characteristics, VOCs are common air, soil, and water contaminants (USGS 2006; USEPA 2008). Automobile exhausts contain VOCs.

TABLE 6.2
Sources and Contribution of NO_x and VOCs

Compound	Source	Percent Contribution
NOx	Motor vehicles	56
	Utilities	22
	Industrial, commercial, and residential fuel consumption	17
	Other	5
VOCs	Industrial and commercial processes	50
	Motor vehicles	45
	Consumer solvents	5

Source: United States Environmental Protection Agency (USEPA). 2024e. *Ground Level Ozone*. www.epa.gov/ground-level-ozone-pollution. (Accessed March 20, 2024).

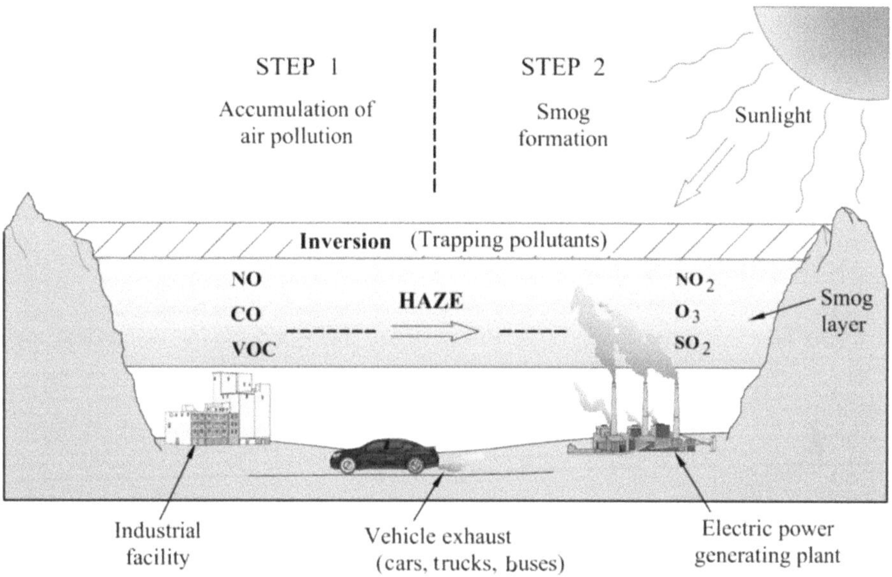

FIGURE 6.15 Formation of photochemical smog. (United States Environmental Protection Agency (USEPA). 2008. *Findings on National Air Quality: Status and Trends through 2006*. EPA454/R-07-007. USEPA. Research Triangle Park. NC.)

When combined with other common air pollutants and sunlight, urban smog will form if atmospheric conditions are favorable – that is, there is an ample supply of the combined sources of VOCs and other smog-forming contaminants. This type of smog formation produces **photochemical smog** and is shown in Figure 6.15. We will further discuss VOCs in Chapter 8.

6.3.4.3 Criteria Pollutants

USEPA has established what are termed **criteria pollutants** which include particulate matter (dust), ground-level ozone, carbon monoxide, sulfur dioxide, nitrogen dioxide, and lead. These pollutants can harm human health as well as the environment (USEPA 2024h, i).

6.3.4.3.1 Dust and Particulate Matter

Dust and particulate matter can be very similar but are not the same. Dust particles are particles suspended in the air and originate from a natural source, such as a dust storm in desert regions. Particulate matter contains small particles that generally originate from anthropogenic sources such as emissions from a coal-fired power plant, a smelter, or automobile exhaust. Other examples of particulate pollution originating from an anthropogenic source can be a wildfire which is started from an anthropogenic source and the Dust Bowl of the 1930s, which was heavily influenced by poor agricultural practices.

6.3.4.3.1.1 Dust

Dust originates from several types of natural sources. Dust may be heavy enough to see but light enough to be carried by the wind. Dust is generally defined as microscopic particles of material that can be carried in air by wind. Dust can be made up of pollen, bacteria, smoke, ash, salt crystals, and small bits of soil or rock, including sand. Significant sources of dust are from desert regions such as the Sahara Desert in northern Africa. In fact, it is not uncommon for dust originating from the Sahara Desert to reach North and South America. Another significant source of dust can be volcanic eruptions which can emit millions of metric tons of ash high into the atmosphere (USGS 2023). Significant accumulation of dust in the atmosphere can influence climate. In fact, a strong correlation exists between aeolian activity and the decadal droughts related to the Pacific Decadal Oscillation in the southern Pacific Ocean (USGS 2023). Aeolian processes are defined as a type of geomorphic process that involves the erosion, transport, and deposition of particles caused by wind at or near the surface of the Earth. An example of an aeolian deposit is dunes composed of sand in desert regions.

Dust originating from a natural source can be a hazard to human health and the environment if the concentrations are high enough and exposure is long enough. Therefore, dust can be considered a geologic hazard and an anthropogenic-induced hazard if the dust originates from an unregulated source, which is dependent upon the circumstances of the release of dust into the air.

Inhalation of dust can cause a range of health problems, including the following (USEPA 2024g):

- Allergic reactions
- Asthma
- Eye irritation
- Sneezing
- Coughing
- Hay fever
- Chronic breathing and lung problems
- Heart disease
- Lung infections
- Reduced life span
- Death

6.3.4.3.1.2 Particulate Matter

Particulate matter (PM) is also known as particle pollution. PM is a complex mixture of very small particles and liquid droplets that generally originates from a regulated anthropogenic source. Particle pollution consists of a number of components, including the following (USEPA 2024g):

• Acids	• Nitrates
• Sulfates	• Organic compounds
• Metals	• Soil or dust particles

The size of the particles is directly related to their potential for causing adverse health effects. Particles less than 10 μm or less are small enough to pass through the nose and throat and enter the lings while breathing (USEPA 2024g). Once inhaled, these particles can affect the heart and lungs and cause substantially adverse health effects.

USEPA groups PM into two categories (USEPA 2024g):

- Particles ranging in size from 2.5 to 10 μm are considered inhalable coarse particles – most likely present near roadways and dust-producing industries
- Particles less than 2.5 μm are considered fine particles and are present in smoke and haze. These particles can be emitted from forest fires or power plants, certain industries, and automobiles.

Adverse health effects caused by inhalation of PM include the following (USEPA 2024g):

- Increased respiratory irritation, coughing, and difficulty breathing
- Decreased lung function
- Aggravated asthma
- Development of chronic bronchitis
- Irregular heartbeat
- Nonfatal heart attacks
- Premature death in people with heart or lung disease

Urban areas have the highest concentrations of PM, which is a significant distributor of contaminants in the atmosphere. Contaminants such as SVOCs, some VOCs, PCBs, and many pesticides and herbicides may sorb to a soil particle and travel a significant distance through wind action (USEPA 1998). The size of PM is significant because the largest sizes tend to settle to the ground surface first. Smaller particles can travel around the globe and remain suspended for years if favorable conditions exist (USEPA 2023d). Here is some good news: there was a 20% to 30% decline in atmospheric PM within the urban areas of the United States between 2001 and 2016 (USEPA 2023d). An additional concern is the potential exposure to other contaminants absorbed onto PM, which are then ingested or inhaled.

6.3.4.3.2 Carbon Monoxide

Carbon monoxide (CO) is a colorless and odorless gas that can be harmful when inhaled in large amounts. CO is released into the atmosphere when something is burned. The largest source of outdoor CO is cars, trucks, and other machinery and vehicles that burn fossil fuels. A variety of items in household can also emit CO in indoor air, which include improperly working chimneys, space heaters, generators, furnaces, and gas stoves (USEPA 2024h).

6.3.4.3.3 Lead

Sources of lead emission vary from one area to another. At the national level, major sources of lead in the air are ore and metal processing plants. Other sources include incinerators and lead-acid battery manufacturers. As a result of USEPA's regulatory efforts including the removal of lead as an additive to gasoline, lead levels in the air have been reduced by 98% (USEPA 2024h).

6.3.4.3.4 Sulfur Dioxide

Sulfur dioxide is one of a group of gases. However, the other gases are much less common in the air. Sulfur dioxides are designated as a group with the symbol SO_x. Sulfur dioxide is the compound of greatest concern and is used as an indicator compound. The largest source of this group is from burning of fossil fuels. Sulfur dioxide can harm humans and the environment. Sulfur dioxide reacts with other compounds in the atmosphere and is a major component of acid rain which harms whole ecosystems (USEPA 2024h).

6.3.4.3.5 *Nitrogen Dioxide*

Nitrogen dioxide (NO_2) is a group of compounds of highly reactive gases known as oxides of nitrogen or nitrogen oxide with a symbol of NO_x. Other nitrogen oxides include nitrous acid and nitric acid. Breathing air with high concentrations of nitrogen oxides can irritate the lungs and aggravate respiratory diseases, particularly asthma. Nitrogen oxides react with water in the environment and other compounds in acid to form acid rain.

6.3.4.3.6 *Ground-Level Ozone*

Ozone is a gas that is composed of three oxygen atoms. Ozone occurs both in the Earth's upper atmosphere and at ground level. Also called stratospheric ozone or good ozone, it occurs naturally in the upper atmosphere, where it forms a protective layer that shields us from the sun's harmful UV light. This beneficial ozone has been partially destroyed by other harmful anthropogenic chemicals (see Section 6.2.3).

Ozone at ground level is a harmful air pollutant to humans and the environment and a main ingredient of smog. Ozone is not emitted directly but reacts with other substances in the atmosphere from VOC emissions from cars, power plants, refineries, chemical plants, and other sources (USEPA 2024h).

6.3.4.4 Nuclear Radiation

Nuclear radiation is treated as an air hazard in this book. However, releases of nuclear radiation can also occur directly to water and land. In addition, historical releases of nuclear radiation eventually impact all three media. Releases of nuclear radiation to the environment are monitored by the International Atomic Energy Agency (IAEA) to streamline communication and safety in the event a release occurs. IAEA has also established a rating scale of nuclear radiation releases that ranges from zero, which represents a deviation, but no release occurs to seven, which represents a major accident. The rating scale with the number of recorded incidents is listed in Table 6.3.

The Chernobyl nuclear accident in Russia in April 1986 and the Fukushima nuclear accident in Japan in April 2011 were both rated a seven. The Three Mile Island nuclear accident in Pennsylvania in March 1979 was rated a number six.

TABLE 6.3
IAEA Rating Scale of Nuclear Radiation Releases

Rating	Meaning	Number of Recorded Incidents Worldwide
0	Deviation	4
1	Anomaly	7
2	Incident	8
3	Serious incident	4
4	Accident	9
5	Significant accident	4
6	Serious accident	1
7	Major accident	2

Source: International Atomic Energy Agency (AIEA). 2024. *International Nuclear Event Scale*. www.IAEA.org./nuclear-event-scale. (Accessed June 23, 2024).

TABLE 6.4
Notable Air Pollution Incidents

Location	Substance Released	Year	Impact
Bhopal, India	Methyl isocyanate	1984	15,000 estimated deaths
Oil Fires, Kuwait	VOCs and particulates	1991	Environmental impacts worldwide
Dust Bowl 1930s Mid-West and Plains States	Primarily particulates and dust	1930s	Premature death, increased erosion, crop loss, widespread wildlife impacts
Chernobyl, Russia	Nuclear radiation	1986	Estimated 4,000 deaths, widespread long-term radiation poisoning
London, England Smog	VOCs from coal and automobiles	1952	Massive air pollution exposing millions of humans
Three Mile Island, Pennsylvania	Nuclear radiation	1979	Potential nuclear radiation exposure to a large human population
Fukushima, Japan	Nuclear radiation	2011	Large release of nuclear radiation to the air and Pacific Ocean.
Donora Pennsylvania	Heavy metals, sulfur dioxide and particulates	1948	Twenty deaths and 7,000 hospitalizations

Source: Earth and Human. 2024a. *Mass Air Pollution Incidents*. www.earthandhuman.org. (Accessed May 26, 2024).

According to USEPA (2022), exposure to radiation does not cause immediate health effects but can cause a small increase in the risk of the probability of injury, disease, or death from the exposure later in life. Risks of injury, disease, or death are dependent on the type, dose, length of exposure, and individual susceptibility to the radioactive agent. Exposure to high levels of radiation over a short period of time can cause symptoms such as nausea, vomiting, and death (USEPA 2022). Approximately one-third of the US population is located within 80 kilometers of a nuclear power plant (Nuclear Regulatory Commission 2023).

6.3.5 Summary of Anthropogenic Impacts to Air

Anthropogenic impacts to air affect all of us with every breath we now take into our bodies. Air contaminants now contribute to premature deaths of millions of humans every year (WHO 2024a). The effects of air pollution on wildlife are likely similar to those being felt by humans. While individual incidents may not have the impact on the environment that accumulation of contaminants have in the long term because of accumulated toxicity effects, they are still significant and signal that, regardless of a lot of prevention efforts put into place or engineered, they still occur and can be devastating. Table 6.4 lists some of the more notable individual incidents of air pollution releases affecting humanity.

6.4 ANTHROPOGENIC IMPACTS TO WATER

The importance of water, as with air and land, on Earth cannot be overstated. Water, along with energy and organic molecules, was the prerequisite for the origin of life here billions of years ago. Today, water plays a central role in maintaining our survival as a species, preserving the natural environment, and therefore achieving a sustainable planet (United Nations 2003). Yet, there are immense challenges facing humans with respect to securing water for their basic needs and long-term quality of life. Although almost three-fourths of Earth's surface is covered by water, most of this water is not potable, and more than 95% of freshwater is either frozen, underground or in a gaseous phase.

Human activities have disrupted the water cycle on Earth. Humans have redirected rivers, built dams to store water, drained water from wetlands, polluted many freshwater bodies and groundwater, and used the oceans as their privy for centuries (USGS 2024d). Figure 6.20 presents a diagram showing the water cycle including anthropogenic influence points and locations (USGS 2024d). Water levels in ponds, lakes, and human-managed and constructed reservoirs rise and fall with season variability. However, the current study indicates that nearly 60% of seasonal variability in Earth's surface water storage now occurs in dam reservoirs and other water bodies managed by humans. This suggests that humans have a dominant effect on Earth's water cycle (Cooley et al. 2021).

6.4.1 Historical Anthropogenic Effects to Water in the United States

In the late 1960s, there were several indications that the surface waters of the United States were heavily impacted and polluted through human activities. Some of these were as follows (USEPA 2024a):

- Pollution in the Chesapeake Bay killed fish resulting in an estimated loss to the fishing industry of $3 million annually in 1968.
- Bacteria levels in the Hudson River were 170 times greater than what was considered the safe limit as measured in 1969.
- An estimated 26 million fish were killed in Lake Thonotosassa, Florida, due to wastewater discharges from four food processing plants in 1969.
- A floating oil slick on the Cuyahoga River, just southeast of Cleveland, Ohio, caught fire causing significant damage to two railroad trestles. The cause of the fire was never determined, but investigations concluded that discharges of volatile organic compounds to the river with low flashpoints provided enough fuel that a passing train, for instance, could have provided an ignition source to start the fire.
- The Department of Health, Education and Welfare's Bureau of Water Hygiene reported that 30% of drinking water samples had detected chemicals exceeding recommended Public Health Service limits.
- A study by the FDA in 1971 concluded that over 70% of samples of swordfish had mercury at concentrations that were considered unfit for human consumption.

These impacts listed above along with pollution in Lake Erie made it clear to the United States Congress to develop regulations to protect the nation's water, which then became the Clean Water Act of 1972.

As first described in Chapter 4, the ability of water to dissolve so many different substances allows cells to use nutrients, minerals, and chemicals in biological processes. In natural systems, wherever water goes – the air, the ground, streams, and lakes, biota, or through our bodies – it takes valuable chemicals, minerals, and nutrients picked up along the way. Water is also unique because it is the only substance on Earth found in all three states (liquid, solid, and gas) within our planet's temperature ranges. Liquid water is essential to life, as it comprises approximately 60% of the human body by weight and 70% of the human brain. Some organisms are 90% liquid water by weight (Hill et al. 1993).

Other historical anthropogenic effects on water just in the United States include the following (USGS 2024e; USEPA 2024j; Fernandes and Pacheco 2021):

- Changing the course of the Chicago in the late 1800s
- Construction of more than 8,100 major dams including the Hoover Dam in Arizona and Nevada border on the Lower Colorado River in the 1930s, Glen Canyon Dam in Utah on the

Lower Colorado River in the mid-20th century, Flaming Gorge Dam in Utah, Garrison Dam in North Dakota, Oahe Dam in South Dakota, Fort Peck Dam in Montana, and the Grand Coulee Dam in Washington

- Reduction of biodiversity of aquatic species from physical reconstruction and introduction of pollutants
- Significant reduction of salmon populations is of special note
- Aquifer mining
- Desertification
- Wetland degradation and destruction
- Deforestation
- Forest stress
- Distortion of continental runoff
- Invasive species
- Species extinction

6.4.2 Water and Climate Change

According to the United Nations (2024b), climate change is primarily a water crisis. Water is experiencing the effects of climate change through worsening floods, rising sea levels, melting ice, increased wildfires, and worsening droughts. Climate change is affecting the global distribution of water.

Only 0.5% of water on Earth is usable and available as freshwater. Climate change is dangerously affecting that supply. Over the past 20 years, terrestrial water storage has dropped significantly, which has major ramifications for water security in the future (World Meteorological Organization (WMO) 2024). By 2050, the number of people at risk for flooding will increase from 1.2 billion to 1.6 billion. In the early to mid-2010s, 1.9 billion people lived in severely water-scarce areas. By 2030, that number is projected to increase to 3.2 billion people (United Nations 2024c). The current Arctic sea ice cover (both annual and late summer) is at its lowest level since at least 1850 and is projected to be practically ice-free by 2050 (IPCC 2021).

6.4.3 Disruptions to the Water Cycle

There is now enough evidence to strongly suggest that human activity has changed or disrupted Earth's water cycle from a physical point of view. This does not include the effects of pollution, which is addressed in the next section. Glaciers are melting, atmospheric water vapor is changing, precipitation patterns are changing, droughts have increased in severity and frequency, flooding has increased in severity, runoff and erosion are changing landscapes, infiltration has decreased, and groundwater levels have decreased (Wild Earth Lab 2024).

6.4.4 Pollutant Effects on Water

The transport ability of water also means that water can carry substances harmful to humans and the environment. If these contaminants are present at a sufficient concentration and the exposure is long enough, potentially harmful effects can occur. The damage can be immediate and obvious, as when oil is washed up on a beach and kills waterfowl – or slow developing and silent – a scenario represented all too often by the benzene that originated from a leaking gasoline tank and carried by groundwater into a drinking water well (Rogers 2020).

The rivers, streams, lakes, wetlands, and groundwater are priceless resources. They provide drinking water, irrigation for crops, provide habitat for aquatic and terrestrial creatures, and numerous recreational opportunities. However, pollution from urban areas, especially from agriculture, continues to pose a threat to water quality (USGS 2024e).

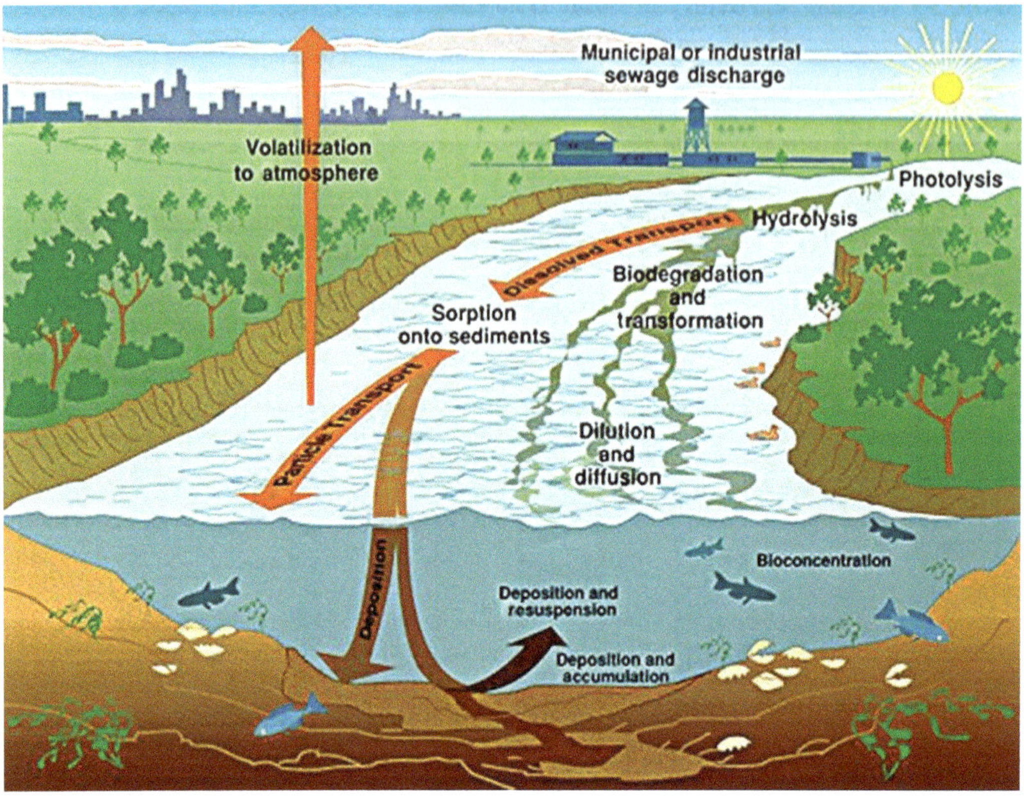

FIGURE 6.16 Some anthropogenic risks to surface water. (From United States Geological Survey (USGS). 1995. *Contaminants in the Mississippi River*. USGS Circular 1133. USGS. Washington, DC.)

Figure 6.16 depicts a municipal or industrial waste water plant discharging to surface water and many of the ensuing contaminant fate and transport processes, including the following (Rogers 2020):

- Transport of discharged wastewater solute downstream
- Mixing due to turbulent advection and turbulent diffusion
- Photolysis
- Hydrolysis
- Biodegradation
- Transformation
- Volatilization to the atmosphere
- Sorption of contaminants in sediment
- Bioaccumulation of contaminants by animal and plant life
- Dilution

The types of pollution in water may not differ that significantly from those present in air but differences in the physical chemistry of pollutants and in the geology of a particular location may have a significant influence on contaminant concentrations and how water may react to the presence of certain pollutants. The general categories of water pollution include the following:

- Suspended matter such as plastic and garbage
- Oil spills which are mixtures of individual chemical compounds that can number in the hundreds

- Microbiological
- Chemicals including VOCs, heavy metals including hexavalent chromium, PCB's dioxins
- Thermal (either very cold or very warm)
- Oxygen-depletion chemicals such as some fertilizers
- Pesticides and herbicides
- Many others as listed in Section 6.1.

6.4.5 Summary of Anthropogenic Impacts to Water

Anthropogenic impacts on water affect the entire planet. Water contaminants and poor sanitation now affect more than 2 billion humans daily and contribute to the premature deaths of millions of humans every year (WHO 2024b).

The effects of water pollution on wildlife are likely similar to those being felt by humans. While individual incidents may not have the impact on the environment that accumulation of contaminants have on long-term health because of accumulated toxicity effects, they are still significant and signal that, regardless of a lot of efforts being put into place or engineered, they still occur with and can be devastating. Table 6.5 lists some of the more notable individual incidents of water pollution releases affecting humanity.

6.5 ANTHROPOGENIC EFFECTS TO LAND

Land change is tightly intertwined with multiple issues of past, present, and future land-use practices, geology, weather patterns, soil and carbon dynamics, and climate change (USGS 2024f). According to the World Economic Forum (2024), 95% of land (excluding those areas covered in permanent ice

TABLE 6.5
Notable Water Pollution Incidents

Location	Substance Released	Year	Impact
Gulf of Mexico	Oil spill	2010	Estimated 560 million liters of oil spilled into the Gulf of Mexico. Widespread ecological impacts
Kuwait Oil Release	Oil	1991	Estimated 907 million liters of oil released into the Persian Gulf. Widespread ecological impacts.
Pacific Garbage Patch	Garbage	Current	Garbage patch covering an area twice the size of Texas between Hawaii and California
Southern Japan	Mercury	1950s	Mercury poisoning caused by mercury releases to water and the ocean. Estimated 2,000 human deaths. Widespread ecological impacts.
Ok Tedi, New Guinea	Two billion tons of mining waste released to surface water	1984	Mining waste released to the Ok Tedi River causing widespread ecological damage and over 30,000 residents evaluated
Valdez, Alaska	Oil spill	1989	Estimated 40 million liters of oil spilled into Prince William Sound in southern Alaska causing widespread ecological damage and wildlife deaths
Gulf of Mexico Dead Zone	Fertilizers	Ongoing	Phosphorous and nitrogen nutrient loading causing algal blooms and oxygen deficiency leading to marine organism deaths

Source: Earth and Human. 2024b. *Mass Water Pollution Incidents*. www.earthandhuman.org (Accessed May 26, 2024).

cover) on Earth shows some kind of human modification, with 85% bearing evidence of multiple forms of human impacts. Some of these activities include (World Economic Forum 2024):

- Human settlements and urban areas
- Agriculture
- Cropland
- Livestock
- Livestock grazing areas
- Transportation including roads, airports, and railroads
- Mining
- Energy production
- Forest farming and mining
- Water storage (reservoirs)
- Electrical infrastructure, including powerlines lights, and substations
- Industrial development
- Stormwater control areas
- Landscaping

6.5.1 Anthropogenic Land Use Effects

Humans have affected the physical surface of land in many ways. Land development, overpopulation, pollution, mining, agriculture, burning fossil fuels, and deforestation are just a few of these physical activities that then trigger other negative impacts that include the following (National Geographic Society 2024b):

- Soil erosion
- Poor air quality
- Poor water quality
- Species extinction
- Ecosystem collapse
- Habitat loss
- Deforestation
- Wildfires
- Noise pollution
- Light pollution
- Chemical pollution
- Waste disposal (landfills)
- Illegal waste disposal or dumping
- Eutrophication
- Contaminating water supplies both surface and groundwater and polluting the oceans

Just in the United States, 880 million acres of land is farmland, which represents 39% of land in the United States and includes 1.9 million farms (United States Department of Agriculture (USDA) 2024a).

If we rewind the clock 1,000 years ago, only 4 million km^2 of land was used for farming (Our World in Data 2024). The current estimate of land used for farming world-wide is 48 million km^2. Figure 6.17 shows the increase in land use for farming since year zero. That is about five times the size of the United States. The vast majority of the world's agricultural land is used to raise livestock and dairy (Our World in Data 2024).

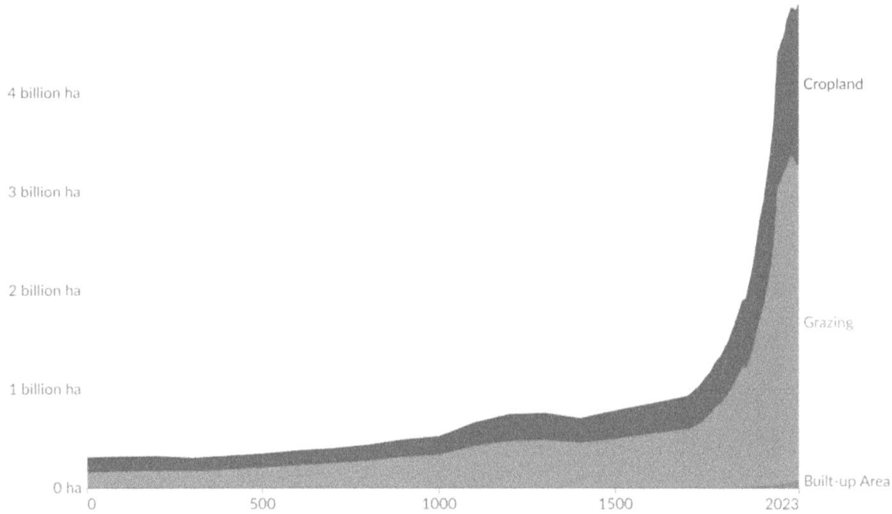

FIGURE 6.17 Growth in agricultural land use in hectares over the past 2023 years. (From World Economic Forum. 2024. *The Extent of Human Impact of Earth's Surface*. www.weforum.org/Human-impact-on-Earth's-surface. (Accessed March 24, 2024).)

According to Hooke and Duque (2024), the human population has doubled in the past 40 years and is expected to double again in the next 40 years. The expansion of infrastructure and agriculture necessitated by this population growth has accelerated the pace of land transformation and degradation. The current rate of land transformation, particularly for agriculture, is unsustainable.

6.5.2 Anthropogenic Land Pollution

Land pollution refers to the deterioration of the Earth's land surfaces, at or below ground level, from anthropogenic sources (NOAA 2024e). Causes of anthropogenic land pollution are numerous and do not just include releases of chemical contaminants but also include human development and activities that disturb the land and often destroy natural habitats.

Land pollution is not a new phenomenon. Land pollution has been a problem for thousands of years and has become increasingly serious as the human population increases. An increasing human population opens the door for exposure to chemical contaminants and to more bacteria and disease through an accumulation of human waste, animal waste, and our garbage (NOAA 2024e; USEPA 2023d). We will discuss more about land pollution in Chapters 7 and 8.

6.5.3 Summary of Anthropogenic Land Impacts

Anthropogenic impacts to land affect the entire planet. Contamination of the land affects large human populations and causes ecological harm and even the collapse of regional ecosystems (WHO 2024c). The effects of land pollution on wildlife are likely similar to the effects being felt by humans. While individual incidents may not have the impact on the environment that accumulation of contaminants have on long-term health because of accumulated toxicity effects, they are still significant and signal that regardless of a lot of prevention efforts being put into place or engineered, they still occur with and can be devastating. Table 6.6 lists some of the more notable individual incidents of land pollution releases affecting humanity.

TABLE 6.6
Notable Land Pollution Incidents

Location	Substance Released	Year	Impact
Dust Bowl 1930s, Central United States	Primarily particulates	1930s	Premature death, increased erosion, crop loss, widespread wildlife impacts
Chernobyl, Russia	Nuclear radiation	1986	Estimated 4,000 deaths, widespread long-term radiation poisoning
Guiyu, China	Lead	Current	Dumping of e-wastes causing lead poisoning in children
Superfund Sites, United States	Numerous	Historical	Over 1,500 superfund sites have caused significant ecological impacts that prevent land re-use. A couple of notable sites include:
Times Beach, MO	Dioxins	1960s and 1970s	Town was abandoned under a declaration from the president
Love Canal, NY	Halogenated VOCs	1940s to 1970s	Increased cancer risks for those exposed, including children

Source: Earth and Human. 2024c. *Mass Land Pollution Incidents*. www.earthandhuman.or. (Accessed May 26, 2024).

6.6 ANTHROPOGENIC IMPACTS ON LIVING ORGANISMS

Human activity is a major threat to Earth's biodiversity. Modifications to the air, water, and land by humans have significantly impacted living organisms indirectly through, if not all, each of the activities we have discussed in the chapter. In addition, there are anthropogenic activities that directly impact biodiversity, such as hunting.

The impacts on biodiversity from human activity have generally taken a long time to occur and to be quantified. In part, this is because the human population growth thus far has been exponential, meaning that its growth rate stays the same regardless of its population size. This makes the population grow faster and faster as it gets larger. Populations may grow exponentially for some period, but they ultimately reach a capacity when it is limited by resource availability. However, humans have continued to work around the carrying capacity concept through technological and medical breakthroughs that assist in supporting an ever-growing population. This threatens biodiversity because the more humans there are, the more this displaces other species and reduces species abundance and richness. Humans affect other species through many activities; some argue that humans impact other species by conducting any activity. Some of these most influential activities include the following (United Nations 2024d; World Wildlife Fund (WWF) 2022):

- Land development
- Pollution
- Deforestation
- Natural resource exploitation
- Extensive overuse of nonrenewable resources
- Over-exploitation (i.e., hunting or extermination)
- Invasive species
- Disease
- Habitat destruction
- Wetland destruction
- Agriculture

6.6.1 Wildlife Population Decline

According to the Living Planet Report published by the WWF (2022), monitored populations of vertebrates including mammals, birds, amphibians, reptiles, and fish have seen a 69% drop in populations since 1970. Several factors are key drivers of biodiversity decline and include habitat loss, species overexploitation, invasive species, climate change, and disease (WWF 2022). Populations in Latin America and the Caribbean have fared worst, with an average decline of 94% in species populations. Global freshwater species have also been disproportionately impacted, declining 83% on average (WWF 2022). The Living Planet Report analyzed almost 32,000 species worldwide. It provides a comprehensive measure of how wildlife is responding to environmental pressures driven by biodiversity loss and climate change, while also allowing a view of human impact on biodiversity (WWF 2022).

According to the WWF study, our future depends on reversing the loss of nature as much as it depends on addressing climate change. The decreases in wildlife populations and abundance can have dire consequences for human health and our economy. This is because when wildlife populations decline to this degree, it means that dramatic changes are impacting wildlife habitat and the food and water that they rely upon to exist. This is significant because it is these natural systems that also sustain human life (WWF 2022).

A couple of examples that may come as a surprise to many that contribute to wildlife population decline are collisions with anthropogenic structures. The next two sections briefly describe seemingly innocent circumstances where anthropogenic behavior contributes to species decline.

6.6.1.1 Wildlife Decline due to Anthropogenic Structures

Birds face many threats in urban areas. Up to 1 billion birds die each year in the United States from building collisions (United States Fish and Wildlife Service 2024; Audubon Society 2024). For example, on October 9, 2023, 960 birds died by colliding with just one building in Chicago. The birds were collected and counted by scientists from the Field Museum and the University of Chicago. The migrating birds were migrating south and struck the building in large groups. The scientists believe that a combination of poor weather, migrating at night, building height, configuration, and poor lighting each contributed to the bird deaths (British Broadcasting Company 2024).

Birds cannot perceive glass as a solid barrier and often fly directly into windows or become entangled in reflective surfaces. Bright lights and unusual reflections can also disorient birds and cause deadly collisions. Three main hazards that buildings create for birds include the (United States Fish and Wildlife Service 2024):

- Glass windows and walls: Birds see the reflection of the sky and trees and try and fly straight through. Transparent or reflective glass can be especially harmful
- Artificial lighting: Bright lights disrupt natural migratory behaviors and circadian rhythms, increasing night-time collisions with buildings
- Building and antenna height: Tall structures interfere with flight paths and migration routes, especially near coastal areas and wetlands.

6.6.1.2 Wildlife Decline due to Certain Types of Renewable Energy Sources

Renewable energy sources are not exempt from causing environmental harm. Wind turbines have been recognized as a favored alternative and clean-energy source with low environmental impact. The selection of sites for wind farms creates serious conservation concerns on biodiversity. Wind turbines are a serious threat to birds and bats, especially if they are located along migratory bird pathways such as the Mississippi flyway (Kumara et al. 2022).

An example of bird deaths in the United States from a wind farm is Altamont Pass in northern California, which was the first large wind in the United States and was constructed in the early 1980s and includes more than 7,000 wind turbines. According to the Audubon Society (2020), raptors including golden eagles, red-tailed hawks, burrowing owls, and bats were hardest hit. The belief is that birds, especially raptors, keep their eyes focused on the ground when searching for prey. They do not notice the wind turbines that may rotate at speeds exceeding100 km/h. The Audubon Society counted an average of 1,300 raptors killed per year within the Altamont Pass wind farm (Audubon Society 2020). An accurate estimate of bird deaths worldwide due to wind turbines has not yet been evaluated but is likely to be significant.

6.6.2 Invasive Species

An **invasive species** is any kind of living organism that is not native to an ecosystem and causes harm that is directly or indirectly introduced by humans or is the result of human activity (USDA 2024b; USEPA 2024k).

Invasive species' mode of invasion is commonly as a stowaway during the transportation of goods. USEPA (2024k) estimated that 30% of invasive species to the Great Lakes were as ballast water in ships carrying cargo. Ballast water is taken into a ship or discharged from a ship as it loads or unloads cargo to accommodate the ship's weight changes. Invasive species are a worldwide problem and have persisted for hundreds of years. Most wildlife and environmental agencies are losing the battle with invasive species for one simple and disturbing reason, this type of pollution breeds.

To make this point clear, the United Nations established the Global Invasive Species Program (GISP) in 1997 to develop a global strategy to deal with invasive species (United Nations 2024e). Worldwide, there are thousands of invasive species (United Nations 2024e). Invasive species are characterized as aggressive when they are transported into areas where they are not naturally occurring usually because they do not have a natural predator or other organisms have not evolved defenses (United Nations 2024e).

The number of different types of invasive species in North America alone total the following (USDA 2024a and 2024b):

- Plants – 1,911 species
- Insects – 6,647 species
- Pathogens – 1,278 identified
- Animals – 220 species

The number of invasive species and diseases in North America totals more than 10,000 (USDA 2024a and 2024b).

6.6.3 Species Extinction

Extinction is the dying out of a species (National Geographic Society 2024c). Extinction plays an important role in the evolution of life because it opens up opportunities for new species to emerge. Over long periods of time, the number of species can be relatively stable at times, meaning the number of species going extinct roughly matches the number of new species (National Geographic Society 2024c). However, there are instances when species' extinction occurs rapidly to the point that a large percentage of species are lost to extinction. The causes of mass extinction, as we leraned from Chapter 2, vary. Currently, many scientists believe that we might be entering a time of mass extinction caused by humans (National Geographic Society 2024c).

As we have learned, 40% of land has been converted for food production. Agriculture is also responsible for 90% of global deforestation and accounts for 70% of Earth's freshwater use. The result has a devastating effect on species' habitat and has placed most species on Earth in decline

(WWF 2022). The list below encompasses some North American species that have gone extinct just since 1500 AD and include the following (International Union on the Conservation of Nature (IUCN) 2024):

- Great Auk – 1852
- California Golden Bear – 1922
- Labrador Duck – 1878
- Cascade Mountain Wolf – 1940
- Mexican Grizzly Bear – 1964
- Sea Mink – 1860
- South Rocky Mountain Wolf – 1935
- Carolina Parakeet – 1918
- Heath Hen – 1932
- Bachman's Warbler – 1988
- Passenger Pigeon – 1914
- Ivory-Billed Woodpecker – 1942
- Eastern Cougar – 2011 declared
- Eastern Elk – 1887
- Newfoundland Wolf – 1911
- Caribbean Monk Seal – 1952
- Southern California Kit Fox – 1903
- Dusky Seaside Sparrow – 1987
- Eskimo Curlew – 1981
- Tacoma Pocket Gopher – 1970

According to the WWF (2022) and the IUCN (2024), hundreds of species of animals are now extinct directly due to humans in several different ways, including hunting, habitat destruction, poisoning, development, and pollution. According to the United Nations, more than 1 million species of animals and plants are threatened with extinction as a direct result of human activities (United Nations 2024d).

6.6.4 Summary of Anthropogenic Impacts to Living Organisms

Many scientists believe that stresses placed upon the natural world from anthropogenic sources are equal to or more significant than climate change (WWF 2022). Like climate change, impacts on biodiversity take time to occur and be noticed by the scientific community. However, in some instances, once it is noticed, it may be too late to save certain species from extinction. In addition, once one species is identified as endangered, there are likely many others that are also endangered but have not been evaluated or are not the focus of scientific inquiry. To combat species loss will require proactive legislative measures of global proportion to build a nature-positive society (WWF 2022). As we have discussed in this chapter and according to the WWF (2022), species loss is connected to the destruction of habitat which then may isolate species populations to levels that are not sustainable. Additional factors are our lack of understanding and knowledge on recognizing the warning signs and the speed at which humans are changing the landscape of Earth.

6.7 SUMMARY AND CONCLUSION

Natural geological and environmental risks and associated impacts are of key concern to lives and livelihoods across the Earth. Understanding and monitoring geological and environmental processes and using this information to identify and quantify risk potential are critical to improving risk reduction methods so that sustainability potential for survival is considerably improved (British Geological Survey 2024). However, as this chapter has introduced us to the risks and associated impacts caused by humans, we have discovered through the course of this chapter that risks and impacts are not equal. They depend on many variables that are a complex interplay between the geological environment and human civilization.

We will go into more detail on wildlife population decline, invasive species, and their extinction in Chapter 9. In the next chapter, we shall further explore anthropogenic impacts on the environment through natural resource exploitation, including urban land development, agricultural development, forest management, mining and petroleum development, and water and the oceans.

REFERENCES

Audubon Society. 2020. *Audubon Society. Analysis: Is It Possible to Have Wind Power While Keeping Birds Safe?* www.audubon.org./news/birds-and-wind-turbines. (Accessed June 22, 2024).

Audubon Society. 2024. *Hundreds of Millions of Birds Are Killed Annually from Building Collisions*. www.audubon.org/news/Birds-and-building-collisions. (Accessed June 22, 2024).

Bannon, C. 2024. Fresh Water in Roman Law: Rights and Policy. *Journal of Roman Studies*. Vol. 107. www.cambridge.org/core/journals/journal-of-roman-studies/article/fresh-water-in-roman-law-rights-and-policy/548B1C559B3D6ACEDF50C4576DD14603/core-reader#fns01. (Accessed March 24, 2024).

Berg, F. K., and Gentine, P. 2017. The Impact of Anthropogenic Land Use and Land Cover Change on Regional Climate. *Nature*. Article Number 989. 8pp. (Accessed March 22, 2024).

British Broadcasting Company. 2024. *Nearly 1,000 Birds Die After Striking Chicago Building*. www.bbc.com/news/world-us-canada/brid-deaths-in-chicago. (Accessed June 22, 2024).

British Geological Survey. 2025. *Environmental Policy and Sustainability Strategy*. www.bgs.ac.uk/about-bgs/our-work/environmental-policy. (Accessed February 4, 2025).

Carson, R. 1962. *Silent Spring*. Houghton Mifflin. Boston, MA.

Concern USA. 2022. *Ten Solutions to Climate Change That Will Actually Make a Difference*. https://concernusa.org/news/solutions-to-climate-change. (Accessed May 5, 2024).

Cooley, S. W., Ryan, J. C., and Smith, L. C. 2021. Human Alteration of Global Surface Water Storage Variability. *Nature*. Vol. 591, pp. 78–81.

Earth and Human. 2024a. *Mass Air Pollution Incidents*. https://earthandhuman.org/mass-air-pollution-incidents. (Accessed May 26, 2024).

Earth and Human. 2024b. *Mass Water Pollution Incidents*. https://earthandhuman.org/mass-water-pollution-incidents. (Accessed May 26, 2024).

Earth and Human. 2024c. *Mass Land Pollution Incidents*. https://earthandhuman.org/mass-land-pollution-incidents. (Accessed May 26, 2024).

Environmental Education Associates. 2021. *The Famous Benjamin Franklin Letter on Lead Poisoning*. http://environmentaleducation.com/wp-content/uploads/userfiles/Ben%20Franklin%20Letter%20on%20EEA%281%29.pdf. (Accessed March 23, 2024).

Fernandes, L. F., and Pacheco, F. A. 2021. Impacts of Anthropogenic Activities on Watersheds in a Changing Climate. *Special Edition of Journal Water*. MDPI. Basil, Switzerland. 244pp. www.mdpi.com/journal/water/special.issues/watersheds. (Accessed March 22, 2024).

Gassert, F., Reig, P., Luo, T., and Maddocks, A. 2013. *A Weighted Aggregation of Spatially Distinct Hydrological Indicators*. World Resources Institute. Washington, DC. 28pp.

Haynes, W. 1954. *American Chemical Industry – A History*. Vols. I–IV. Van Nostrand Publishers. New York. NY.

Hemond, H. F., and Fechner-Levy, E. J. 2000. *Chemical Fate and Transport in the Environment*. Academic Press. London, England.

Hill, J. W., Feigl, D. M., and Baum, S. J. 1993. *Chemistry and Life*. Macmillan Publishing Company. New York, NY.

Hooke, R. L., and Martin-Duque, J. F. 2024. Land Transformation by Humans: A Review. *Geological Society of America (GSA) Today*. Vol. 22, no. 12, pp. 4–10.

Intergovernmental Panel on Climate Change (IPCC). 2021. *Climate Change: The Physical Science Basis*. IPCC AR6 WGI. IPCC. 3949pp.

International Atomic Energy Agency (AIEA). 2024. *International Nuclear Event Scale*. www.IAEA.org./nuclear-event-scale. (Accessed June 23, 2024).

International Union on the Conservation of Nature (IUCN). 2024. *Extinct Species in North America since 1500 AD*. www.iucn.org. (Accessed March 27, 2024).

Kumara, H. N., Babu, S., Rao, G. B., Mahato, S., Bhattacharya, M. Rao, N. V., Tamilinyan, D., Parengal, H., Deepak, D., Balakrishnan, A., and Bilaskar, M. 2022. Responses of Birds and Mammals to Long-Established Wind Farms. *Nature*. Vol 12, Article Number 1339. 22pp.

National Association of Clean Air Agencies. 2013. *Background and History of EPA Regulations of Greenhouse Has (GHG) Emissions Under the Clean Air Act*. Washington, DC. p. 19. Microsoft Word – Background and History of EPA Regulation of Greenhouse Gas-Aug 2013-FINAL (4cleanair.org). (Accessed March 20, 2024).

National Atmospheric and Space Administration (NASA). 2024. *Paleoclimatology: Explaining the Evidence.* https://earthobservatory.nasa.gov/features/Paleoclimatology_evidence. (Accessed March 17, 2024).

National Geographic Society. 2024a. *Paleoclimatology.* https://education.nationalgeographic.org/recource/paleoclimatology. (Accessed March 15, 2024).

National Geographic Society. 2024b. *Humans Impact on the Environment.* www.nationalgeographic.org/topic/resource-library-human-imapcts-environment. (Accessed March 24, 2024).

National Geographic Society. 2024c. *Extinction.* https://education.nationalgeographinc.org/resource/extinction. (Accessed March 24, 2024).

National Oceanic and Atmospheric Administration (NOAA). 2024a. *Global Monitoring Laboratory – Carbon Cycle Greenhouse Gases.* https://gml.noaa.gov/ccgg/data. (Accessed March 16, 2024).

National Oceanic and Atmospheric Administration (NOAA). 2024b. *Monthly Average Mauna Loa CO_2.* https://gml.noaa.gov/ccgg/trends/nlo.html. (Accessed March 16, 2024).

National Oceanic and Atmospheric Administration (NOAA). 2024c. *Ice Core.* https://ncei.noaa.gov/products/paleoclimatology/ice-cores. (Accessed March 17, 2024).

National Oceanic and Atmospheric Administration (NOAA). 2024d. *Climate at the Core.* https://climate.gov/news-features/climate-tech/climate-cores. (Accessed March 17, 2024).

National Oceanic and Atmospheric Administration (NOAA). 2024e. *A Brief History of Pollution.* https://A Brief History: Pollution Tutorial (noaa.gov). (Accessed March 23, 2024).

National Snow and Ice Data Center (NSIDC). 2024. *What Do Ice Cores Reveal About The Past?* www.nsidc.org/ask-scientist/ice-core-climate-history. (Accessed March 17, 2024).

Nuclear Regulatory Commission. 2023. *Radiation Dispersion Maps and Information.* www.nrcgov/radiatio-dispersion-maps. (Accessed June 24, 2024).

Ohio State University. 2024. *University Libraries. We Have Met the Enemy and He Is Us.* www.library.osu.edu/sitewe=have-met-the-enemy-and-he-is-us. (Accessed May 19, 2024).

Our World in Data. 2024. Research and Data to Make Progress Against the World's Largest Problems. www.ourworldindata.org. (Accessed February 4, 2025).

Pennsylvania Historical Association. 2024. *The Vision of William Penn.* http://explorepahistory.com/story.php?storyId=1-9-3&chapter=1. (Accessed March 23, 2024).

Rogers, D. T. 2020. *Urban Watersheds: Geology, Contamination, Environmental Regulations, and Sustainability.* CRC Press. Boca Raton, FL. 606pp.

Rogers, D. T., and Rogers, M. H., 2022a. *Integrating Future Anthropogenic Climate Impacts into a Geological and Chemical-Based Sustainability Model.* Paper No. 381709. Geological Society of America. Denver, Colorado.

Rogers, M. H., and Rogers, D. T. 2022b. Incorporating Climate Change Modeling to Parcel Level Sustainability Efforts Globally. *International Journal of Sustainable Development.* Vol. 11, no. 4.

Smithsonian. 2024a. *Air Pollution Goes Back Further than You Think.* www.smithsonianmag.com/science-nature/air-pollution-goes-back-way-further-you-think-180957716/. (Accessed March 23, 2024).

Smithsonian. 2024b. *The Deadly Donora Smog of 1948. The Deadly Donora Smog of 1948 Spurred Environmental Protection – But Have We Forgotten the Lesson?* | History | Smithsonian Magazine. (Accessed March 23, 2024).

Smithsonian. 2024c. *A 500 Million Year Survey of Earth's Climate.* www.science.org/content/article/500-million-year-history-of.earth. (Accessed March 24, 2024).

Tarlock, D. A. 2001. *History of Environmental Law. Encyclopedia of Life Support Systems.* University of Chicago Press. Chicago, IL. 1450pp.

United Nations. 2003. *International Decade for Action; Water for Life 2005–2015.* United Nations. New York, NY. www.un.org/waterforlifedecade/background.html. (Accessed March 21, 2024).

United Nations. 2024a. *The United Nations Environment Programme.* www.unenvironment.org/environment-you. (Accessed March 19, 2024).

United Nations. 2024b. *Water and Climate Change.* www.unwater.org/water-facts/water-and-climate-change. (Accessed March 21, 2024).

United Nations. 2024c. *Climate Change and Water.* www.un.org/climate-change/water. (Accessed March 21, 2024).

United Nations. 2024d. *Global Biodiversity: Status of the Earth's Living Resources.* www.un.org/global-biodiversity-status. (Accessed March 23, 2024).

United Nations. 2024e. *Biodiversity: Invasive Species.* United Nations System-Wide Earth Watch. United Nations. www.un.org/earthwatch/biodiversity/invasivespecies.html. (Accessed March 27, 2024).
United States Advisory Council. 2024. *Keep America Beautiful Campaign.* www.adcouncil.org/Our-Campaigns/The-Classics/pollution-keep-america-beautiful-iron-eyes-cody. (Accessed March 24, 2024).
United States Census Bureau. 2024. *United States Population through Time and World Populations Clock.* www.UScensus.gov/USpopluation (Accessed March 19, 2024).
United States Department of Agriculture (USDA). 2024a. *Land Use for Farming. National Agriculture Statistics Service.* https://nass.usda.gov/farmland. (Accessed March 23, 2024).
United States Department of Agriculture (USDA). 2024b. *National Invasive Species Information Center (NISIC).* www.invasivespeciesinto.gov. (Accessed March 27, 2024).
United States Environmental Protection Agency (USEPA). 1991. *U.S. Efforts to Address Global Climate Change. Report to Congress.* USEPA Government Printing Office 21P-3001. USEPA. Washington, DC. 87pp.
United States Environmental Protection Agency (USEPA). 1998. *Taking Toxics Out of the Air.* EPA451/K-98-001. USEPA. Washington, DC.
United States Environmental Protection Agency (USEPA). 2008. *Findings on National Air Quality: Status and Trends through 2006.* EPA454/R-07-007. USEPA. Research Triangle Park, NC.
United States Environmental Protection Agency (USEPA). 2021. *EPA Report Shows Disproportionate Impacts of Climate Change on Socially Vulnerable Population in the United States.* www.epa.gov/newsrelease/epa-report-shows-disproportionate-Impacts-of-climate-change. (Accessed May 5, 2024).
United States Environmental Protection Agency (USEPA). 2022. *Radiation Health Effects.* www.epa.gov/radiation-health-effects. (Accessed June 2024).
United States Environmental Protection Agency (USEPA). 2023a. *National Environmental Protection Act.* www.epa.gov/NEPA. (Accessed December 26, 2023).
United States Environmental Protection Agency (USEPA). 2023b. *Summary of Environmental Regulations of the United States.* https://epa.gov/summary-environmental-laws. (Accessed December 26, 2024).
United States Environmental Protection Agency (USEPA). 2023c. *Donora Smog of 1948.* EPA Alumni Association: News and Views. www.epaalimni.org/examples/donora/index.html (Accessed December 26, 2023).
United States Environmental Protection Agency (USEPA). 2023d. *EPA's Report on the Environment (ROE).* www.epa.gov/report-environment. (Accessed December 26, 2023).
United States Environmental Protection Agency (USEPA). 2024a. *Global Greenhouse Gas Emissions Data.* www.epa.gov/ghgemissions/globaldata. (Accessed March 15, 2024).
United States Environmental Protection Agency (USEPA). 2024b. *Sources of Greenhouse Gas Emissions.* www.epa.gov/ghgemissions/sources. (Accessed March 16, 2024).
United Stated Environmental Protection Agency (USEPA). 2024c. *Global Greenhouse Gas Emissions Data.* Global Greenhouse Gas Emissions Data | Greenhouse Gas (GHG) Emissions | US EPA. www.epa.gov/air-emissions-inventories/national-emissions-inventory-nei. (Accessed March 16, 2024).
United States Environmental Protection Agency (USEPA). 2024d. *Regulations for Greenhouse Gas (GHG) Emissions.* www.epa.gov/regulations-emissions-vehicles-and-engines/regulations-greenhouse-gas-ghg-emissions. (Accessed March 20, 2024).
United States Environmental Protection Agency (USEPA). 2024e. *Ground Level Ozone.* www.epa.gov/ground-level-ozone-pollution. (Accessed March 20, 2024).
United States Environmental Protection Agency (USEPA). 2024f. *Upper Atmosphere Ozone.* www3.epa.gov/region1/eco/uep/ozone.html. (Accessed March 20, 2024).
United States Environmental Protection Agency (USEPA). 2024g. *Particulate Matter (PM) Pollution.* www.epa.gov/pm-pollution. (Accessed March 20, 2024).
United States Environmental Protection Agency (USEPA). 2024h. *Criteria Pollutants.* www.epa.gov/criteria-air-pollutants. (Accessed March 20, 2024).
United States Environmental Protection Agency (USEPA). 2024i. *Contaminants of Emerging Concern.* www.epa.gov/wqc/contaminants-emerging-concern-including-pharmaceuticals-and-personal-care-products. (Accessed March 21, 2024).
United States Environmental Protection Agency (USEPA). 2024j. *Contaminants of Emerging.* www.epa.gov/wqc/contaminants-emerging-concern-including-paharmaceutials. (Accessed February 4, 2025).

United States Environmental Protection Agency (USEPA). 2024k. *Invasive Species*. www.epa.gov/greatlakes/invasive-species. (Accessed March 27, 2024).

United States Fish and Wildlife Service. 2024. *Buildings and Birds*. www.fws.gov/story/threats-birds-collisi ons-buildings-glass. (Accessed June 21. 2024).

United States Geological Survey (USGS). 1995. *Contaminants in the Mississippi River*. USGS Circular 1133. USGS. Washington, DC.

United States Geological Survey (USGS). 2006. *Volatile Organic Compounds in Nation's Ground Water and Drinking-Water Supply Wells. National Water-Quality Assessment Program*. USGS Circular 1292. USGS. Washington, DC.

United States Geological Survey (USGS). 2023. *Sources, Compositions, and Effects of Atmospheric Dust*. www.usgs.gov/cernters/geoscience-and-environmental-science/dust. (Accessed December 28, 2023).

United States Geological Survey (USGS). 2024a. *Paleoclimate*. www.usgs.gov/science/science-explore/clim ate/paleoclimate. (Accessed March 15, 2024).

United States Geological Survey (USGS). 2024b. *Ice Dynamics, Paleoclimate, and Ice Cores*. www.usgs.gov/centers/geosciences-and-environmental-change. (Accessed March 17, 2024).

United States Geological Survey (USGS). 2024c. *Important Paleoenvironmental Perspectives on Modern Climate Change*. www.usgs.gov/programs/climate-research-and-development-program. (Accessed March 18, 2024).

United States Geological Survey (USGS). 2024d. *The Water Cycle*. www.usgs.gov/science/water-cycle. (Accessed March 22, 2024).

United States Geological Survey (USGS). 2024e. *Water Quality in the Nation's Streams and Rivers – Current Conditions and Long-Term Trends*. www.usgs.gov/mission-areas/water-resources. (Accessed March 22, 2024).

United States Geological Survey (USGS). 2024f. *Land Change*. www.usgs.gov/news/featured-story/land-cha nge. (Accessed March 23, 2024).

United States National Park Service (NPS). 2024. *John Muir*. www.nps.gov//John-Muir. (Accessed June 14, 2024).

United States Research Archives. 2024. *Industrial Emissions in the Late 1800's*. www.UnitedStates.research.archives.gov (Accessed March 22, 2024).

Wild Earth Lab. 2024. *How Humans Affect the Water Cycle*. https://wildearthlab.com/human-impacts-on-th-ewatercycle. (Accessed March 22, 2024).

Wilson, R., and Spengler, J. D., editors. 1996. *Particles in Our Air: Concentration and Health Affects*. Harvard University Press. Cambridge, MA.

World Economic Forum. 2024. *The Extent of Human Impact of Earth's Surface*. www.weforum.org/Human-impact-on-Earth's-surface. (Accessed March 24, 2024).

World Health Organization (WHO). 2024a. *Nine Out of Ten People Breathe Unhealthy Air*. WHO. Geneva, Switzerland. wwwwho.int/news-room/detail/02-022018. (Accessed March 19, 2024).

World Health Organization (WHO). 2024b. *Drinking Water and Pollution*. www.who.int/news-room/fact-she ets/drinking-water-pollution. (Accessed May 26, 2024).

World Health Organization (WHO). 2024c. *Land Pollution*. www.who.int/news-room/fact-sheets/land-pollut ion. (Accessed May 26, 2024).

World Meteorological Organization (WMO). 2024. *Weather, Climate, Water*. https://wmo.int/wake-looming-for-water. (Accessed March 21, 2024).

World Wildlife Fund (WWF). 2022. *Living Planet Report – Building a Nature Positive Society*. Almond, R. A., Grooten, M., Bignoli, J., and Petersen, D., editors. WWF. Gland, Switzerland. 114pp.

7 Natural Resource Exploitation

7.1 INTRODUCTION

Exploitation of the natural environment goes back thousands of years in human history. Natural resource exploitation began in earnest approximately 11,000 years ago when humans transitioned from hunter-gatherer or foraging lifestyle to a farming and more sedentary lifestyle. This transition to a new era in human history is termed **Neolithic Era** and is often interpreted as one of the most rapid and significant transitions in human history (University of Cambridge 2024; Harari 2018). By producing and storing food, humans began mastering the natural world and took the first significant steps toward thousands of years of rapid technological development. This transition greatly increased fertility, permitted the opportunity for the development of specialist craftsmen, and the construction of permanent structures (University of Cambridge 2024; Harari 2018). These changes in human history opened the door to the other subject areas that will be discussed in this chapter that include how natural resource exploitation began and how it developed. The types of natural resource exploitation described include agriculture, urban land development, forest management, mining and petroleum development, and finally, water and the oceans.

This chapter examines aspects of human innovation and engineering as they relate to natural resources. It highlights the incredible power of our collective intelligence to solve complex challenges during our brief history. Creating culture and civilization has unleashed the opportunity of imaginative thought. Instead of focusing our efforts on individual survival, technological advancements have given us the opportunity and time so that we can explore how life on Earth has evolved since its origin billions of years ago. It has also made it possible for us to explore fundamental subatomic particles, explore and learn about our universe, and travel into space. Our curious nature and the power of our brains permit us the ability of awareness and self-reflection, consciousness, and questioning our existence. Now we find ourselves at a crossroad where our success as a species may no longer be an advantage for our future survival but may be leading us down a path where it threatens the existence of many living species on Earth, including our own.

7.2 HISTORICAL PERSPECTIVE

The transition to a farming lifestyle had a cascading effect and made civilization a reality. Farming gave rise to civilization, and, in so doing, required certain inputs to promote and support an ongoing civilization that included (National Geographic Society 2024a):

- Building permanent homes and constructing the first cities
- Harvesting forests for building materials and fuel

 DOI: 10.1201/9781003463153-9

- Expanding the need for mining and petroleum for tools, fuel, and construction of roads and buildings
- Settling near water sources that included freshwater for potable water sources and irrigation and oceans and large waterways for the transport of goods and commerce

There is evidence that the transition to agriculture may have started as far back as 19,000 years in what is now the Jordanian desert region of the Middle East. However, the point in time from approximately 14,500 years ago is still generally considered the time in which the first indications appear in cultural change associated with agriculture and included planting wild grains and construction of stone buildings. This region where farming is believed to have begun is known as the **Fertile Crescent** in the Levant region, which is located in the Middle East from northern Egypt through Israel and Jordan to the Persian Gulf. Even though the current climate of this region is a rather inhospitable desert, during the time of the transition to farming, this area was one that included marshlands and grasslands with gazelle and antelope. In addition, given that this location is at the crossroads of Africa, Europe, and Asia, it is not a surprise that it would be a site of regional agricultural innovation (University of Cambridge 2024). Agriculture emerged at several locations around the world during this transitional period and included the following:

- Africa: rice, millet, and sorghum
- Levant: wheat and barley
- China: millet and rice
- North America: maize, sunflower, and squash
- South America: potatoes and other crops
- Papua New Guinea: bananas and yams

Associated with this transition to farming was also the domestication of goats, cows, sheep, and pigs, which all have their origins in the Fertile Crescent. Agriculture and animal domestication provided a more stable supply of food and enabled humans to change their lifestyle and become more sedentary, which then permitted them to build more permanent buildings and form communities. This shift from hunter-gatherer to a farming society afforded humans more security and time to innovate, which included pottery, jewelry, tools, and cosmetics. Lastly, this dynamic shift changed human culture forever and altered the human relationship with nature and paved the way for the development of civilization. From this time onward, humans have modified the Earth through agriculture, building cities, resource exploitation, and other related activities, which gave humans the advantage over most all other living creatures and subsequently resulted in a seemingly relentless explosive increase in the human population (National Geographic Society 2024b).

7.3 AGRICULTURAL DEVELOPMENT

According to the United Nations (2024a), agriculture is the largest land use on Earth and now occupies 40% of the land surface on the planet. In addition, agriculture and associated activities are the largest pollution source on Earth and are also the largest single factor responsible for species population decline. **Agriculture** is defined as the cultivation and exploitation of animals, plants (including fungi), and other forms of organic life for human use, including food, fiber, medicines, fuel, and any other human use (Environmental Science 2024).

7.3.1 Agricultural Development in the United States

The development of agriculture in the United States and throughout the world generally required the clearing of land so that domesticated crops could be planted. In the United States, agricultural conservation techniques date as far back as 1682, when William Penn ordered that one acre of forest be preserved for every five acres cleared for settlement (Pennsylvania Historical Association 2024). In addition, Benjamin Franklin widely wrote about the adverse effects of treating the land, the effects of lead, and the dumping of waste (Environmental Education Associates 2024).

It was not until the Industrial Revolution, beginning in the 1750s, that signaled a more rapid transition from making products by hand to making products with machines. The stationary steam engine invented by James Watt led to widespread use of steam power that used coal. New chemical manufacturing methods, iron production processes, the development of machine tools, and the rise of the factory system all came about in relatively short time, which sparked increased agricultural activities and the rise of the industrial age and resulted in the discharge of large amounts of waste emitted into the air, into waterways, and discarded onto the land. In 1862, then president, Abraham Lincoln signed into law an act of the United States Congress that formed the United States Department of Agriculture (USDA) (Crop Life in America 2024).

It was at about this time in the history of the United States that agriculture transitioned from subsistence farming to business farming. Innovations in food production because of the Industrial Revolution made the transition from subsistence farming to farming as a business possible. This transition considerably increased food production, and excess food was sold as a commodity. An additional transition occurred in agriculture through the rise of bioengineering beginning in the late 20th century. During this transition, bioengineered organisms, also called **genetically modified organisms** (GMO), were developed that had genetically favored DNA to resist drought and other desired characteristics. The aim was to produce more food (Crop Life in America 2024).

Some believe that humans are entering a fourth stage of agricultural development that is labeled the age of artificial intelligence, or AI. The influence of AI on agriculture is not yet understood but will likely include machine learning to operate farm machinery and conduct other tasks once the responsibility of humans (Crop Life in America 2024).

In 1940, 18% of the labor force in the United States were farmers. By 1970, the percentage of the labor force as farmers declined to 4.6%. By the year 2000, farmers made up 1.5% of the United States labor force (Crop Life in America 2024).

7.3.2 Ecological Effects of Agriculture

Ecological effects of agriculture in just the United States are profound. The amount of land used for agriculture is staggering. The amount of land in the United States is estimated at 2.26 billion acres. The amount of land used for agriculture is 900 million acres or 40% of all the land in the United States (United States Environmental Protection Agency (USEPA) 2024a). This does not include infrastructure such as roads and secondary processing locations, including grocery stores.

The negative ecological effects of agriculture on the natural environment include the following (USEPA 2024b):

- Land development. Land that is no longer available for native plant and animal species
- Wetland destruction
- Changing natural drainage patterns. Agricultural development increases runoff and decreases groundwater recharge
- Increased erosion by removing native species
- Deforestation
- Releasing carbon dioxide into the atmosphere affects climate change
- Decreases water quality
- Introduces animal waste to stormwater runoff

- Livestock grazing in riparian zones (areas immediately adjacent to running freshwater) can change landscape conditions by reducing stream bank vegetation and increasing water temperatures, sedimentation, and nutrient levels
- Enhancing invasive species presence and growth by removing native species
- Decrease native aquatic species in nearby freshwater rivers, streams, and lakes
- Spreading of disease. Some studies to elucidate the connection between land-use changes and infectious disease may be influenced by land use or other environmental factors
- Decreased native species populations. This is especially significant if no wildlife corridors are created that connect patches of native vegetation (Darke and Tallamy 2019)

7.3.3 Chemical Contaminant Effects of Agriculture

As recently as the first half of the 20th century, environmental and ecological awareness and legislation were generally lacking not only in the United States but also worldwide. There were a few international agreements that primarily focused on boundary waters, navigation, and fishing rights along shared waterways between countries. However, they ignored pollution and ecological issues. In fact, the most convenient and least costly method of waste disposal at the time was "up the stack or down the river" (Haynes 1954). It was not until the publishing of the Rachel Carson book, *Silent Spring* in 1962, which described the environmental effects of the pesticide compound dichlorodip henyltrichloroethane, or DDT, on birds, that the environmental movement in the United States was catapulted into the public eye (Carson 1962).

7.3.3.1 Fertilizers

Fertilizers are chemical compounds designed to promote plant and fruit growth when applied (USEPA 2024b). The most common fertilizers are nitrogen (N), phosphorus (P), and potassium (K). Nitrogen is found primarily in an organic form in soils but can also occur as nitrate. Because nitrate is very soluble and mobile, it is transported by surface water to rivers, lakes, and streams where it can promote algal growth. In many cases the algal growth is extensive. Nitrate can also contaminate drinking water. Phosphorus occurs in soil in organic and inorganic forms, but being more soluble than nitrate, it can also be depleted in soil through surface water runoff. Phosphorus can also promote algal growth in rivers, lakes, and streams, because it is a limiting nutrient in fresh water. Potassium (K) in fertilizers is commonly incorporated as potash – an oxide form of potassium that includes the compounds potassium chloride, potassium sulfate, potassium nitrate, and potassium carbonate. The term "potash" comes from the historical practice of extracting potassium carbonate (K_2CO_3) by leaching wood ashes and evaporating the solution in large iron pots. Fertilizers containing potassium generally do not promote algal growth (USEPA 2024b).

The USDA has tracked fertilizer use in the United States since 1960. According to the USDA, approximately 63.5 kg (140 pounds) of fertilizer containing N–P–K are applied each year per acre of land farmed, and this amount has increased more than 200% since 1960 (USDA 2024a). In 2018, USDA estimated that 1.41 million km^2 (350 million acres) were planted for crops, which means that approximately 22.2 trillion kg (49 trillion pounds) of fertilizers were applied to farmland in the United States (USDA 2024a). In urban areas, fertilizers containing N–P–K are common and routinely applied to residential lawns and golf courses to help maintain green, healthy grass and gardens (USEPA 2024c). According to USEPA (2024d), nitrogen, phosphorus, and potassium fertilizers are not currently known to cause cancer. Exposure to high concentrations may cause nausea and vomiting (USEPA 2024d).

A major sink for fertilizers is surface water because they are applied to the soil surface and are considered soluble in water and mobile – especially in the case of nitrates and phosphorus. Once in surface water, nitrate and phosphorus can promote excessive algal growth. Significant algal growth can deplete the dissolved oxygen in surface water and cause suffocation and death to aquatic organisms. The solubility of some fertilizers, combined with the relationship between surface water and groundwater, may lead to groundwater contamination. The natural process of enrichment of surface waters with plant nutrients is termed **eutrophication**. When anthropogenic activities such as fertilization or sewage discharges accelerate this natural process, **cultural eutrophication** occurs (McGucken 2000).

Perhaps the most significant example of cultural eutrophication in North America is commonly called the **Red Tide** in the Gulf of Mexico. The cause is from *Karenia brevis*, which is a type of algae that produces potent neurotoxins. The toxins can be suspended in the air near beaches and cause human respiratory irritation and illness. They can also accumulate in shellfish and cause neurotoxic shellfish poisoning (National Oceanic and Atmospheric Administration (NOAA) 2024a). The algae can block sunlight and deplete oxygen, resulting in fish die-offs. In the Gulf of Mexico, red tides can occur anytime but are most common between August and December. Historical records of red tides indicate that date back at least to the 1840s. However, they appear to be more common and larger in size over the last few decades. Scientists are not completely sure of what triggers and sustains red tides, but there appears to be a connection with nutrient loading within the Gulf of Mexico, exacerbated by overuse of fertilizers in the upper Mississippi Valley. Fertilizers are soluble in water and, therefore may be transported and deposited in the gulf via the Mississippi River. This provides the algae with a food source. There may also be a connection with hurricanes that mix sediment rich in nutrients and become churned up by the hurricane and are released into the water column (Smithsonian 2024a).

7.3.3.2 Pesticides and Herbicides

Simply put, pesticides and herbicides are manufactured to kill things. USEPA (2024e) defines a **pesticide** as preventing, destroying, repelling, or mitigating any pest. A **herbicide** is defined as a substance that is used to kill unwanted plants commonly referred to as weeds. **Pests** are defined as living organisms occurring where they are not wanted or causing damage to crops or humans or other animals (USEPA 2024e). Examples include the following:

- Insects
- Mice or other animals
- Unwanted plants (weeds)
- Fungi
- Microorganisms, such as bacteria and viruses
- Invasive species that include many plants and animals

Perhaps the most famous of all pesticides is the now-banned substance called **DDT**. It became famous as an environmental contaminant after the book *Silent Spring*, written by Rachel Carson, was published in 1962 (Carson 1962). The book cataloged the environmental impacts of the indiscriminate spraying of DDT in the United States and questioned the logic of releasing large amounts of chemicals into the environment without fully understanding their effects on ecology or human health. DDT was widely used in agriculture and by consumers in the 1940s and 1950s. DDT (CAS Registry Number 50-29-3) is now considered a probable carcinogen (Group: A2) by USEPA (2024c) and is no longer used as a pesticide in the United States after it was banned in 1972. The 2001 United Nations Environmental Program meeting held in Stockholm, Sweden (and put into effect in 2004) permitted its use for "vector control" – organisms that produce pathogens, such as mosquitoes (USEPA 2024f).

Pesticides and herbicides are released into the environment for specific purposes and are most commonly detected in near-surface soils and surface water rather than groundwater (USGS 2006). The highest concentrations of pesticides exist in the nation's streams and sediments within urban areas where they have been detected in 83% of streams and 70% of sediments (USGS 2009). The total tonnage of pesticide and herbicide use in the United States has remained constant during the

1980s and early 1990s but has increased by nearly 60% in the last 20 years mostly for agriculture and home and garden purposes (USEPA 2024e).

7.3.4 Effects of Agriculture on Water and the Oceans

There is a common saying in geological circles that states, *whatever nature makes eventually ends up in the ocean*. Agricultural activities on land affect the oceans in many ways. Some of these activities result in negative effects to the oceans, some of which include the following (United Nations 2024b):

- *Increased erosion and sedimentation:* Clearing land and exposing soil increases the amount of erosion. Those eroded sediments from agriculture are transported by streams and rivers and eventually are deposited in the ocean. The sedimentation of these particles can smother coral reefs, seagrass beds, and other marine habitats, disrupting their delicate ecosystems.
- *Nutrient flux:* Included in the runoff from agricultural land are nutrients. Excessive use of fertilizers leads to nitrogen and phosphorus runoff and upon reaching the oceans, can cause or exacerbate algal blooms, oxygen depletion, and eventually reach a state where large areas of the oceans become a dead zone, where marine organisms struggle to survive.
- *Pesticides and marine life:* The use of pesticides in agriculture poses a direct threat to marine life. When pesticides are applied excessively, they can migrate to water bodies, streams, and rivers, which transport them eventually to the oceans. Along the way, they can impact aquatic life, including fish, crustaceans, and marine mammals.
- *Aquaculture and oceanic impact:* Fish farming is called **aquaculture**. Aquaculture is an important industry for meeting the world's growing demand for seafood. However, poorly managed aquaculture operations can negatively impact ocean health. Issues such as disposal of fish waste into the ocean can lead to polluting the ocean, resulting in habitat destructions and disease outbreaks.
- *Climate change and ocean acidification:* Agriculture contributes to climate change through greenhouse gas emissions directly from tilling soil, deforestation to increase cropland, and other related land-use changes. These factors contribute to ocean acidification and rising sea levels. Acidic oceans pose a threat to marine life, particularly shell-forming organisms like corals and mollusks.

7.4 URBAN LAND DEVELOPMENT

Urbanization refers to the concentration of human populations into discrete areas. Urbanization dates back more than 6,000 years ago (National Geographic Society 2024b). The earliest city is considered by modern-day scholars to be Uruk and was located near or along the banks of the Euphrates River in a region known as Mesopotamia in the Middle East. By around 4,500 years ago, the city was a walled city, most likely for protection, and had an estimated population of 65,000 residents. Other areas where cities were created include Egypt, Greece, and Italy.

Currently, there are an estimated 512 cities with a population that exceeds 1 million residents (United Nations 2024c). In 1950, there were only 83. Currently, there are 103 cities in China alone that have a population of greater than 1 million residents. India is next with 46 and then Brazil with a total of 10 (United Nations 2024c).

7.4.1 Negative Ecological Effects of Urban Land Development

Historical urban areas, including Rome, London, Athens, Cairo, and others, offer examples of the negative effects of living in high-density cities that result in the development of disease, famine, and pollution. An example response to pollution concerns, the Roman Senate passed legislation to protect the city's supply of water, which addressed concern over freshwater and sewage disposal

(Bannon 2024). Another example, this time to address air and water pollution in the 14th century, England, is when the king proclaimed the prohibition of the burning of coal in London and the disposal of waste into waterways, specifically the Thames River (Smithsonian 2024b).

This increased concentration of people into cities leads to the transformation of land for residential, commercial, industrial, and transportation purposes (USEPA 2024g). The conversion of Earth's land surface to urban use is one of the most irreversible human impacts on the global biosphere. The environmental impacts of urban expansion reach far beyond urban areas themselves. Some of the effects include the following (Seto et al. 2017):

- Increases the loss of highly productive farmland, which then increases the loss of forest land to farming
- Affects energy demand
- Alters the climate
- Modifies hydrologic and biochemical cycles
- Fragments natural habitats
- Reduces biodiversity
- Introduces invasive species
- Greatly increases pollutant loading to the land, air, and water
- Creates light pollution
- Creates noise pollution
- Increases contaminant exposures
- Creates artificial environments

7.4.2 Environmental Effects of Population Density Differences

In rapidly urbanizing areas, agriculture intensifies on remaining undeveloped land and is likely to expand to new areas, putting pressure on land resources (Jiang et al. 2013). Despite its small land coverage, urban land and its expansion have exhibited profound impacts on global environments. For instance, urban areas hold approximately 50% of the world's population but is responsible for approximately 70% of global greenhouse gas emissions (Chen et al. 2020).

In the United States, urban areas are of a lower density than other parts of the world, such as Europe and Asia. This occurs because the United States primarily practices single-family homes on rather large areas of land. An additional factor is that there is generally a greater distance of travel to places of work and commerce, which increases travel distance by the use of automobiles. This further decreases urban density because of the amount of increased land required for transportation, namely roads and parking lots (USEPA 2024g).

As an example, let us compare Atlanta in the United States with Barcelona in Spain. Atlanta has a population of 2.5 million residents and occupies an area of 4,280 km^2, has an urban density of 6 persons per hectare, and has an energy demand of 80 MJ per person. Barcelona has a similar population of 2.8 million residents but occupies an area of 162 km^2, which translates into an urban density of 173 persons per hector, and has an energy demand of 9 MJ per person. This example demonstrates the difference of a low-density urban area that depends on the automobile for transportation and discourages other forms of transportation such as walking. Atlanta residents display more of a sedentary lifestyle, which negatively affects health with increased risk of overall mortality, cardiovascular disease, and some types of cancer compared to the residents of Barcelona.

7.4.3 Chemical Contaminant Effects of Urbanization

Urbanization increases the variety and volume of pollutants in the air, water, and land. In urban areas, much of the land surface is covered by buildings, pavement, and compacted landscapes. These surfaces do not allow rain and snowmelt to soak into the ground, and because of this, there is an increase in the amount of runoff. Typically, pollutants commonly present in runoff include the following (USEPA 2024h):

- Sediment
- Oil and grease
- Volatile organic compounds from automobiles
- Viruses, bacteria, and nutrients from pet waste and unmaintained or faulty septic systems
- Road salts
- Heavy metals from roof shingles, cars, fences, gutters, and other sources
- Thermal pollution from impervious surfaces such as streets and rooftops

These pollutants can harm fish and wildlife populations, kill native vegetation, pollute drinking water supplies, and make recreational areas unsafe to enjoy (USEPA 2024h).

7.4.4 The Birth of Megacities

One effect of the increase in people living in urban areas is the rise of the megacity, which is a city that has more than 10 million residents. There are 31 megacities in the world, with China being home to 6 megacities and India being home to 4 megacities. Some of the largest in the world include the following (United Nations 2024c):

- Tokyo, Japan – estimated population of 37.8 million
- Shanghai, China – estimated population of 34.8 million
- Jakarta, Indonesia – estimated population of 31.7 million
- Delhi, India – estimated population of 26.4 million
- Seoul, Korea – estimated population of 25.5 million
- Beijing, China – estimated population of 24.9 million
- New York City, USA – estimated population of 23.8 million
- Mexico City, Mexico – estimated population of 21.6 million
- Sao Paulo, Brazil – estimated population of 21.2 million
- Cairo, Egypt – estimated population of 20.5 million

An effect of these large cities is urban sprawl, which often occurs in large cities and is characterized as unchecked population increase and land occupation in less populated areas. These areas typically have less infrastructure and thereby have higher environmental impacts (National Geographic Society 2024b).

7.5 FOREST MANAGEMENT

Humans have been cutting down trees and forests for thousands of years. **Deforestation** is defined as the intentional cutting down and clearing of forested land. The main driver in forest loss is agriculture. One-third of forests on Earth have been cut down, and the rate of deforestation is still accelerating (National Geographic Society 2024c). There are many uses of wood and wood pulp products that include the following (National Geographic Society 2024c; USEPA 2024i):

- Building materials
- Furniture
- Landscaping
- Wood pallets

- Packaging
- Paper and pulp products
- Home products
- Fuel
- Musical instruments
- Tools
- Carving
- Veneer
- Marine applications such as boats

7.5.1 Deforestation

Much of Earth's farmland was once forest. Europe once was 80% forest; now it is 34%. Today, the greatest amount of deforestation is occurring in tropical rainforests. Building and upgrading roads into forests make them more accessible for exploitation. Slash-and-burn agriculture is a significant factor and contributor to deforestation in the tropics (see Figure 7.1). Throughout the history of the United States, there have been numerous legal battles and conflicts between environmental groups and the timber industry. The demand for timber for home building, farming, and other uses in the United States has always seemed to be high and has placed significant pressure on the environment through habitat destruction. As mentioned in the beginning of this chapter, William Penn was aware of the importance of healthy forests in 1681 when he required that for every 5 acres of forest cleared, 1 acre remain preserved.

FIGURE 7.1 An example of slash-and-burn deforestation in Brazil. (Photograph by Daniel T. Rogers.)

7.5.2 Deforestation in the United States

Recognized forestland in the United States is currently estimated at 3.0 million km^2 and represents an increase over the low mark of 1926. The National Forest Management Act of 1976 was enacted to establish guidelines for the management of renewable resources on national forest land. The Act specifically addresses timber harvesting within national forests. The purpose of the Act was to protect national forests from permanent damage from excessive logging and clear-cutting practices (United States Forest Service 2024).

Throughout the history of the United States, there have been numerous legal battles and conflicts between environmental groups and the timber industry. The demand for timber for home building, farming, and other uses in the United States has always seemed to be high and has placed significant pressure on the environment through habitat destruction. As mentioned in the beginning of this chapter, William Penn was aware of the importance of healthy forests in 1681 when he required that for every 5 acres of forest cleared that 1 acre remain preserved.

One of the more contentious conflicts was over the Northern Spotted Owl in the Pacific Northwest. The Northern Spotted Owl experienced a decrease in population levels that threatened its extinction. A major study was undertaken to evaluate the causes of the population decrease, and the results indicated that there were three major threats to the owl population decrease that involved timber harvesting that included (1) variability of birth and death rates through time, (2) loss of genetic variation, and (3) random catastrophes (United States Forest Service 2024).

The passage of the Act required the National Forest Service to conduct an inventory of all national forests and to use a systematic and interdisciplinary approach to resource management that also included monitoring biological effects of timber harvesting (United States Forest Service 2024). Although deforestation in the United States has not been as much as in other countries of the world, the decrease is still significant. In 1620, approximately 4.1 million km^2 of the United States was forest. The low point was in 1926 at 2.9 million km^2. Since 1926, the amount of forested land in the United States is 3.0 million km^2 (USDA 2024b).

Figure 7.2 shows the amount of virgin forest in the United States in 1620 and in 2020 (USDA 2024b). Figure 7.3 shows numerous areas where the forest has been harvested in Stanislaus National Forest near Yosemite National Park in California. Figure 7.4 shows a patchwork of forest harvesting, called **fragmented forest harvesting**.

7.5.3 Ecological Impacts of Forest Farming

The United States boasts the world's fourth largest amount of forests, encompassing approximately 8% of global forests. With almost 310 million hectares (760 million acres) of land, forests cover nearly one-third of the country's total land area. From the boreal forested land in Alaska to the deciduous forests in the eastern states, the pine plantations in the southern states, the temperate rainforests of the northwest, and even the tropical rainforests of Puerto Rico and Hawaii, the diversity of the forests in the United States is truly remarkable. What is also amazing is that 69% of the forested areas within the United States are privately owned. The remaining 31% of forested areas are managed under local, tribal, state, and federal government (United Nations 2024d). The ecological impacts of forest loss are staggering. The following describes a few of the most important effects (United Nations 2024d):

- Loss of biodiversity
- Loss of species abundance
- Habitat destruction
- Increased carbon dioxide emissions

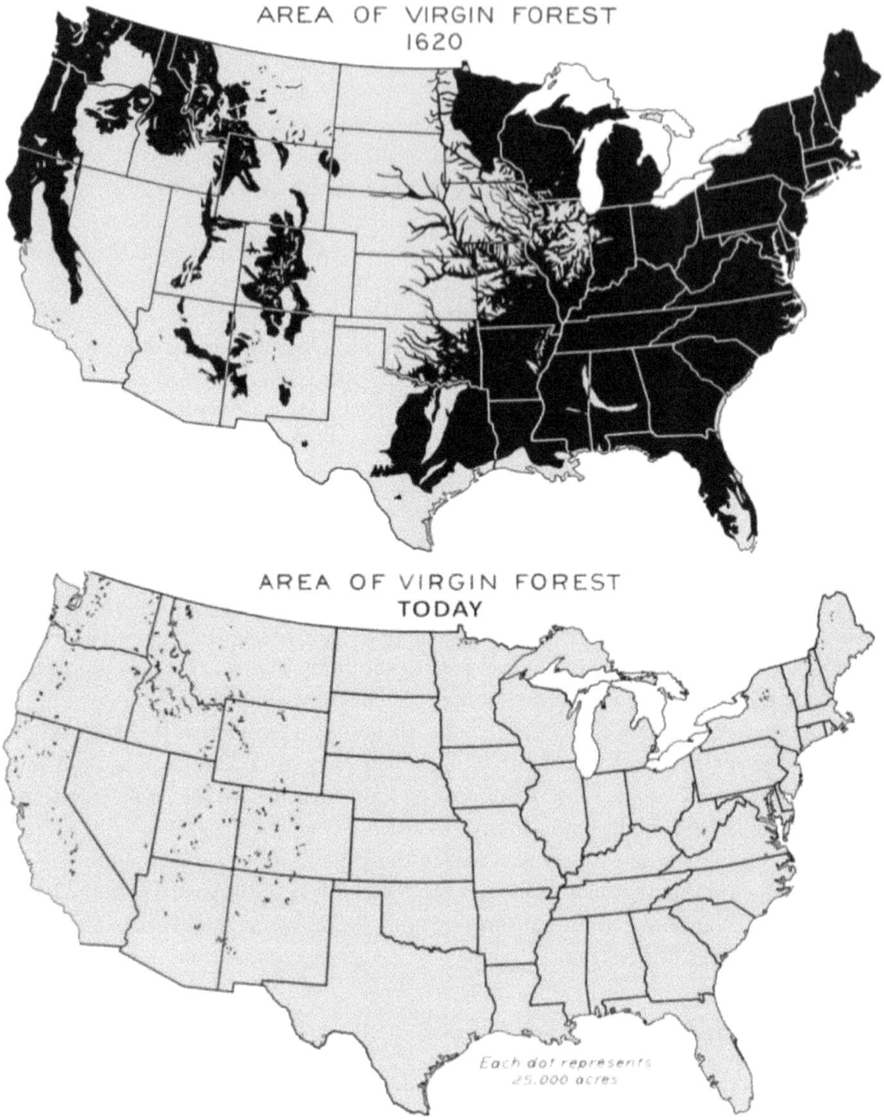

FIGURE 7.2 Amount of virgin forest in the United States in 1620 and 2020. (From United States Department of Agriculture. 2024. *U.S. Forest Resource Facts and Historical Trends*. www.fia.fs.fed.us/library/brochures/docs/2012/ForestFacts_1952-2012_English.pdf. (Accessed March 29, 2024).)

- Soil erosion
- Soil degradation
- Water cycle disruption
- Increased flooding
- Increased pollutant discharge
- Increased pollutant transport
- Increased localized temperature
- Cultural heritage loss

FIGURE 7.3 Forest harvesting in California near Yosemite National Park. (Photograph by Daniel T. Rogers.)

7.5.4 Wildfires in the United States

Wildfires affect all biomes, from forests and savannas to grasslands and tundra. Globally, over 370 million hectares (914 million acres) of land burn every year. This burning is estimated to have released 1.8 billion metric tons of greenhouse gases into the atmosphere. Wildfires in forests account for 5% of the total land burned but account for 80% of greenhouse gas emissions (United Nations 2024d).

The total amount of land lost to wildfires in the United States has increased in recent decades. The average for the decade from 2011 to 2020 was just over 3 million hectares (7.5 million acres), which exceeds the amount of forest harvested for wood products (Rainforest Alliance 2024). In the last 30 years, climate change is estimated to have doubled the total forest area burned in the western United States. Unsustainable land use and environmental degradation reduce the resilience of natural ecosystems to wildfires, while deforestation and drainage of peatlands worsen drought conditions and increase flammability (United Nations 2024d). The United Nations (2024d) has estimated that wildfire-induced air pollution is associated with 340,000 premature deaths from respiratory and cardiovascular issues. The negative economic effects from wildfires are mounting. The 2018 California Camp Fire is estimated to have caused around 19 billion dollars damage in direct costs, while the 2019–2020 wildfires in Australia resulted in a 23-billion-dollar damage in direct costs. It is now a realization that up to 6 million homes in the United States cannot obtain fire insurance due to the risk of wildfires (United Nations 2024d).

7.6 MINING AND PETROLEUM DEVELOPMENT

Mining and petroleum development are building blocks of our current society. Mining provides humans with the resources to assist in economic growth and improve standards of living. Petroleum provides humans with its main source of energy and also consumer goods made from plastics. Comparatively speaking, mining and petroleum development directly impact much less land than farming and forest management.

For organization purposes, we will discuss mining and petroleum development separately in the following sections.

7.6.1 Mining Development

Mining extracts useful material from the Earth. Although mining provides many valuable minerals, it can also harm people and the environment. Some examples of substances that are mined include the following (National Geographic Society 2024d):

- Gold
- Silver
- Copper
- Aluminum
- Titanium
- Lithium
- Zinc
- Chrome
- Cadmium
- Arsenic
- Nickel
- Molybdenum
- Manganese
- Magnesium
- Zircon
- Iron
- Phosphorus
- Talc
- Potash
- Rare earth and other minerals, including tin, cobalt, vanadium, tungsten, and niobium
- Limestone and dolomite
- Sand
- Gravel
- Salt
- Clay
- Dimension stone, such as granite and marble
- Coal (covered in Section 7.6.2)
- Gems such as diamonds, rubies, and sapphires
- Other materials and substances

7.6.1.1 Mining Purposes and Techniques

Mining is the process of extracting useful materials from the Earth. Mining dates back thousands of years. Some of the first substances mined were flint and obsidian for the construction of arrows, which were used in hunting and for weapons. The mining of gold and silver was typically done for decorative and ceremonial purposes. There are two main categories of modern mining techniques, and can be simply stated as above-ground mining and below-ground mining (National Geographic Society 2024d). There are two other types of mining that are less common and are called placer mining and in situ mining. Placer mining involves the removal of valuable mineral deposits from loose river deposits. In situ mining involves in situ leaching or solution mining. In situ mining involves the application of a liquid into the ore body that dissolves the targeted mineral and is subsequently recovered (American Mining Services 2024).

Typically, above-ground mining is the method of choice when the product of choice is near the surface. Below-ground mining is typically conducted through the construction of vertical and horizontal shafts sometimes extending thousands of meters underground. Typical, below-ground mining is more expensive and more dangerous than above-ground mining. Both mining techniques generate high volumes of waste materials. This is typical because the desired material targeted for mining may only be at a concentration of 10% or less within mined material. The process of extracting the desired material generates high volumes of waste in both techniques (National Geographic Society 2024d).

Mining revenue from the top 40 mining companies had a total revenue of 544 billion dollars in 2022. The most mined products in tons mined in 2022 included the following:

- Iron ore – 2,600,000,000 tons
- Industrial metals – 181,579,892 tons
- Technology and precious metals – 1,474,899 tons

7.6.1.2 Ecological Impacts from Mining

Ecological impacts occur with both above-ground and below-ground mining techniques. Surface mining can be destructive to the surrounding landscape, leaving huge open pits behind. Mining can also harm the environment through a byproduct of mining called acid mine waste, especially if the minerals being mined contain sulfides. The process of mining exposes materials higher in sulfide concentration, and when rainwater or streams dissolve the sulfides, they form acids. The acidic water damages aquatic plants and animals. Mine waste can also cause water pollution from other toxic metals that may be contained at low concentrations in the water, including mercury and arsenic (National Geographic Society 2024d).

An incident in 2015 in Colorado is a good example of the potential negative environmental impacts from mining. In 2015, USEPA was conducting an investigation of the Gold King Mine near Silverton, Colorado. While conducting the investigation, pressurized water began leaking above the mines tunnel, which resulted in spilling an estimated 12 million L (more than 3 million gallons) of mine waste into Cement Creek, which is a tributary of the Animas River USEPA 2024. The mine waste, containing heavy metals and low pH water, traveled many tens of kilometers downstream of the mine and turned the water in the Animas River a bright yellow. The USEPA has since placed the Bonita Peak Mining District, which includes the Gold King Mine, on the National Priorities Site, commonly known as Superfund (USEPA 2024i).

Other types of ecological impacts from mining include the following:

- Destruction of natural habitat by actual mining process but also from road building and other land destruction activities associated with mining, such as the process of mineral exploration
- Drainage pattern disruption
- Increased erosion
- Creation of sinkholes
- Degraded water quality
- Water pollution
- Species disruption
- Noise pollution
- Chemical pollution
- Light pollution

7.6.2 Petroleum Development

Petroleum development includes oil, natural gas, and coal. Petroleum accounts for the supply of 80% of the energy needs of the United States (United States Department of Energy (DOE) 2024a). The petroleum industry in the United States employs 12.3 million people as of the end of 2023.

Between 2012 and 2025, the petroleum industry in the United States is projected to provide 1.6 trillion dollars in tax revenue to the federal and state governments.

7.6.2.1 Petroleum Formation and Benefits

As we have discussed earlier in this book, petroleum, natural gas, and coal are formed from the remains of ancient organisms, such as plants, algae, and bacteria. Over millions of years of intense heat and pressure, these organic remains (fossils) transform into carbon-rich substances (National Geographic Society 2024e). Petroleum is also called crude oil. Petroleum is refined to make gasoline, diesel fuel, and aviation fuel. In addition, there are thousands of other products where petroleum is a component, including some of the following (United States Department of Energy (DOE) 2024b):

- Tires
- Refrigerators and other appliances
- Life jackets
- Plastics
- Pharmaceuticals
- Anesthetics
- Petroleum jelly
- Lip balm
- Computers
- Adhesives
- Antiseptics
- Asphalt
- Clothing
- Toothpaste
- Water pipes
- Yarn
- Solar panels
- Spacesuits
- Lipstick
- Heart valves
- Crayons
- Cell phones
- Thousands of other products

7.6.2.2 Petroleum Exploration and Extraction

In some places, petroleum bubbles to the surface of the Earth. An example of where crude oil is at the surface is the Le Brea Tar Pits in Los Angeles, California. The total amount of petroleum in a reservoir is termed **oil-in-place**. Many petroleum liquids that make up a reservoir are unable to be extracted. The part of the reservoir's oil-in-place that can be extracted is termed a reservoir's **oil reserves**. Oil is typically discovered by drilling at locations where the geology favors the presence of oil and is called exploratory drilling. Once discovered, additional drilling is conducted to evaluate the size and producibility of the oil reserve. The largest oil reserve in the United States is Prudhoe Bay, Alaska. Other large reserves are located in Texas and other locations (National Geographic Society 2024e).

Oil reserves are also located offshore. In the United States, offshore oil reserves are located in the Gulf of Mexico and off the southern California coast (National Geographic Society 2024e). Offshore oil is drilled from offshore drilling platforms. Some are very large and can be tethered to the ocean floor and float or they can be a rigid structure that is fixed to the ocean floor with concrete or steel legs (National Geographic Society 2024e).

7.6.2.3 Ecological Impacts from Petroleum Development

Petroleum products such as gasoline, diesel fuel, aviation fuels, and oil are some of the most significant pollutants on Earth. As stated earlier, 80% of greenhouse gases originate from petroleum combustion. However, many of the compounds are manufactured using petroleum compounds and include some compounds listed below (USEPA 2024j):

- Benzene
- Toluene
- Ethyl benzene
- Xylene compounds
- PCBs
- Tetrachloroethene
- Trichloroethene
- Vinyl chloride
- Dichlorodiphenyltrichloroethane

Other ecological impacts, mainly from petroleum spills, include the following (USEPA 2024h):

- Ocean contamination
- Water contamination
- Ecological destruction
- Greenhouse gas emissions and climate change impacts
- Disruption of wildlife behavior
- Microplastic ingestion
- Plastic pollution
- Coastline and beach pollution impacts

One of the largest environmental disasters involving petroleum products in the world occurred in April 2010 in the Gulf of Mexico. Figure 7.4 shows a photograph of the Deepwater Horizon engulfed in flames. Some of the ecological damage from the spill, which discharged millions of liters of oil into the Gulf of Mexico, include the following (NOAA 2024b):

- Spilled an estimated 300,000,000 L of oil into the ocean
- Impacted 1,800 km of coastline
- Impacted 150,000 km^2 of surface water
- Plant cover and vegetation mass reduced along 350 to 720 miles of shoreline
- Amphipods, periwinkles, shrimp, forage fish, red drum, fiddler crabs, and insects killed
- 4 to over 8 billion harvestable oysters lost
- Between 51,000 and 84,000 thousand birds killed
- Between 56,000 and 166,000 small juvenile sea turtles killed
- Up to 50% decrease in Barataria dolphin population
- An estimated 2 to 5 million newly hatched fish were killed
- 400 to 700 square miles of coral reefs damaged

In March 1989, the oil tanker Exxon Valdez spilled an estimated 40 million liters of crude oil into Prince William Sound in Alaska. At the time, it was the worst oil spill in United States history until

FIGURE 7.4 Deepwater Horizon engulfed in flames. (From National Oceanic and Atmospheric Administration (NOAA). 2024b. Deep Water Horizon Incident Response. https://response.restoration.noaa.gov.deep-water-horizon. (Accessed March 30, 2024).)

the Deepwater Horizon spill in 2010 (USEPA 2024k). The oil slick impacted more than 2,500 km of coastline and killed hundreds of thousands of seabirds, otters, seals, and whales. As a direct result of the oil spill, the Oil Spill Prevention Act of 1990 was enacted. Cleanup of the oil was mainly achieved through mechanical means and included the following (USEPA 2024k):

- Applying oil booms to concentrate the oil and pump the oil into containers aboard recovery ships
- Washing shorelines
- Physically recovering oil from shorelines and beaches
- Excavating shorelines and beaches

Burning the oil and applying a dispersant were both attempted but had little or no effect. Billions of dollars were spent recovering oil and cleaning tens of miles of shoreline and saving and cleaning affected wildlife (USEPA 2024k; NOAA 2024c). Figure 7.5 shows the Exxon Valdez in Prince William Sound after the spill occurred and recovery efforts were underway (USEPA 2024k). Figure 7.6 shows a photograph of some affected wildlife (USEPA 2024k). Immediate effects of the spill included estimated deaths of 100,000 to as high as 250,000 seabirds, nearly 3,000 sea otters, 300 harbor seals, nearly 250 bald eagles, 22 orcas, and an unknown number of salmon and herring (NOAA 2024c).

The Oil Spill Prevention Act resulted in instrumental changes in oil production, transportation, and distribution that included the following (USEPA 2024k, 2024l):

- Establishing a fund for future cleanup of spills
- Increasing spill response preparedness

FIGURE 7.5 Exxon Valdez oil spill. (From United States Environmental Protection Agency (USEPA). 2024c. *USEPA Exxon Valdez Spill Profile*. www.epa.gov/emergency-response/exxon-valdez-spill-profile. (Accessed March 31, 2024).)

FIGURE 7.6 Example of affected wildlife from the Exxon Valdez oil spill. (From United States Environmental Protection Agency (USEPA). 2024c. *USEPA Exxon Valdez Spill Profile*. www.epa.gov/emergency-response/exxon-valdez-spill-profile. (Accessed March 31, 2024).)

- Increasing reliability in mapping and location of ships
- Improving ship designs
- Instituting measures and controls to minimize spills during loading and unloading
- Increasing financial assurance requirements

7.7 WATER AND OCEANS

Water as well as energy and organic molecules were the prerequisites for the origin of life on Earth billions of years ago. Today, water plays a central role in maintaining our survival as a species, preserving the natural environment, and therefore achieving a sustainable planet (United Nations 2003).

Oceans cover 70% of the surface of the Earth or 360 million km^2. In many respects, the oceans of the world are the last resting place for most anthropogenic pollution. As an example, an estimated 7 million metric tons of plastic is dumped into the world's oceans every year (NOAA 2024d). This is just a portion of the pollution that enters the oceans every year. The NOAA (2024d) estimates that billions of pounds of trash and other pollutants enter the oceans each year. The following sections offer discussion points on anthropogenic impacts to water and the ocean that go beyond plastic pollution.

7.7.1 Disruptions to the Water Cycle

The water cycle on Earth has been disrupted by anthropogenic activities. Our actions as a species have effects on every step of the water cycle. Some of those impacts to the water cycle include the following (USGS 2024):

- *Glaciers and ice sheets:* Glaciers and ice sheets are the largest storage of freshwater on Earth. However, water stored in glaciers and semi-permanent snowfields has decreased every year for more than 30 years. The main reason for this melting is climate change, which is caused by humans.
- *Sea-level rise from melt water:* Meltwater from glaciers and ice caps results in a rise in sea level.
- *Sea-level rise from ocean heating:* Heating of the oceans adds to sea-level rise due to thermal expansion
- *Precipitation:* Changes to the global climate impact where and how much precipitation occurs on land.
- *Runoff.* Increased or decreased runoff due to disrupted climate patterns leads to more severe drought and extreme precipitation events.
- *Infiltration:* In general terms, surface infiltration is lowered in urban areas and leads to more stormwater runoff and can result in increased erosion and decreased groundwater recharge.
- *Groundwater:* Groundwater provides potable water for approximately 50% of the population of the United States. In addition, groundwater is used significantly for agriculture and livestock. Overuse of groundwater has significant impacts related to land subsidence and decreased water quality in surface waters.
- *Transpiration:* Transpiration is the process of plants taking up liquid water through their roots and releasing water vapor to the atmosphere through their leaves. Transpiration is an important part of the water cycle because it moves water from the soil back into the atmosphere. Tropical rainforests rely on the process to maintain plant life. Human activities through many activities, including deforestation, have disrupted the transpiration cycle.

7.7.2 Water Pollution

Water pollution from anthropogenic sources has now resulted in polluting most, if not all, water systems on Earth through the introduction of pollutants, including chemical contaminants and microorganisms, to the air, water, and land (Rogers 2020). Water pollution impacts now affect every human and kill more people each year than war and all other forms of violence combined. The most significant anthropogenic activity resulting in water degradation is agriculture. In large part, this is because agriculture uses up to 70% of freshwater utilized by humans. Contaminants from agricultural sources include pesticides (including DDT), herbicides, fertilizers, volatile organic compounds, polynuclear aromatic hydrocarbons, sewerage, road salt, pathogens, oil, gasoline, heavy metals, and many others (Natural Resources Defense Council (NRDC) 2024a).

An additional pollutant source to water is wastewater, including sewerage. NRDC (2024a) estimates that more than 80% of wastewater flows back into the environment without being treated or reused. In some least-developed countries, the figure exceeds 95%. The United States also has difficulty treating sewerage and wastewater, especially during extreme precipitation events when large volumes of untreated wastewater and sewerage are released into the Great Lakes, Mississippi and Ohio Rivers, other surface water bodies, and the ocean. Impacts from extreme precipitation events and aging infrastructure of water treatment systems are a couple of the main causes for the release of more than 850 billion gallons of untreated wastewater and sewerage each year in the United States (NRDC 2024a).

7.7.3 Water for Energy and Storage

Humans have learned early in our history to use water for energy and storage. This was accomplished through constructing aqueducts to bring freshwater to Rome 2,000 years ago, where windmills were used to move water and provide energy to conduct other tasks. In the 20th century, dams were built to generate electricity. Dams are massive structures that retain water for domestic use, irrigation, generating electricity, and industrial use. Dams are also constructed to assist in flood mitigation.

However, dams have negative ecological effects as well. They trap enormous amounts of sediments in their reservoirs, interrupt the flow of nutrients, disrupt the natural flow of rivers and streams, impede migratory routes of many fish, disrupt the flow of carbon between continents and the ocean, cause the loss of biodiversity of aquatic species, release greenhouse gases from decomposing vegetation, and cause habitat destruction for birds and other animals that once lived in floodplains (Scientific American 2024).

7.7.4 Ocean Pollution

Water and the ocean impact every human life. The ocean supplies freshwater to the Earth and nearly all of Earth's oxygen. The ocean moderates Earth's climate, influences our weather, and affects human health. The ocean provides food, medicines, minerals, and energy resources. Much of the human population lives near the ocean. However, humans have not treated the ocean with kindness and respect. The oceans have been the dumping ground for not just our garbage but for some of our most toxic wastes for hundreds of years. The oceans have been treated as the human privy and garbage disposal.

Only in the past few decades has the dumping of wastes in the oceans come under some environmental regulations (United Nations 2024f). Figure 7.7 is a photograph showing a historical example of liquid chemical disposal directly into the ocean from a ship. Figure 7.8 is a photograph of what has come to be known as the garbage barge, in which garbage from New York City would be loaded on a barge and then dumped into the ocean several miles offshore. However, polluting the oceans continues to this day. The oceans receive pollution from direct discharge or dumping, from deposition from the atmosphere, from rivers and streams that discharge into the oceans, and directly from the land itself.

The oceans were once believed to be so vast and deep that the effects of pollution would never directly threaten the oceans. This has been proven false. The rate of accumulation of pollution is proportional to human population and increasing energy requirements. The more people, the more pollution ends up in the oceans. The most common types of pollution in the oceans include plastic and other forms of garbage, heavy metals, organic pollutants such as polychlorinated biphenyls (PCBs) and oil, radioactive substances and waste, biological pollutants, pesticides, herbicides, fertilizers, and untreated raw sewage (NOAA 2024d).

FIGURE 7.7 Example of chemical disposal directly in the ocean from the rear of a ship. (From United States Environmental Protection Agency (USEPA). 2024m. *Summary of the Oil Pollution Act of 1990*. Summary of the Oil Pollution Act | US EPA; www.usepa.gov/summary-or-oil-pollution-act-of-1990. (Accessed May 11, 2024).)

FIGURE 7.8 Solid waste disposal into the ocean from the garbage barge in 1988. (From United States Environmental Protection Agency (USEPA). 2024m. *Summary of the Oil Pollution Act of 1990*. Summary of the Oil Pollution Act | US EPA; www.usepa.gov/summary-or-oil-pollution-act-of-1990. (Accessed May 11, 2024).)

Pollution in the oceans does not all originate from direct discharge but also is delivered through the atmosphere in the form increased concentrations of carbon dioxide in the atmosphere that is partially adsorbed by the oceans and causes acidification of the surface layers (NOAA 2024d). Absorption of carbon dioxide by the oceans has been estimated to be as high as 25%, which represents millions of tons of carbon dioxide absorbed by the oceans every year. The additional carbon dioxide absorbed by the oceans reacts with ocean water and forms an acid called carbonic acid and is lowering the pH of the upper portions of the oceans worldwide. In fact, the oceans are acidifying faster than they have in some 300 million years. The effect of the acidification has been linked to the bleaching of coral reefs, including the Great Barrier Reef off the northeastern coast of Australia (Natural Resources Defense Council 2024b).

An additional form of pollution in the ocean that is not commonly mentioned is noise. Sound waves travel farther and faster in the oceans than they do in the atmosphere. Increased human-induced noise in the ocean is harming numerous marine species because they use sound to navigate, find food, and mate. These crucial activities for marine life survival are having widespread impacts (NRDC 2024b).

Some impacts to the ocean from anthropogenic sources include the following (NOAA 2024d; NRDC 2024b):

- Habitat destruction
- Ocean acidification
- Ocean heating
- Nitrogen enrichment
- Chemical pollution
- Plastic pollution
- Garbage disposal
- Destructive and overfishing
- Surface water runoff
- Deoxygenation
- Deep-sea mining
- Noise pollution
- Skincare chemical pollution

Even though the oceans cover over 70% of the Earth's surface, they are poorly understood. Many geologists and other scientists state that, "Whatever Nature Creates Eventually Ends Up in The Ocean," which could not be more true. However, this saying can also be reinterpreted to state that "Anything Humans Discard Eventually Ends Up in the Ocean." Because the oceans are not well understood, it is likely that anthropogenic impacts to the oceans are also not well understood, meaning that the list below is likely incomplete (NOAA 2024f).

7.7.4.1 Habitat Destruction

Ocean habitat has been affected by human activities from three primary types of activities that include physical destruction, climate change, and contamination, including chemical and solid waste. Physical destructive activities include drilling, mining, dredging, destructive anchoring, removal and destruction of coral reefs, and land reclamation. Climate change habitat destructive activities include ocean heating, marine heat waves, acidification, and sea-level rise. Contaminant destructive factors include chemical contaminants, garbage, and plastic pollution (NOAA 2024d, 2024e).

7.7.4.2 Ocean Acidification

Human activities have increased carbon dioxide concentration in the atmosphere, which has also caused ocean acidification through carbon dioxide reacting with ocean water to create carbonic acid.

Since the Industrial Revolution began more than 200 years ago, the concentration of carbon dioxide in the atmosphere has increased by more than 30%. The ocean absorbs carbon dioxide, and from this absorption, creates carbonic acid (H_2CO_3), and during this time since the Industrial Revolution, the acidity of the oceans has increased by 30% (NOAA 2024f).

7.7.4.3 Ocean Heating

When sunlight reaches Earth's surface, the world's oceans absorb some of this energy and store it as heat. This heat is initially absorbed at the surface, but some of it eventually spreads to deeper waters. Currents also move this heat around the planet. Water has a much higher heat capacity than air, meaning the oceans can absorb larger amounts of heat energy with only a slight increase in temperature. The total amount of heat stored by the oceans is called **ocean heat content**, and measurements of water temperature reflect the amount of heat in the water at a particular time and location. Ocean temperature plays an important role in the Earth's climate system, particularly sea surface temperatures, because heat from ocean surface waters provides energy for storms, such as hurricanes, thereby influencing weather patterns (USEPA 2024n).

Increasing GHG concentrations are trapping more energy from the sun. The oceans have not yet warmed as much as the atmosphere, even though they have absorbed more than 90% of the Earth's extra heat since 1955 (USEPA 2024n).

Figure 7.9 shows changes in the heat content of the top 700 meters of the world's oceans between 1955 and 2020. Ocean heat content is measured in joules, a unit of energy, and is compared against the 1971 to 2000 average, which is set to zero for reference. In the figure, an increase of 1 unit on this graph (1×10^{22} J) is equal to approximately 17 times the total amount of energy used by all the people on Earth in a year. The lines were independently calculated using different methods by government organization in four countries that include the US NOAA, Australia's Commonwealth Scientific and Industrial Research Organisation, China's Institute of Atmospheric Physics, and the Japan Meteorological Agency's Meteorological Research Institute (USEPA 2024n).

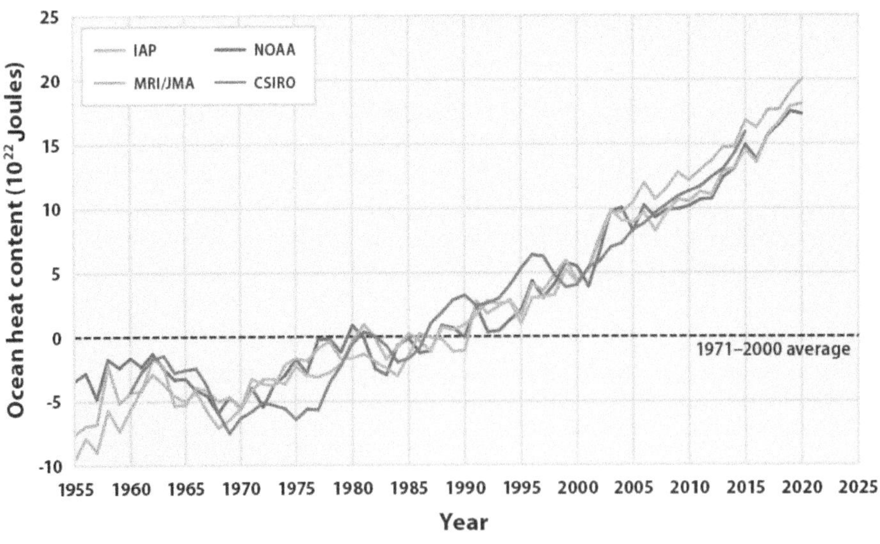

FIGURE 7.9 Change in heat content of the ocean since 1955. (From United States Environmental Protection Agency (USEPA). 2024a. *Climate Change Indicators: Ocean Heat.* Climate Change Indicators: Ocean Heat | US EPA; www.epa.gov/climate-indicators/cliamte-change-indicators-ocean-heat. (Accessed April 25, 2024).)

The heating of the oceans is contributing to several negative effects that are now being realized and include sea-level rise, coral bleaching, marine heatwaves, increasing the rate of ocean-terminating glacial melting, and loss of marine biodiversity (United Nations 2024e).

The increased heat stored in the ocean because of climate change will eventually be released back into the atmosphere. This means that the Earth will experience even more warming in the future. A way to visualize or think of the relationship between the atmosphere and ocean is that the atmosphere is nature's short-term memory and the ocean is nature's long-term memory.

7.7.4.4 Nitrogen Enrichment

Nitrogen enrichment from anthropogenic sources reduces the diversity of marine bacteria, especially in intertidal zones and other locations where nitrogen enters the ocean. The impacts and significance of nitrogen enrichment to the role of biogeochemical cycles are not fully understood. In addition, the role of phosphorus does not appear to be as significant in reducing the diversity of marine bacteria (Xu et al. 2022).

7.7.4.5 Chemical Pollution

Major oil spills have impacted the oceans, coastline, and wildlife. These incidents tend to raise our collective awareness and, in some instances, trigger legislative action to prevent these types of disasters from occurring in the future. However, there are thousands of chemical pollutants that are released into the air, water, and land that make their way to the ocean. There are other impacts that are long-term impacts that have affected reproductive cycles and other biological processes. Contaminant loading from chemical pollution to the ocean is unknown (National Geographic Society 2024f).

7.7.4.6 Plastic Pollution

Plastic pollution is a recent occurrence but has made significant impacts on the ocean and wildlife. The number adds up to trouble for the oceans and humans. There are an estimated 5.25 trillion pieces of plastic debris in the ocean. Of that mass, nearly 270,000 metric tons float on the surface, while some four billion plastic microfibers per square kilometer litter the depths of the ocean (National Geographic Society 2024g).

7.7.4.7 Garbage Disposal

For centuries, the ocean was a dumping ground and landfill of the Earth. Not until the 1980s did disposal of waste into the oceans become regulated (USEPA 2024m). The amount of garbage in the ocean is unknown but is certainly a staggering amount. Scientists have known for decades that the accumulating mass of discarded debris in the ocean was creating negative impacts on seabirds, fish, coral reefs, plankton, and marine animals (National Geographic Society 2024g).

7.7.4.8 Destructive and Overfishing

Certain fishing practices, called **destructive fishing**, not only contribute to the overfishing of target species, but also damage the environment by dredging the seabed or catching unintended species. The term originated approximately 70 years ago and is defined as any type of fishing technique or fishing practice that reduces the fishing stock in an unsustainable way and destroys fish and invertebrate habitats and ecosystems (National Geographic Society 2024h).

7.7.4.9 Surface Water Runoff

As mentioned earlier, fertilizers, pesticides, and thousands of other chemicals transported by surface water contribute to the decline in ocean water quality. Surface water runoff contains eroded materials from the land that often contain harmful chemicals that can bioaccumulate in the tissues of marine organisms, such as PCBs, and many other substances (NOAA 2024f).

7.7.4.10 Deoxygenation

Fertilizers, namely phosphorus and nitrogen, and other chemicals tend to reduce oxygen levels locally through the development of algal blooms and sometimes regionally in the ocean so much so that they are labeled dead zones (NOAA 2024f).

7.7.4.11 Deep-Sea Mining

This is a relatively new issue that involves mining the seafloor for rare earth and other metals that are required for laptop computers, phones, and other commercial products. The mining process, which involves scraping or dredging the ocean floor, creates large sediment plumes and destroys habitat for fish and many invertebrates (NOAA 2024f).

7.7.4.12 Noise Pollution

Sound waves travel farther and faster in the oceans than they do in the atmosphere. Increased human-induced noise in the ocean is harming numerous marine species because they use sound to navigate, find food, and mate. These crucial activities for marine life survival are having widespread impacts (NRDC 2024b).

7.7.4.13 Skincare Chemical Pollution

Common chemicals used in thousands of products to protect against harmful effects of ultraviolet light radiation threaten corals and other marine life. Healthy coral reefs are one of the most valuable ecosystems on Earth. Chemicals contained in sunscreen and other personal healthcare products threaten reefs. The following is a short list of the effects of these products on marine ecosystems (NOAA 2024f):

- *Green algae:* Can impair growth and photosynthesis.
- *Coral:* Accumulates in tissues. Can induce bleaching, damage DNA, deform young, and even kill coral life.
- *Mussels;* Can induce genetic defects in young.
- *Fish:* Can decrease fertility and reproduction and cause female characteristics in male fish.
- *Dolphins:* Can accumulate in tissue and be transferred to young.

Chemicals in sunscreen that are suspected of causing harm to marine life include the following (NOAA 2024g):

- Oxybenzone
- Benzophenone-1
- Benzophenone-8
- OD-PABA
- 4-Methylbenzlidene camphor
- 3-Benzylidene camphor
- Nano-titanium dioxide
- Nano-zinc oxide
- Octinoxate
- Octocrylene

7.8 SUMMARY AND CONCLUSION

In this chapter, we discussed a few of the most impactful human activities that negatively impact Earth. They have included agriculture, forest management, urban development, mining and petroleum development, and water and the oceans.

Human advances in science have created many technologies that have improved our standard of living and, in some circumstances, have improved our relationship with nature. But our technological advancements have also created many unintended negative side effects that question the benefit of some technologies. An example involves the creation of synthetic compounds that nature has difficulty breaking down. Some of these compounds include PCBs, chlorinated hydrocarbons, and many pesticides, including DDT, which began the environmental movement in the United States in earnest over 50 years ago. This is to say nothing about plastics, which we all should know about as we commonly use and discard huge amounts of plastic in our everyday lives.

In the next chapter, we shall discuss these synthetic compounds and other forms of chemical pollution and learn about their behavior in nature. We shall also learn why humans need to understand the geology and subsurface environment of any area because it is the geological environment that is important to understand if these impacts are to be understood fully and reduce risks posed by these substances so that human health and the environment can be better protected.

REFERENCES

American Mining Services. 2024. *Different Types of Mining*. www.americanminingservices.com/different-types-of-mining. (Accessed March 30, 2024).

Bannon, C. 2024. Fresh Water in Roman Law: Rights and Policy. *Journal of Roman Studies*. Vol. 107. www.cambridge.org/core/journals/journal-of-roman-studies/article/fresh-water-in-roman-law-rights-and-policy/548B1C559B3D6ACEDF50C4576DD14603/core-reader#fns01. (Accessed March 29, 2024).

Carson, R. 1962. *Silent Spring*. Houghton Mifflin. Boston, MA.

Chen, G, Li, X., Liu, X., Chen, Y., Liang, X., Leng, J., Xu, X., Liao, W., Qiu, Y., Wu, Q., and Huang, K. 2020. Global Projections for Future Urban Expansion Under Shared Socioeconomic Pathways. *Journal of Nature Communications*. Vol. 11, Article no. 537, pp. 1–12.

Crop Life in America. 2024. *History of Agriculture in the United States*. www.croplifeamerica.org/history. (Accessed March 28, 2024).

Darke, R., and Tallamy, D. 2019. *The Living Landscape*. Timber Press. Portland, Oregon. 392pp.

Environmental Education Associates. 2024. *The Famous Benjamin Franklin Letter on Lead Poisoning*. http://environmentaleducation.com/wp-content/uploads/userfiles/Ben%20Franklin%20Letter%20on%20EEA%281%29.pdf. (Accessed March 28, 2024).

Environmental Science. 2024. *History of Agriculture*. www.environmentalscience.org/history-of-agriculture. (Accessed March 28, 2024).

Harari, Y. N. 2018. *Sapiens: A Brief History of Humankind*. Harper Collins Publishers. New York, NY. 443pp.

Haynes, W. 1954. *American Chemical Industry – A History*. Vols. I–IV. Van Nostrand Publishers. New York, NY.

Jiang, L., Deng, X., and Seto, K. C. 2013. The Impact of Urban Expansion on Agricultural Land Use intensity in China. *Journal of Land Use Policy*. Vol. 35, pp. 35–39.

McGucken, W. 2000. *Lake Erie Rehabilitated: Controlling Cultural Eutrophication, 1960s–1990s*. University of Akron Press. Akron, OH.

National Geographic Society. 2024a. *The Farming Revolution*. https://education.nationalgeographic.org/resource/development-of-farming. (Accessed March 28, 2024).

National Geographic Society. 2024b. *Urbanization*. https://education.nationalgeographic.org/resource/urbanization. (Accessed March 29, 2024).

National Geographic Society. 2024c. *Deforestation*. https://education.nationalgeographic.org/resource/deforestation. (Accessed March 29, 2024).

National Geographic Society. 2024d. *Mining*. https://education.nationalgeographic.org/resource/mining. (Accessed March 30, 2024).

National Geographic Society. 2024e. *Petroleum*. https://education.nationalgeographic.org/recource/petroleum. (Accessed March 30, 2024).

National Geographic Society. 2024f. *Marine Pollution*. www.nationalgeographic.org/marine-pollution. (Accessed May 13, 2024).

National Geographic Society. 2024g. *Ocean Trash*. www.nationalgeographic.org/recource/ocean-trash. (Accessed May 13, 2024).

National Geographic Society. 2024h. *Destructive Fishing*. www.nationalgeographic.org/resource/destructive-fishing. (Accessed May 13, 2024).

National Oceanic and Atmospheric Administration (NOAA). 2024a. *Gulf of Mexico/Florida Harmful Algal Blooms*. https://oceanservice.noaa.gov/hazards/hab/gulf-mexico. (Accessed March 29, 2024).

National Oceanic and Atmospheric Administration (NOAA). 2024b. *Deep Water Horizon Incident Response*. https://response.restoration.noaa.gov.deep-water-horizon. (Accessed March 30, 2024).

National Oceanic and Atmospheric Administration (NOAA). 2024c. *Exxon Valdez Oil Spill. Office of Response and Restoration*. www.noaa.gov/exxon-valdez-oil-spill-restoration. (Accessed March 31, 2024).

National Oceanic and Atmospheric Administration (NOAA). 2024d. *Ocean Pollution*. www.noaa.gov/resource-collections/ocean-pollution. (Accessed March 31, 2024).

National Oceanic and Atmospheric Administration (NOAA). 2024e. *Climate Change: Ocean Heat Content*. www.noaa.gov/news-features/understanding-climate/climate-change-impacts. (Accessed May 12, 2024).

National Oceanic and Atmospheric Administration (NOAA). 2024f. *Ocean Acidification*. www.noaa.gov/education/resource-collections/ocean-coast/ocean-acidification. (Accessed May 13, 2024).

National Oceanic and Atmospheric Administration (NOAA). 2024g. *Skincare Chemicals and Coral Reefs*. www.noaa.gov/news/suscreen-coral-reefs. (Accessed May 12, 2024).

Natural Resources Defense Council (NRDC). 2024a. *Water Pollution*. www.nrdc.org/stories/water-pollution. (Accessed May 12, 2024).

Natural Resources Defense Council (NRDC). 2024b. *Ocean Pollution: The Dirty Facts*. https://nrdc.org/stories/ocean-pollution-dirty-facts. (Accessed March 31, 2024).

Pennsylvania Historical Association. 2024. *The Vision of William Penn*. http://explorepahistory.com/story.php?storyId=1-9-3&chapter=1. (Accessed March 28, 2024).

Rainforest Alliance. 2025. Why Our Forest Are Burning. www.rainforest-alliance.org/insights/why-our-forests-are-burning. (Accessed February 4, 2025).

Rogers, D. T. 2020. *Urban Watersheds: Geology, Contamination, Environmental Regulations, and Sustainability*. CRC Press. Boca Raton, FL. 606pp.

Scientific American. 2024. *The Downside of Dams*. www.scientificamerican/article/doenside-of-dams. (Accessed May 13, 2024).

Seto, K. C. Golden, J. A., Alberti, M., and Turner, B. L. 2017. Sustainability in an Urbanizing Planet. *Proceedings of the National Academy of Sciences*. Vol 114, no. 34, pp. 8935–8938.

Smithsonian. 2024a. *Toxic Red Tides*. www.smitsonianmag.com/smart-news/toxic-red-tides. (Accessed March 2, 2024).

Smithsonian. 2024b. *Air Pollution Goes Back Further than You Think*. www.smithsonianmag.com/science-nature/air-pollution-goes-back-way-further-you-think-180957716/. (Accessed March 29, 2024).

United Nations. 2003. *International Decade for Action; Water for Life 2005–2015*. United Nations. New York, NY. www.un.org/waterforlifedecade/backgroundhtml. (Accessed May 12, 2024).

United Nations. 2024a. *Human Development Report*. United Nations Development Programme. www.hdr.undr.org/sites/2024. (Accessed March 28, 2024).

United Nations. 2024b. *Pollution from Farming Imperils Coastal Habitats*. www.unep.org.news-and-stories/pollution-from-farming. (Accessed March 29, 2024).

United Nations. 2024c. *World Population Prospects*. United Nations Department of Economic and Social Affairs. Population Division. https://esa.un.org/unpd/wpp/data. (Accessed March 29, 2024).

United Nations. 2024d. *As Wildfires Increase, Integrated Strategies for Forest, Climate, and Sustainability Are Ever More Urgent*. www.un.org/en/un-chronicle/wildfires-increase. (Accessed March 30, 2024).

United Nations. 2024e. *How Is Climate Change Impacting the World's Ocean*. www.un.org/en/climatechange/science/climateissues/climateimpacts. (Accessed April 25, 2024).

United Nations. 2024f. *Oceans and the Law of the Sea*. www.un.org/en/sections/issues-depth/oceans-and-law-sea. (Accessed April 16, 2024).

United States Department of Agriculture (USDA). 2024a. *United States Fertilizer Use and Cost*. USDA. Washington, DC. www.ers.usda.gov/data-products/fertilizer-use-and-cost. (Accessed March 29, 2024).

United States Department of Agriculture (USDA). 2024b. *U.S. Forest Resource Facts and Historical Trends*. www.fia.fs.fed.us/library/brochures/docs/2012/ForestFacts_1952-2012_English.pdf. (Accessed March 29, 2024)

United States Department of Energy (DOE). 2024a. *The Economic Benefits of Petroleum*. www.energy.gov/articles/economic-impact-oil-and-gas. (Accessed March 30, 2024).

United States Department of Energy (DOE). 2024b. *Products Made From Oil and Natural Gas*. www.energy.gov/articles/products-made-from-oil-and-natural-gas. (Accessed March 30, 2024).

United States Environmental Protection Agency (USEPA). 2024a. *Land Use: What Are the Trends in Land Use and Their Effects on Human Health and the Environment*. www.epa.gov/report-environment/land-use. (Accessed March 28, 2024).

United States Environmental Protection Agency (USEPA). 2024b. *EPA's Report on the Environment (ROE)*. www.epa.gov/report-environment. (Accessed March 29. 2024).

United States Environmental Protection Agency (USEPA). 2024c. *Integrated Risk Information System*. www.epa.gov/iris. (Accessed March 24, 2024).

United States Environmental Protection Agency (USEPA). 2024d. *Agriculture Nutrient Management and Fertilizer*. www.epa.gov/agriculture/agriculture-nutrient-management-and-fertilizer. (Accessed March 29, 2024).

United States Environmental Protection Agency (USEPA). 2024e. *Pesticides*. www.epa.gov/pesticides. (Accessed March 29, 2024).

United States Environmental Protection Agency (USEPA). 2024f. *DDT – A Brief History and Status*. www.epa.gov/ingredients-used-pesticide-products/ddt-brief-history-and-status. (Accessed March 29, 2024).

United States Environmental Protection Agency (USEPA). 2024g. *Urbanization – An Overview*. www.epa.gov/caddis/urbanization-overview. (Accessed March 29, 2024).

United States Environmental Protection Agency (USEPA). 2024h. *Nonpoint Source Pollution: Urban Areas*. www.epa.gov/nps/nonpoint-source-pollution. (Accessed March 29, 2024).

United States Environmental Protection Agency (USEPA). 2024i. *Emergency Response to August 2015 Release from Gold King Mine*. www.epa.gov/goldkingmine. (Accessed March 30, 2024).

United States Environmental Protection Agency (USEPA). 2024j. *Petroleum Hydrocarbons and Chlorinated Hydrocarbons*. www.epa.gov/petroleumhydrocarbons. (Accessed March 30, 2024).

United States Environmental Protection Agency (USEPA). 2024k. *USEPA Exxon Valdez Spill Profile*. www.epa.gov/emergency-response/exxon-valdez-spill-profile. (Accessed March 31, 2024).

United States Environmental Protection Agency (USEPA). 2024l. *Summary of the Clean Water Act*. https://epa.gov/laws-regulations/summary-clean-water-act. (Accessed March 31, 2024).

United States Environmental Protection Agency (USEPA). 2024m. *Summary of the Oil Pollution Act of 1990*. www.usepa.gov/summary-or-oil-pollution-act-of-1990. (Accessed May 11, 2024).

United States Environmental Protection Agency (USEPA). 2024n. *Climate Change Indicators: Ocean Heat*. www.epa.gov/climate-indicators/cliamte-change-indicators-ocean-heat. (Accessed April 25, 2024).

United States Forest Service. 2024. *Forest Conservation and Management Act*. www.nfs.gov/FCMA. (Accessed March 30, 2024).

United States Geological Survey (USGS). 2006. *Volatile Organic Compounds in the Nation's Ground Water and Drinking-Water Supply Wells*. USGS Circular 1292. USGS. Reston, VA.

United States Geological Survey (USGS). 2009. *Toxic Substance Hydrology Program*. DNAPL. www.usgs.gov/definitions/dnapl_def.html. (Accessed March 29, 2024).

United States Geological Survey (USGS). 2024. *The Water Cycle*. www.usgs.gov/the-water-cycle. (Accessed May 12, 2024).

University of Cambridge. 2024. *From Foraging to Farming: The 10,000 Year Revolution*. www.cam.ac.uk/research/news/from-foraging-to-farming. (Accessed March 28, 2024).

Xu, Y. F., Dong, X. M., Lou, C., Ma, S. N., Xu, J. L., and Cui, Y. D. 2022. Nitrogen Enrichment Reduces the Diversity of Bacteria and Alter Their Nutrient Strategies in Intertidal Zones. *Journal of Aquatic Microbiology*. Vol. 9, pp. 99–108.

8 The Science of Chemical Contaminants and Pollution

8.1 INTRODUCTION

Would there be a need for environmental laws and regulations or the United States Environmental Protection Agency if there were no harmful chemicals released directly into the environment from anthropogenic sources? The answer is yes because there are many other impacts from anthropogenic sources that affect the environment. This is where sustainability takes over and plays a critical role. This may seem like a paradox but as we proceed deeper into the meaning of sustainability, it will become more than clear that there are numerous activities that impact our environment that go beyond what is commonly termed pollution or contamination. Many of these other forms do not introduce chemicals into the natural world. For instance, any number of different land uses where humans inflict enormous harm on our environment. Some of these land uses include mining, petroleum development, urban land development, wetland destruction, deforestation, forest management techniques, and agriculture. As we are discovering, these activities do not just affect a centralized area where the activity took place. These types of activities often have a rippling effect that negatively impacts surrounding areas and, in some cases, spreads like a disease that eventually can affect the entire planet. This is where sustainability plays an important and central role.

In this chapter, we will concentrate our discussion on chemical contaminants since they have formed the cornerstone in the development of environmental regulations and restoration of their harmful effects. We begin by defining a pollutant or contaminant. We then will explore how geology influences the behavior of contaminants once they are released. We conclude with a more detailed review of the more common types of contaminants.

8.2 DEFINING POLLUTION AND CONTAMINATION

Pollution and **contamination** are often used synonymously to mean the introduction into the environment by humans of substances that are harmful or poisonous to human health and ecosystems (van der Perk 2006). Over time, the term pollution evolved to include not only substances, but also other anthropogenic activities such as energy wastes such as heat, light, noise, plastic, dust, garbage, pollen from agricultural trees, paper products, or any waste that is discarded into the environment from any human activity that impacts the air, land, or water. Therefore, pollution generally is considered a broader term, and a chemical contaminant can then be considered a subset of pollution. For our purposes in this chapter, we will concentrate on chemical contaminants as this term is most often associated with chemical releases.

DOI: 10.1201/9781003463153-10

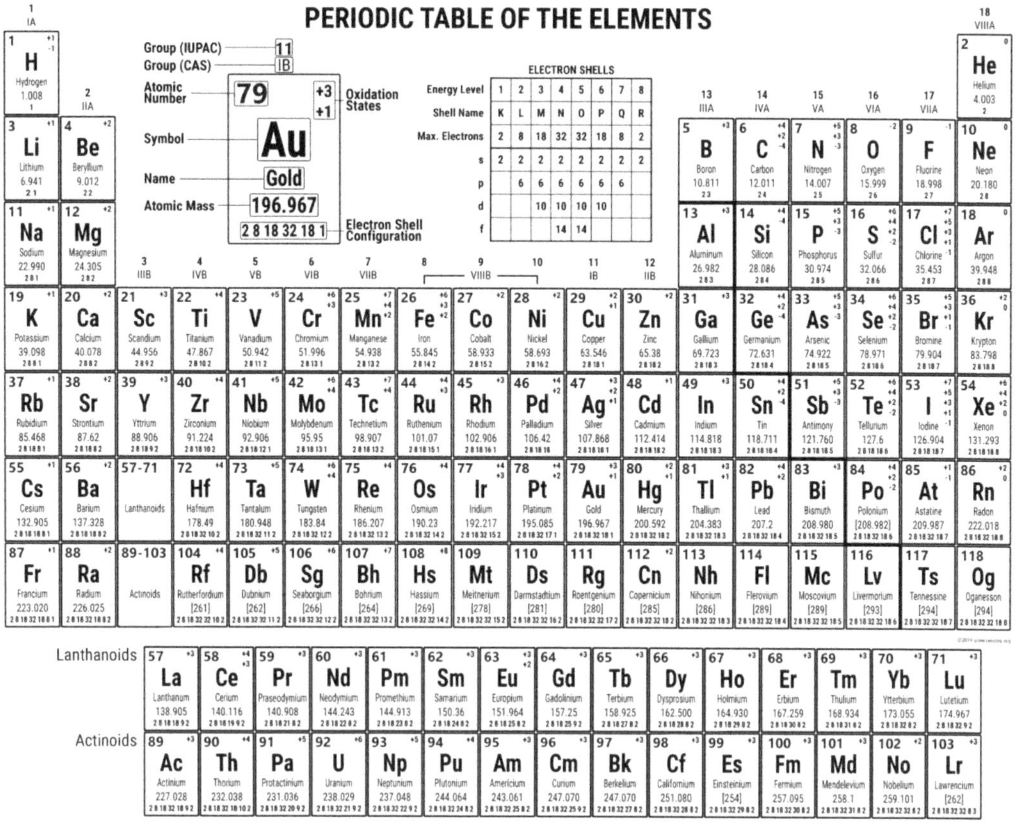

FIGURE 8.1 Periodic table of elements. (From United States Institute of Health. 2024a. *Periodic Table of Elements*. https://pubchem.ncbi.nlm.nih.gov/periodic-table. (Accessed May 11, 2024).)

A chemical or substance becomes a contaminant when it is released into the environment either inadvertently or improperly – at the wrong place or in the wrong amounts. For example, milk becomes a contaminant when large quantities are released into a stream. In urban areas, contaminants are everywhere – in the air, soil, water, inside buildings, and in our homes. For reference purposes and as we will be discussing chemical contaminants, we will often need to refer to the Periodic Table of Elements for reference and educational purposes in this chapter. Figure 8.1 presents the Periodic Table of Elements (United States Institute of Health 2024a). Most households contain chemicals that would be considered contaminants and hazardous if they were released into the environment or disposed of improperly (see Figure 8.2). These chemicals include the following (Rogers 2020):

• Cleaners	• Solvents	• Gasoline
• Pesticides	• Herbicides	• Lawn fertilizers
• Some paints	• Plastic	• Oil
• Grease	• Dirt	• Electronic equipment

In addition, the list of common household products shown below contains contaminants if they were improperly disposed of and include computer equipment, televisions, some electrical equipment, most batteries, some building products, and wood with certain applied preservatives or

FIGURE 8.2 Household items that can become contaminants. (From United States Geological Survey (USGS). 2006a. *Volatile Organic Compounds in the Nation's Ground Water and Drinking-Water Supply Wells*. USGS Circular 1292. USGS. Reston, VA.)

coatings (Rogers 2020). Thousands of environmental pollutants exist, and the following categories of contaminants are commonly present in the environment (Rogers 2020):

- Volatile organic compounds (VOCs)
  - Halogenated VOCs
  - Trihalomethane VOCs
  - Non-halogenated VOCs
- Polynuclear aromatic hydrocarbons (PNAs or PAHs)
- Semi-volatile organic compounds (SVOCs)
- Polychlorinated biphenyls (PCBs)
- Pesticides and herbicides
- Heavy metals
- Fertilizers
- Others including:
  - Greenhouse gases
  - Chlorofluorocarbons (CFCs)
  - Carbon dioxide
  - Carbon monoxide
  - Particulates (dust)
  - Ozone
  - Pathogens
  - Invasive species
  - Pharmaceuticals
  - Cyanide
  - Asbestos

- Acids
- Bases
- Radioactive compounds
- Dioxins
- Polyfluoroalkyl substances (PFAS)
- Emerging contaminants

Where do contaminants end up when released into the environment and, how do humans and the environment become exposed to contaminants? The answer to this question is obtained through an understanding of the process called **contaminant fate and transport;** it is defined as the sequence of anthropogenic and natural events involving a pollution source, its mobilization or transport, and its ultimate fate or resting place, termed a **sink** (Rogers et al. 2007). Sources of pollution include many human activities performed primarily at the surface, resulting in the release of toxic substances into the environment. These toxic substances may be transported over time or remain relatively close to their sources before they are degraded, transformed, or destroyed, because: (1) pollutants released into the environment are often mixtures; (2) each contaminant is unique chemically; and (3) the geologic environment in which the pollutants are released is also unique, understanding the physical chemistry of each, the microbiologic environment, and the geologic environment into which they are released becomes necessary for characterizing their fate and transport (Rogers et al. 2007).

During their transport and before reaching their final sink, certain pollutants may reside at multiple intermediate sinks for different periods of time. Intermediate sinks include surface water, groundwater, and the atmosphere because the contaminants held within these water-containing sinks will flow and ultimately reach the oceans. Aquifers with a very low hydraulic conductivity are for practical purposes considered final sinks, as are inland bogs and some wetlands. Sediment and soil can function as intermediate or final sinks, as erosion may move both of these unconsolidated materials. The oceans are usually a final sink of pollution, but wave action occasionally brings pollutants onshore. To a large extent, the level of human health and/or environmental risk is a function of two fundamental concepts introduced in this chapter: *mobility* and *persistence*. **Mobility** is a measure of a substance's potential to migrate. **Persistence** is a measure of a substance's ability to remain in the environment before being degraded, transformed, or destroyed (Rogers et al. 2007). Substances with higher toxicity pose greater health risks, but the risk to humans and the environment grows exponentially if the chemical is both mobile and persistent. For example, a highly toxic, but immobile chemical may affect a few people in a warehouse through inhalation, whereas a mobile and persistent chemical of moderate toxicity can contaminate a public water supply, or migrate to a different sink where the potential for widespread human exposure and ecosystem damages is much higher.

The following sections discuss how and where a contaminant is released, how its migration through the soil, groundwater, and atmosphere proceeds, and how it ends up in sinks. We conclude with a brief description of the fate and transport of each pollutant group.

8.3 RELEASE OF CONTAMINANTS INTO THE ENVIRONMENT

The fate and transport of pollutants begin with their release into the environment. Pollutant releases originate from numerous sources and under different circumstances; they vary in degree, concentration, duration, mass, volume, and whether a single contaminant is released or if a mixture of contaminants is released. Each of these factors influences their fate and transport. For instance, some releases may be very small and avoid detection. On the other hand, some releases are so large and sudden (e.g., a tanker spill) that they become the leading story of the next newscast. Sudden and large releases increase the probably of severe environmental impairment or destruction, especially if they occur in or near a sensitive ecological area.

Historically, there was no regulation of the disposal of wastes potentially containing pollutants. Until the mid-20th century, the most convenient but least costly method of waste disposal was "up the stack or down the river" (Haynes 1954). We now know the perception of a pollutant leaving your immediate vicinity and being gone forever is not true. Today, many releases of pollutants are permitted but carefully monitored, such as wastewater discharges from industrial and municipal sources and those to the atmosphere from industrial and commercial sources and power plants. The legislation (such as the Clean Water Act of 1972 and the Clean Air Act of 1990) governing such release permits a point source such as a wastewater plant or a smokestack to release specific compounds at low concentrations to ensure there are no adverse health and environmental effects detectable on the media they are released into (water or air). Some releases, as we shall discover and discuss in detail, are not regulated and include the majority of household wastes and releases of pollution from many agricultural sources. A majority of the other pollutant releases into the environment are unintentional or accidental and occur from numerous sources under a multitude of circumstances, and some of these include the following (Fetter 1993; Rogers 1996; United States Geological Survey (USGS) 2006a):

- Permitted releases of contaminants in wastewater and air emissions
- Spills and leaks from several types of containers or operations, including:
  - Drums of various sizes and shapes
  - Pipelines
  - Above-ground storage tanks
  - Underground storage tanks
  - Tanker trucks and other transport vehicles
  - Tanker ships
  - Railroad tanker cars
  - Aircraft
- Accidents or collisions involving automobiles where gasoline and oil may be released, railroad derailments, etc.
- Septic systems
- Sewer leaks
- Landfills and dumps
- Automobile exhaust
- Sumps and dry wells
- Former disposal lagoons
- Animal feed lots
- Pesticide and herbicide application
- Fertilizer application
- Injection wells
- Application of pesticides, herbicides, and fertilizers
- Application of de-icing compounds (i.e., road salts)

Exploring and evaluating the fate and transport of pollutants in the environment is a complex interplay between the geological environment and the contaminant. Figure 8.3 shows that fate and transport are also influenced by the method and location, volume, mass, and duration of the contaminant release. We will break down and describe these factors influencing the migration of contaminants, degradation of contaminants, and then describe how the fate and transport of each group of substances is influenced by the group's specific chemistry.

8.4 PRINCIPLES OF CONTAMINANT FATE AND TRANSPORT

Interaction with the environment begins immediately after a pollutant has been released. Once released, a pollutant can do three things (Hemond and Fechner-Levy 2000):

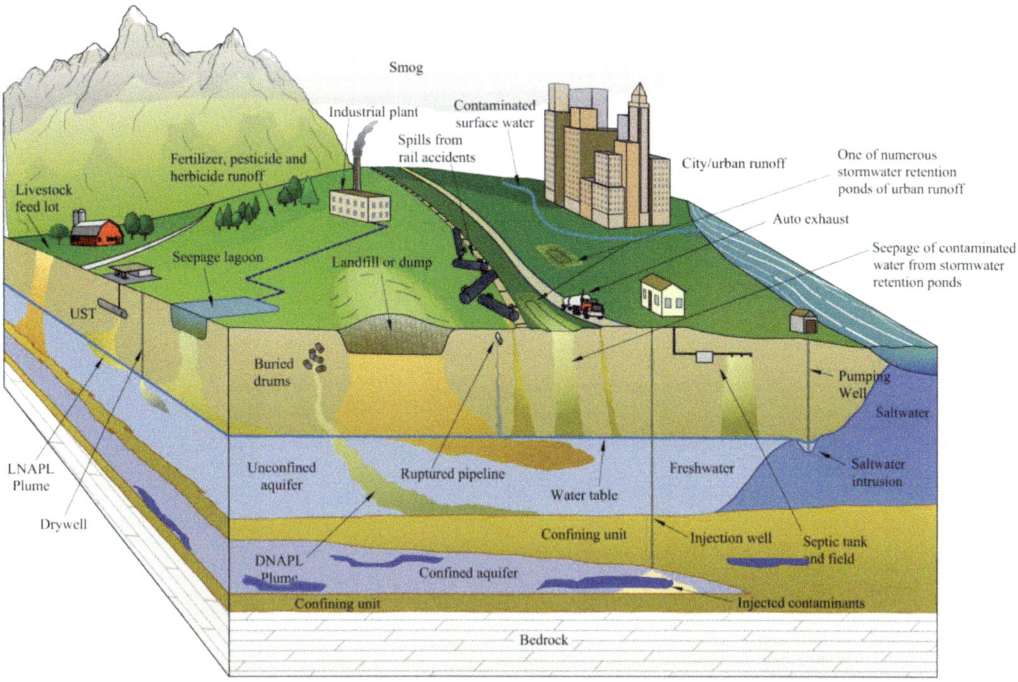

FIGURE 8.3 Sources and locations of many contaminant releases.

- Stay put
- Migrate in soil, water, air, or a combination of media
- Degrade, transform, or get destroyed

The factors controlling these three outcomes depend upon the physical chemistry of the contaminant and the characteristics of the receiving environment (USGS 2006a; Rogers et al. 2007). For example, some pollutants may change physical states. In one such example, one of more than 250 chemical compounds that comprise gasoline begins to immediately evaporate after exposure to the atmosphere. Other pollutants may degrade in a manner of minutes after being released, while some may last for thousands of years or sometimes longer, as with certain radioactive compounds. Therefore, computing a mass balance should be the first action when assessing any particular release, as this will ensure that the mass or volume of contaminant released is accurately measured (Equation 8.1). Conducting a mass balance also serves to validate or refute our current understanding of the behavior of different environmental elements influencing the pollutants once they have been released.

A simple mass balance is expressed as:

$$\begin{aligned}\text{Amount released} = \text{Amount recovered} + \text{Amount lost to the environment}\\ (\text{Air} + \text{Water} + \text{Soil})\end{aligned} \tag{8.1}$$

An additional factor is what media the contaminant is released into first: air, water, or soil. This also influences its behavior and interaction with the environment. Figure 8.4 presents an excellent example of contaminant release into all three environmental media at the same time. Once in the environment, pollutants can and often move between soil, surface water, groundwater, and the atmosphere, and they can also degrade. Many pollutants degrade quickly if conditions are favorable,

FIGURE 8.4 Aerial photo showing releases to the air, water, and land. (From New Jersey Department of Environmental Protection (NJDEP). 2015. *Preliminary Assessment Report*. NJDEP. Trenton, NJ.)

yet others can persist and last for years or decades if conditions are unfavorable (USGS 2006a). Factors degrading contaminants fall into two broad categories: biotic degradation and abiotic degradation. **Biotic** degradation involves microorganisms or fungi, and occurs when an organism, such as a bacterium, uses a contaminant as a source of food and either degrades or transforms the contaminant (USGS 2006a). **Abiotic** degradation involves other processes not including microorganisms. Examples of abiotic degradation include photolysis – the degradation as a result of exposure to sunlight (USGS 2006a) – and hydrolysis. Hydrolysis involves cleaving a molecule into two parts by the addition of a molecule of water.

When examining pollutant degradation in the environment, it is important not to confuse dilution with degradation. If given enough time, contaminants may become diluted, and this process results in a decrease in the pollutant concentrations per unit volume of the media being measured. However, dilution is not degradation, as it does not involve a chemical transformation of the pollutant.

8.4.1 Contaminant Transport Concepts

Pollutant transport in the environment is dominated by three physical transport mechanisms (USEPA 1996a; USGS 2006a): advection/convection, molecular diffusion, and dispersion. **Advection** is the horizontal transport of any property of the atmosphere and water. Common examples are the transfer of heat by wind and sediment transport within a flowing stream. **Convection** is the vertical advection of air, water, or other fluid as a result of thermal differences. We introduced the concept of convection in Chapter 2 as the driving force behind plate tectonics. **Molecular diffusion** is the movement of a chemical from an area of higher concentration to an area of low concentration due to the random motion of the chemical molecules. **Dispersion** (also referred to as hydrodynamic dispersion) is the tendency for pollutants to spread out from the path of the expected advective flow (USGS 2006a). Occasionally, the effects of diffusion and dispersion are treated together, but for the purposes of this book, we treat them separately.

The rate of advective transport of a pollutant is often expressed in terms of flux density. **Flux density** is the mass of a chemical transported across an imaginary surface of a unit area per unit of time. Equation 8.2 shows this relationship (Hemond and Fechner-Levy 2000). This relationship is independent of the media involved (soil, surface water, groundwater, or the atmosphere):

$$J = CV \tag{8.2}$$

where J = flux density, J = (Mass/[Length × Width] × Time) or $[M/L^2 \times T]$
C = concentration of the chemical per cubic liter or meter of media $[M/L^3]$
V = velocity [length/time] or $[L/T]$

An example of molecular diffusion is shown in Figure 8.5 (Payne et al. 2008). From the time of release to infinity, a pollutant released into a fluid such as air or water will diffuse throughout the fluid at random locations.

Fick's First Law of diffusion (Equation 8.3) can be used to predict the diffusive flux of a pollutant (solute) across an imaginary plane as a function of the rate of change in concentration with distance (Hemond and Fechner-Levy 2000):

$$J = D(dC/dx) \text{ (one dimension)} \tag{8.3}$$

where J = the flux density $(M/L^2 \times T)$
D = the Fickian mass transport coefficient (L^2/T)
C = chemical concentration (M/L^3)
X = the distance over which a concentration change is being considered (L)

Note: In simple calculations, the minus sign is often omitted if the direction of Fickian transport is clear.

General factors that influence or control the rate of migration of a contaminant include the following (Hornsby 1990):

- Physical and chemical properties of the contaminants themselves
- The geological environment where the release occurs
- Climatological factors
- Vegetation factors

Specific physical properties affecting migration of contaminants in soil include the following (USGS 2006a; Rogers et al. 2007; Payne et al. 2008):

- *Solubility in water:* The more soluble a contaminant is in water, the more mobile it will be in the subsurface environment.
- *Vapor pressure:* As vapor pressure increases, affinity to volatilize increases, and the more likely a contaminant will be present in the gas phase.
- *Molecular weight:* The higher the molecular weight, the greater the energy requirement to transport the contaminant in the horizontal direction. Increased molecular weight may induce the contaminant to migrate downward in areas of steep slopes.
- *Chemical stability and persistence:* Stable compounds have more time to migrate and may migrate further if conditions are favorable before they are degraded, transformed, or destroyed.
- *Sorptive properties:* Sorptive processes include adsorption and absorption. The lower the sorptive properties, the higher the migration potential. For this relationship to exist, the number of sorption sites within the geological materials has not been exceeded (Payne et al. 2008).

Specific geologic factors affecting the migration of pollutants in soil include the following (Rogers 1996; USGS 2006a; Rogers et al. 2007, 2012):

- *Composition:* Soil type has a significant influence on the migration of contaminants. In general, coarse-grained soils, such as sand and gravel, do not impede the migration of contaminants nearly as much as soils composed of clay.

- *Porosity and permeability:* Direct evidence of migration potential is hydraulic conductivity. Soils composed of sand and gravel have a hydraulic conductivity of 100 to 1,000 times higher than soils composed of clay. Although soils or sediments composed of clay may have a high relative porosity, they are usually not as permeable because the porosity is generally not interconnected.

In some cases, however, as shown in Chapter 3, secondary porosity such as root fragments and vertical fractures can make a seemingly impervious clay deposit much more permeable than expected.

- *Organic carbon content:* As just discussed, an increased total organic content in soil may impede the migration of certain types of contaminants as long as the contaminant mass does not exceed the holding capacity of the soil. Soils or sediments composed of sand and gravel generally have lower organic carbon content compared to soils composed of clay.
- *Soil chemistry:* The pH, redox potential, and other soil chemistry factors influence the contaminant migration of many different types of compounds. Many metals, for instance, are particularly sensitive to pH differences in soil, and these differences – along with the characteristics of each metal – influence their migration patterns in the environment.
- *Stratigraphy:* In stratigraphy, heterogeneity and the anisotropic nature of the geologic sediments play a significant role at both micro- and macro-scales. The geology beneath the surface can and does change dramatically in just a few feet in any direction. The result is differing sediment types and chemical composition, including pH and redox potential, of subsurface layers acting to impede or enhance contaminant migration.
- *Unconformities:* The presence of unconformities influences the migration of groundwater as well as the migration of contaminants. Hydrogeologically, the presence of an unconformity indicates there is a surface or plane in the subsurface geologic environment. This surface or plane often produces a significant difference in the hydraulic conductivities within the soils or sediments above and below the unconformity, especially if these units are fine-grained sediments such as silts or clays. As a result, contaminants released in this type of location use the unconformity as a sink and migrate much further than expected.

Specific climatological factors affecting the migration of pollutants in soil include the following (Freeze and Cherry 1979; Rogers 2020):

- *Freeze–thaw cycles:* Freeze–thaw cycles can lead to the development of vertical fractures in the soil to depths approaching 10 meters. These vertical fractures are a type of secondary porosity, and, if present, can greatly increase the migration potential of contaminants vertically through the soil column. In addition, the freezing of near-surface soils may trap contaminants at the surface and lead to increased contaminant loading during warmer periods when the ice melts.
- *Liquid precipitation:* Because water is the universal solvent, geographic locations receiving abundant rainfall play a significant role in enhancing the migration of contaminants, especially if they are soluble. Rainfall also enhances the migration of contamination through the physical transport of particles with sorbed contamination on their surfaces.
- *Snowfall:* Airborne deposition of contaminants may become temporarily trapped in seasonal snowpack and released when the snowpack melts. Increased contaminant loading to the environment may occur during warmer periods when the snowpack melts (Wania et al. 1998).
- *Wind:* Many locations within the United States contain significant amounts of wind-blown deposits, especially in the southwest. Contaminants with a high sorption potential may become attached to fine wind-blown sediment grains and transported over long distances. In addition, volatile contaminants released as a gas are routinely transported by wind.

- *Water vapor:* Humidity plays a significant role in the water cycle by affecting air movement, reacting with contaminants in the gas phase and contaminants released into the atmosphere attached or sorbed onto particulate matter. Contamination goes along for the ride when precipitation formed around contaminated condensation nuclei is transported from the atmosphere to the lithosphere and then entrained by surface runoff.
- *Fog:* Fog can be an effective agent for the transfer of acid rain by transferring it to vegetation or other surface materials through direct contact.
- *Flood events and hurricanes:* Due to their catastrophic nature and magnitudes, floods and hurricanes may not only increase contaminant migration but also cause significant releases. During the 1993 floods of the Mississippi River, numerous barrels containing hazardous waste were swept away and deposited in the Gulf of Mexico.
- *Solar energy:* Sunlight breaks down some contaminants through a process, called photolysis. In toxic microorganisms, ultraviolet light passes easily through cell walls, cytoplasm, and nuclear membranes and prevents DNA replication.

Specific vegetative factors that affect migration of pollutants in soil include the following (Nudunuri et al. 1998; Rogers 2020):

- *Roots:* These pathways within shallow subsurface geological materials can enhance the migration of contaminants.
- *Certain plants can uptake contaminants:* Many types of plants and trees have the capability with their root systems to assist in the removal of contaminants from shallow subsurface soil. Contaminants may be stored in plant tissues or are transformed through biologic processes.
- *Microorganisms:* Microorganisms in the soil often biodegrade many different types of contaminants by using the contaminants themselves as a source of food.

8.4.2 Contaminant Degradation Concepts

The degradation of specific compounds in the environment is expressed in terms of their half-lives. **Half-life** is the average amount of time required to degrade half or 50% of a specific pollutant population (USEPA 1998a, 1998b). Contaminants degrade through biotic or abiotic processes, and the processes controlling their rate of decay are dependent upon the following factors (USEPA 1996a, 1996b; USGS 2006a):

- *The nature of the release:* It contains the following factors:
  - Media receiving the release.: It depends on whether the release was into the atmosphere, surface or subsurface soil, surface water, ocean, directly to groundwater, or a combination of media.
  - Amount (volume or mass) of the release.
  - Number of contaminants released. Was it a single contaminant, or a mixture of several contaminants?
  - Physical state of the release (i.e., liquid, solid, or gas)
  - Time duration of the release
- *Geologic environment:* The most significant geologic factors include the following:
  - Soil composition and other physical characteristics such as permeability, porosity, moisture content, composition, extent and distribution, thickness, total organic carbon content, pH, redox potential, dissolved oxygen (if saturated), and other parameters.
  - Depth to bedrock, type, composition, distribution, fractures, permeability, porosity of bedrock and other parameters
  - Terrain and topography

- Potential surface and subsurface migration pathways.
- *Climatic factors:*
- Release Location: Different climates such as deserts, mountains, humid areas, or temperate regions can influence the type and rate of degradation.
- *Surface water features:*
  - *Distance to surface water bodies and their type:* Immature streams, mature streams, rivers, lakes, wetlands, and bogs can differ in their pH levels due to the rock composition of their channels and bottoms, and the amount of organic matter they receive from outside (called allochthonous inputs).
- *Weather conditions at the time of the release:* Weather conditions are often important and sometimes overlooked as potentially significant. Those conditions affecting degradation and migration include the following:
  - Temperature
  - Humidity
  - Precipitation
  - Wind speed and direction
- *Biological factors:*
  - The type, distribution, and amount of microorganisms will influence the rate and can determine if degradation can occur.
  - *Anthropogenic factors:* Anthropogenic factors are often overlooked and frequently significant. These include the following:
  - Physical landscape alteration (i.e., buildings, roads, parking lots, etc.)
  - Surface water drainage modifications including stormwater control and wetland destruction
  - Alteration of native vegetation
  - Introduction of invasive vegetation
  - Regional contaminant loading including sources, duration, type, release points, and physical state of contaminants (i.e., solid, liquid, or gas)
  - Developmental history of the area and region

8.4.2.1 Biotic Degradation

Microbes have the ability to oxidize a variety of organic contaminants including many VOCs, PAHs, and other compounds. This capability arises from their enormous variety, populations, rapid growth, and diversity of environmental niches. Soluble organic compounds with low molecular weights such as alcohols and organic acids are metabolized and degraded rapidly by microbes, perhaps because these compounds also occur naturally and microbes have evolved to degrade them more efficiently (Hemond and Fechner-Levy 2000). Halogenated synthetic or anthropogenic compounds, however, are not easily degraded by microbes (USEPA 1998b), and some contaminants that escape biotic degradation bioaccumulate in the bodies of organisms. The rate of biodegradation by microorganisms generally slows if the organic contaminants possess the following (Hemond and Fechner-Levy 2000):

- Increasing molecular weight
- Decreasing water solubility
- Presence of benzene or aromatic rings
- A large amount of branching within the molecular structure
- Presence of halogen atoms in the structure (chlorine, fluorine, bromine, or iodine)

8.4.2.2 Abiotic Degradation

Abiotic degradation applies to degradation processes accomplished without microorganisms. Common abiotic degradation processes include the following:

- Photolysis
- Hydrolysis
- Reduction – oxidation
- Radioactive decay

Photolysis (sometimes referred to as photodegradation or photochemical degradation) occurs in the presence of sunlight. Common examples of this process include the fading of colored and dyed objects and the transformation of plastic object textures from pliable to brittle. Photolysis is most common in the atmosphere, surface water, and at Earth's surface. The degree of degradation caused by sunlight depends on the wavelength spectrum of the light, intensity of light exposure, and duration (Hemond and Fechner-Levy, 2000). If the energy per photon is sufficient to break a specific chemical bond photolysis can be initiated. Once begun, increased light intensity will result in a faster rate of degradation. Failure to achieve the energy level required to break a bond means degradation will not occur. Ultraviolet light is especially effective at degrading many organic contaminants (USEPA 1996a).

Degradation by photolysis occurs within compounds capable of absorbing light energy and is often observed in organic compounds with double bonds between their carbon atoms. Many VOCs, PAHs, and SVOCs characterized by a benzene ring fit this pattern. Under favorable conditions, the degradation of many organic compounds by photolysis may occur in a short period of time – from a few hours to a few days (Lyman et al. 1990).

The process of **hydrolysis** occurs when a water molecule breaks. Contaminant degradation by hydrolysis also destroys a molecule of contaminant. Two types of chemical compounds are susceptible to degradation by hydrolysis (Schwartzenbach et al. 1993):

- Alkyl halides, straight-chained or branch-chained hydrocarbons where one or more hydrogen atoms have been replaced by a chlorine, fluorine, bromine, or iodine atom. Using an X to represent a halogen atom and an R to represent the hydrocarbon group, the basic hydrolysis reaction is shown in Equation 8.4:
- Esters, compounds containing a modified carboxylic acid group (–COOH), where the acid hydrogen atom has been replaced by a different organic functional group. The process of hydrolysis within this group converts the ester compound into the "parent" organic acid and an alcohol.

In reduction–oxidation degradation reactions (redox), electrons are transferred from one atom to another. **Chemical reduction** is defined as the addition of electrons and **chemical oxidation** is defined as the loss of electrons (Hemond and Fechner-Levy 2000). In a reaction involving atoms A and B, if atom A gains an electron, it is reduced, and Atom B, having donated an electron is the **reductant**. Because atom B loses an electron, B is oxidized and atom A is the **oxidant**. Each reaction involving the loss or gain of an electron is termed **half-reaction**. The oxidation of contaminants can occur very rapidly through combustion or incineration. Here, fire transforms the contaminants through oxidation at greatly elevated temperatures and is used in cooking, heating, and transportation applications (Hemond and Fechner-Levy 2000).

8.4.3 Transport and Fate of Contaminants in Soil

Folklore holds that the presence of soil protects groundwater quality by filtering contaminants before they reach and impact groundwater (Hornsby 1990). Soil does have a limited ability to filter contamination; however, it does not do a perfect job of holding, filtering, degrading, transforming, or destroying contaminants. These capabilities also depend upon a number of factors related to the chemistry of the contaminant and the geological environment where the contaminant is released.

Soil is defined as the unconsolidated mineral matter on the immediate surface of the Earth (Soil Science Society of America 1987). Basic to an understanding of soil are the factors affecting its development and ultimate physical structure. The composition, texture, and thickness of soil are influenced by its source material, plant growth, micro- and macro-organisms, climate, topography, the process of formation (e.g., alluvial, fluvial, and glacial), and physical and chemical weathering since original formation (Brady and Well 1999). Structurally, soil is composed of three phases: soil gases, soil water, and organic and inorganic solids (Sawhney and Brown 1989). The gas and water phases may comprise 25% to 50% of the total volume of surface soil, especially at shallow depths (USEPA 1999).

Pollutants released into the soil can migrate within all three phases. Once a pollutant is resident in soil, these factors determine its migration rate (Schnoor 1996; USEPA 1999; Rogers 2020):

- Contaminant mass released
- Duration of the release
- Physical chemistry of the contaminant
- Physical chemistry of the soil (e.g., pH, redox potential and mineralogy)
- Amount of water present
- Permeability of the soil
- Retention capacity of the soil
- Distribution of plant matter
- Biological interaction between the contaminant and indigenous microorganisms

All of these factors must be well understood before an accurate assessment of the fate and transport of a contaminant in soil can be made.

Contaminants migrate through the soil by two basic processes: diffusion and mass flow. The rates of diffusion and mass flow greatly depend upon the local geology and the physical chemistry of the contaminant. Diffusion of substances through soil and aquifer materials occurs in response to differences in energy from one point to another. These energy gradients may be caused by differences in temperature or chemical concentrations within the contaminated area. In most cases, however, the principal process of moving a contaminant through soil is mass flow or advection, because contaminants generally want to move downward through the soil under the force of gravity (USGS 2006a). Contaminant-specific physical and chemical attributes affecting the migration of contaminants in soil include the following (USEPA 1996a; Wiedemeier et al. 1999):

- Solubility
- Vapor pressure
- Density
- Chemical stability
- Persistence
- Adsorption potential

In soil, solid-phase contaminants migrate much more slowly than liquid-phase contaminants, and tend to remain relatively close to their point of release or deposition (USEPA 1999). Before they can migrate a significant distance, solid-phase contaminants must change phase or undergo a transformation process. For example, heavy metals – a solid-phase contaminant – typically remain at their point of release or deposition. If they undergo oxidation, however, their solubility and other properties enabling migration may increase (Lindsay 1979). Also, once a contaminant begins to dissolve in water, it may also be subject to further transformation reactions induced by indigenous bacteria present in the surface soil (Sutherson and Payne 2005).

If a source continues to emit contaminants in solid or liquid form that dissolve in water, the underlying soil will eventually become saturated. The leading edge of contamination will migrate either horizontally or vertically or both as long as the retention capacity of the soil is exceeded (USEPA 1999). When the contaminant release stops, the migration of the liquid contaminant will significantly decrease as the soil regains its retention capacity (Rogers 2020). As soil is also composed of gas, contaminant migration through the vapor phase is often observed with contaminants having higher relative vapor pressures. VOCs are frequent participants in this type of migration. Capillary forces can also induce the migration of liquid-phase contaminants.

8.4.4 Transport and Fate of Contaminants in Surface Water

The transport of contaminants in surface water is dominated by turbulent advective flow. Because the rate of flow in a river or stream varies significantly by location and over time, estimating the contaminant flow involves averaging the streamflow variations and contaminant concentrations over a specified time interval. Conducting measurements at multiple locations also provides a more accurate measure of the rates of streamflow and contaminant transport.

Molecular diffusion also influences contaminant migration in surface water. Turbulent flow is characterized by water moving in constantly changing and unpredictable patterns. The swirls resulting from turbulent flow are called **eddies,** and they appear in many sizes, volumes, and velocities. Random mixing of the water within eddies creates turbulent diffusion and also influences mass transport. Wave action can create similar eddying effects in lakes and other non-flowing water bodies. Analytically, the transport rates for chemicals in surface water are expressed in terms of flux density. Flux density is the mass of a chemical transported across an imaginary surface of a unit area per unit of time (Hemond and Fechner-Levy 2000). Fick's First Law is also used to describe the flux density of mass transport by turbulent dispersion (Hemond and Fechner-Levy 2000).

As most urban areas of the United States obtain their potable water from surface sources, wastewater discharges are a concern. Treatment costs rise when the source of supply is contaminated, and the risks of biological contamination also increase. Moreover, a majority of urban areas are located along rivers and streams and other surface water bodies, and many of these water bodies have already been degraded – some significantly (USGS 1995a, 1995b; Rogers 2020).

When certain contaminants are released to surface waters through overland flow, stormwater runoff, or wastewater discharge they may accumulate in sediments. Compounds with a higher likelihood of accumulation have low solubility, high molecular weight, low potential to degrade, and high sorption potential. Contaminants having these physical chemistry attributes typically do not sustain themselves in surface water unless the rate of flow is substantial. Even then, they may be carried along the bottom of the stream or river until the carrying capacity of surface water is insufficient and the contaminants settle to the bottom (Rogers 2020).

If the source of contamination persists, greater amounts of the contaminant will be deposited. The accumulation of contaminants in sediments increases the exposure risk to aquatic and terrestrial plant and animal life. If any of the contaminants exhibit bioaccumulation properties, they may proceed up the food chain from bottom-dwelling macroinvertebrates to small fish and eventually to larger fish, predatory birds, and other organisms. Humans are situated at the top of the food chain, and the potential risks to human health must be considered when evaluating the fate and transport of contaminants in surface water (USEPA 2024a).

Contaminants considered to be bioaccumulative include mercury, PCBs, chlordane, dioxins, and DDT (USEPA 2024a).

The number of surface water bodies under advisory includes the following (USEPA 2024a):

- 44% of the Nation's total lake acres (excluding the Great Lakes), representing approximately 18 million acres of surface water
- 35% of the Nation's total river miles, or approximately 1.3 million miles

- 45% of the Nation's contiguous coastal waters
- 100% of the Great Lakes and their connecting waters

8.4.5 Transport and Fate of Contaminants in Groundwater

Transport of contaminants in groundwater is dominated by three factors: advection, dispersion, and molecular diffusion. When applied to groundwater**, advection** is the movement of contaminants by the bulk motion of groundwater flow, dispersion is the tendency for contaminants to spread out from the path of the expected advective flow, and diffusion is the action of spreading of molecules from areas of high concentration to areas of low concentration at the molecular level. Groundwater flow lacks turbulent diffusion because velocities are typically much slower. In some instances, however, groundwater does display turbulent dispersion, especially in karst topography where water beneath the surface flows and behaves much like a stream at the surface. Figure 8.5 shows a spill from an underground storage tank (USGS 2006a). Here, advective transport of contaminants in groundwater is occurring at the water table boundary. Diffusion, biodegradation, volatilization, and recharge from surface precipitation affecting the contaminant migration are also shown. The effects of dispersion and diffusion are represented by the spread of the contaminant plume as it migrates from a hole or rupture at the bottom of the tank (USGS 1998).

8.4.5.1 Effects of Dispersion and Diffusion

The representation of dispersion in Figure 8.5 is overly simplistic, because the geology of unconsolidated sediments is very complex and typically displays a high degree of heterogeneity and anisotropic distribution patterns. As a result, contaminants migrating in unconsolidated deposits do not

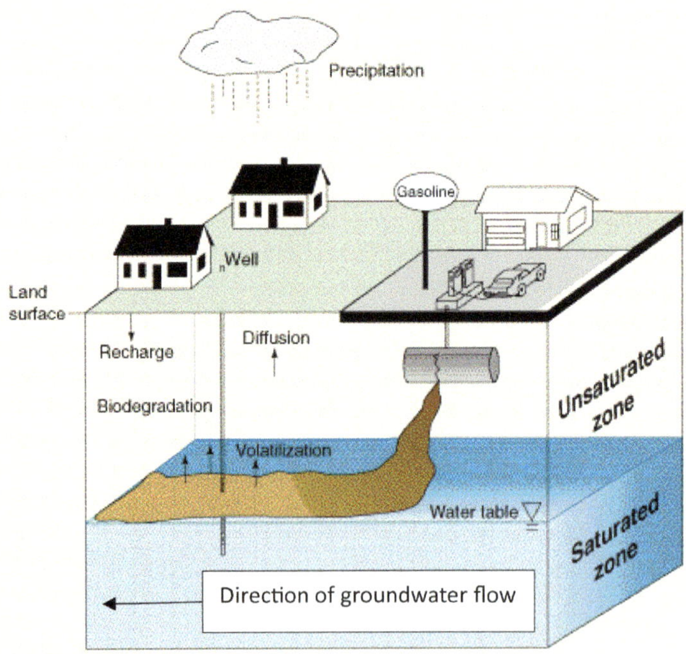

FIGURE 8.5 Advective transport and other processes affecting the migration of contaminants in groundwater. (From United States Geological Survey (USGS). 1998. *Simulating Transport of Volatile Organic Compounds in the Unsaturated Zone using the Computer Model R-UNSAT*. USGS Fact Sheet 019-08. USGS. Washington, DC.)

migrate uniformly but migrate within the physical parameters of advection dictated by the particular subsurface geology. More highly permeable layers have a higher hydraulic conductivity and behave as preferred groundwater and contaminant migration pathways. These layers are essentially super highways for groundwater and contaminant transport, and, in some instances, the hydraulic conductivity is from 100 to sometimes 1,000 times greater over distances of just a few centimeters (Payne et al. 2008).

Zones with higher permeability move water more quickly and have a higher flux density. If more water moves through these higher permeability zones, then a potentially greater contaminant mass also moves through. As contamination continues to migrate along these flow paths of higher permeability dictated by the subsurface geology, diffusion of contaminants into less permeable zones occurs. The top portion of the figure represents the flow paths of contaminants in the early stages of migration, and the later stages of migration are shown in the figure's bottom portion. Over time, the contaminant (solute) has diffused into the less permeable, lower hydraulic conductivity geologic materials (Payne et al. 2008). The transport of contaminants in groundwater is also influenced by many of the same factors affecting the migration of contaminants in unsaturated soil or the vadose zone:

- Physical chemistry of the contaminants:
  - Solubility
  - Molecular weight
  - Vapor pressure
  - Stability and persistence
- Sorption potential
- Type, distribution, and amount of microorganisms
- Tendency to biodegrade
- Dissolved oxygen content of groundwater
- Geological factors:
  - Stratigraphy (including thickness and distribution of geological units down to micro-stratigraphic scales at the centimeter or even millimeter)
  - Presence of unconformities
  - Soil or sediment chemistry
  - Organic carbon content
  - Porosity and permeability
  - Composition
- Climate factors:
  - Freeze and thaw cycles
  - Recharge from surface precipitation
  - Flood events
  - Seasonal climatic variations
- Vegetative factors including types and distribution of surface vegetation, root networks, and water requirements

Sorption potential has a significant effect on the migration of contaminants in groundwater because it slows the migration of contaminants even as the flow rate of the transporting groundwater remains constant. This effect is termed **retardation** (USGS 2006a), and the degree of retardation present depends upon the specific contaminant's sorptive affinity and the amount of total organic carbon in the aquifer matrix.

Many contaminants are captured by pumping wells or migrate to surface water if the travel times and/or distances are short enough before they degrade. Figure 8.6 shows an example of travel times and capture zones in groundwater beneath urban areas. Any contaminant reaching groundwater

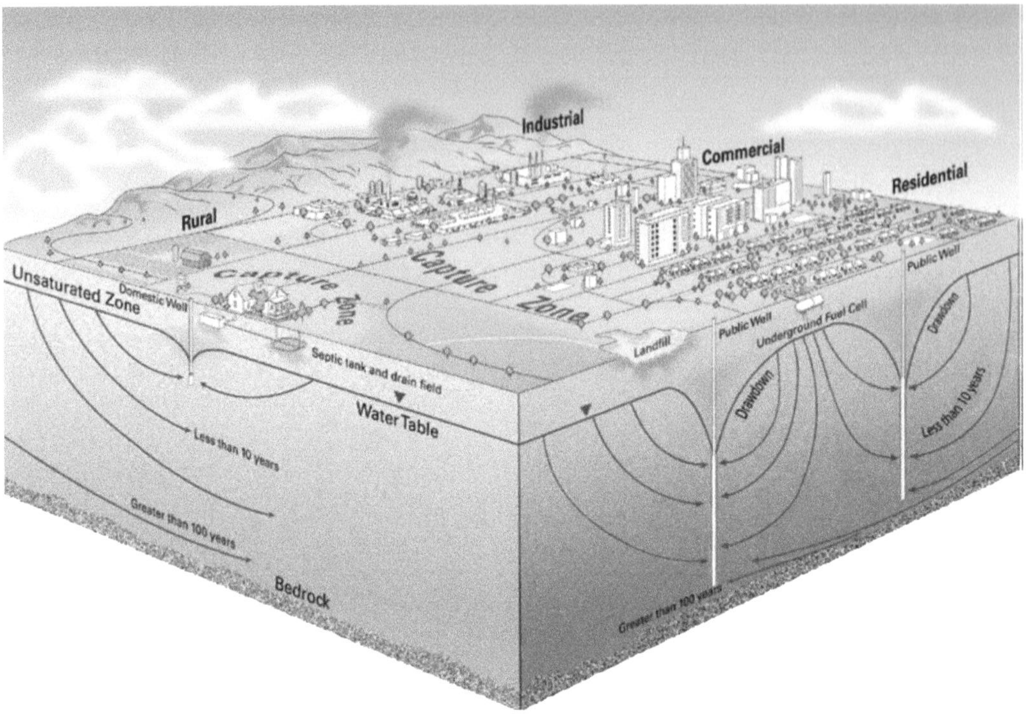

FIGURE 8.6 Potential groundwater travel time and capture zone beneath an urban area. (From United States Geological Survey (USGS). 2006a. *Volatile Organic Compounds in Nation's Ground Water and Drinking-Water Supply Wells. National Water-Quality Assessment Program.* USGS Circular 1292. USGS. Washington, DC.)

within the area marked *capture zone* has the potential to enter the public water supply if the contaminant does not degrade before reaching a public water supply well. Several sources of contamination listed in Section 8.1 are also shown: septic tanks, underground storage tanks (USTs), landfills, industrial facilities, and power plants. These contaminant sources are typical for any urban area within the United States and pose distinct threats to contaminate a public or private water supply.

8.4.5.2 Karst Topography

In karst topographical settings, the transport of groundwater contamination may behave similarly to surface water (Ford and Williams 2007). Some karst formations may exhibit turbulent advective flow because they have flow rates approaching the velocities observed in surface water flow (Heath 1983). Figure 8.7 shows an example of contaminant flow in a karst aquifer (USGS 1995a).

8.4.6 Transport and Fate of Contaminants in the Atmosphere

Different contaminants affect different portions of the atmosphere. For instance, CFCs affect the protective ozone layer in the upper portions of the atmosphere. Contaminant behavior in the atmosphere is very similar to the behavior of contaminants observed in surface water. Advective transport, turbulent diffusion, and molecular diffusion also influence contaminant migration in the atmosphere (Hemond and Fechner-Levy 2000). Figure 8.8 shows smoke from a fire billowing up (turbulent diffusion) into the atmosphere, and the horizontal movement of the smoke by advective transport. We also see a type of advective transport called convection; in this process, air rises due to thermal differences in the atmosphere. Turbulent advective mixing by wind and convection of

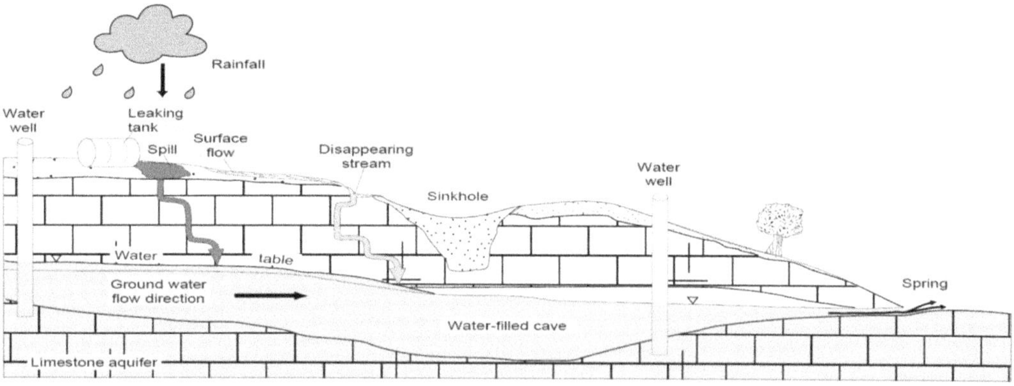

FIGURE 8.7 Contaminant migration in a karst aquifer. (From United States Geological Survey (USGS). 1995a. *Ground-Water Quality Protection.* Open-File Report 95-376. USGS. Nashville, TN. www.pubs.usgs.gov/of/1995/ofr-95376. (Accessed March 25, 2019).)

FIGURE 8.8 Smoke rising into the atmosphere from a wildfire. (Photograph by Daniel T. Rogers.)

the atmosphere is most significant within the troposphere (Schlatter 2009). These forces are very effective at transporting contaminants in the gas phase, and also excel at moving solid particulate matter in the atmosphere (USEPA 2008).

Anthropogenic sources of atmospheric contaminants are significant, and present themselves as an array of different contaminants released in high volumes annually (USEPA 1998a). Most contaminants are released into the atmosphere from anthropogenic sources at or near the surface, with most of the impacts occurring in the troposphere, and to a lesser degree the stratosphere. Contaminants released near the ground can mix throughout the troposphere in a few weeks, but it can take years or decades for them to reach the stratosphere (Hemond and Fechner-Levy 2000). Records of atmospheric contaminants and their effects can be traced to the 13th century when King Edward I banned the burning of kiln coal in London due to its impacts on air quality (Wilson and Spengler 1996).

Temperature and *pressure* are two important factors affecting the migration of contaminants into the atmosphere. Temperature and pressure play a much more significant role in atmospheric fate and transport than in soil and water because of the ideal gas laws, which relate to a combination of the relationships between temperature, density, and pressure. There are two ideal gas laws which are referred to by the scientists who first proposed each. What became Boyle's Law was proposed by Robert Boyle in 1662 who stated that for a fixed amount of an ideal gas, kept at a fixed temperature, pressure and volume are inversely proportional. This means that if the pressure of such a system is doubled, the volume of that system becomes half of its original value. Boyle's law is expressed numerically as follows:

$$PV = K \tag{8.6}$$

where P = pressure
V = volume K = a constant

Charles law is named after Jacques Charles who proposed that in a closed system, the ratio between the temperature and the volume must be a constant and is numerically expressed as the following:

$$V/T = K \tag{8.7}$$

where V = volume T = temperature in Kelvin scale K = a constant

Temperature and pressure typically decrease with increasing altitude in the troposphere. The temperature in the lower portion of the stratosphere is relatively constant, and helped give rise to its name meaning "stratified".

Oxygen is a recent addition to the atmosphere in geological terms. The origin of oxygen began with algae production approximately 2.45 billion years ago (Farquhar et al. 2000; Raub and Kirschvink 2008). We do not have a definitive explanation of oxygen's atmospheric origins, but one hypothesis states oxygen levels rose when the volcanism providing large amounts of hydrogen to the atmosphere declined. Methane and carbon dioxide were also gradually displaced by oxygen until their levels rose rapidly about 2.3 billion years ago. The presence of oxygen in the atmosphere plays a significant role and affects contamination in the environment through oxidation reactions and rates of combustion (USEPA 1991).

Contaminants initially released into the atmosphere often do not remain in the air; they settle out and contaminate the soil or surface water. Some contaminants, however, remain in the atmosphere for long periods of time, and other contaminants initially released into soil or water may volatilize and contaminate the air. In some cases, air contaminants may settle out of the atmosphere and adsorb onto a soil grain on the land surface only to be picked up later by the wind and sent airborne again. Factors controlling whether a contaminant remains in the atmosphere include the following:

- Physical and chemical factors of the contaminants including vapor pressure, molecular weight, solubility, reactivity
- Geography and local topography
- Climate and weather conditions

The average person inhales approximately 20,000 liters of air per day (USEPA 2008). The World Health Organization (WHO 2024) estimates that 2.4 million people die every year from causes directly attributable to air pollution, with the elderly and young children at the most risk. Specific diseases caused by prolonged exposure to air contaminants are chronic and often do not immediately appear after exposure. These diseases include heart disease, lung cancer, and bronchitis. The burning of fossil fuels in power plants and automobile, truck, and bus exhaust account for 90% of all air pollution in the United States (USEPA 2008).

Deposition of contaminants onto the land from the atmosphere occurs in two different ways:

1. *Dry deposition:* Dry deposition is typically dust or particulate matter settling out of the air. The amount of dry deposition depends upon the amount of suspended particles, wind speed and duration, and particle size. Contaminants are often present within dry deposition events, especially in urban areas where they may be sorbed onto the surfaces of particulate matter in the air (USEPA 2008).
2. *Wet deposition:* Occurs when snow, fog, or a rain droplet forms and then dissolves or carries a contaminant to the surface. Acid rain is a good example of wet deposition.

8.5 FATE AND TRANSPORT OF CONTAMINANTS IN THE ENVIRONMENT

Now that we have been introduced to how contaminants move around the natural environment after being released, let us turn our discussion to how individual contaminants groups.

8.5.1 Volatile Organic Compounds

VOCs are organic compounds that generally volatilize or evaporate readily under normal atmospheric pressure and temperatures. They usually have a high vapor pressure, low-to-medium solubility, and low molecular weight. As a result of these chemical characteristics, VOCs are common air, soil, and water contaminants (USGS 2006a; USEPA 2008). A subgroup of VOCs are halogenated VOCs. This subgroup will be discussed separately in the sections below.

8.5.1.1 Non-Halogenated Volatile Organic Compounds

Many VOCs have a molecular structure that is centered around what is termed, the **benzene ring** (see Figure 8.9) (Jensen 2009). Table 8.1 is a list of common VOCs. Automobile exhaust contains VOCs. When combined with other common air pollutants and sunlight, urban smog will form if atmospheric conditions are favorable – that is, there is an ample supply of the combined sources of VOCs and other smog-forming contaminants. This type of smog formation produces **photochemical smog**.

VOCs are also released directly onto the ground surface through leaks or spills at, or near the surface. Sources of these leaks include underground storage tanks, service stations, refineries, and pipelines. Because of these surface and shallow subsurface releases, VOCs are common groundwater contaminants and have been detected in the groundwater of numerous aquifers in the United States (USGS 2006a).

A study of groundwater in the United States detected VOCs at a concentration of 0.02 µ/L in more than 50% of approximately 3,500 samples collected from 100 different groundwater aquifers across the country (USGS 2006a). The VOCs detected most often included bromoform, bromodichloromethane, chloroform, chloromethane, 1,1-dichloroethane (DCE),

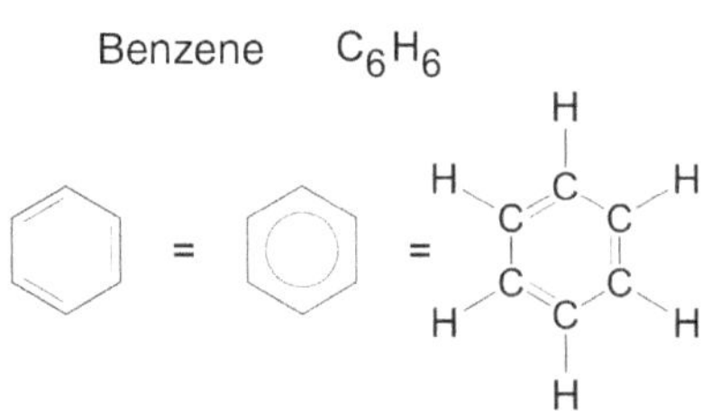

FIGURE 8.9 Benzene and the benzene ring.

TABLE 8.1
Common Uses of VOCs

Compound (Listed Alphabetically)	CAS Registry Number	Common Uses
Acetone [a]	67-64-1	Nail polish remover, paint thinner, laboratory chemical
Benzene[b,c]	71-43-2	Gasoline component, solvent, pharmaceutical, dyes, and plastics
Bromodichloromethane	75-27-4	Trihalomethane (byproduct of water purification)
Bromoform	75-25-2	Trihalomethane (byproduct of water purification)
Bromomethane	75-83-9	Fumigant
Carbon disulfide[d]	75-15-0	Cellulose manufacturing, soil fumigant
Carbon tetrachloride[f]	56-23-5	Solvent, dry cleaning, agriculture, formerly used in fire extinguishers
Chlorobenzene	108-90-7	Solvent, insecticides
Chloroethane	75-00-3	Solvent, pharmaceutical manufacturing
Chloroform	67-66-3	Trihalomethane (byproduct of water purification)
Chloromethane	74-87-3	Solvent, agricultural chemicals, and cellulose
Dibromochloromethane	124-48-1	Trihalomethane (byproduct of water purification)
1,1-Dichloroethane (DCE)	75-34-3	Solvent, manufacture of plastic wrap and adhesives
1,2-DCE	107-06-2	Solvent, paint ingredient, fumigant, vinyl chloride production
1,1-Dichloroethene	75-35-4	Manufacturing of plastics, adhesives, refrigerants
cis-1.2-Dichloroethene	156-59-2	Fat extraction from meat and fish, refrigerants, pharmaceuticals
trans-1,2-Dichloroethene	156-60-5	Fat extraction from meat and fish, refrigerants, pharmaceuticals
Dichloromethane	79-09-2	Paint and stain remover, aerosol propellant, caffeine extraction
1,2-Dichloropropane	78-87-5	Solvent, stain remover, former soil fumigant
cis-1,3-Dichloropropene	10061-01-5	Solvent
trans-1,3-Dichloropropene	10061-02-06	Solvent
Ethyl benzene[g]	100-41-4	Gasoline component, solvent, styrene manufacturing
2-Hexanone[h]	591-78-6	Paint and paint thinner, used to dissolve oil and waxes
Methyl ethyl ketone (MEK)	78-93-3	Solvent
Methyl isobutyl ketone (MIBK)	108-10-1	Solvent, used in metal extraction
Methyl-*tert*-butyl ether (MTBE)	1634-04-4	Former gasoline additive, formerly used to dissolve gallstones
Naphthalene[l]	91-20-3	Gasoline component, mothballs, insecticide
Styrene	100-42-5	Gasoline component, coating, paint, rubber, adhesives
1,1,2,2-Tetrachloroethane	79-34-5	Solvent, ingredient in paints and pesticides
Tetrachloroethene (PCE)[e]	127-18-4	Solvent, dry cleaning, textile processing
Toluene[i]	108-88-3	Gasoline component, solvent
1,1,1-Trichloroethane (TCA)[e]	71-55-6	Solvent, cosmetic ingredient, aerosol products, textile processes
1,1,2-Trichloroethane	79-00-5	Solvent, manufacturing of plastic wraps
Trichloroethene (TCE) [e]	79-01-6	Solvent, caffeine extraction, dry cleaning, paint and ink, rubber
Vinyl acetate	108-05-4	Polymer
Vinyl chloride[j]	75-01-4	Rubber, paper, and glass industries, PVC manufacturing
Xylenes[k]	1330-20-7	Gasoline component, paint thinner ingredient, plasticizer

Source: USGS. 2006a. *Volatile Organic Compounds in the Nation's Ground Water and Drinking-Water Supply Wells.* USGS Circular 1292. USGS. Reston, VA; USEPA. 2024a. *EPA's Report on the Environment (ROE).* www.epa.gov/report-environment. (Accessed May 11, 2024).)

a Levy (2009).
b Lide (2008).
c ATSDR (2007a).
d Holleman and Wiberg (2001).
e Doherty (2000).
f ATSDR (2007b).
g ATSDR (1996)
h ATSDR (2005a)
i ATSDR (2001a).
j ATSDR (2006a)
k ATSDR (2007c)
l ATSDR (2005b)

dichlorodifluoromethane, methylene chloride, dibromochloromethane, MTBE, trichloroethene (TCE), tetrachloroethene (PCE), 1,1,1-trichloroethane (TCA), trans-1,2-dichloroethene, toluene, and trichlorofluoromethane.

The same study indicates the vulnerable nature of many aquifers in the United States, and their location corresponds with many urban areas. In fact, of the 28 major urban areas of the United States, 27 have detectable concentrations of VOCs (only Kansas City is missing) and represent a population of 125.4 million – over 40% of the entire population of the United States.

VOCs exist as Light non-aqueous Phase Liquids (LNAPLs) and Dense non-aqueous Phase Liquids (DNAPLs). Because LNAPLs are lighter than water, they tend to float on top of groundwater, whereas the heavier-than-water DNAPLs tend to sink through the water column in an aquifer if conditions are favorable (USGS 2006a).

Contaminant degradation rates vary widely and depend on many factors including: (1) the nature of the release; (2) the physical chemistry of the contaminants themselves; (3) the geological environment where the contaminants are released; and (4) the presence, type, and distribution of microorganisms (Howard et al. 1997; USEPA 1998a; McKone and Enoch 2002; USGS 2006a).

Degradation rates also vary by media. In general, organic compounds in the atmosphere, including VOCs, degrade more quickly than the same organic compounds released and migrate to subsurface soil and groundwater (USEPA 1998a, 1998b).

8.5.1.2 Halogenated Volatile Organic Compounds

A group of VOCs, called halogenated VOCs, are very toxic and do not degrade easily. Halogenated VOCs have also been commonly referred to as chlorinated solvents or DNAPL compounds (USEPA 1996a, 1996b). Halogenated VOC compounds include compounds such as PCE, TCE, 1,1,1-TCA, DCE, vinyl chloride, PFAS, methyl bromide, several pesticides, and many other compounds. They are considered very toxic and persistent in the environment if they are released and can remain in the environment for decades and can even increase in toxicity with time (USEPA 1996a; Suthersan et al. 2017).

Halogens, meaning salt producers, are elements that form a group called Group 7A and are located on the Periodic Table in column 17. The six halogenated elements are fluorine, chlorine, bromine, iodine, astatine, and tennessine. These six elements have special characteristics and properties that make them unique and different from other elements. They are not present in their elemental form in nature, but rather they are present in combination with other nonmetals in molecular form or with several metals in the molecular structure of minerals. The primary reason why these elements are not present in elemental form in nature is that they are highly reactive and strongly attach to other molecules, especially organic molecules (International Union of Pure and Applied Chemistry (IUPAC) 2024; United Nations 2024; USGS 2006a). Halogenated elements of Group 7A form salts when they react with other molecules, including metals. Halogens are reactive nonmetal elements and have seven valence electrons as pure elements, one short of a stable octet. They form diatomic molecules with atoms by nonpolar covalent bonds when joined. Some properties of halogenated elements include the following (IUPAC 2024; United Nations 2024):

- Halogens are highly electronegative and highly reactive with alkali metals and alkaline soil.
- Halogens have a relatively weak intermolecular force and can gain an electron by reacting with other atoms.
- Halogens have low melting and boiling points and are poor conductors of electricity and heat.
- Halogens are very toxic and potentially lethal because of their high reactivity. As halogens get heavier down column 17 in Group 7A in the Periodic Table, they become less reactive, and their degree of toxicity also decreases. However, astatine is considered dangerous because it is radioactive (United States Institute of Health 2024b).

- Halogens are elements that are naturally present in all three basic states of solid, liquid, and gas. Fluorine and chlorine most often occur in the gas phase. Bromine is naturally present as a liquid, and iodine and astatine are naturally present in the solid phase.

Fluorine is highly toxic and reactive and is visible as a pale-yellow gas. It is so toxic that many scientists who studied fluorine died because of the elements' high toxicity. Fluorine is abundant in the Earth's crust and is the 13th-most abundant element on Earth. Fluorine is the most reactive element and is the only element that is more electronegative than oxygen. The high reactivity of fluorine allows some of the strongest bonds possible, especially to carbon. For example, Teflon is fluorine bonded with carbon, is extremely resistant to thermal and chemical attacks, and has a high melting point (IUPAC 2024).

Chlorine is highly toxic and reactive and forms a yellow-to-green gas. Chlorine is considered too reactive to be present in elemental form in nature. Chlorine is commonly present in synthetic pesticides and VOCs. Chlorine is abundant in the Earth's crust and is commonly used in bleaches, cleaners, and pesticides.

Bromine is the only nonmetal substance that is present in liquid form at standard pressure and temperature and is moderately abundant in the Earth's crust. Bromine is used in fire extinguishers and pesticides.

Iodine is a solid but is also volatile and may sublimate when heated and form a violet-colored gas. Earlier, iodine was used as a disinfectant, mainly for skin wounds. Compounds of iodine are used as catalysts, drugs, and dyes. Silver iodide (AgI) is an important component of the photographic industry and is used in seeding clouds to increase precipitation yield.

Astatine is only present in trace amounts in the Earth's crust and is radioactive. Astatine is commonly present in association with some ores of uranium in nature.

Tennessine is an element whereby characteristics are generally unknown due to its rarity and it was first created in 2010 by a team of scientists from Oak Ridge National Laboratory in Tennessee, Vanderbilt University, and Lawrence Livermore National Laboratory. Its properties and behavior are greatly dependent on theoretical expectations.

An example of the resistance of halogenated compounds degradability is presented in Figure 8.10, which is the degradation sequence of the halogenated VOC tetrachloroethene. The degradation sequence follows a modified Dominico (1987) analytical solution for mass transport in a consecutive

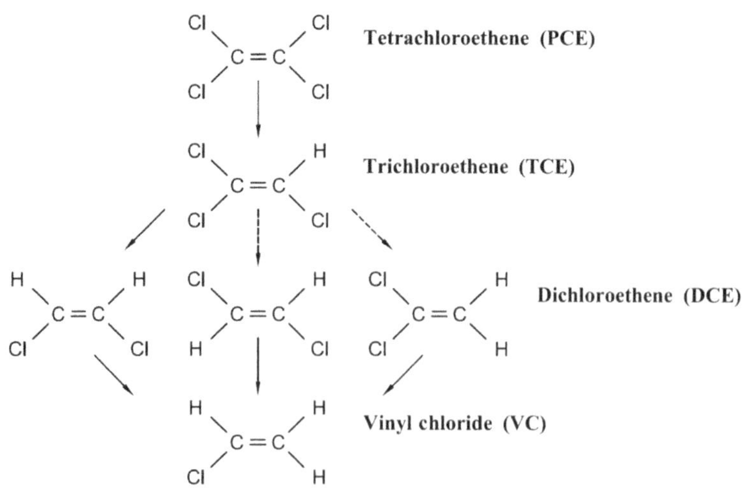

FIGURE 8.10 Degradation sequence of tetrachloroethene.

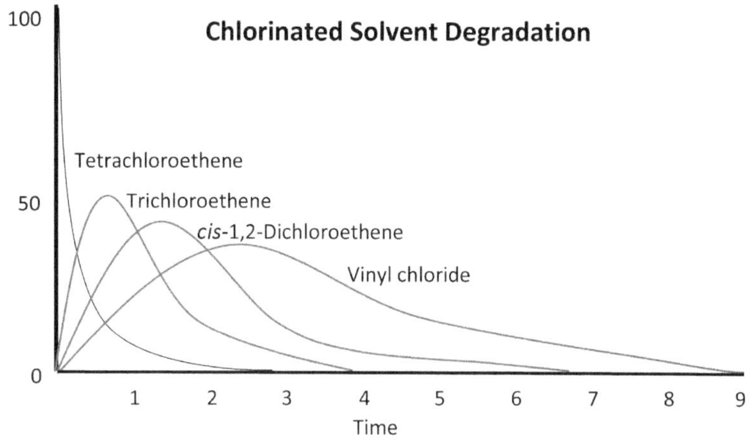

FIGURE 8.11 Chlorinated solvent degradation. (Modified from Payne, F. C., and Rogers, D. T. Chlorinated Solvent Degradation – A Consecutive, Irreversible Reaction Sequence. In: *Remediation of Recalcitrant Chlorinated Volatile Organic Compounds in Groundwater*. Groundwater Management Book 21. National Groundwater Association. Dublin, OH. pp. 97–109.)

$$\frac{C_x}{C_0} = \exp\left\{\frac{x}{2\alpha_x}\left[1-\left(1+\frac{4\lambda\alpha_x}{\upsilon}\right)^{\frac{1}{2}}\right]\right\} erf\left[\frac{Y}{4(\alpha_y x)^{\frac{1}{2}}}\right] erf\left[\frac{Z}{4(\alpha_z x)^{\frac{1}{2}}}\right]$$

Where,

C_x - contaminant concentration in a downgradient well along the plume centerline at a distance x (mg/l),
C_0 - contaminant concentration in the source well (mg/l),
x - centerline distance between the downgradient well and source well (feet),
α_x, α_y and α_z - longitudinal, transverse, and vertical dispersivity (feet), repectively,
$D_x = \alpha_x \times v$, $D_y = \alpha_y \times v$, $D_z = \alpha_z \times v$,
λ- degradation rate constant (1/day),
$\lambda = 0.693/t_{1/2}$ (where $t_{1/2}$ is the degradtion half-life of the compound).
v - groundwater velocity (feet/day),
Y - source width (feet),
Z - source depth (feet),
erf - error function,
exp - exponential function.

FIGURE 8.12 Dominico steady-state model for multidimensional transport of decaying contaminant species analytical transport equation. (From Domenico, P. A. 1987. An Analytical Model for Multidimensional Transport of Decaying Contaminant Species. *Journal of Hydrology*. Vol. 91, pp. 49–58.)

and irreversible fourth-order differential equation with default persistence of 2, 4, 6, and 8 years for each compound (see Figures 8.11 and 8.12) (Payne and Rogers 1997).

As shown in Figure 8.11, the maximum vinyl chloride concentration is achieved approximately 25 years after being released. According to Agency for Toxic Substances and Disease (ATSDR 2006a), vinyl chloride is approximately 60 times more potent than its parent compound tetrachloroethene. Therefore, the degradation sequence of tetrachloroethene demonstrates that peak toxicity is achieved approximately 25 years after being released. The example shown in the figure uses an initial concentration of tetrachloroethene of 1,000 μ/L and the time scale is presented in decades (Payne and Rogers 1997).

8.5.1.3 Trihalomethane Volatile Organic Compounds

Trihalomethane VOCs are chemical compounds in which three of the four hydrogen atoms of methane (CH_4) are replaced by halogen atoms. The most common trihalomethane VOCs include chloroform, bromoform, bromodichloromethane, and dibromochloromethane. These compounds have been detected in the groundwater of many US aquifers (USGS 2006a). Trihalomethanes have high relative vapor pressures and commonly evaporate quickly when in contact with the atmosphere. Therefore, exposure to trihalomethanes is of special concern during showering and washing (ATSDR 1997a, 2005a). They degrade by photolysis when exposed to direct sunlight and can also be degraded by microorganisms (ATSDR 1997a, 2005a).

8.5.2 Polynuclear Aromatic Hydrocarbons

PAHs are LNAPL compounds and being lighter than water they tend to float on surface water and groundwater. PAH compounds are formed synthetically or naturally during incomplete combustion of coal, tar, gasoline, wood, and garbage, and may also be present in tobacco, medicines, dyes, plastics, pesticides, and charbroiled meat (United States Department of Health and Human Services (USDHH) 1995). PAHs do not readily dissolve in water and have low vapor pressures compared to most VOCs. PAHs are common constituents of automobile exhaust, especially from diesel fuel (USGS 2006a), and are common air contaminants in urban areas. In addition, PAHs have a high sorptive affinity and can therefore attach to particulate matter (ATSDR 1996). Many PAH compounds biodegrade under favorable conditions. The half-life of PAHs is shortest in the atmosphere due to photochemical degradation and lasts just a few days or weeks (ATSDR 1996). The half-life of PAH compounds in soil and groundwater is longer and may last for several years or decades. When released to soil, PAHs tend to migrate more slowly than VOCs because of their higher molecular weight and sorption to soils with high organic content.

8.5.3 Polychlorinated Biphenyls

PCB compounds are synthetic halogenated semi-volatile organic compounds. There are 209 individually distinct PCB compounds, each termed a PCB congener, that contain 1 to 10 chlorine atoms on a biphenyl molecule (see Figure 8.13). Once in the environment, PCBs do not readily degrade – they remain in the environment for long periods of time and often cycle between air, water, and soil. PCBs can be carried long distances attached to particulate matter and have been detected in snow and seawater far away from any known point of release (USEPA 1996a). This transport capability is confirmed by their worldwide detection. The lighter the PCB compound (fewer number of chlorine atoms in its structure), the farther it can potentially be transported from its release point.

PCBs are not very soluble, have high adsorption potential, and low vapor pressures. This combination of physical properties largely determines their environmental distribution, as they are only present at high concentrations in water or air when sorbed to particulate matter. Sinks where PCBs are frequently detected include soil near release points and sediments in rivers, streams, and lakes (USEPA 1996a, 1996b). PCBs bioaccumulate in the leaves and above-ground parts of plants and

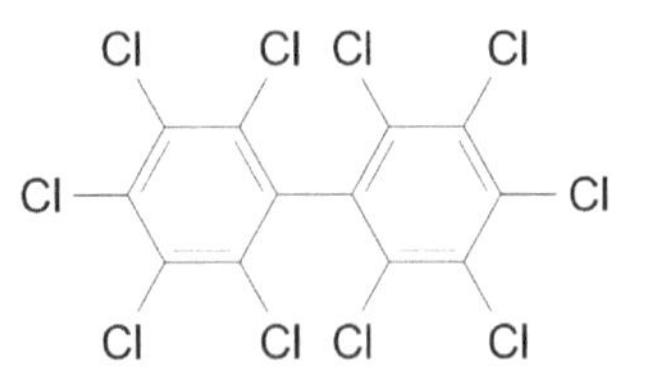

FIGURE 8.13 Basic PCB structure.

food crops, and in aquatic organisms and fish where PCBs are present in sediments (USEPA 2024b). As a result, humans and other organisms ingesting impacted plant material, food crops, or organisms containing PCBs may bioaccumulate PCBs in their body tissues (ATSDR 2001b; USEPA 2009a).

8.5.4 Semi-Volatile Organic Compounds

SVOCs are much less volatile than VOCs, but notable exceptions here are the amine compounds that exist as a gas at room temperature and standard pressure (ATSDR 2002a). In terms of solubility, phthalates and phenols do not readily dissolve in water, whereas amines and esters may dissolve, become mobile, and reach groundwater (ATSDR 2002a, 2002b, 2002c, 2002d). When released into the environment SVOCs are commonly detected in soil because they have high adsorption potentials. Many SVOCs, including pentachlorophenol, are degraded by microorganisms under favorable conditions and are also susceptible to photolysis and hydrolysis (ATSDR 2001c, 2002a). Amines and esters degrade in minutes when exposed to direct sunlight (ATSDR 1999a, 2002a, 2002b, 2002c, 2002d).

8.5.5 Heavy Metals

Heavy metals are released directly into air, water, and soil. In most cases, these contaminants do not remain in the atmosphere for long periods of time because they have high specific gravity and become deposited onto the land surface shortly after being emitted. Mercury, however, has been detected as far away as 50 miles from its source after being released into the atmosphere (USEPA 1997). Lead is also considered a common air pollutant (USEPA 2008). Major sources of lead include metal melting facilities, battery manufacturing, and leaded gasoline and fuels. Table 8.2 lists many heavy metals and their common uses.

TABLE 8.2
Common Uses for Each of the Select Heavy Metals

Heavy Metal	Atomic Number	Uses
Arsenic	33	Wood preservative, poison, insecticide, pigments, chemical weapons
Barium	56	Superconductors, pigments, fireworks, lubricants, optics
Cadmium	48	Batteries, plastic stabilizer, pigments, metal plating, coatings, alloys
Chromium III	24	Pigments, inks, glass, steel additive, dyes, leather tanning, refractory, alloys
Chromium VI	24	Metal plating, corrosion resistance additive, wood preservative
Copper	29	Wire, building products, piping, jewelry, electromagnetics, brass, and other alloys
Lead	82	Batteries, paint, ceramics, firearms, industrial coolant, electrodes, solder, construction materials, alloys, formerly a gasoline additive
Mercury	80	Electrical switches, thermometers, manometers, medical and dental applications, cosmetics, mercury-vapor lamps, formerly used in hat making. Mercury is the only metal evaluated that is a liquid and has a measurable vapor pressure at standard pressure and temperature.
Nickel	28	Batteries, steel additive, magnets, coins, alloys
Selenium	34	Photocells, electronics, semiconductors, steel additive, alloys, copier and printing drums, glass manufacturing, pigments
Silver	47	Coins, electronics, circuit boards, alloys, mirrors, decorative items, jewelry, photographic films, batteries
Zinc	30	Metal plating, rust inhibitor (galvanization), brass alloy, batteries, cathodic protection, paint pigment, fire retardant, propellant, photocopying products, medical applications

Source: Pradyot, P. 2003. *Handbook of Inorganic Chemical Compounds*. McGraw Hill. New York, NY.; Krebs, R. E. 2006. *The History and Use of Earth's Chemical Elements: A Reference Guide*. Greenwood Publishing Group. Oxford, England.

The good news is that there has been a 92% decrease in atmospheric lead concentrations over the period from 1980 through 2018. Contributing to the observed decrease has been the removal of lead from gasoline and fuels, increased efficiency in air pollution control equipment, and the regulation of lead emissions sources (USEPA 2024a).

The solubility of heavy metals in water is very low, except for arsenic and chromium VI at a neutral pH. Due to their inability to form a solution, the preferred sinks for heavy metals are soil and sediments. Many metals undergo some transformation such as oxidation after being released into the environment, but are not destroyed and remain in the environment (ATSDR 1999b, 1999c, 2003, 2004a, 2005c, 2005d, 2007d, 2007e, 2007f, 2008a, 2008b). Due to their low solubility and high specific gravities, they tend to remain near release points. This is why increased concentrations of heavy metals are present in the near-surface soil of urban areas as a result of releases from anthropogenic sources (Murray et al. 2004).

Some heavy metals such as mercury accumulate in the sediment of lakes, rivers, and streams. Mercury may undergo a transformation through a process known as methylation after being released into the environment, typically when it reaches a surface water body (USEPA 1997). The process of methylation transforms elemental mercury to methyl mercury (CH_3Hg). Methyl mercury is the most toxic form of mercury and has the potential to bioaccumulate in aquatic organisms, including fish (USGS 2000).

On land and in sediments, other heavy metals such as barium, cadmium, chromium, lead, copper, and zinc can accumulate in plant matter if they are present at sufficient concentrations in soil within the root zone of plants (Nudunuri et al. 1998). Removal of contaminants from near-surface soil is not uncommon and is sometimes the preferred alternative. Suitable locations for removal include closed landfill sites requiring a vegetative cap to minimize erosion potential, or sites where contamination is shallow, relatively static, and has become a chronic problem (Singer et al. 2003). **Phytoremediation** – the removal of contamination using plants – has been applied to other contaminants including organic compounds, but with limited success.

Murray et al. (2004) evaluated heavy metals in urban soil. This study represents an initial effort to characterize the metal concentrations in surface and near-surface soil in an urban environment in southeast Michigan that includes the City of Detroit. The results of the study discovered that:

1. Heavy metal concentrations in an urban environment are greatest at the surface
2. A directional increase of metal concentrations in a west-to-east trend across the metropolitan area commensurate with a general west-to-east increase in urbanization and industrial activity
3. Identified lead as the heavy metal with the highest surface concentration with mean levels present at more than 16 times background at residential sites and 15.5 times greater than background at industrial sites. More importantly, lead concentrations in surface soil at specific industrial sites and adjacent residential neighborhoods were detected at levels hundreds of times the level that may occur naturally in the soil.
4. Heavy metal concentrations beneath the surface (at depths between 0.5 and 10 m), although elevated, were in general agreement with those of previous studies conducted in Michigan and the United States.
5. Heavy metal concentrations in soils at depths greater than 10 m are in complete agreement with previous studies of heavy metal concentrations in rural areas of Michigan and the United States. This suggests that soil at depths greater than 10 m is generally only affected to a minor extent by anthropogenic heavy metal sources.

The results of the Murray et al. (2004) study of soil metals have important implications for land-use planning and for future site investigations in any urban area when heavy metal contamination is suspected and include the following:

- Evaluation of human and ecological risk can be achieved by concentrating investigation efforts on heavy metals in the surface soils. The results of the study have demonstrated that heavy metal concentrations are highest at the surface and quickly decrease with increasing depth. Therefore, concentrating investigative efforts at the surface will efficiently identify the most elevated heavy metal concentrations derived from anthropogenic sources.
- Innovative regional strategies for the cleanup of sites with metal contamination need to be developed.
- Identification of potential hydrologic and ecological impacts of metals in the near-surface zone is essential for the protection of groundwater. As this study has identified that the occurrence of heavy metals in an urban environment is highest at the surface, the results can be integrated into evaluations of interflow, base flow, increased leaching from acid rain, infiltration/inflow to public water supplies, runoff to surface water systems, and the potential revegetation capacity at redeveloped sites.
- Mapping of the surface geology in urban areas can pinpoint soils particularly sensitive and prone to contamination.
- Conducting future studies related to the occurrence of heavy metals in other urban areas should consider the effects of historic land-use activities and the potential human health issues and risks associated with heavy metals in the surface soils.

With respect to dissolved metals in groundwater, results from the Murray et al. (2004) study indicate urbanization and its accompanying industrial activities have contributed to heavy metal impacts across all land-use categories (industrial, commercial, and residential) and throughout the varied surface geology of the metropolitan area. Chromium is prevalent at industrial sites, and the high levels of Cr in groundwater strongly suggest the chromium is in a hexavalent form – which is much more mobile and soluble than its trivalent form – and has greater potential to damage the environment and impact human health. This is especially significant as hexavalent chromium has a high contaminant risk factor. Significantly high contamination ratios were demonstrated by Cr, Cd, and Pb within the study area's sand unit, but there were also several metals detected at concentrations exceeding their MCLs in clay-rich soil units, indicating the existence of some mobility within a soil medium previously considered virtually immobile for metals.

8.5.6 Pesticides and Herbicides

Pesticides and herbicides are released into the environment for specific purposes and are most commonly detected in near-surface soils and surface water rather than groundwater (USGS 2006b). The highest concentrations of pesticides exist in the nation's streams and sediments within urban areas where they have been detected in 83% of streams and 70% of sediments (USGS 2006b). The total tonnage of pesticide and herbicide use in the United States has remained constant during the 1980s and early 1990s but has increased by nearly 60% in the last 20 years mostly for agriculture and home and garden purposes (USEPA 2017).

Permethrin and toxaphene are two other widely used compounds in insecticides and pesticides, respectively. The fate and transport concern with permethrin centers around its use as an insect repellent, application to crops, and flea treatments for pets. All of these activities involve human exposure (ATSDR 2005e), and with a half-life of approximately 28 days, there is often adequate time for human contact. Toxaphene strongly sorbs to soil particles and is not very soluble in water (SRC 2024). Common sinks where toxaphene may be present include sediments and soil, from where it bioaccumulates in fish and mammals. Toxaphene is also in the atmosphere as it evaporates when in a solid form or dissolved (ATSDR 1997b). It is estimated the half-life of toxaphene is more than 10 years in soil (SRC 2024), so there is a high likelihood it is still present at appreciable concentrations in the environment.

8.5.7 Fertilizers

The most common fertilizers include nitrogen (N), phosphorus (P), and potassium (K). A major sink for fertilizers is surface water because they are applied to the soil surface and are considered soluble in water and mobile, especially in the case of nitrates and phosphorus. Once in surface water, nitrate and phosphorus can promote excessive algal growth. Significant algal growth can deplete the dissolved oxygen in surface water and cause suffocation and death to aquatic organisms. The solubility of some fertilizers, combined with the relationship between surface water and groundwater, may lead to groundwater contamination. The natural process of enrichment of surface waters with plant nutrients is termed **eutrophication**. When anthropogenic activities such as fertilization or sewage discharges accelerate this natural process, **cultural eutrophication** occurs (McGucken 2000).

8.5.8 Dioxins

Dioxins are halogens and are considered very toxic (ATSDR 2006b). The dioxin compound, or congener, considered to be the most toxic is 2,3,7,8-tetrachlorodibenzodioxin (TCDD). As one method of dioxin formation is through incineration and combustion, dioxin compounds are present in the atmosphere and have been detected around the globe (ATSDR 2006b). Dioxin and dioxin-like compounds (DLCs) are a group of compounds that are persistent organic pollutants (POPs). In many instances, dioxins are formed through incomplete combustion or intense thermal industrial processes using PCBs or PCB-like chemicals. Dioxins typically have chlorine in their atomic structure and their toxicity varies depending on the position of the chlorine atoms (USEPA 2024b). In fact, the toxicity of individual dioxin compounds can vary by a factor of approximately 30,000 times (ATSDR 2006b).

When dioxins are released into the environment, they typically sorb to soil particles and are often detected in sediments in lakes, rivers, and streams acting as sinks for dioxins (ATSDR 2006b). Dioxin compounds are considered bioaccumulative contaminants, with the potential to build up in the food chain and yield detectable concentrations in the tissues of many animals (ATSDR 2006b; USEPA 2024b). Due to the presence of chlorine in their atomic structure, dioxins do not readily degrade once they are formed and released into the environment (USGS 2006b).

Dioxins are essentially insoluble in water but have a high relative solubility in lipids. Dioxins tend to associate themselves with organic matter such as plankton, plant leaves, and animal fat. In addition, they tend to sorb to inorganic particles such as ash and soil with high clay content (USEPA 2006b). Dioxins tend to bioaccumulate in animals (ATSDR 2006b).

Times Beach in Missouri is one of the first Superfund sites in the United States and is listed as a Superfund Site because of dioxin contamination, specifically TCDD (USEPA 1988).

8.5.9 Other Pollutants

Other pollutants described below include dioxins, cyanide, asbestos, acids and bases, radioactive compounds, greenhouse gases, carbon dioxide, carbon monoxide, ozone, sulfur dioxide, particulate matter, pathogens, invasive species, polyfluoroalkyl substances, and emerging contaminants.

8.5.9.1 Cyanide

In the atmosphere, cyanide is most often present as hydrogen cyanide. When present in surface water, cyanide compounds usually will form hydrogen cyanide and then enter the atmosphere through evaporation. When released into the soil, cyanide compounds are considered fairly mobile when the retention capacity of the soil is exceeded and may migrate and contaminate groundwater. Cyanide compounds are degraded by microorganisms when present at low concentrations.

When concentrations of cyanide compounds are elevated, they tend to be toxic to microorganisms and resist degradation (ATSDR 2006c). The half-life of cyanide in the atmosphere ranges between 1 and 3 years (ATSDR 2006c). In soil and water, the half-life of cyanide compounds is much more difficult to estimate because the concentration, distribution, and presence of microorganisms available to degrade the cyanide compounds varies.

8.5.9.2 Asbestos

Asbestos fibers do not degrade, evaporate, or dissolve in water and remain virtually unchanged in the environment (ATSDR 2001d). Asbestos originates from naturally occurring minerals and is therefore present in the environment. Average background concentrations of asbestos in air range from 0.00001 to 0.0001 fibers per milliliter of air and are highest in urban areas (ATSDR 2001d). Small-diameter asbestos fibers can remain suspended in the atmosphere for a long period of time compared to larger fibers (those larger than 10 μm) (ATSDR 2001d). As asbestos is widely used in building materials, it is most common in urban areas and in places where natural deposits are present. Asbestos can become airborne through the disturbance of asbestos-containing materials during demolition or remodeling activities.

8.5.9.3 Acids and Bases

When released into soil, acids and bases neutralize rapidly. They are diluted when they come into contact with water if a difference in pH levels exists. Therefore, if environmental impairment occurs, it must be realized rapidly before the acid or base becomes neutralized. This impairment occurs with the majority of sudden and accidental releases but does not hold true for acid rain which generates adverse effects with slower onsets. Acids and bases may migrate a significant distance – sometimes more than one mile – when released in the atmosphere and may cause significant impairment to living organisms exposed to their vapors (ATSDR 2004b, 2004c).

8.5.9.4 Radioactive Compounds

Two types of radioactive compounds will be addressed in this section. One type is the naturally occurring compound radon. The other are radioactive active compounds that originate from nuclear power generation plants. A third type is solar radiation, which is addressed separately.

8.5.9.4.1 Radon

Radioactive compounds occur naturally, with the most common being radon. Radon is produced from the decay of uranium (ATSDR 2008c) and is present in air, water, and soil. Radon may build up in basements, especially if cracks exist, or other subsurface structures located above natural deposits having higher relative uranium levels. The half-life of radon is approximately 4 days (ATSDR 2008c). Most of the human exposure attributed to other radioactive compounds results from medical devices, diagnostic treatments, testing equipment such as X-ray machines, and cancer therapy (Kathren 1991; ATSDR 2000).

8.5.9.4.2 Nuclear Radiation

Radionuclides are created during nuclear explosions, irradiation in nuclear reactors, particle X-ray bombardment of materials, and decay and fission of parent isotopes (Deblonde et al. 2020). Natural radionuclides are created by stellar nucleosynthesis and cosmic-ray spallation processes. However, with the exception of uranium and its isotopes, natural radioactive isotopes do not pose a significant health threat to humans or the environment. Anthropogenic radionuclides are produced under extreme pressures, temperature, and radiation fields, as well in non-equilibrium physiochemical conditions, and this may subsequently influence the surrounding geology and the radionuclides' fate in the environment. As an example, radionuclides released into the environment from an underground nuclear detonation may exceed temperature of 1,000,000 K, vaporize approximately 70 metric tons of rock, and melt approximately 700 tons of rock for every kiloton of explosive yield (Deblonde et al. 2020).

Understanding how radionuclides behave in the environment is essential for effective isolation of nuclear waste, management of contaminated sites, ensuring long-term protection of ecosystems, and limiting potential impacts on human health. Radioactive compounds from nuclear detonations and nuclear power generating plants present a unique set of challenging circumstances in minimizing environmental risk. One challenge is the potency of the radiation to human health and the other is that in many circumstances, the half-life of some compounds is greater than 10,000 years (Deblonde et al. 2020). An additional challenge is the difficulty of studying the fate and transport behavior of these compounds because of the risk of exposure to the hazard and the exceptionally long half-lives of the compounds. Lastly, many of the radionuclides of concern either do not naturally occur on Earth or are present in small quantities. Some of the compounds include technetium, neptunium, americium, and others. Therefore, the behavior of nuclear waste is intrinsically difficult to evaluate because of the wide range of elements they contain and cover a wide-spread media realm including gas, solids, soluble, cations, anions, oxo-species, and others (Deblonde et al. 2020).

8.5.9.5 Greenhouse Gases

Greenhouse gases decay very slowly and are primarily atmospheric contaminants. Some quantities of these gases are naturally removed from the atmosphere, such as the removal of carbon dioxide during photosynthesis. However, the anthropogenic addition of carbon dioxide and other greenhouse gases into the atmosphere has greatly exceeded the capacity of the natural environment to remove them (USEPA 2009b). As a result, greenhouse gas concentrations have been increasing (USEPA 2018).

8.5.9.6 Carbon Dioxide

As we have discussed at length, carbon dioxide is a greenhouse gas but it is worth mentioning again here that there is another important method in which carbon dioxide is absorbed and that is by the oceans. The rate of accumulation of pollution is proportional to human population. The more people, the more carbon dioxide ends up in the oceans. However, carbon dioxide is not evenly distributed in the oceans vertically (National Atmospheric and Space Administration (NASA) 2021). In addition, there is not a clear understanding of how the oceans react to increasing concentrations of carbon dioxide in the atmosphere (NASA 2021).

Absorption of carbon dioxide by the oceans has been estimated to be as high as 25% which represents millions of tons of carbon dioxide absorbed by the oceans every year. The additional carbon dioxide absorbed by the oceans reacts with ocean water and forms an acid called carbonic acid, lowering the pH of the oceans worldwide. In fact, the oceans are acidifying faster now than they have in some 300 million years. The effect of acidification has been linked to the bleaching of coral reefs, including the Great Barrier Reef off the northeastern coast of Australia (Natural Resources Defense Council (NRDC) 2024).

8.5.9.7 Carbon Monoxide

Carbon monoxide is released when fuel is not burned completely. Carbon monoxide is formed naturally and anthropogenically. The most significant source of carbon monoxide is from automobile exhaust (USEPA 2025b). Inside homes, significant sources of carbon dioxide emissions are natural gas and oil furnaces, hot water heaters, appliances, wood-burning stoves, and fireplaces. Carbon monoxide is also created as a byproduct of several industrial processes including metal melting and chemical synthesis. From 1980 to 2016 there has been a decrease of greater than 80% in carbon monoxide in the United States. This decrease was attributed to improved air pollution control equipment for stationary and mobile sources of air pollution (USEPA 2021a).

8.5.9.8 Ozone

Ozone is a gas occurring in Earth's upper atmosphere and at ground level. As noted in Chapter 7, ozone in the upper atmosphere is greatly beneficial to life on Earth because it filters UV radiation, but ozone occurring at ground level is considered an air pollutant (USEPA 2024a). Ground-level ozone is not emitted directly into the air – it is created by chemical reactions between oxides of nitrogen (NO_x) and VOCs in the presence of sunlight.

Emissions from automobile exhausts, gasoline vapors, chemical solvents, electrical generating facilities, and some factories trigger the production of ground-level ozone (USEPA 2009b). This variety of ozone is a concern in urban regions of the United States during the summer because strong sunlight and hot weather can generate higher levels (USEPA 2009b). Since the passing of the Clean Air Act of 1990, atmospheric ozone concentration in the United States has declined by 25% (USEPA 2024a). Better control of stationary and mobile sources of air pollution such as automobile exhaust is behind this improvement (USEPA 2024a).

The ozone layer in Earth's upper atmosphere filters much of the ultraviolet radiation from the sun and is beneficial to all life on Earth. However, anthropogenic activity from the use of some aerosols has eroded some of the ozone layer. Depletion of the ozone layer means that there is less protection from certain ultraviolet wavelengths of solar radiation from the sun. Studies have linked the depletion of the ozone layer to certain types of skin cancer. Depletion of the ozone layer has also been linked to causing negative effects on plants, marine ecosystems, biogeochemical cycles, and solid earthen materials (USEPA 2024e).

8.5.9.9 Sulfur Dioxide

Sulfur dioxide is a component of smog and also combines with nitrous oxide compounds to eventually form sulfuric acid which is commonly referred to as acid rain (USEPA 2024c). It is removed from the atmosphere during precipitation and is typically neutralized quickly in soil if the pH of the soil is greater than 7. Some areas of the northeastern United States have soils lacking the ability to effectively neutralize the effects of acid rain, and there have been adverse effects on aquatic life and vegetation in the region. Efforts to reduce sulfur dioxide emissions from stationary and mobile sources have resulted in a decrease of 75% since 1980 in the United States (USEPA 2024c). Nevertheless, the pH of rain in the eastern United States remains acidic.

8.5.9.10 Particulate Matter

Particulate matter (PM) is also known as particle pollution, more commonly known as dust. PM is a complex mixture of very small particles and liquid droplets that generally originates from a regulated anthropogenic source. Urban areas have the highest concentrations of anthropogenic-created particulate matter, which is a significant distributor of contaminants in the atmosphere. Contaminants such as SVOCs, some VOCs, PCBs, and many pesticides and herbicides may sorb to a soil particle and travel a significant distance through wind action (USEPA 1998a). The size of particulate matter is significant because the largest sizes tend to settle to the ground surface first. Smaller particles can travel around the globe and remain suspended for years if favorable conditions exist (USEPA 2024a). USEPA regulates two varieties of PM. One variety is PM less than 10 microns in diameter and the other is PM less than 2.5 microns in diameter. Here is some good news: there was a 20% to 30% decline in atmospheric particulate matter within the urban areas of the United States between 2001 and 2016 (USEPA 2024a).

8.5.9.11 Pathogens

Bacteria, parasites, and viruses are considered pathogens and may be present in large numbers in the environment. They are in and on the food we eat, in and on our bodies, in the air we breathe and the water we drink, in soil, and at depths within the Earth. Many are beneficial, but some have the potential to adversely affect our health and wellbeing and are commonly referred to as *germs* (Madigan et al. 2008).

8.5.9.12 Invasive Species

Invasive species have been around for a few hundred years and now number in the tens of thousands of different species. Invasive species are difficult to impossible to contain once released, so prevention is key. The main reason invasive species are difficult to contain is that the mode of migration is heavily influenced by breeding, hitching a ride, or as a travel companion with humans. With an increasing human population, increased standard of living, a warming climate, and increased human travel, the prospects of increased invasive species challenges are almost certain.

8.5.9.13 Polyfluoroalkyl Substances

PFAS chemicals are persistent, mobile, and very toxic. They have been widely produced and used in numerous consumer products for decades and have just recently been identified as a contaminant (Bell et al. 2019). Due to its soluble nature, PFAS chemicals are most often detected in surface and groundwater and have also been detected in fish tissues. PFAS compounds have also been detected in foam that may form on lakes and may be concentrated along windward shorelines. In addition, because of their widespread use, potency, high relative toxicity, and mobility and persistence in the environment, many PFAS compounds are found in human blood and many animals, fish, and other organisms. PFAS chemicals have also been detected at low concentrations in a variety of food products. Reliable estimates of their degradation have not yet been quantified (USEPA 2024f).

8.5.9.14 Chlorofluorocarbons

A hundred years ago in the 1920s, toxic and flammable compounds such as ammonia, chloromethane, propane, and sulfur dioxide were used as refrigerants because they produced a cooling effect while expanding or vaporizing. In the late 1930s, an alternative was developed called chlorofluorocarbon, or CFC for short, composed of dichlorofluoromethane and was called freon. CFCs are now a class of compounds called CFCs that are gaseous and contain carbon, fluorine, and chlorine. Almost immediately after the development of CFCs, they became very popular in many commercial products such as refrigerants, spray can solvents, propellants, and in air conditioners. Although CFCs are non-reactive in the troposphere, they are broken down in the stratosphere by ultraviolet radiation from the sun where they then react with and destroy the ozone layer. CFCs are now commonly known as **ozone-depleting substances** (ODS). In addition, CFCs are strong greenhouse gases (USEPA 2019).

Since the 1970s, CFCs have been heavily regulated because of their destructive effects on the ozone layer. In 1978, the United States banned the use of CFCs. As a replacement for CFCs, compounds called hydrochlorofluorocarbons (HCFCs) were developed. HCFCs also deplete the ozone layer and are greenhouse gases but to a much lesser extent (USEPA 2019). HCFCs are also being phased out. While new production of these refrigerants has ended in the United States, large volumes still exist in older cooling systems. According to USEPA (2019), preventing the release of these harmful refrigerants is one of many ways to mitigate climate change.

8.5.9.15 Emerging Contaminants

Emerging contaminants including many pharmaceuticals, 1,4-dioxane, 1,2,3-trichloropropane, PFAS compounds, phenol isopropylated phosphate (PIP), and perchlorates are resistant to degradation and can remain in the environment for long periods of time (Bell et al. 2019; USEPA 2024d, 2024g, 2024h). Emerging contaminants have been detected in groundwater where they can migrate long distances due to their relatively high solubility and resistance to degradation. Emerging contaminants are difficult to investigate because many have entered the environment from non-traditional sources such as residential septic systems and agricultural locations, as opposed to industrial sources (Bell et al. 2019).

Emerging contaminants include a wide variety of compounds. In addition, thousands of new chemicals are manufactured every year. Therefore, the likelihood of additional chemicals being added is almost certain. A partial list of emerging contaminants includes the following (Bell et al. 2019; Barnes et al. 2008; USEPA 2024h):

- PIP
- PFAS
- 1,4-Dioxane
- 1,2,3-Trichloropropane
- Perchlorates
- Pharmaceuticals and drugs including:
  - Antibiotics
  - Steroids
  - Progesterone
  - Testosterone
  - Antibacterial chemicals
  - Hormones
  - Narcotics
  - Sildenafil citrate, commonly known as Viagra
  - Benzoylmethylecgonine, commonly known as cocaine
  - Many other legal and illegal drugs
- Insect repellants
- Solvents
- Detergents
- Plasticizers
- Fire retardants
- Veterinary antibiotics
- Pesticides
- Herbicides
- Fertilizers
- Makeup
- Skincare chemicals
- Sunscreen
- Body lotions
- Many others

8.6 SUMMARY AND CONCLUSION

There are tens of thousands of pollutants, and they are everywhere. After being released into the environment, they migrate in the air, water, soil, or in some combination of media. Some are persistent, while others are not. Some dissolve in water and some do not. Some are transported around the globe in the atmosphere while others do not. The physical chemistry of each contaminant combined with the geography, geology, hydrogeology, and atmospheric conditions all play a role in affecting the fate and transport of contaminants and determine their final disposition.

Some contaminants can change form, such as a gas to a liquid or a liquid to a gas, and cycle between the soil, air, and water if they last long enough (e.g., some VOCs). Other contaminants not changing form can be found in sinks or areas where they accumulate (e.g., river and lake sediment). Risk only occurs when an exposure pathway exists. Toxicity, therefore, is not the only factor to consider when evaluating risk. Mobility and persistence of a contaminant must also be considered. If they last long enough, the oceans are the final resting place for many contaminants.

Of special note within this chapter are a group of contaminants called halogenated compounds (those containing chlorine, fluorine, bromine, iodine, astatine, and tennessine within their structure). Contaminants containing halogen atoms within their structure generally remain in the environment for long periods of time. This is because, in large part, they are molecularly stable and synthetic compounds, and the natural environment has difficulty in degrading them. Several of these compounds have the ability to accumulate in tissues of living organisms if exposed (e.g., PCBs), whereby possibly exposing humans after they work their way up the food chain.

REFERENCES

Agency for Toxic Substances and Disease Registry (ATSDR). 1996. *Polycyclic Aromatic Hydrocarbons*. General Contaminant Class. ATSDR ToxFAQs. ATSDR. Atlanta, GA.

Agency for Toxic Substances and Disease Registry (ATSDR). 1997a. *Chloroform*. CAS Registry Number 127-18-4. ATSDR ToxFAQs. ATSDR. Atlanta, GA.

Agency for Toxic Substances and Disease Registry (ATSDR). 1997b. *Toxaphene*. CAS Registry Number 8001-35-2. ATSDR ToxFAQs. ATSDR. Atlanta, GA.

Agency for Toxic Substances and Disease Registry (ATSDR). 1999a. *Dimethylamine*. CAS Registry Number 124-40-3. ATSDR ToxFAQs. ATSDR. Atlanta, GA.

Agency for Toxic Substances and Disease Registry (ATSDR). 1999b. *Mercury*. CAS Registry Number 7439-97-6. ATSDR ToxFAQs. ATSDR. Atlanta, GA.

Agency for Toxic Substances and Disease Registry (ATSDR). 1999c. *Silver*. CAS Registry Number 7440-22-4. ATSDR ToxFAQs. ATSDR. Atlanta, GA.

Agency for Toxic Substances and Disease Registry (ATSDR). 2000. *Radon Toxicity: Who Is At Risk?* ATSDR. Atlanta, GA.

Agency for Toxic Substances and Disease Registry (ATSDR). 2001a. *Toluene*. CAS Registry Number 108-88-3. ATSDR ToxFAQs. ATSDR. Atlanta, GA.

Agency for Toxic Substances and Disease Registry (ATSDR). 2001b. *Polychlorinated Biphenyls*. ATSDR ToxFAQs. ATSDR. Atlanta, GA.

Agency for Toxic Substances and Disease Registry (ATSDR). 2001c. *Pentachlorophenol*. CAS Registry Number 87-86-5. ATSDR ToxFAQs. ATSDR. Atlanta, GA.

Agency for Toxic Substances and Disease Registry (ATSDR). 2001d. *Asbestos*. CAS Registry Number 1332-21-4. ATSDR ToxFAQs. ATSDR. Atlanta, GA.

Agency for Toxic Substances and Disease Registry (ATSDR). 2002a. *Di(2-ethylhexyl) phthalate*. CAS Registry Number 117-81-7. ATSDR ToxFAQs. ATSDR. Atlanta, GA.

Agency for Toxic Substances and Disease Registry (ATSDR). 2002b. *Toxicological Profile for Flame Retardant Ester Compounds*. ATSDR ToxFAQs. ATSDR. Atlanta, GA.

Agency for Toxic Substances and Disease Registry (ATSDR). 2002c. *Benzyl Acetate*. CAS Registry Number 140-11-4. ATSDR ToxFAQs. ATSDR. Atlanta, GA.

Agency for Toxic Substances and Disease Registry (ATSDR). 2002d. *Ethyl Acetate*. CAS Registry Number 141-78-6. ATSDR ToxFAQs. ATSDR. Atlanta, GA.

Agency for Toxic Substances and Disease Registry (ATSDR). 2003. *Selenium*. CAS Registry Number 7782-49-2. ATSDR ToxFAQs. ATSDR. Atlanta, GA.

Agency for Toxic Substances and Disease Registry (ATSDR). 2004a. *Copper*. CAS Registry Number 7440-50-8. ATSDR ToxFAQs. ATSDR. Atlanta, GA.

Agency for Toxic Substances and Disease Registry (ATSDR). 2004b. *Ammonia*. CAS Registry Number 7664-41-7. ATSDR ToxFAQs. ATSDR. Atlanta, GA.

Agency for Toxic Substances and Disease Registry (ATSDR). 2004c. *Hydrochloric Acid*. CAS Registry Number 7647-01-0. ATSDR ToxFAQs. ATSDR. Atlanta, GA.

Agency for Toxic Substances and Disease Registry (ATSDR). 2005a. *Bromoform*. CAS Registry Number 75-25-2. ATSDR ToxFAQs. ATSDR. Atlanta, GA.

Agency for Toxic Substances and Disease Registry (ATSDR). 2005b. *Naphthalene*. CAS Registry Number 91-20-3. ATSDR ToxFAQs. ATSDR. Atlanta, GA.

Agency for Toxic Substances and Disease Registry (ATSDR). 2005c. *Nickel*. CAS Registry Number 7440-02-0. ATSDR ToxFAQs. ATSDR. Atlanta, GA.

Agency for Toxic Substances and Disease Registry (ATSDR). 2005d. *Zinc*. CAS Registry Number 7440-66-6. ATSDR ToxFAQs. ATSDR. Atlanta, GA.

Agency for Toxic Substances and Disease Registry (ATSDR). 2005e. *Permethrin: Toxicologic Information about Pesticides*. CAS Registry Number 52645-53-1. ATSDR. ATSDR. Atlanta, GA.

Agency for Toxic Substances and Disease Registry (ATSDR). 2006a. *Vinyl Chloride*. CAS Registry Number 75-01-4. ATSDR. Atlanta, GA.

Agency for Toxic Substances and Disease Registry (ATSDR). 2006b. *Dioxins: Chemical Agent Briefing Sheet*. ATSDR. Atlanta, GA.

Agency for Toxic Substances and Disease Registry (ATSDR). 2006c. *Cyanide*. CAS Registry Number 74-90-8, 143-33-9, 151-50-8, 592-01-8, 544-92-3, 506-61-6, 460-19-5, 506-77-4. ATSDR ToxFAQs. ATSDR. Atlanta, GA.

Agency for Toxic Substances and Disease Registry (ATSDR). 2007a. *Benzene*. CAS Registry Number 71-43-2. ATSDR ToxFAQs. ATSDR. Atlanta, GA.

Agency for Toxic Substances and Disease Registry (ATSDR). 2007b. *Ethybenzene*. CAS Registry Number 100-41-4. ATSDR ToxFAQs. ATSDR. Atlanta, GA.

Agency for Toxic Substances and Disease Registry (ATSDR). 2007c. *Xylene*. CAS Registry Number 1330-20-7. ATSDR ToxFAQs. ATSDR. Atlanta, GA.

Agency for Toxic Substances and Disease Registry (ATSDR). 2007d. *Lead*. CAS Registry Number 7439-92-1. ATSDR ToxFAQs. ATSDR. Atlanta, GA.

Agency for Toxic Substances and Disease Registry (ATSDR). 2007e. *Barium*. CAS Registry Number 9440-39-3. ATSDR ToxFAQs. ATSDR. Atlanta, GA.

Agency for Toxic Substances and Disease Registry (ATSDR). 2007f. *Arsenic*. CAS Registry Number 7440-38-2. ATSDR ToxFAQs. ATSDR. Atlanta, GA.

Agency for Toxic Substances and Disease Registry (ATSDR). 2008a. *Cadmium*. CAS Registry Number 7440-43-9. ATSDR ToxFAQs. ATSDR. Atlanta, GA.

Agency for Toxic Substances and Disease Registry (ATSDR). 2008b. *Chromium*. CAS Registry Number 7440-47-3. ATSDR ToxFAQs. ATSDR. Atlanta, GA.

Agency for Toxic Substances and Disease Registry (ATSDR). 2008c. *Radon*. CAS Registry Number 14859-67-7. ATSDR ToxFAQs. ATSDR. Atlanta, GA.

Barnes, K. K., Kolpin, D. W., Furlong, E. T., Zaugg, S. D., Meyer, M. T., and Barber, L. B. 2008. A National Reconnaissance of Pharmaceuticals and Other Organic Wastewater Contaminants in the United States. *Journal of Science in the Total Environment*. Vol. 402, no. 2–3, pp. 192–200.

Bell, C. A., Gentile, M., Kalve, E., Ross, I., Horst, J., and Suthersan, S. 2019. *Emerging Contaminants*. CRC Press. Baoc Raton, FL. 439pp.

Brady, N. C., and Well, R. R. 1999. *The Nature and Properties of Soils*. 12th Edition. Prentice Hall. Upper Saddle River, NJ.

Deblonde, G. P., Kersting, A. B., and Zavarin, M. 2020. Open Questions on the Environmental Chemistry of Radionuclides. *Nature*. Vol. 3, no. 167, pp. 1–10.

Doherty, R. E. 2000. A History of the Production and Use of Carbon Tetrachloride, Tetrachloroethylene, Trichloroethylene, and 1,1,1-Trichloroethane in the United States: Part 1 – Historical Background; Carbon Tetrachloride and Tetrachloroethylene. *Journal of Environmental Forensics*. Vol. 1, no. 2, pp. 69–81.

Domenico, P. A. 1987. An Analytical Model for Multidimensional Transport of Decaying Contaminant Species. *Journal of Hydrology*. Vol. 91, pp. 49–58.

Farquhar, J., Huiming, B., and Thiemens, M. 2000. Atmospheric Influence of Earth's Earliest Sulfur Cycle. *Science*. Vol. 289, no. 5480, pp. 756–758.

Fetter, C. 1993. *Contaminant Hydrogeology*. 2nd Edition. Prentice Hall. Upper Saddle River, NJ.

Ford, D. and Williams, P. 2007. *Karst Hydrogeology and Geomorphology*. John Wiley & Sons. New York, NY. 475p.

Freeze, R. A., and Cherry, J. A. 1979. *Groundwater*. Prentice Hall. Englewood Cliffs, NJ.

Haynes, W. 1954. *American Chemical Industry – A History*. Vols. I–IV. Van Nostrand Publishers. New York. NY.

Heath, R. C. 1983. *Basic Ground-Water Hydrology*. United States geological Survey. Water Supply Paper 2220. United States Government Printing Office, Alexandria, VA.

Hemond, H. F., and Fechner-Levy, E. J. 2000. *Chemical Fate and Transport in the Environment*. Academic Press. London, England.

Holleman, A. F., and Wiberg, E. 2001. *Inorganic Chemistry*. Academic Press. San Diego. CA.

Hornsby, A. G. 1990. *How Contaminants Reach Groundwater*. University of Florida Institute of Food and Agriculture. Gainsville, FL. 102pp.

Howard, P. H., Michalenko, E. M., Basu, D. K., and Aronson, D. 1997. *Handbook of Environmental Fate and Exposure Data for Organic Chemicals. Volume 5. Solvents 3*. Lewis Publishers. Chelsea, MI.

International Union of Pure and Applied Chemistry (IUPAC). 2024. *Nomenclature of Organic Chemistry, Halogenated Volatile Organic Compounds*. www.IUPAC.org/halogenatedvolatileorganiccompounds. (Accessed September 10, 2024).

Jensen, W. B. 2009. The Origin of the Circle Symbol for Aromaticity. *Journal of Chemical Education*. Vol. 86, no. 4, pp. 423–425.

Kathren, R. 1991. *Radioactivity and the Environment*. Taylor & Francis Publishers. Lieden, The Netherlands.

Krebs, R. E. 2006. *The History and Use of Earth's Chemical Elements: A Reference Guide*. Second Edition. Greenwood Press. Westport, Connecticut. 456p.

Levy, A. B. 2009. Acetone. In: *Ullmann's Encyclopedia of Industrial Chemistry*. 7th Edition. John Wiley & Sons–VCH. New York. NY.

Lide, D. R. 2008. Physical Constants of Organic Compounds. In*: Handbook of Physics and Chemistry*. 89th Edition. CRC Press. Boca Rotan, FL.

Lindsay, W. L. 1979. *Chemical Equilibria in Soils*. John Wiley & Sons. New York, NY.

Lyman, W. J., Reehl, W. F., and Rosenblatt, D. M. 1990. *Handbook of Chemical Property Estimation Methods*. American Chemical Society. Washington, DC. 459pp.

Madigan, M. T., Martinko, J. M., Dunlap, P. V., and Clark, D. P. 2008. *Brock Biology of Microorganisms*. 12th Edition. Prentice Hall. New York, NY.

McGucken, W. 2000. *Lake Erie Rehabilitated: Controlling Cultural Eutrophication, 1960s–1990s*. University of Akron Press. Akron, OH. 320pp.

McKone, T. E., and Enoch, K. G. 2002. CalTox™, a Multimedia Total Exposure Model Spreadsheet Users Guide. In: *Lawrence Berkeley National Laboratory*. LBNL 47399. University of California. Berkeley, CA.

Murray, K. S., Rogers, D. T., and Kaufman, M. M. 2004. Heavy Metals in an Urban Watershed in Michigan. *Journal of Environmental Quality*. Vol. 33, pp. 163–172.

National Atmospheric and Space Administration (NASA). 2021. *The Ocean's Carbon Balance*. The Ocean's Carbon Balance (nasa.gov). www.noaa.gov/education/resource-collections/climate/carbon-cycle (Accessed March 27, 2021).

Natural Resources Defense Council (NRDC). 2024. *Ocean Pollution: The Dirty Facts*. https://nrdc.org/stories/ocean-pollution-dirty-facts. (Accessed May 11, 2024).

New Jersey Department of Environmental Protection (NJDEP). 2015. *Preliminary Assessment Report*. NJDEP. Trenton, NJ.

Nudunuri, K. V., Erickson, L. E., and Govindaraju, R. S. 1998. Modeling the Role of Active Biomass on the Fate and Transport of Heavy Metals in the Presence of Root Exudates. *Journal of Hazardous Waste Research*. Vol. 1, no. 9, pp. 1–25.

Payne, F. C., Quinnan, J. A., and Potter, S. T. 2008. *Remediation Hydraulics*. CRC Press. Boca Raton, FL.

Payne, F. C., and Rogers, D. T. 1997. Chlorinated Solvent Degradation – A Consecutive, Irreversible Reaction Sequence. In: *Remediation of Recalcitrant Chlorinated Volatile Organic Compounds in Groundwater*. Groundwater Management Book 21. National Groundwater Association. Dublin, OH. pp. 97–109.

Pradyot, P. 2003. *Handbook of Inorganic Chemical Compounds*. McGraw Hill. New York, NY.

Raub, T. D., and Kirschvink, J. L. 2008. A Pan-Precambrian Link between Deglaciation and Environmental Oxidation. In: Cooper, A. K., Barrett, P. J., Stagg, H., Stump, E., and Wise, W., editors. *Antarctica: A Keystone in a Changing World*. The National Academic Press. Washington, DC.

Rogers, D. T. 1996. *Environmental Geology of Metropolitan Detroit*. Clayton Environmental Consultants. Novi, MI.

Rogers, D. T. 2012. Environmental Risk Analysis through Integration of Geologic Vulnerability and Air, Soil, and Water Contaminant Risk Factor Derivation. *International Geological Congress*. Vol. 1, pp. 3003.

Rogers, D. T. 2020. *Urban Watersheds: Geology, Contamination, Environmental Regulations, and Sustainability*. CRC Press. Boca Raton, FL. 606pp.

Rogers, D. T., Murray, K. S., and Kaufman, M. M. 2007. Assessment of Groundwater Contaminant Vulnerability in an Urban Watershed in Southeast Michigan, USA. In: Howard, K. W. F., editor. *Urban Groundwater – Meeting the Challenge*. Taylor & Francis, London, England.

Sawhney, B. L., and Brown, K., editors. 1989. *Reactions and Movement of Organic Chemicals in Soils*. Special Publication No 22. Soil Science of America. Madison, WI.

Schlatter, T. W. 2009. *Atmospheric Composition and Vertical Structure*. National Oceanic and Atmospheric Administration (NOAA). Boulder, CO.

Schnoor, J. L. 1996. *Environmental Modeling: Fate and Transport of Pollutants in Water and Soil*. John Wiley & Sons. New York, NY.

Schwartzenbach, R. P., Gachwend, P. M., and Imoden, D. M. 1993. *Environmental Organic Chemistry*. John Wiley & Sons. New York, NY. 767pp.

Singer, A. C., Crowley, D. E., and Thompson, I. P. 2003. Secondary Plant Metabolites in Phytoremediation and Biotransformation. *Trends in Biotechnology*. Vol. 21, no. 3, pp. 123–130.

Soil Science Society of America. 1987. *Glossary of Soil Science Terms*. www.soils.org/publications/soils-glossary/q=publications/soil-glossary. (Accessed February 4, 2025).

SRC. 2024. *Environmental Fate Data Base (EFDB)*. CHEMFATE Chemical Search. http://srcinc.com/what-we-do/efbd.aspx. (Accessed May 11, 2024).

Suthersan, S. S., Horst, J., Schnobrich, M., Welty, N., and McDonough, J. 2017. *Remediation Engineering: Design Concepts*. Second Edition. CRC Press. Boca Raton, FL. 603p.

United Nations. 2024. *Halogenated Volatile Organic Compounds*. www.un.org/halogenatedvolatileorganiccompounds. (Accessed September 10, 2024).

United States Department of Health and Human Services (USDHH). 1995. *Toxicological Profile for Polycyclic Aromatic Hydrocarbons*. ATSDR. Atlanta. GA.

United States Environmental Protection Agency (USEPA). 1988. *Times Beach Record of Decision*. www.epa.gov/timesbeachrecordofdecision. (Accessed September 19, 2024).

United States Environmental Protection Agency (USEPA). 1991. *Air Pollution and Health Risk USEPA*. EPA/450/3-90-022. USEPA. Washington, DC.

United States Environmental Protection Agency (USEPA). 1996a. *Transport and Fate of Contaminants in the Subsurface*. EPA/625/4-89/019. USEPA. Washington, DC.

United States Environmental Protection Agency (USEPA). 1996b. *Bioscreen. Natural Attenuation Decision Support System*. Office of Research and Development. Washington, DC.

United States Environmental Protection Agency (USEPA). 1997. *Mercury Study Report to Congress: Volume III: Fate and Transport of Mercury in the Environment*. EPA-454/R-97-005. USEPA. Washington, DC.

United States Environmental Protection Agency (USEPA). 1998a. *Taking Toxics Out of the Air*. EPA451/K-98-001. USEPA. Washington, DC.

United States Environmental Protection Agency (USEPA). 1998b. *Chemical Fate Half-Lives for Toxics Release Inventory (TRI) Chemicals*. USEPA. Washington, DC.

United States Environmental Protection Agency (USEPA). 1999. *Fundamentals of Soil Science as Applicable to Management of Hazardous Wastes*. EPA/540/S-98/500. Office of Research and Development. Washington, DC.

United States Environmental Protections Agency (USEPA). 2025a. Dioxin. www.epa.gov/dioxin. (Accessed February 4, 2025).

United States Environmental Protection Agency (USEPA). 2025b. Basic Information About Carbon Monoxide (CO) Outdoor Air Pollution. www.epa.gov/co-pollution/basic-information-about-carbon-monoxide-co-outdoor-air-pollution. (Accessed February 4, 2025).

United States Environmental Protection Agency (USEPA). 2008. *Findings on National Air Quality: Status and Trends through 2006*. EPA454/R-07-007. USEPA. Research Triangle Park, NC.

United States Environmental Protection Agency (USEPA). 2009a. *Biennial National Listing of Fish Advisors for 2008*. EPA-823-F-09-007. USEPA. Washington, DC.

United States Environmental Protection Agency (USEPA). 2009b. *Ozone – Good Up High Bad Nearby. Air Quality Planning and Standards*. USEPA. Washington, DC.

United States Environmental Protection Agency (USEPA). 2017. *Pesticide Industry Usage*. USEPA Biological and Economic Analysis Division, Office of Pesticide Programs. Washington, DC. 24pp.

United States Environmental Protection Agency (USEPA). 2018. *Global Greenhouse Gas Emissions Data*. https://epa.gov/ghgemissions/global-greenhouse-gas-emissions-data. (Accessed May 11, 2024).

United States Environmental Protection Agency (USEPA). 2019. *Ozone Layer Depletion*. www.epa.gov/ozonelayerdepletion. (Accessed September 16, 2024).

United States Environmental Protection Agency (USEPA). 2024a. *EPA's Report on the Environment (ROE)*. www.epa.gov/report-environment. (Accessed May 11, 2024).

United States Environmental Protection Agency (USEPA). 2024b. *Advisories for Fish and Shellfish Consumption*. www.epa.gov/fish-tech. (Accessed May 11, 2024).

United States Environmental Protection Agency (USEPA). 2024c. *What Is Acid Rain*. www.epa.gov/acidrain/what/index.html. (Accessed May 11, 2024).

United States Environmental Protection Agency (USEPA). 2024d. *Integrate Risk Information System (IRIS)*. www.epa.gov/ncea/iris/intro.htm. (Accessed May 11, 2024).

United States Environmental Protection Agency (USEPA). 2024e. *Health and Environmental Effects of Ozone Layer Depletion*. www.epa.gov/ozone-layer-protection/health. (Accessed June 10, 2024).

United States Environmental Protection Agency (USEPA). 2024f. *PFAS Explained*. www.epa.gov/pfas-explained. (Accessed June 2, 2024).

United States Environmental Protection Agency (USEPA). 2024g. *Preliminary Information on Manufacturing, Processing Distribution, USE and Disposal of Phenol, Isoproplated, Phosphate (3:1) (PIP)*. www.epa.gov/sites/production/files/2017-08/. (Accessed May 11, 2024).

United States Environmental Protection Agency (USEPA). 2024h. *Contaminants of Emerging Concern*. www.epa.gov/wqc/contaminants-emerging-concern-including-pharmaceuticals-and-personal-care-products. (Accessed May 11, 2024).

United States Geological Survey (USGS). 1995a. *Ground-Water Quality Protection*. Open-File Report 95-376. USGS. Nashville, TN. www.pubs.usgs.gov/of/1995/ofr-95376. (Accessed December 2009).

United States Geological Survey (USGS). 1995b. *Contaminants in the Mississippi River*. USGS Circular 1133. USGS. Washington, DC.

United States Geological Survey (USGS). 1998. *Simulating Transport of Volatile Organic Compounds in the Unsaturated Zone Using the Computer Model R-UNSAT*. USGS Fact Sheet 019-08. USGS. Washington, DC.

United States Geological Survey (USGS). 2000. *Mercury in the Environment*. Fact Sheet 146-00. USGS. Washington, DC.

United States Geological Survey (USGS). 2006a. *Volatile Organic Compounds in Nation's Ground Water and Drinking-Water Supply Wells. National Water-Quality Assessment Program*. USGS Circular 1292. USGS. Washington, DC.

United States Geological Survey (USGS). 2006b. *Pesticides in the Nation's Streams and Groundwater. 1992-2001 – A Summary*. USGS Fact Sheet 2006-3028. USGS. Washington, DC.

United States Institute of Health. 2024a. *Periodic Table of the Elements*. https://pubchem.ncbi.nlm.nih.gov/periodic-table. (Accessed May 11, 2024).

United States Institute of Health. 2024b. *A Summary of Toxicity and Health Risks due to Exposure to Halogen Compounds*. www.ncbi.nlm.nih.gov/hallogentoxicity. (Accessed September 10, 2024).

Van der Perk, M. 2006. *Soil and Water Contamination: From Molecular to Catchment Scale*. Balkema: Proceedings and Monographs in Engineering, Water and Earth Sciences. Taylor and Francis. London, UK.

Wania, F., Hoff, J. T., Jia, C. Q., and Mackey, D. 1998. The Effects of Snow and Ice on the Environmental Behavior of Hydrophobic Organic Chemicals. *Journal of Environmental Pollution*. Vol. 102. pp. 79–95.

Wiedemeier, T. H., Rifai, H. S., Newell, C. J., and Wilson, T. J. 1999. *Natural Attenuation of Fuels and Chlorinated Solvents in the Subsurface*. John Wiley & Sons. New York, NY.

Wilson, R., and Spengler, J. D., editors. 1996. *Particles in Our Air: Concentration and Health Affects*. Harvard University Press. Cambridge, MA.

World Health Organization (WHO). 2024. *Health Statistics and Health Information Systems*. Mortality Database Tables. www.who.int/healthinfo/morttables/en/index/html. (Accessed May 11, 2024).

9 Invasive and Endangered Species and Extinction

9.1 INTRODUCTION

In this chapter we address anthropogenic impacts on living organisms. This will include discussing invasive species, endangered species, extinct species, and pathogens that include viruses, bacteria, and parasites. We will also discuss some select species that have been affected by climate-related impacts. Invasive species and extinction of species in the United States are nothing new. The rate of invasive species and extinction of species is increasing and is caused by humans (National Geographic Society 2024a). We will first briefly discuss life on Earth and then examine invasive species and then move on to endangered species, species extinction, and finally we will address pathogens. We conclude the chapter with an examination of the outlook at the horizon and what the future may hold for many species on Earth given certain scenarios.

9.2 LIFE ON EARTH

Life on Earth exists in various environments, including air, water, soil, and even exists in rock at significant depths beneath the surface. The region where life exists on Earth is commonly referred to as the biosphere and comprises many different ecosystems. While scientists have suggested that Earth may have developed favorable conditions for the existence of life as far back as 4.3 billion years, the oldest fossils discovered so far date to 3.7 billion years old. During the 600-million-year time period between 4.3 and 3.7 billion years ago, life may have emerged and existed repeatedly but evidence of life may have been erased by geological processes and catastrophic collisions with asteroids and comets (University of Chicago 2024). The details of those early events are not well preserved in Earth's oldest rocks, but some hints do exist. Those hints come from the oldest zircons, which are very durable minerals that form in magma. Scientists have discovered traces of a form of carbon within the oldest zircons, one of which is 4.1 billion years old. As carbon is a necessary building block of life on Earth, some scientists have suggested that perhaps there is a link between the carbon in zircons and the origin of life on Earth. However, the evidence is meager at best and does not prove life's existence at that early date.

The origin of life on Earth stands as one of the greatest mysteries of science. Various hypotheses have been proposed, but all remain unverified. Two theories as to the origin of life on Earth involve volcanically active hydrothermal environments on land and on the ocean floor, and a third suggests the possibility of life's origin within rocks and hydrothermal fluids at depths perhaps of a few kilometers beneath the surface of the ground. A fourth theory involves the transport of the amino acids, considered the building blocks of life, on comets or asteroids, which also suggests that life may exist on other planets (University of Chicago 2024; Jabr 2024; Moody et al. 2024).

DOI: 10.1201/9781003463153-11

Some organisms thrive in scalding, highly acidic hydrothermal fluids, on land at locations in Iceland, Norway, and Yellowstone National Park. Other organisms have been discovered in deep-ocean hydrothermal vents on the ocean floor where ocean water comes into contact with magma that results in superheated plumes that form chimney-like vents. Organisms have also been discovered within the Earth's crust 3 km beneath the surface. These organisms live within fluid-filled pores of rock at high-relative temperatures compared to the surface. These organisms do not appear to require light or oxygen to exist and thrive, but seem to use methane, hydrogen, and metals to exist and have been referred to as extremophiles because they exist in such extreme environments (University of Chicago 2024; Jabr 2024). Until the discovery of extremophiles, the essential ingredients for life were thought to be light as an energy source, water, and the elements, carbon, hydrogen, oxygen, nitrogen, and phosphorus. Therefore, the search for the origin of life continues.

Life now likely numbers more than 100 million species and exists everywhere on Earth and even within the Earth since it first appeared billions of years ago. Life has had many barriers to overcome such as asteroid and comet collisions, changing chemistry and physical properties, isolation on continents by mountains ranges, deserts, and climate. Some of the more significant barriers for terrestrial life are the separation of continents and islands by vast expanses of ocean. However, an additional factor has recently been introduced that some species on Earth have exploited and is a result of our species' success. Human technological advancements have permitted the human population to exceed 8 billion people and humans now routinely travel the globe. Because of these advancements, some species have hitched a ride with humans to new habitats and, in some instances, have invaded and taken over new regions of Earth. As we will see in the following sections, the negative aspects of this unintentional consequence of anthropogenic influence have led to results that threaten many species.

9.3 INVASIVE SPECIES

We begin by discussing invasive species because, in many instances, invasive species are responsible for a significant portion of the problem of habitat destruction and loss of native species abundance and contribute to forcing native species down a dangerous path to endangerment and perhaps even extinction. The sad reality of this journey is that it is caused by human behavior.

9.3.1 What Is an Invasive Species

An invasive species is an organism that is not indigenous, or native, to a particular area. An expanded and generally accepted definition of an **invasive species** is any kind of living organism that is not native to an ecosystem and causes harm that is directly or indirectly introduced by humans or is the result of human activity (United States Department of Agriculture (USDA) 2024a; USEPA 2024a).

Invasive species can and have caused significant environmental and economic harm to the new area. Not all non-native species are considered invasive. For example, most of the food crops throughout the world, including rice, wheat, corn, tomatoes, and a host of others, are not considered invasive. To be invasive, a species must adapt to the new area easily. It must also reproduce quickly and cause harm to the physical property of the new area, or native plants and animals of the region.

9.3.2 Origins of Invasive Species

The origin of an invasive species generally occurs from one or more of the following (National Geographic Society 2024a):

- *Transported to a new area on purpose:* This type is often introduced as a form of pest control. For example, in 1949, five cats were brought to Marion Island, a part of South Africa in the southern Indian Ocean. The cats were introduced as a method of pest control for mice on the island. By 1977, there were estimated 3,400 cats on the island and were endangering the local bird populations.

- *Pets that are released or escape into the environment:* An example of a pet being released is the Burmese python. Some were released into the Everglades in southern Florida. The pythons are native to the jungles of southeast Asia and have few predators in the Everglades. As a result, the pythons feast on many native species, including the White Idis, small deer, racoons, rabbits, and other small- to medium-sized mammals and even the American Alligator (National Geographic Society 2024a).
- *Accidental introduction:* Some species get introduced to a new area accidently. A good example is the zebra mussel. Zebra mussels are native to the Black Sea and Caspian Sea in central Asia. They arrived in the Great Lakes of North America accidentally, either stuck to large ships or in ballast water that was released once arriving to the Great Lakes. Zebra mussels are now so prevalent that they threaten many native species. Another example of an accidental introduction is the Emerald Ash Borer Beetle. The beetle is smaller than a penny but has caused significant ecological, economic, and cultural devastation. The beetle is native to Asia and is now one of the most destructive invasive species in North America. They have killed hundreds of millions of Ash trees, especially in the northeastern states (USDA 2024b). They have also been confirmed in Oregon in 2022. They are believed to have entered North America in an infested wood pallet from Asia to the Great Lakes. A method of spread now is the transport of dead Ash trees used as firewood (USDA 2024b).
- *Introduced through climate change:* This method is more recent and due predominantly through climate change. For example, let's examine bark beetles. There are approximately 600 identified species of bark beetles in the United States with an estimated 200 species only in California. Many species are very small and are routinely just a few millimeters long (United States Forest Service 2024). Bark beetles live beneath the bark of host trees in what are termed **galleries** that are unique to each species. They are opportunistic and prefer to infest trees weakened by other agents or factors, which include disease, infestation by other insects, mechanical damage (such as a storm or fire), high soil compaction, air pollution, drought, or high density. Bark beetles produce aggregating attractants that ensure mass attack of suitable host material that lead to group-killing of trees. A high reproductive potential allows bark beetles to multiply rapidly when conditions are favorable (United States Forest Service 2024).

This is key because over the last couple of decades, bark beetle destruction has occurred throughout many states, especially western states, and Canada. Winters have not been as cold and the bark beetles avoid dying off in massive numbers and because of climate change, many tree species the bark beetles prefer, namely pine and other related species, are stressed which has greatly increased bark beetle abundance and damage. In fact, Jasper National Park in Alberta Canada is so infested that an estimated 80% of trees have been affected by the bark beetle. With such damage and the killing of millions of trees in Canada, this has added to the fuel for wildfires that have ravaged many parts of the boreal forest of Canada in the last few years.

The bark beetle is not as traditional for a typical invasive species. However, because of a human-induced reason, the bark beetle has taken advantage of a warming atmosphere through more favorable habitat due to the warming. In addition, an examination of the most exploited habitats favor northern latitudes where climate change has been more severe.

9.3.3 Select Invasive Species

From a certain perspective, an invasive species is much like a form of pollution. The concept of a living organism being a type of pollution is not a stretch, after all, we do label some invasive species as an infestation, which is pollution-like. It is important to note a couple items of fundamental difference: one is that pollution does not breed and reproduce, pollution does not adapt, and pollution does not evolve. However, pollution does degrade, and, in some instances, pollution degrades or combines with other natural substances or phenomenon and becomes worse. A couple of examples are (1) the degradation of chlorinated volatile organic compounds , which typically reaches peak human toxicity 20 years after being released into the environment, and (2) the formation of smog which requires certain pollutants to be released and atmospheric condition to be met to form smog. Just as pollution, there is no intentional act to cause harm, but the result can be just as harmful.

Typically, invasive species' mode of invasion is as a stowaway during the transportation of goods. USEPA (2024a) estimated that 30% of invasive species to the Great Lakes was contained in ballast water in ships or attached to ships carrying cargo. Ballast water is taken into a ship or discharged from a ship as it loads or unloads cargo to accommodate the ship's weight changes.

Invasive species is a worldwide problem and has persisted for hundreds of years. Most wildlife and environmental agencies are losing the battle with invasive species, for one simple and disturbing reason: this type of pollution breeds and can evolve.

To make this point clear, the United Nations established the Global Invasive Species Program (GISP) in 1997 to develop a global strategy to deal with invasive species (United Nations 2024a). Worldwide, there are thousands of invasive species (United Nations 2024a). Invasive species are characterized as aggressive when they are transported into areas where they are not naturally occurring usually because they do not have a natural predator or other organisms have not evolved defenses (United Nations 2024a). The top 16 invasive species currently causing the most harm in the United States include the following (USEPA 2024a):

1. *Burmese python:* Introduced in the Florida Everglades and were considered established in 2002. The python has reduced deer, racoon, marsh rabbits, bobcats, possum, and alligators by as much as 99%.
2. *Emerald ash borer:* Originally from Russia, China, and Japan, they were accidently introduced in southeastern Michigan in 2002 in infected wood crates. The Emerald Ash has infected and killed hundreds of millions of Ash trees in the United States.
3. *Nutrua:* Commonly called swamp rats. They were imported from South America into the southern United States for their fur but now have grown out of control and are in the wild. Their endless digging along the banks of rivers and streams kills plants and fish and causes soil erosion. They not only damage crops and other property, but they also gnaw prodigiously through wetland plants turning swamps into open waters. Their extensive burrow systems destabilize roads, bridges, levees, and golf courses. They are also known to transmit diseases, such as tuberculosis.
4. *European starling:* Since their introduction into the United States in the 1890s, they have become the most successful foraging bird in the United States with a population now estimated at 400 million. They compete with native species and destroy crops such as grains and pitted fruits. They swarm agricultural troughs, contaminating food and water, and are linked with diseases like histoplasmosis, a lung condition afflicting agricultural workers.
5. *Northern snakehead:* The most feared fish in the Chesapeake Watershed. A sharp-toothed fish from Korea and China that originally infested the Chesapeake Watershed has now spread to other parts of Virginia and Delaware.
6. *Brown marmorated stink bug:* Stink bugs destroy fruits and vegetables to such an extent that they have been known to increase the price of produce. They first showed up in Allentown Pennsylvania in 1998 after crawling out of a cargo ship. Originally, they are from China.
7. *Feral hogs:* Essentially, farm pigs gone wild. They are established in 47 states with massive populations in Texas, Florida, Georgia, Virginia, and North Carolina. The USDA (2024a) estimates that there are likely more than 6 million feral swine in the United States.

8. *Lionfish:* Lionfish are native to the Pacific Ocean. They were first spotted off the Miami coast in the mid-1980s. They are characterized as eating everything that they can stuff into their mouths. That's how they obtained their name. They are destroying life on the coral reefs of the Atlantic Ocean.
9. *Norway rat:* First introduced in the United States in 1775 and now they are everywhere, including Hawaii and Alaska.
10. *Tegu:* Tegus are brown lizards that are muscular, fast, and love to eat eggs. They are known to harass pets and invade homes where they were released in Florida. Game officials have largely given up on eradicating them from the wild.
11. *Asian citrus psyllid:* The Asian Citrus Psyllid carries a bacterium that attacks and kills fruit trees, especially orange trees. Estimates are that 80% of the orange trees in Florida are infected.
12. *Brown tree snake:* The brown tree snake has infected the island of Guam and has decimated the local bird population. They are so out of control on the Island that they are even responsible for power outages.
13. *Asian carp:* In the United States, there are four species of Asian Carp: bighead carp, black carp, grass carp, and silver carp. In the 1970s, the United States imported them to assist in keeping aquatic farms clean, control weeds in canal systems, and help with sewerage treatment. However, they escaped into the Mississippi River basin and the rest was history. They have voracious appetites and can eat as much as 20% to 100% of their body weight every day. They have been outbreeding native species and pushing them out of their own ecosystem while expanding into new waterways. Silver carp, which have a disconcerting habit of leaping out of the water, have been seen as a hazard to boaters and anglers. There is great concern that if the Asian Carp invades the Great Lakes, many native species will be eliminated.
14. *Domestic cats:* Cats are one of the most beloved pets in the United States. It is perhaps because they are so beloved that it may surprise many that they are also considered a destructive invasive species and not just in the United States but in many other parts of the world. Historians believe that the ancient Egyptians were the first to domesticate cats, and that these pets then spread to Europe during the time of the Roman Empire. When Europeans colonized North America, they brough cats with them as pets. Since then, cats have flourished and have, in turn, harmed many native species. The harm caused by cats is alarming, in that cats may be responsible for killing as many as 1.3–4.0 billion birds, and 6.3–22.3 billion mammals annually. In fact, it is so significant that the American Bird Conservancy estimates that cats have contributed to the extinction of more than 60 species of birds, mammals, and reptiles.
15. *Gypsy moths:* French artist and astronomer Etienne Leopold Trouvelot first brought gypsy moths to the United States in the late 1860s to see if they could produce textiles better than silkworms. Unfortunately, they escaped and spread into the countryside of Massachusetts. Since then, they have spread throughout the United States and threaten more than 300 species of trees and shrubs, reducing animal habitats and presenting an increased fire hazard. They are very fond of oak trees. The United States spends approximately 30 million dollars a year on gypsy moth eradication, suppression, and research.
16. *Kudzu:* First introduced into the United States in 1877, Kudzu is an invasive plant species that grows very fast, smothering other plants in the southeastern United States to as far north as Indiana. Kudzu is now growing and killing other plant species in the United States at a rate of up to 1,000 km^2/year.
17. *Dandelions:* Dandelions are non-native plants that originated in Europe and Africa. Dandelions spread easily and may take over gardens and lawns. They are one of the most successful plants on Earth and have also been spotted in Antarctica. They have a deep root system and produce many seeds.

The number of different types of invasive species in North America alone totals the following (USDA 2024b):

- Plants – 1,911 species
- Insects – 6,647 species
- Pathogens – 1,278 identified what defines a pathogen?
- Animals – 220 species

The number of invasive species and pathogens in North America alone totals more than 10,000 (USDA 2024b).

9.4 ENDANGERED SPECIES

In 2024, the list of endangered species surpassed 1 million worldwide according to the International Union for Conservation of Nature (IUCN 2024a). The sections below describe the meaning of endangered species, the response strategies and assessments and legislation efforts to combat extinction of endangered species, and lists some of the endangered species in the United States and worldwide.

9.4.1 What Is an Endangered Species

An **endangered species** is an organism that is threatened with extinction. Species become endangered for two main reasons: loss of habitat and loss of genetic variation (National Geographic Society 2024b).

A loss of habitat can happen naturally. Non-avian dinosaurs lost their habitat about 65 million years ago. As discussed in Chapter 2, the rather hot, dry climate of the Cretaceous changed very quickly, most likely when an asteroid slammed into the Earth. The impact of that event forced debris into the atmosphere, reducing the amount of heat and light that reached Earth's surface. The dinosaurs were unable to adapt and the non-avian dinosaurs quickly went extinct (National Geographic Society 2024b).

Human activity is the major cause of habitat loss. Urban development and agriculture have considerably reduced natural habitat worldwide. In fact, agriculture alone has engulfed an estimated 40% of habitable land on Earth (United Nations 2024a).

9.4.2 Worldwide Response to Endangered Species

Conservation efforts for saving critically endangered species involve a multi-faceted approach that addresses various threats to the species and their habitats. These efforts are usually coordinated through collaboration with governments, conservation organizations, local communities, and the public. The following is a generalized list of key components of the general strategy methods (IUCN 2024a):

- Habitat protection and restoration
- Community engagement
- Anti-poaching methods
- Legislation and Policy Advocacy
- Captive Breeding and Reintroduction
- Public Awareness and Education

Conservation efforts to save endangered species from extinction often require combing more than one strategy listed above to meet the specific needs to save a species. Continuous monitoring and adaptive management are crucial for assessing the effectiveness of efforts and conservation interventions.

Critically endangered species is an indication of the broader biodiversity crisis on Earth. Humans can save many species but need the collective efforts and commitments from everyone.

9.4.3 United States Endangered Species Act of 1973

The Endangered Species Act of 1973 was designed to protect animal species from extinction from a consequence of human interference from economic growth and development un-tempered by adequate concern and conservation. The Act was established to reverse a trend of those identified endangered species from extinction at whatever the cost (United States Fish and Wildlife Service 2024). The Act is administered by the Fish and Wildlife Service (FWS) and the National Oceanic and Atmospheric Administration (NOAA). Listing status is a sliding criterion that indicates the degree to which a certain species is threatened with extinction. The listing status used in the United States is as follows (United States Fish and Wildlife Service 2024):

- *E = Endangered:* Endangered is the highest of alerts meaning the closest to extinction in the United States and generally means any species that is in danger of extinction throughout all or a significant portion of its natural habitat range.
- *T = Threatened:* Threatened is any species which is likely to become endangered within the foreseeable future throughout all or a significant portion of its natural habitat range.
- *C = Candidate:* Candidate is a species that is under consideration due to a drop in population or because of loss of a significant reduction in its natural habitat.

9.4.4 Endangered Species in the United States

The first list of endangered species in the United States included 14 mammals, 36 birds, 6 reptiles and amphibians, and 22 fish. According to the United States FWS, there are currently a total of 711 animal species on the list and 941 plant species for a total of 1,652 on the endangered species list just in the United States (United States Fish and Wildlife Service 2024). Some of the animal species in the United States that are or were on the endangered species list include the following (United States Fish and Wildlife Service 2024):

- Bald Eagle – removed in 2007
- California Condor
- Peregrine Falcon – removed in 1999
- Mexican Wolf
- Grey Whale
- California Southern Sea Otter
- Hawaiian Goose
- Black-Footed Ferret
- Whooping Crane
- Kirkland's Warbler
- Gray Wolf
- Red Wolf
- Grizzly Bear – removed in 2007
- Florida Key Deer
- Virginia Big-Eared Bat
- Florida Grasshopper Sparrow

9.4.5 Worldwide Endangered Species

Similar to the United States, other countries have enacted laws that identify and protect endangered species within their borders. For instance, Canada enacted the Canadian Species at Risk Act (SARA) and Europe enacted the Habitat's Directive (Council Directive 92/43/EEC). In addition, one of the most recognized independent international systems of species assessment is the Red List of Threatened Species, created by the IUCN. Other international agreements include the Convention on the Conservation of Migratory Species of Wild Animals (CMS), Convention on International Trade in Endangered Species of Wild Fauna and Flora (CITES).

Worldwide, there are thousands of species of animals and plants that are in danger of extinction because of direct human involvement that includes development, habitat loss and destruction, poaching, pollution, and other factors. The World Wildlife Fund (WWF) maintains a general list of those animals that are in danger and also includes an additional category different from that of the United States by the addition of another category called **critically endangered**. Some of the animals currently listed by the WWF as critically endangered or endangered include the following (WWF 2024):

- Amur Leopard
- Cross River Gorilla
- Malayan Tiger
- South China Tiger
- Sumatran Rhino
- Asian Elephant
- Bluefin Tuna
- Chimpanzee
- Hectors Dolphin
- Indochinese Tiger
- Sea Lion
- Tiger
- Saola
- Hawksbill Turtle
- Kakapo
- Philippine Eagle
- Fruit Dove
- Black Rhino
- Eastern Lowland Gorilla
- Sumatran Orangutan
- Sumatran Elephant
- Sumatran Tiger
- Western Lowland Gorilla
- Blue Whale
- Borneo Pygmy Elephant
- Ganges River Dolphin
- North Atlantic Wright Whale
- Snow Leopard
- Black Spider Monkey
- Northern White Rhino
- Sao Tome Fiscal
- Javan Hawk Eagle
- Northern Bald Ibis
- Philippine Cockatoo
- Yellow-eyed Penguin
- Bornean Orangutan
- Javan Rhino
- Orangutan
- Mountain Gorilla
- Amur Tiger
- African Elephant
- Bonobo
- Galapagos Penguin
- Indian Elephant
- Red Panda
- Sri Lanka Elephant
- African Giraffe
- Vaquita
- Yangtze Finless Porpoise
- Forest Owlet
- Spoon-billed Sandpiper

The above list is just to name a few of the more familiar animals that are now considered threatened with extinction. Unless major efforts from the world community are conducted immediately, the loss of these animals and countless others are likely inevitable. In fact, a United Nations (2024d) study estimates that there are over one million species at risk of extinction.

9.5 SPECIES EXTINCTION

In this section, we address species extinction, which is where many endangered species end up if intervention efforts fail and continued genetic variation pressures continue because of loss of habitat and over factors discussed in the previous section.

9.5.1 What Is Species Extinction

Species extinction occurs when a species no longer lives anywhere on Earth because the last members of the species have died. Extinction can occur due to environmental changes or evolutionary changes in the species. Extinction drives the evolution of life by making room for new species. The loss of a species impacts food chains, food webs, and pollination activities (National Geographic Society 2024b).

9.5.2 Recent Extinct Species in North America

The list below encompasses some North American species that have gone extinct just since 1800 AD and include the following (IUCN 2024b):

- Great Auk – 1852
- California Golden Bear – 1922
- Labrador Duck – 1878
- Cascade Mountain Wolf – 1940
- Mexican Grizzly Bear – 1964
- Sea Mink – 1860
- South Rocky Mountain Wolf – 1935
- Carolina Parakeet – 1918
- Heath Hen – 1932
- Bachman's Warbler – 1988
- Blue Walleye – 1980
- Harelip Sucker – 1859
- Tecopa Pupfish – 1970
- Yellowfin Cutthroat Trout – 1950
- Rocky Mountain Locust – 1902
- Passenger Pigeon – 1914
- Ivory-Billed Woodpecker – 1942
- Eastern Cougar – 2011 declared
- Eastern Elk – 1887
- Newfoundland Wolf – 1911
- Caribbean Monk Seal – 1952
- Southern California Kit Fox – 1903
- Dusky Seaside Sparrow – 1987
- Eskimo Curlew – 1981
- Tacoma Pocket Gopher – 1970
- Blackfin Cisco – 1960
- Silver Trout – 1930
- Thicktail Chub – 1960
- Pigtoe or Pearly Mussel – 1970
- Xerces Blue – 1950

9.5.3 Recent Extinct Species Worldwide

The list below includes some species outside of North America that have gone extinct that are attributed to human influences of habitat loss (United Nations 2024d):

- Sri Lanka Tree Frog – Sri Lanka
- Harlequin Toad – Central America
- Yunna Lake Newt – China
- Indian Caecilian – India
- Gastric-Brooding Frog – Australia
- Torrent Frog – Australia
- Gunther's Streamlined Frog – Australia
- Bali Tiger – Bali
- Elephant Bird – Madagascar
- Eastern Moa – New Zealand
- Pyrenean Ibex – Spain
- Atlas Bear – northern Africa
- Auroch – Europe
- Irish Elk – United Kingdom
- Galapagos Damsel – Galapagos Island
- Lake Titicaca Orestias – South America
- Quagga – southern Africa
- Tarpan – Eurasia
- Levuana Moth – Fiji
- Polynesian Tree Snail – Polynesia
- Barbary Lion – Northern Africa
- Cape Lion – Southern Africa
- Cave Lion – Eurasia
- Caspian Tiger – Kazakhstan
- European Lion – Europe
- Javan Tiger – Java
- Stephens Island Wren – New Zealand
- Giant Moa – New Zealand
- Dodo Bird – Mauritius Island
- Moa-Nalo – Hawaii
- Shomburgk's Deer – Thailand
- Bluebuck – southern Africa
- Syrian Elephant – Syria
- Cyprus Dwarf Hippopotamus – Cyprus
- Gravenche – Europe
- Ferghana – China
- Syrian Wild Ass – Syria
- Turkoman – Eurasia
- Lake Pedder Earthworm – Tasmania
- Madeiran Large White – Europe

9.5.4 Extinction Rate in Earth History

Extinctions are a normal part of evolution. There is a natural background rate to the timing and frequency of extinctions. Extinctions occur naturally and periodically over time (Jablonski 1986). In fact, 10% of species are lost to extinction every million years, 30% every 10 million years, and 65% every 100 million years. Evolution occurs through a balance of extinction (meaning the end of a species) and speciation, which is the evolution of new species (Ritchie 2022). More than 99% of all organisms that have ever lived on Earth are extinct (National Geographic Society 2024d). The background rate of species extinction is measured in the number of families that go extinct every million years, which is considered to be five as a background rate (Jablonski 1986).

9.5.5 Mass Extinctions in Earth History

There have been five mass extinction events in the last 500 million years of Earth history. Mass extinction events may have occurred earlier than 500 million years ago, but very little is known about extinction events that far back in Earth history. The five mass extinction events in the last 500 million include the following (National Geographic Society 2024d):

1. 444 million years ago during the late Ordovician Period that resulted in 86% species loss
2. 360 million years ago during the late Devonian Period that resulted in 75% species loss
3. 250 million years ago at the end of the Permian Period that resulted in 96% species loss
4. 200 million years ago at the end of the Triassic Period that resulted in 80% species loss
5. 65 million years ago at the end of the Cretaceous Period that resulted in 76% species loss

A mass extinction is generally defined by the loss of at least 75% of species within a relatively short period of time, generally considered to be within 2 million years (Ritchie 2022). Figure 9.1 shows the extinction rate for the last 500 million years of Earth history. Variation in extinction rates does occur as depicted in Figure 9.1.

Evidence suggests that the five mass extinctions in Earth history were caused by the following (National Geographic Society 2024c; Ritchie 2022):

1. Mass extinction at the end of the Ordovician Period 444 million years ago was a time where intense glacial and interglacial periods created large sea-level variations and resulted in significant shoreline migrations. Tectonic uplift of the Appalachian Mountains at this same time

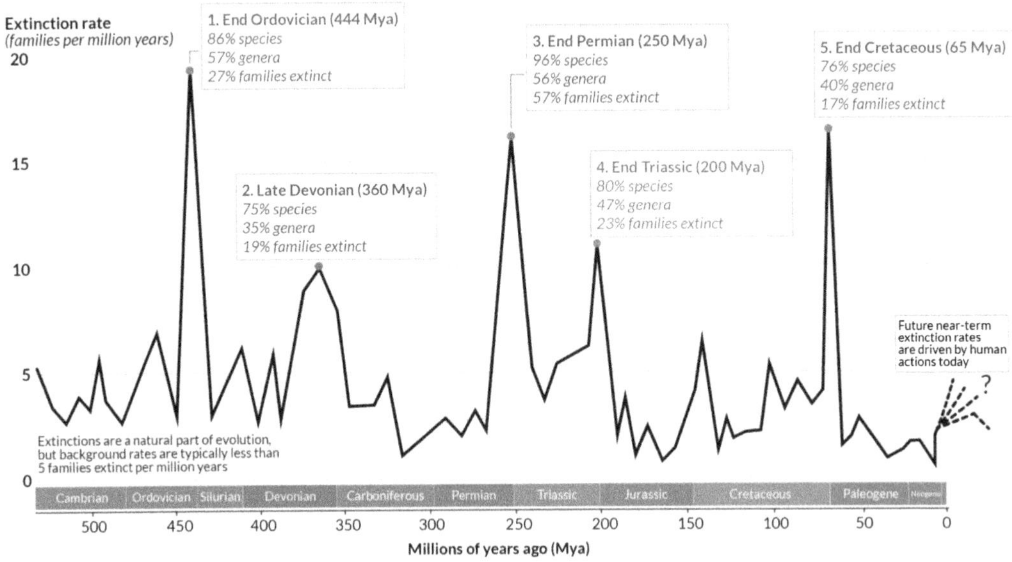

FIGURE 9.1 Extinction rate for the past 500 million years of Earth history. (From Ritchie, H. 2022. *There have Been Five Mass Extinctions in Earth's History*. OurWorldinData.org. https://ourworldindata.org/mass-extinctions. (Accessed May 27, 2024).)

created increased weathering and erosion and sequestration of carbon dioxide, which caused significant alterations in climate and the chemistry of the ocean. The most important factor in reducing carbon dioxide and therefore, global cooling, is the weathering of minerals known as silicates, such as feldspar and mica. Feldspar and mica are common ingredients in granite. The presence or absence of plate tectonics on a given planet significantly affects the rates and efficiency of this atmospheric thermostat. The basic chemical reaction for this to occur is $CaSiO_2 + CO_2 = CaCO_3 + SiO_2$. When the first two molecules in the equation combine, limestone is formed, and carbon dioxide is removed (Ward and Brownlee 2000).
2. Mass extinction at the end of the Devonian Period 360 million years ago was a time when Earth experienced rapid growth and diversification of land plants, which caused rapid and severe global cooling.
3. Mass extinction at the end of the Permian Period 250 million years ago was a time in which intense volcanic activity occurred in what is now Siberia. The volcanic activity emitted large quantities of greenhouse gases into the atmosphere. Carbon dioxide and sulfur emissions from the volcanic activity caused ocean acidification, acid rain, and other changes in ocean and land chemistry.
4. Mass extinction at the end of the Triassic Period 200 million years ago was a time when underwater volcanic activity in the Central Atlantic Magmatic Province (CAMP) caused global warming and dramatic and rapid change in the chemistry of the ocean.
5. Mass extinction at the end of the Cretaceous Period 65 million years ago was a time when an asteroid struck Earth near the Yucatan in Mexico, which caused global cataclysmic events to the atmosphere, land, and ocean.

Each of the five mass extinctions in Earth history possessed a climate element and occurred rapidly. Two of the five mass extinctions related to global cooling and a reduction of atmospheric carbon dioxide. Two others related to global warming and an increase in atmospheric carbon dioxide. The last mass extinction was the result of an asteroid impact and changed the land, ocean, atmosphere, and climate. Therefore, four of the five mass extinction events in the last 500 million years of Earth's history involved fluctuations in atmospheric carbon dioxide concentrations.

9.5.6 Anthropogenic Influences of Species Extinction

What causes a species to go extinct? We have learned that species extinction is a natural part of evolution. However, anthropogenic interference is causing an increase in the baseline extinction rate at a pace that is estimated to be between 1,000 and 10,000 times the natural rate (National Geographic Society 2024d; Barnosky et al. 2011). There are many circumstances that may cause a species to become endangered and ultimately extinct and include the following (USEPA 2024c; National Geographic Society 2024d):

- *Overhunting and overharvesting:* This circumstance has been the fate of many large animals that are slow, are a human food source, or have a medicinal value, especially when humans had migrated to a previously uninhabited area. Many large mammals such as the mammoths, mastodons, and camels of the Americans between 13,300 and 15,000 years ago may have experienced this outcome. A recent example is the Passenger Pigeon which was recently overhunted into extinction. Elephants and rhinoceros are a couple species that are currently in this situation.
- *Habitat loss:* Habitat loss is considered one of the main drivers of species extinction. Deforestation, land development, agriculture, human over-populations, water resource over-exploitation, wetland destruction, and other anthropogenic influences cause habitat loss.

- *Highly specialized species:* Highly specialized species that have specific habitat requirements generally do not do well when faced with a changing environment. A potential example here is the Monarch Butterfly which lays its eggs on a plant called the Milk Weed. If the Milk Weed plants experiences a decline, so does the butterfly (WWF 2023).
- *Pollution:* Effects of pollution on many species has been well documented and began with the book by Rachel Carson called *Silent Spring* (Carson 1962) that documented the effects of dichlorodiphenyltrichloroethane (DDT) on animal species.
- *Invasive species:* Invasive species can outcompete indigenous species, and reduce population numbers and lower genetic diversity within some animal and plant populations.
- *Human–wildlife conflict:* Examples of conflict between humans and certain wildlife species include the wolf, mountain lion, grizzly bear, and other animal species considered apex predators.
- *Disease:* Disease contributes to decline in species abundance. Examples include a deadly fungus that reduced numbers of 30 amphibian species in Panama in the early 2000s.
- *Low birth rate:* Animals that do not reproduce often and have few offspring have difficulty maintaining optimal species diversity in a changing climate. Pandas and elephants are good examples of species that exhibit low birth rates.
- *High genetic vulnerability:* If a species has low genetic diversity, it can have difficulty to evolve quickly enough in a changing environment. The koalas and cheetah are good examples of species with high genetic vulnerability.
- *Rare species:* If a species is already rare, it has a lower probability of survival under changing environmental conditions. The Sumatran Tiger is a good example of an already rare species.

9.5.7 Future Outlook of Species Extinction

According to the WWF (2024) and the IUCN (2024b), hundreds of species of animals are now extinct directly due to humans in several different ways including hunting, habitat destruction, poisoning, development, and pollution. According to the United Nations, more than 1 million species of animals and plants are threatened with extinction as a direct result of human activities worldwide (United Nations 2024a). The more than 200 scientists who participated in the study concluded that biodiversity loss poses a global threat at the same level as the risks posed by climate change (United Nations 2024d).

According to the report by the WWF (2022), such a collapse in biodiversity would wreak havoc on interconnected ecosystems of the planet, putting human communities at risk by compromising food sources, contaminating clean water and air, and eroding natural defense against extreme weather such as hurricanes and floods. In addition, the study concluded that species extinctions will become increasingly common in the coming decades with the driving factors included continued urban and agricultural development, deforestation, pollution, overfishing, plastic pollution, and poaching (see Figure 9.2) (United Nations 2024d).

The basic message of the report is the same as what the scientific community has been saying for more than 30 years and that is biodiversity is important for human wellbeing, and humans are destroying it. To produce the report included 145 biodiversity scientific experts plus thousands of other contributors who took more than three years to compile data from over 15,000 sources (United Nations 2024d).

There has been much concern and debate on the increase in the rate of extinctions caused by anthropogenic activities. The debate has not focused on whether human activities have resulted in the extinction of species but the magnitude of the extinction rate. Could it be that human activities trigger the sixth mass extinction event? (Barnosky et al. 2011).

FIGURE 9.2 Dead seal with a frisbee around its neck. (From United Nations. 2024d. *UN Report: One million Species at Risk of Extinction*. www.un.org/sustainable/one-million-species-at-risk-of-extinction. (Accessed April 7, 2024).)

9.6 PATHOGENS

Pathogens are what most people call germs. Simply put, a **pathogen** is a microorganism that may cause disease within its host. Pathogens are different in many ways, but they have one main similarity: once they enter the body, they can damage cells or interfere with the body's normal activities (National Institute of Health (NIH) 2024).

9.6.1 What Is a Pathogen

Pathogens are taxonomically widely diverse and comprise viruses, bacteria, parasites, fungus, as well as unicellular and multi-cellular eukaryotes (National Institute of Health 2024). Depending on the type of pathogen in your body, you can get a minor illness or a deadly disease. Pathogens are different in many ways, but they have one main similarity: once they enter the body, they can damage cells or interfere with the body's normal activities (NIH 2024). Although we do not always think of bacteria, parasites, fungi, and viruses as everyday invasive species or environmental contaminants, they are common and can cause many diseases and adverse health effects. There are four main routes of exposure of a pathogen and include the following (NIH 2024):

- *Inhalation:* If an infected person sneezes or coughs, pathogens can be released into the local air, which could be inhaled by another individual.
- *Contact with blood:* Infected blood that is transferred to another individual through sexual contact, touching a person's wound, or from blood-sucking insects.
- *Food and water:* Eating or drinking something infested with pathogens can make an individual ill.
- *Physical contact:* If an individual touches someone or touches something that contains a pathogen such as a handrail, kitchen counter, toilet, etc., and then touches their nose, mount, eyes, ears, or a cut, that pathogen can enter your body.

Therefore, considering the above list, pathogens can contaminate water supplies and are also present in the air, in soil, on and within food, and most other surfaces. Not all viruses, bacteria, fungi, and parasites are pathogens. Instead of causing disease, some actually help the body. An example would be yogurt and other fermented foods such as kimchi may lower the risk of type 2 diabetes and heart disease (NIH 2024).

According to the NIH (2024), continued increase in human population will likely result in increasing the risk of pathogen spread.

9.6.2 Types of Pathogens

Bacteria are organisms made up of just one cell. They are capable of multiplying themselves through a process called **binary fission**, whereby a single bacterium grows to approximately twice its normal size and then splits into two daughter cells that are copies of the original bacterium. Bacteria live everywhere, even inside most organisms. Most bacteria are harmless, and some are beneficial by destroying other harmful bacteria within our bodies (Madigan et al. 2008). However, some may cause diseases, such as tuberculosis. One of the more common harmful bacteria is a group called *E. coli*, which is short for *Escherichia coli*. Most *E. coli* are harmless, but a strain called serotype O157:H7 can cause food poisoning in humans. Three basic shapes of bacteria exist: rounded, rod shaped, and spirals. *E. coli* bacteria is rod shaped. The ability for *E. coli* to survive for a brief period outside the body creates the potential for the bacteria to spread and infect other people. The spread of *E. coli* usually occurs when there is poor sanitation or when untreated sewage is discharged from municipal wastewater treatment plants, which occasionally occurs during flood events (USEPA 2009). Adverse health effects from exposure to *E. coli* bacteria typically include gastroenteritis, urinary tract infections, skin rashes, and neonatal meningitis (USEPA 2009).

A **virus** is a subcellular infectious agent capable of replicating itself inside the cells of another organism (Madigan et al. 2009). Viruses are typically 100 times smaller than a bacterium. They consist of two parts: (1) DNA or RNA molecules carrying genetic information; and (2) a protein coat protecting the genes. Some viruses may also have an outside layer of fat surrounding them while they are outside a cell. Viruses spread in many ways, including aerosol routes (coughing and sneezing), through infected or contaminated water, and exchange of body fluids. Viruses cause diseases such as the common cold, influenza, chickenpox, mumps, Ebola, and HIV (Madigan et al. 2008).

A **parasite** is an organism living on or within a different organism (the host) at the expense of the host organism. Common examples of parasites causing adverse health effects in humans are Cryptosporidium and Giardia; both may cause severe intestinal disorder (United States Center for Disease Control 2009). Exposure to these two parasites occurs by consuming infected water (United States Center for Disease Control 2009). In humans, Giardia creates uncomfortable, but curable gastrointestinal symptoms, whereas Cryptosporidium can create lifelong gastrointestinal symptoms in persons with weak immune systems and may result in death.

A **fungal** pathogen is a type of fungus that causes diseases in humans or other organisms. These fungi are eukaryotic microorganisms that can affect various parts of the body, such as skin, nails, lungs, blood, and brain. There are about 300 species of fungi that are known to be pathogenic to humans. Some examples of pathogenic fungi are Candida, Aspergillus, Cryptococcus, Histoplasma, Pneumocystis, and Stachybotrys (United States Center for Disease Control (CDC) 2024).

9.7 ACTIONS TAKEN IN THE UNITED STATES FOR SPECIES PROTECTION AND HABITAT PRESERVATION

The United States has not been silent when it comes to invasive species and extinction. However, it has taken time to foster attention at the national level to organize and take effective action. The action taken to assist with the eradication of invasive species and extinction prevention has generally been to evaluate and identify the human actions that constitute the main cause of species influence and then test what methods are the most productive at solving the problem. Much of this scientific research, eradication methods, and methods to save species from extinction has been conducted through the enactment of environmental laws. The aim of enacting these laws was to understand why and how some species either became an invasive species or why and how a native species became threatened with extinction. There are many environmental laws in the United States and the aims of most, if not all, environmental laws are to protection human health and the environment.

Laws that are commonly attributed to protecting human health and the environment include the following (USEPA 2024b):

- Rivers and Harbors Act of 1899
- Antiquities Act of 1906
- Atomic Energy Act of 1946
- Atomic Energy Act of 1954
- Brownfield Revitalization Act of 2002
- Clean Air Act (CAA) of 1970
- Clean Water Act (CWA) of 1972
- Coastal Zone Management Act of1972
- Comprehensive Environmental Response, Compensation, and Liability Act (CERCLA) of 1980
- Emergency Planning and Community Right-to-Know Act of 1986
- Endangered Species Act of 1973
- Energy Policy Act of 1992
- Energy Policy Act of 2005
- Federal Power Act of 1935
- Federal Feed, Drug, and Cosmetic Act (FFDCA) of 1938
- Federal Insecticide, Fungicide, and Rodenticide Act (FIFRA) of 1947
- Fish and Wildlife Coordination Act of 1968
- Food Quality Protection Act of 1996
- Fisheries Conservation and Management Act of 1976
- Global Climate Protection Act of 1987
- Hazardous Materials Transportation Act (HMTA) of 1975
- Insecticide Act of 1910
- Lacey Act of 1900
- Low-Level Radioactive Waste Policy Act of 1980
- Marine Protection Act of 1972
- Marine Mammal Protection Act of 2015
- Medical Waste Tracking Act of 1980
- Migratory Bird Treaty Act 1916
- Mineral Leasing Act 1920
- National Environmental Policy Act of 1969
- National Forest Management Act of 1976
- National Historic Preservation Act of 1966
- National Parks Act of 1980
- Noise Control Act 1974
- Nuclear Waste Policy Act of 1982
- Ocean Dumping Act of 1988
- Occupational Safety and Health Act (OSHA) of 1970
- Oil Spill Prevention Act of 1990
- Pollution Prevention Act (PPA) of 1990
- Refuse Act of 1899
- Resource, Conservation, and Recovery Act (RCRA) of 1976
- Rivers and Harbors Act of 1899

- Safe Drinking Water Act (SDWA) of 1974
- Superfund Amendments and Reauthorization Act (SARA) of 1986
- Surface Mining Control and Reclamation Act of 1977
- Toxic Substance Control Act (TSCA) of 1976
- Wild and Scenic Rivers Act of 1968

One thing to ponder is to ask the question, with so many environmental regulations that address protecting the environment and preventing species extinction, how can it be that there are so many threatened and endangered species?

9.8 SUMMARY AND CONCLUSION

Invasive species, endangered species, extinct species, and pathogens discussed in this chapter each have connections to human behavior and geology. The geology of a given area influences habitat for species and when human activity occurs in those areas, the outcome can be devastating. One of the most alarming aspects of this chapter is the conclusion from the scientific community that biodiversity loss poses a global threat at the same level as the risks posed by climate change (United Nations 2024d). It is important for each one of us to realize that we are but one of about 4,500 species of mammals that exist on Earth and that we are but one species in an estimated 30 to perhaps more than 100 million species that form Earth's biosphere. We are totally dependent on the biosphere for the food we eat, the air we breathe, and other requirements to maintain life. We find ourselves on Earth after billions of years of Earth's existence by chance, not by design. Our species has become so successful and with such speed at modifying our environment that many other life forms have been overcome and dominated by technological and cultural innovations and wars and revolutions.

Now we come to the point in the next chapter of assessing the anthropogenic impacts on a global scale by examining how countries of the world have responded to environmental pressures and we will examine how efforts have been effective or not by examining the environmental health of the oceans and Antarctica.

REFERENCES

Barnosky, A. D., Matzke, N., Tomiya, S., Wogen, G. O., Swartz, B., Quental. T. B., and Ferrer, E. A. 2011. Has the Earth's Sixth Mass Extinction Already Arrive? *Nature*. Vol. 471, no. 7336, pp. 51–57.

Carson, R. 1962. *Silent Spring*. Houghton Mifflin. Boston, MA.

International Union for Conservation of Nature (IUCN). 2024a. *Endangered Species*. www.iucn.org/endangered-species. (Accessed April 8, 2024).

International Union for Conservation of Nature (IUCN). 2024b. *Extinct Species in North America*. www.iucn.org/extinct-species. (Accessed April 8, 2024).

Jablonski, D. 1986. Mass and Background Extinctions: The Alternation of Macrobiodiversity Regimes. *Science*. Vol. 231, pp. 129–133.

Jabr, F. 2024. *Becoming Earth: How the Planet Came to Life*. Random House. New York, NY. 269pp.

Madigan, M. T., Martinko, J. M., Dunlap, P. V., and Clark, D. P. 2008. *12th Edition. Brock Biology of Microorganisms*. Prentice-Hall, Inc. New York, NY.

Moody, R. E., Alvarex-Carretero, S., Mahendrarjah, T. A., Clark, J. W., Betts, H. C., Dombroski, N., Szánthó, L. L., Boyle, R. A., Daines, S., Chen, X., Lane, N., Yang, Z., Shields, G. A., Szöllősi, G. J., Spang, A., Pisani, D., Williams, T. A., Lenton, T. M., and Donoghue, P. C. J. 2024. The Last of the Universal Common Ancestor and Its Impact on the Early Earth System. *Nature Ecology and Evolution*. Vol. 8, 1654–1666.

National Geographic Society. 2024a. *Invasive Species*. https://education.nationalgeographic.org/esource/invasive-species. (Accessed April 2, 2024).

National Geographic Society. 2024b. *Endangered Species*. https://education.nationalgeographic.org/resource/endangered-species. (Accessed April 7, 2024).

National Geographic Society. 2024c. *The Farming Revolution*. https://education.nationalgeographic.org/resource/development-of-farming. (Accessed April 7, 2024).
National Geographic Society. 2024d. *Species Extinction*. https://education.nationalgeographic.org/resource/species-extinction. (Accessed April 7, 2024).
National Institute of Health (NIH). 2024. What are Pathogens. https://pmc.ncbi.nlm.nih.gov/articles/PMC5648414. (Accessed April 7, 2024).
Ritchie, H. 2022. *There Have Been Five Mass Extinctions in Earth's History*. OurWorldinData.org. https://ourworldindata.org/mass-extinctions. (Accessed May 27, 2024).
United Nations. 2024a. *Biodiversity: Invasive Species*. United Nations System-Wide Earthwatch. www.un.org/earthwatch/biodiversity/invasivespecies.html. (Accessed April 2, 2024).
United Nations. 2024d. *UN Report: One million Species at Risk of Extinction*. www.un.org/sustainable/one-million-species-at-risk-of-extinction. (Accessed April 7. 2024).
United States Center for Disease Control (CDC). 2009. *Cryptosporidium Fact Sheets. Department of Health and Human Services*. CDC. Atlanta, GA.
United States Department of Agriculture (USDA). 2024a. *National Invasive Species Information Center (NISIC)*. www.invasivespeciesinto.gov. (Accessed April 2, 2024).
United States Department of Agriculture (USDA). 2024b. *National Invasive Species Information Center (NISIC). Emerald Ash Borer Beetle*. www.invasivespeciesinfo.gov/terrestrial/invertebrates/emerold-ash-boror-beetle. (Accessed April 2, 2024).
United States Environmental Protection Agency (USEPA). 2009. *E. Coli Bacteria. Total Coliform Rule-Basic Information*. USEPA. Washington, D. C.
United States Environmental Protection Agency (USEPA). 2024a. *Invasive Species*. www.epa.gov/greatlakes/invasive-species. (Accessed April 2, 2024).
United States Environmental Protection Agency (USEPA). 2024b. *Summary of Environmental Regulations of the United States*. https://epa.gov/summary-environmental-laws. (Accessed April 7, 2024).
United States Environmental Protection Agency (USEPA). 2024c. *Threatened and Endangered Species*. https://epa.gov/endangered-and-threatened-species. (Accessed May 27, 2024).
United States Fish and Wildlife Service. 2024. *Endangered Species Act*. www.fws.gov/endangered/laws-policies/. (Accessed April 7, 2024).
United States Forest Service. 2024. *Bark Beetles*. www.fs.usda.gov/internet/fse_documents/fseprd563458. (Accessed April 2, 2024).
University of Chicago. 2024. *The Origin of Life on Earth*. https://Universityofchicago.edu/theoriginoflifeonearth. (Accessed August 20, 2024).
Ward, P. D., and Brownlee, D. 2000. *Rare Earth: Why Complex Life Is Uncommon in the Universe*. Springer-Verlag. New York, NY. 333pp.
World Wildlife Fund (WWF). 2022. *Living Planet Report 2022: Building a Nature-Positive Society*. WWF. Gland, Switzerland, 115pp.
World Wildlife Fund (WWF). 2023. *Troubling News for Monarch Butterfly Populations*. www.worldwildlifefund.org/stories/troubling-news-for-monarch-butterfly-populations. (Accessed May 27, 2024).
World Wildlife Fund (WWF). 2024. *Species List of Endangered, Vulnerable, and Threatened Animals*. www.wwf.org/species/endangered-list. (Accessed April 7, 2024).

10 Assessing Global Anthropogenic Impacts

10.1 INTRODUCTION

According to the United Nations (UN 2024a), anthropogenic activities that include habitat destruction, pollution, ecosystem impacts due to climate change, introduction of invasive species, and over-exploitation and hunting have affected the entire Earth and have directly led to extinction of thousands of species. In this chapter, we will evaluate and assess global anthropogenic impacts. We have already described numerous impacts upon the Earth caused either directly or indirectly by humans. The aim of this chapter is to step back and take a holistic view of Earth to assess the anthropogenic impacts in a cumulative manner. To view Earth holistically, we will also evaluate impacts to the oceans and Antarctica.

The reason why we must examine the oceans is that all life depends on the oceans and the oceans make up approximately 70% of the surface of the Earth. Antarctica is a continent that has no permanent human settlements. As a first thought you might ask yourself why we should examine Antarctica, as there should not be any significant environmental issues in Antarctica, correct? To most of you, the answer will be humbling and perhaps even disturbing. We shall have to wade through the information and data presented in this chapter to begin to understand and appreciate the depth and magnitude of the challenges that are confronting humans from a geological and environmental perspective and that are very evident even in Antarctica.

This chapter will also address current conditions of pollution on Earth in the air, water, and land. We shall also evaluate how countries of the world have chosen to address anthropogenic impacts. This will be accomplished by examining the effectiveness of environmental protection in more than 50 countries of the world. We will rank each country from most effective to least effective and discuss each country's main impediments preventing them from achieving an effective and healthier natural environment. The countries that will be evaluated account for 70% of the human population and 75% of the land area on Earth. Finally, this chapter will evaluate countries that are important or unique to all humans on Earth. The evaluation will provide us with special insights into unique issues, situations, and problem-solving challenges due to a host of complex circumstances that some countries are facing. These countries have special challenges that, the combination of which, are unique and will require us to explore potential reasons behind some of the challenges faced by these nations because of social concerns, political unrest, lack of action, and economic, geographic, geological, or other factors that result in either poor or excellent environmental performance. Some of these countries include Australia, New Zealand, China, Tanzania, Egypt, Brazil, Argentina, Korea, Peru, Malaysia, and Chile.

DOI: 10.1201/9781003463153-12

10.2 CLIMATE REGIONS ON EARTH

The climate on Earth is influenced by many factors including solar radiation, local and regional geology, plate tectonics, position of the continents and ocean currents, and atmospheric carbon dioxide concentration, just to name a few. We have also discovered in the previous chapter that climate, specifically the atmospheric concentration of carbon dioxide, was a significant influencing factor in four of the five mass extinctions in the last 500 million years of Earth history. Figure 10.1 is a map that shows the climate regions of the world that will be referred to throughout this chapter (National Oceanic and Atmospheric Administration (NOAA) 2024a).

Note that the climate regions outlined on the map represented as Figure 10.1 are also heavily influenced by the geology of specific regions and, in some cases, geology influences the entire country, a phenomenon that is more common than many may think. One of the most dramatic examples of how topography can impact precipitation patterns occurs in the Pacific Northwest region in the United States. The Cascade Mountain range runs from south to north in northern California, Oregon, and western Washington. The mountains create a barrier to air moving eastward for the Pacific Ocean. When the moist, oceanic air encounters the mountains, it begins to rise. As the air rises and moves over the mountains, it cools, causing much of its moisture to condense, forming clouds and precipitation. As the air continues moving to the east, it descends the other side of the mountains, warms up, and dries out.

This phenomenon causes areas on the west side of the mountains to be much wetter than areas on the east side. Meteorologists call this contrast the **orographic effect** (NOAA 2024b). Let's take this a bit further and compare precipitation on either side of the Cascade Mountains at a couple locations to illustrate this phenomenon. Packwood and Yakima are both located in Washington. Both locations are at a similar elevation of approximately 330 m, only about 90 km apart, and

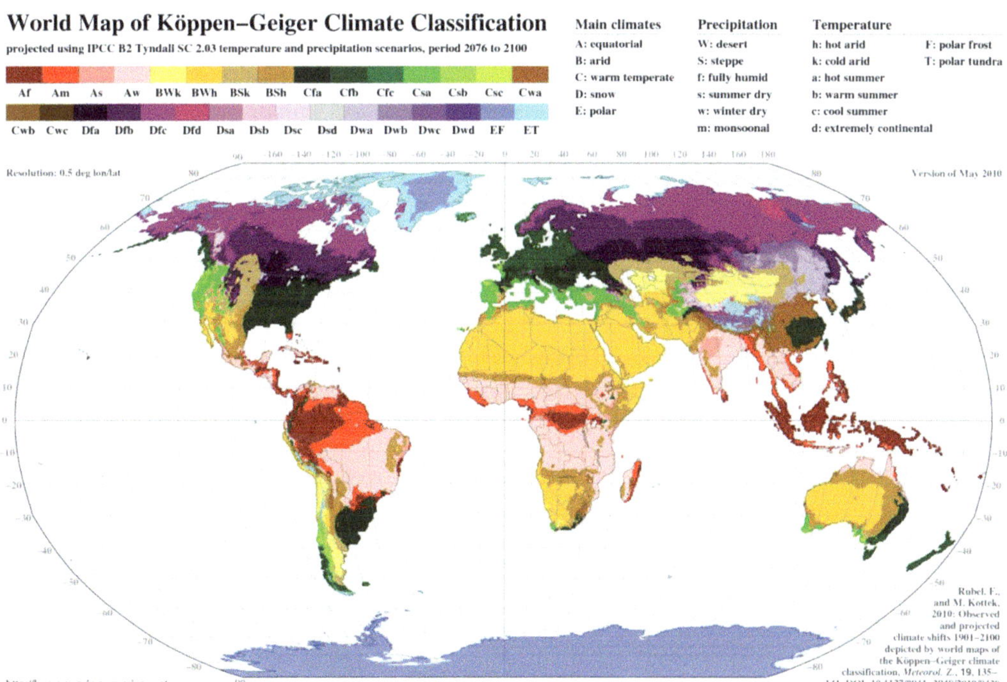

FIGURE 10.1 Climate regions of the world. (From National Oceanic and Atmospheric Administration (NOAA 2024a). Koppen-Geiger Climate Changes. https://sos.noaa.gov/datasets/koppen-geiger-climate-changes. (Accessed April 12, 2024), 2024a.)

share a similar latitude of approximately 46.6°N. According to NOAA's 1981 to 2010 climate data, Packwood receives an average of 142 cm (56 inches) of precipitation per year while Yakima receives 21 cm (8.3 inches). The difference is because Packwood is located on the west side of the Cascade Mountains while Yakima is located on the west side (NOAA 2024b).

Another example is how the Rocky Mountains in the western United States and southern Canada influence the climate regions directly to the east of the mountains, which is also an orographic effect that is partially responsible for the formation of the Great Plains (NOAA 2024b). Another example of this phenomenon is the Himalayan Mountain Range in Asia and the Andes Mountain Range in South America. The examples explained above relate directly to the geology and its influence on the climate and precipitation patterns of the western United States and the world.

To start our exploration of global anthropogenic impacts, let us first look at some global assessments that have been implemented by the UN and the World Health Organization (WHO) that describe the current status of anthropogenic impact on Earth.

10.3 GLOBAL ASSESSMENTS AND STANDARDS

The Earth's pollution encompasses numerous aspects of the land, air, and water. However, there are some of what are known as **sinks** or final resting places of pollution, with the oceans being the largest sink on Earth. In fact, in the geological sciences, there is a common reference that states "whatever is created eventually ends up in the ocean." Other sinks include the land and in some instances the atmosphere.

International organizations such as the WHO and the UN have begun to assess, on a global scale, the quality of freshwater and the air on Earth. A global assessment of the land has not yet been completed. Both organizations have published guidelines for assessing, improving, or maintaining acceptable air and water quality standards as far back as 1983 (WHO 2013; UN 2024b, 2024c). According to WHO (2013, 2024a, 2024b, 2024c) and the UN (2024a, 2024b), approximately 2.8 billion humans lack access to basic sanitation and improved drinking water. While this number is improving, it still represents nearly 30% of the planet's human population.

In addition, according to WHO (2024d), 9 out of 10 people on Earth breath air that is considered unacceptable because it contains high levels of pollution. The UN estimates that most land pollution is caused by agricultural activities, such as grazing, use of pesticides and fertilizers, irrigation, plowing, and confined animal facilities (UN 2024a).

Significant effort was made to include data tables for air pollution emission standards, water quality, remediation goals, and other relevant information for every chemical compound in each evaluated country. However, this was just not possible while keeping this book manageable. Therefore, only select tables listing specific chemicals are provided, and website addresses are provided for those not included.

10.3.1 Air

As stated above and according the WHO, 9 out of 10 people on Earth breath air containing high levels of pollution (WHO 2024d). Updated estimates indicate that nearly 7 million deaths occur each year from breathing unhealthy air (WHO 2024d). Most of the deaths occur in low- to middle-income countries in Asia and Africa. The most problematic pollutant is fine particulate matter which are particles less than 2.5 microns, and they penetrate deep into the lungs causing diseases that include the following (WHO 2024d):

- Stroke
- Pulmonary diseases
- Heart disease
- Respiratory infections
- Lung cancer
- Pneumonia

The major sources of air pollution worldwide include (WHO 2024d):

- Inefficient use of energy in households
- Industry
- Agriculture
- Transportation
- Use of coal

The inefficient use of energy in households is directly related to access to clean cooking fuels, which is the case for an estimated population of 3 billion people worldwide (WHO 2024d). The most significant types of air pollutants worldwide include the following (WHO 2024d):

- Particulate matter (PM10)
- Fine particulate matter (PM2.5)
- Carbon monoxide (CO)
- Ozone (O_3)
- Sulfate
- Nitrates
- Black carbon
- Lead
- Volatile organic compounds (VOCs)
- Benzo(a)pyrene
- Benzene

The WHO has ranked the top ten cities with the most air pollution in the world as the following (WHO 2024e):

1. Cairo, Egypt
2. Delhi, India
3. Beijing, China
4. Moscow, Russia
5. Istanbul, Turkey
6. Guangzhou, China
7. Shanghai, China
8. Buenos Aires, Argentina
9. Paris, France
10. Los Angeles, United States

The WHO published ambient air quality guidelines for ozone, particulate matter, sulfur dioxide, nitrogen dioxide, and carbon monoxide (see Table 10.1).

The guideline limits listed above are not enforceable and are based on human health risk posed by air pollutants. WHO examined 37 air pollutants but has only set guidelines for the five listed in Table 10.1. WHO includes guidance for countries in the form of statistical modeling related to exposure and risk for those compounds that are of concern but did not establish a guideline. This is

TABLE 10.1
WHO Ambient Air Guidelines

Parameter	Units of Measure	Guideline Limit
Ozone (O_3)	Parts per billion	50
Particulate matter (PM10)	Micrograms per cubic meter	25
Sulfur dioxide (SO_2)	24 hour, parts per billion	8
Nitrogen dioxide (NO_2)	Yearly, parts per million	21
Carbon monoxide (CO)	8 hour, parts per million	9

Source: World Health Organization (WHO). 2024e. *World Health Organization Global Urban Ambient Air Pollution Database*. www.who.int/phe/health_topics/outdoorair/database.htm. (Accessed April 13, 2024).

due to many factors that relate to exposure, risk, geography, and other factors at specific locations (WHO 2024d).

Many cities throughout the world exceed WHO guidelines for air quality. Figure 10.2 shows the major cities of the world that exceed the WHO guidelines for particulate matter (PM2.5) on an annual basis (WHO 2024e). As you can see from examining Figure 10.2, most cities in the United States, Australia, Canada, and Europe do not exceed the WHO guidelines. However, many cities in southern Asia, central Africa, and Mexico City exceed the WHO guidelines. The city with the most air pollution in the United States ranks tenth worst in the world and it is Los Angeles, California. Many cities in China also exceed current WHO standards (WHO 2024e, 2024f)

The United States Department of State has an air quality monitoring program for many large cities of the world that is updated daily and is termed **air quality index** (AQI) (United States Department of State 2024). The United States Department of State air quality guidelines for AQI for fine particulate matter (PM2.5) are listed in Table 10.2.

10.3.2 Water

Global driving forces for issues related to water include climate change, increasing water scarcity, population growth, demographic changes, and urbanization. These driving forces are expected to place additional stress on freshwater supplies and sanitation systems and services. As climate change scenarios become increasingly reliable, existing infrastructure will need to be adapted and planning of new systems and services will need to be updated. Extreme weather conditions are also reflected in the increased frequency and intensity of natural disasters and will become the new normal (WHO 2013). Figure 10.1 shows the climate regions of the world.

According to the UN (2024a, 2024b), the most prevalent water quality problem globally is eutrophication, which is a result of high-nutrient loads (mainly phosphorus and nitrogen). The most significant source of water pollution on Earth is agriculture. Other sources include domestic sewerage, industrial effluents, and atmospheric inputs from fossil-fuel burning and wildfires. A significant issue related to water is sanitation and the spread of diseases through untreated water contaminated with human excrement and from livestock and wildlife called **zoonotic pathogens** (WHO 2012, 2013).

As stated above, the WHO (2013, 2024b, 2024c, 2024d) and the UN (2024a, 2024b) estimate that 2.8 billion people do not have access to proper sanitation, while Baum et al. (2013) estimate that the actual number could be as high as 4.1 billion because most evaluations do not account for all biological disease-causing pathogens.

The WHO lists the top ten cities with the most water pollution in the world as the following (WHO 2024b):

1. Sumgavit, Azerbaijan	2. Linfen, China
3. Tianjing, China	4. Sukinda, India
5. Vapi, India	6. La Oroya, Peru
7. Dzerzhinsk, Russia	8. Norilisk, Russia
9. Chernobyl, Russia	10. Kabwa, Zambia

Of the countries we will evaluate, only one city is located in a country that we will not evaluate, which is Zambia. Of the countries we will evaluate, four cities are located in the former Soviet Union, two are located in China, two in India, and one in Peru. Contaminants that have affected the cities with the most water pollution in the world include the following (WHO 2024b):

• VOCs	• Biological contaminants including *Escherichia coli*
• Oils and fuel	• Pesticides
• Radioactive materials	• Fertilizers
• Garbage	• Plastic
• Zoonotic pathogens	• Other pathogens

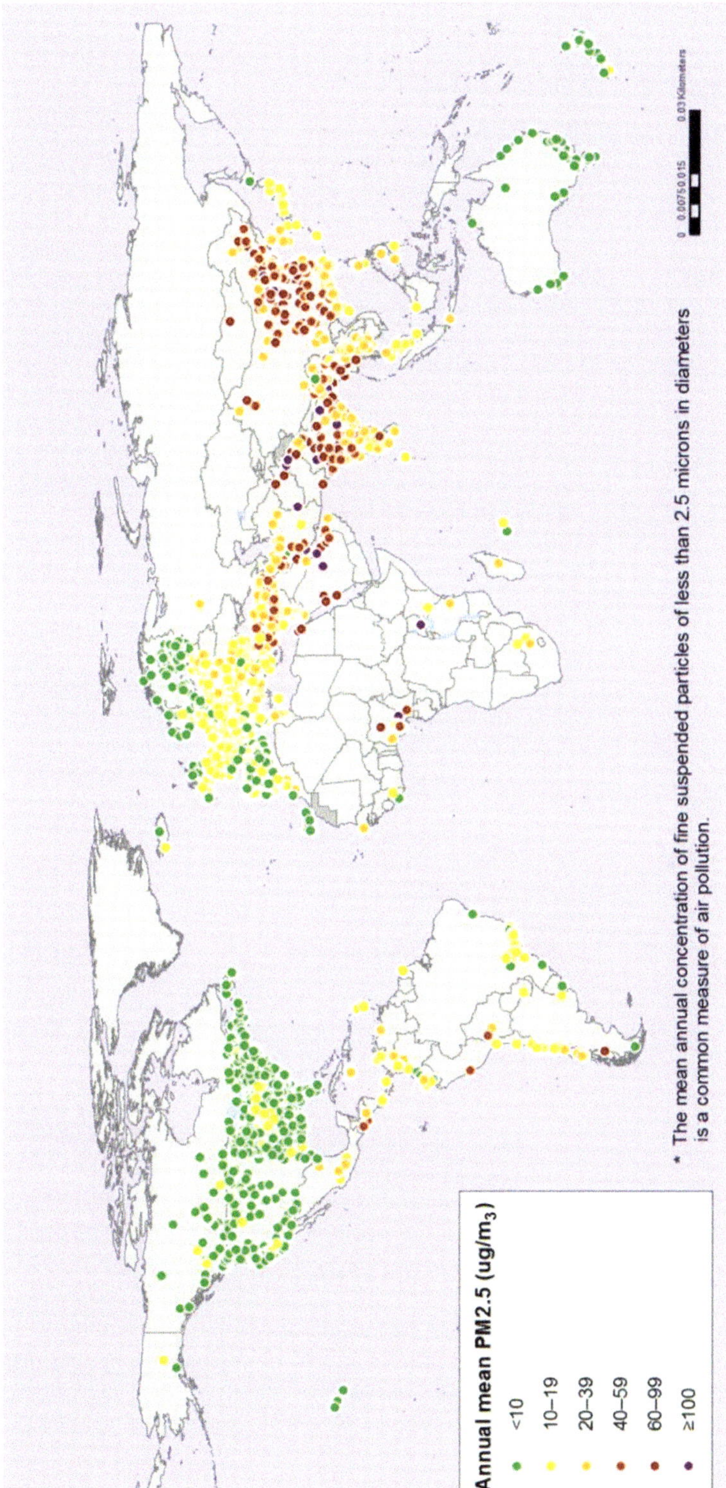

FIGURE 10.2 Annual mean PM2.5 at select cities worldwide. (From World Health Organization (WHO). 2024e. World Health Organization Global Urban Ambient Air Pollution Database. www.who.int/phe/health_topics/outdoorair/database.htm. (Accessed April 16, 2024).)

TABLE 10.2
Air Quality Index

AQI for PM2.5	Value (μg/m³)	Health Effects Statement	Cautionary Statement
Good	0–50	Little or no risk	None
Moderate	51–100	Unusually sensitive individuals may experience respiratory symptoms	Active children and adults, and people with respiratory disease, such as asthma, should limit prolonged outdoor exertion
Sensitive groups	101–150	Increased aggravation of heart and lung disease and premature mortality in persons with cardiopulmonary disease and the elderly; significant increase in respiratory effects in the general population	Active children and adults, and people with respiratory disease, such as asthma, should avoid prolonged outdoor exertion; everyone else, especially children, should limit prolonged outdoor exertion
Unhealthy	151–200	Increased aggravation of heart and lung disease and premature mortality in persons with cardiopulmonary disease and the elderly; significant increase in respiratory effects in the general population	Active children and adults, and people with respiratory disease, such as asthma, should avoid all outdoor exertion; everyone else, especially children, should limit outdoor exertion
Very unhealthy	201–300	Significant aggravation of heart and lung disease and premature mortality in person with cardiopulmonary disease and the elderly; significant increase in respiratory effects in the general population	Active children and adults, and people with respiratory disease, such as asthma, should avoid all outdoor exertion; everyone else, especially children, should limit outdoor exertion
Hazardous	301–500	Serious aggravation of heart and lung disease and premature mortality in person with cardiopulmonary disease and the elderly; serious risk of respiratory effects in the general population	Everyone should avoid all outdoor exertion
Beyond index	501 and greater	Extremely high levels and health risk	Take immediate steps to reduce exposure

Source: United States Department of State. 2024. *United States Department of State Air Quality Monitoring Program.* www.stateair.net/web/post/1/5.html. (Accessed April 16, 2024).

μg/m³ = micrograms per cubic meter.

In 2011, the WHO published minimum standards for water quality. The WHO Water Quality Standards are intended as a benchmark for those countries that have not established standards of their own and are considered minimum standards (WHO 2024b). Table 10.3 lists the WHO Water Quality Standards.

The WHO guidelines for drinking water are not mandatory limits. They are intended to provide a scientific point of departure for development of national or regional numerical drinking water standards for those countries that do not have standards or may have their own standards but are incomplete or, in some cases, may not be satisfactory. The WHO guidelines were established using the following criteria (WHO 2024b):

- The probability of exposure
- Chemical concentration likely to present risk of observable health risks
- Evidence of health effects and relative ease of control of the different sources of exposure

TABLE 10.3
WHO Water Quality Standards

Parameter	Standard (μg/l)	Parameter	Standard (μg/l)	Parameter	Standard (μg/l)
Acrylamide	0.5	Alachlor	20	Aldicarb	10
Aldrin and dieldrin	0.03	Antimony	20	Arsenic	10
Atrazine	100	Barium	700	Benzene	10
Benzo(a)pyrene	0.7	Boron	2,400	Bromate	10
Bromodichloromethane	60	Bromoform	100	Cadmium	3
Carbon tetrachloride	4	Carbofuran	7	Chlorate	700
Chlordane	0.2	Chlorine	5,000	Chlorite	700
Chloroform	300	Chlorotoluron	30	Chlorpyrifos	30
Chromium	50	Copper	2,000	Cyanzine	0.6
2,4-D	30	2,4-DB	90	DDT	1
Dibromoacetonrile	20	1,2,3-Chloropropane	1	1,2-Dibromoethane	0.4
Dibromochloromethane	100	Dichloroacetate	50	Dichloroacetonitrile	20
1,2-Dichlorobenzene	1,000	1,2-Dichloroethane	30	1,2-Dichloroethene	5-
1,4-Dichlorobenzene	300	Dichloromethane	20	Dichloroprop	20
1,2-Dichloroporpane	40	Dimethoate	6	1,4-Dioxane	50
1,3-Dichloropropene	20	Edetic acid	600	Endrin	0.6
Di(ethylhexyl)phthalate	8	Epichlorohydrin	0.4	Ethylbenzene	300
Fenoprop	9	Fluoride	1,500	Hydroxyatrazine	200
Hexachlorobutadiene	0.6	Isoproturon	9	Lead	10
Lindane	2	MCPA	2	Mecoprop	10
Mercury	6	Methoxychlor	20	Metolachor	10
Microcystin-LR	1	Molinate	6	Monochloroacetate	20
Nickel	70	Nitrate (as NO_3^-)	50,000	Nitrite (as NO_2^-)	3,000
Nitrilotriacetate acid	200	Pendimethalin	20	Pentachlorophenol	9
N-Nitrodimethylamine	0.1	Selenium	40	Simazine	2
Sodium dichloroisocyanurate	40,000	Styrene	20	2,4,5-T	9
Terbuthylazine	7	Tetrachloroethene	40	Simazine	2
Toluene	700	Trichloroacetate	200	Trichloroethene	20
Trifluralin	0.02	Trihalomethanes	1	Uranium	30
Vinyl chloride	0.3	Xylenes	500		

Source: World Health Organization (WHO). 2024b. *Guidelines for Drinking-Water Quality*. WHO. Geneva, Switzerland. 516pp.

μg/l = microgram per liter.

The WHO lists many other compounds, but it does not provide numerical values in the guideline due to mitigating factors such as geography, environmental, social, cultural, economic, dietary, and other conditions affecting potential exposure. Therefore, WHO recommends that it may be necessary to undertake a drinking water assessment before establishing standards. Chemical compounds and microbes not listed in Table 10.3 but, in many instances, should be evaluated on a case-by-case basis after a drinking water assessment is completed, include the following (WHO 2024a, 2024b):

- *E. coli* and other microbes
- Chlorothalonil
- Diazinon
- Fenamiphos
- Amitraz
- Cypermethrin
- Dinoseb
- Formothion
- Chlorobenzilate
- Deltramethrin
- Ethylene thiourea
- 4-chloro-o-tolyloxy butyric acid (MCPB)

- Hexachlorocyclohexanes
- Phorate
- Quintozene
- Tributyltin oxide
- Ammonia
- Beryllium
- Bromochloroacetate
- *Bacillus thuringiensis*
- 2-Chlorophenol
- Cyanogen chloride
- 1,1-Dichloroethene
- 1,3-Dichlorobenzene
- 1,3-Dichloropropane
- Di(2-ethylhexyl)adipate
- Formaldehyde
- Hexachlorobenzene
- Inorganic tin
- Malathion
- Methyl parathion
- Monobromoacetate
- Nitrobenzene
- Permethrin
- Petroleum products
- Propanil
- Sodium
- 1,1,1-Trichloroethane
- Trichloroactonitrile
- Methamidophos
- Propoxur
- Toxaphene
- Trichlorfon
- Asbestos
- Bromide
- Carbaryl
- Chlorine dioxide
- Chloropicrin
- Dialkyltins
- 1,1-Dichloroethane
- 2,4-Dichlorophenol
- Diflubenzuron
- Fenitrothion
- Glyphosate
- Heptachlor
- Iodine
- Manganese
- Methyl tertiary butyl ether (MTBE)
- Monochlorobenzene
- Navaluron
- pH
- Pirimiphos-methyl
- Pyriproxfen
- Spinosad
- Total dissolved solids (TDS)
- Trichlorobenzenes
- Oxamyl
- Pyridate
- Triazophos
- Aluminum
- Bentazone
- Chloride
- Chloral hydrate
- Chloroacetones
- Cyanide
- Dibromoacetate
- Dichloramine
- Diquat
- Endosulfan
- Fluoranthene
- Hardness
- Hydrogen sulfide
- Iron
- Methoprene
- Molybdenum
- Mxnes (MX)
- Parathion
- 2-Phenylphenol
- Potassium
- Silver
- Sulfate
- Trichloramine
- Zinc

10.3.3 Land

Land pollution is defined as a degradation or even destruction of the Earth's surface, soil, rock, or subsurface as a result of anthropogenic activities. The expansion of housing developments, businesses, industry, infrastructure, and agriculture all in response to an increasing human population over the last several decades accounts for humans modifying over 50% of the Earth's topsoil (UN 2024a). Land also becomes contaminated from the migration of contaminants through the soil column or when contaminants change media such those carried by the atmosphere, perhaps hundreds or even thousands of miles, and are deposited on the land or water, possibly in a different country.

The surface of the Earth is approximately 510,100,000 km^2 (Coble et al. 1987). Approximately 70% is covered by water or 361,800,000 km^2 and 30% is land or 148,300,000 km^2. The amount of land on Earth that is used for farming is estimated at 40% or 59,320,000 km^2 (UN 2024a). This is a huge number, especially because the amount of urbanized land on Earth is estimated to be only 2.7% and is home to roughly half the current human population (UN 2024a). The amount of land currently used for farming becomes an even larger value as 10.4% of Earth's land or 15,600,000 km^2 is covered by ice and is not farmable, 20% or 29,660,000 km^2 is mountainous, and another 20% or 29,660,000 km^2 is desert and has little or no topsoil (Coble et al. 1987). When factoring out those land areas that are not suitable for farming, only 7% of land remains available on Earth.

The WHO and United States Environmental Protection Agency (USEPA) (2012) have not established guidelines for soil or groundwater. However, most countries that have not developed standards for the remediation of land that includes soil, sediment, or groundwater commonly refer to a set of standards established by the Dutch in 1987 and were revised in 1994, 2001, and 2015, commonly referred to as the Dutch Intervention Values (DIVs) (Lijzen et al. 2001; Netherlands Ministry of Infrastructure and the Environment 2015). Table 10.4 presents a partial list of DIVs. A complete

TABLE 10.4
Dutch Intervention Values for Select Compounds for Soil and Groundwater

	Soil		Groundwater	
Compound	**Target Value (mg/kg)**	**Intervention Value (mg/kg)**	**Target Value (μg/l)**	**Intervention Value (μg/l)**
Arsenic (As)	29.0	55.0	10	60
Barium (Ba)	160	625	50	625
Cadmium (Cd)	0.8	12	0.4	6
Chromium (Cr)	100.0	380	1	30
Copper (Cu)	36.0	190	15	75
Nickel (Ni)	35.0	210	15	75
Lead (Pb)	85.0	530	15	75
Mercury (Hg)	0.3	10.0	0.05	0.3
Silver (Ag)	None	15	None	40
Selenium (Se)	0.7	100	None	40
Zinc (Zn)	140	720	65	800
Chloride	None	None	100,000	None
Cyanide	None	20	5	1,500
Cyanide (complex)	None	50	10	1,500
Thiocyanate	None	20	None	1,500
Benzene	None	1.1	0.2	30
Ethyl benzene	None	110	4	150
Toluene	None	320	7	1,000
Xylenes (sum)	None	17	0.2	70
Styrene (vinylbenzene)	None	86	6	300
Phenol	None	14	0.2	2,000
Cresols (sum)	None	13	0.2	200
Naphthalene	None	None	0.01	70
Phenanthrene	None	None	0.003	5
Anthracene	None	None	0.0007	5
Vinyl chloride	None	0.1	0.01	5
Dichloromethane	None	3.9	0.01	1,000
Trichloroethene (TCE)	None	2.5	24	500
Tetrachloroethene (PCE)	None	8.8	0.01	40
Pentachlorophenol	None	12	0.04	3
Polychlorinated biphenyl (PCB) (sum)	None	1	0.01	0.01
Monochloroanilines (sum)	None	50	None	30
Dioxin (sum)	None	0.00018	None	None
Chloronaphthalene (sum)	None	23	None	6
Chlordane	None	4	0.00002	0.2
DDT (sum)	None	1.7	None	None
DDE (sum)	None	2.3	None	None
DDD (sum)	None	34	None	None
DDT/DDE/DDD sum	None	None	0.000004	None
Aldrin	None	0.32	0.000009	None
Dieldrin	None	None	0.0001	None
Lindane	None	1.2	0.009	None
Atrazine	None	7.1	0.00029	150
Carbofuran	None	0.017	0.0009	100
Asbestos	None	100	None	None

Source: Netherlands Ministry of Infrastructure and the Environment. 2015. Dutch Pollutant Standards. www.government.nl/ministry-of-infrastructure-and-the-environment. or at www.sanaterre.com/guidelines/dutch.html. (Accessed April 16, 2024), 2015.

mg/kg = milligram per kilogram; μg/l = microgram per liter.

list can be viewed at www.sanaterre.com/guidelines/dutch.html (Lijzen et al. 2001; Netherlands Ministry of Infrastructure and the Environment 2015).

DIVs are based on potential risks to human health and ecosystems and are technically evaluated. The values presented usually have two integers for each chemical listed. One is a target level that is considered optimal and will likely result in limited or no health or ecological effect, and the other is a level or concentration above which some intervention is necessary to lower risk posed by site-specific exposure pathways. The DIVs have been used worldwide as a benchmark for cleanup of contaminated sites in many areas of the world where cleanup values have not been established. In addition, even if there are established values, the DIVs are used for comparison purposes. Therefore, in many instances, the DIVs represent the benchmark for cleanup standards worldwide.

10.3.4 Summary of Global Assessments and Standards

Although it may not be clearly noticeable at this point, there is a general lack of enforcement of international standards for dealing with pollution. As we explore the environmental regulations of select countries of the world, many countries struggle with meeting existing standards, in part, because enforcement is either absent or substandard. This is significant because, as we already know, pollution does not respect political boundaries. Next, we shall examine the continent of Antarctica. Antarctica is a continent with no permanent human settlements and the vast majority of the continent is covered in ice.

10.4 ANTARCTICA

Antarctica is a continent at the southernmost part of the world with a total land area of 14 million km^2 with 98% of its surface covered in ice. It shares no border with any other country (see Figure 10.3). There are no permanent settlements in Antarctica. In 1959, Antarctica was established as an international area dedicated to science and research. Several nations including Argentina, France, Japan, United Kingdom, and the United States signed the Antarctic Treaty in 1959. The treaty contains 14 articles which establish key elements related to the peaceful international coordination on the continent, some of which include the following (Antarctic and Southern Ocean Coalition 2024):

- Banning military intervention
- Peaceful scientific research
- International information exchange
- No territorial sovereignty

10.4.1 Pollution in Antarctica

Some pollution issues in Antarctica include the following (McConnell et al. 2014; National Aeronautics and Space Administration (NASA) 2014):

- Ozone depletion
- Increased atmospheric CO_2 concentrations
- Solid and hazardous waste generated from research activities and tourism
- Marine pollution from anthropogenic sources such as spills from ships and cargo vessels
- Litter
- Lead
- Invasive species such as dandelions

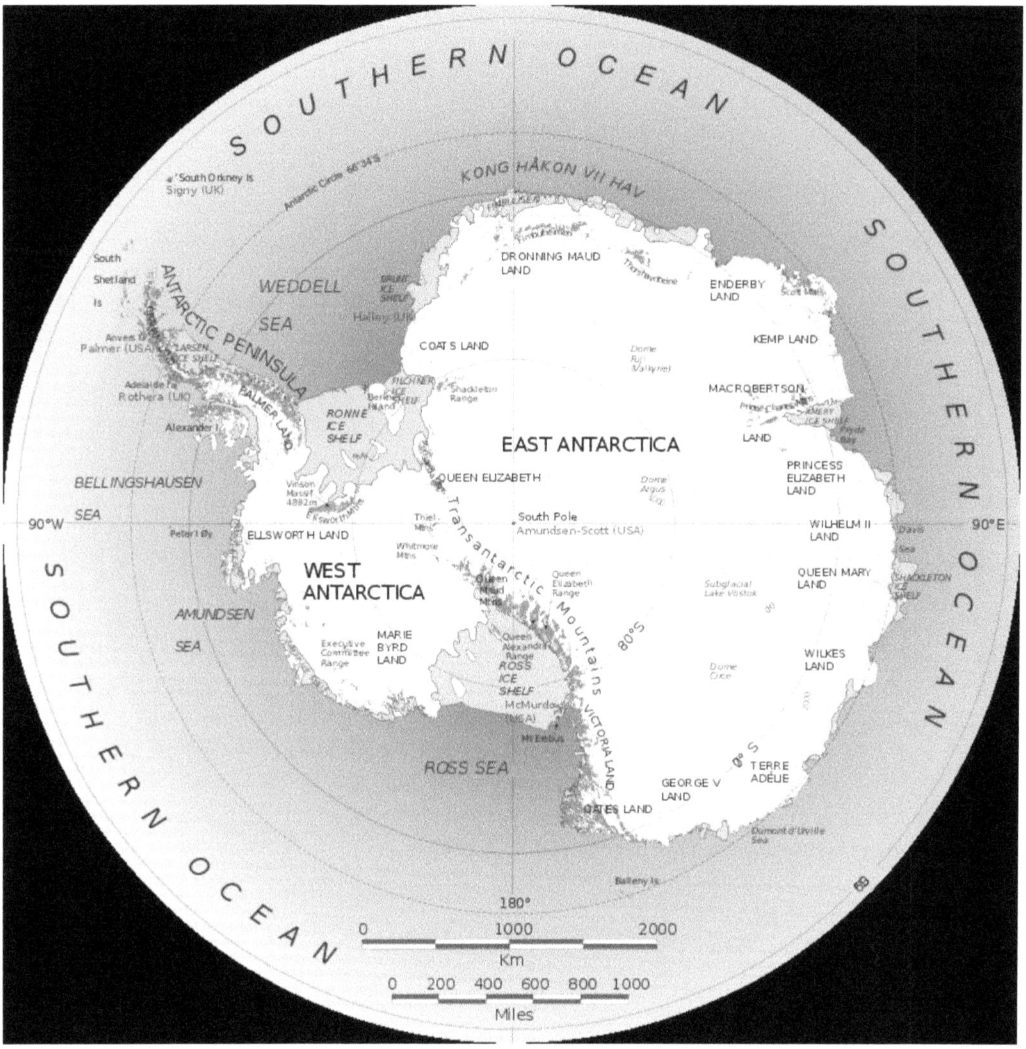

FIGURE 10.3 Map of Antarctica. (From National Aeronautics and Space Administration (NASA). 2024. *Map of Antarctica*. https://nasa.gov/images/contnet/600827main_24.jpg. (Accessed April 16, 2024).)

10.4.2 Environmental Regulations of Antarctica

Environmental regulations in Antarctica do exist and are traced back to the Antarctic Treaty of 1959. All activities in Antarctica must be conducted in accordance with the regulations set forth by the Norwegian Polar Institute. For any activity, the following must be conducted (Norwegian Polar Institute 2024):

- Notification of planned activity
- Preparation of an environmental impact assessment
- Preparation of contingency plans
- Obtaining insurance
- List of experience and equipment

- Preparation of environmental emergency plan
- Waste handling and disposal plan
- Provisions for preventing introduction of alien species
- Preparation of a final report

10.4.3 Summary of Environmental Regulations in Antarctica

It should not come as a surprise after reading this chapter that there are environmental impacts even in Antarctica from anthropogenic sources. This presents an interesting and perhaps conflicting phenomenon to some of you because Antarctica is predominantly devoid of human settlement. It is essentially a wilderness of continental size. So how can Antarctica be polluted, and if Antarctica is polluted what does that say about the rest of the Earth? Antarctica is perhaps the best example that pollution does not respect country borders. Evaluating Antarctica has placed a spotlight on the migration of pollution on Earth. One migration pathway that is common for pollution follows Fick's First Law of Molecular diffusion, which is in full display for us to observe and quantify in Antarctica.

10.5 OCEANS

As discussed in Chapter 7, oceans cover 70% or 360 million km^2 of the Earth's surface. In many respects, the oceans of the world are the last resting places for most anthropogenic pollution. As an example, an estimated 7 million metric tons of plastic is dumped into the world's oceans every year (NOAA 2024c). This is just a portion of the pollution that enters the oceans every year. NOAA (2024c) estimates that billions of pounds of trash and other pollutants enter the oceans each year. Figure 10.4 presents an example of ocean pollution washing ashore in Hawaii. The photograph depicted in Figure 10.4 is especially disturbing and presents some important information concerning the amount of pollution in the ocean as Hawaii is located in the geographic center of the Pacific Ocean and more than 2,500 km from a continent.

FIGURE 10.4 Plastic and other garbage on a beach in Hawaii. (From National Oceanic and Atmospheric Administration (NOAA). 2024c. *Ocean Pollution*. www.noaa.gov/resource-collections/ocean-pollution. (Accessed April 16, 2024).)

10.5.1 Ocean Pollution

The oceans have been the dumping ground not only for our garbage but also for some of our most toxic wastes for hundreds of years. The oceans have been treated as the human privy and garbage disposal site. Only in the past few decades have the dumping of wastes in the oceans come under some environmental regulations. However, polluting the oceans continues to this day. The oceans receive pollution from direct discharge or dumping, deposition from the atmosphere, rivers and streams that discharge into the oceans, and directly from the land itself.

The oceans were once believed to be so vast and deep that the effects of pollutions would never directly threaten the oceans. This has been proven false. The rate of accumulation of pollution is proportional to human population and increasing energy requirements. As the population increases, the more pollution ends up in the oceans. The most common types of pollution in the oceans include plastic and other forms of garbage, heavy metals, organic pollutants such as polychlorinated biphenyls (PCBs) and oil, radioactive substances and waste, biological pollutants, pesticides, herbicides, fertilizers, heavy metals, and untreated raw sewerage (NOAA 2024c; Natural Resources Defense Council (NRDC) 2024).

Pollution in the oceans does not all originate from direct discharge but also is delivered through the atmosphere from increased concentrations of carbon dioxide in the atmosphere that is partially adsorbed by the oceans and causes acidification of the surface layers (NOAA 2017). Absorption of carbon dioxide by the oceans has been estimated to be as high as 25%, which represents millions of tons of carbon dioxide absorbed by the oceans every year. The additional carbon dioxide absorbed by the oceans reacts with ocean water and forms an acid called **carbonic acid** and is lowering the pH of the oceans worldwide. In fact, the oceans are acidifying faster than they have in the past 300 million years. The effect of the acidification has been linked to bleaching of coral reefs, including the Great Barrier Reef off the northeastern coast of Australia (Natural Resource Defense Council 2024; UN 2024d).

The following is a list of anthropogenic impacts on the oceans that are currently known to the scientific community (USEPA 2024a and NOAA 2024c):

- Habitat destruction
- Ocean acidification
- Ocean heating
- Nitrogen enrichment
- Chemical pollution
- Plastic pollution
- Garbage disposal
- Destructive fishing
- Surface runoff
- Deoxygenation
- Deep sea mining
- Noise pollution
- Skincare chemical pollution

10.5.2 Environmental Regulations of the Ocean

The United States has enacted several laws over the last several decades aimed at protecting the Great Lakes and the oceans from pollution. Federal laws addressing waterways were first enacted in the United States as far back as 1899. The Act passed in 1899 and others that address waterways include the following (USEPA 2024b):

- Rivers and Harbors and Refuse Act of 1899
- Coastal Zone Protection Act of 1972

- Fisheries Conservation and Management Act of 1976
- Global Climate Protection Act of 1987
- Marine Protection Act of 1972
- Marine Mammal Protection Act of 2015
- Ocean Dumping Act of 1988
- Oil Spill Protection Act of 1990

10.5.3 Summary of Environmental Regulations of the Oceans

For centuries, the oceans have been governed by the freedom of the seas doctrine which is a principle recognized in the 17th century that limited national rights and jurisdiction over the oceans to a narrow belt of sea, typically 19.3 km, surrounding a nation's coastline (UN 2024e). Since 1982, the UN has been actively involved in establishing protections for the oceans through the adoption of the Law of Sea Convention which established the following (UN 2024f):

- Freedom of navigation rights
- Set territorial sea boundaries 19.3 km offshore
- Set exclusive economic zones of up to 321 km offshore
- Set rules for extending continental shelf rights up to 563 km offshore
- Created the International Seabed Authority
- Created other conflict-resolution mechanisms

The UN is also involved in protecting the marine environment and biodiversity and reducing pollution from marine shipping, especially shipping in the polar regions with the adoption of the Polar Code in 2014. As of 2022, nearly 100,000 commercial shipping vessels that carry cargo were registered worldwide, transporting more than 10 billion metric tons of cargo (UN 2024f).

10.6 EVALUATING ENVIRONMENTAL PROTECTION OF THE WORLD

Environmental protection on a global scale is not an easy task and, for the most part, does not exist. Each country desires its own autonomy and responsibility to create and enforce protecting human health and the environment within their borders. This has led to many inconsistencies and disruptions, mainly because pollution does not respect country borders. In response to this, the UN and the WHO have developed minimum standards. Each country has the option to establish lower standards and create new standards if they so desire.

In this section, we will limit our discussion to how each country focuses on the fundamentals of environmental protection, which include the following:

- Protecting and improving air quality
- Protecting and improving water quality
- Protecting and improving the land
- Protecting and improving habitat and other factors influencing the health of living organisms

We will compare how environmental protection of these countries differs and examine how they compare to the United States. The countries to be evaluated are listed in Table 10.5.

Some findings could be rather surprising, in that it will become apparent that some countries are more advanced compared to the United States in how they have chosen to organize and implement their environmental protection and where they focus their internal resources. In addition, the governments of some countries have different attitudes and relationships with industry.

TABLE 10.5
Countries and Areas in Which Protecting the Environment through Regulations Are Examined

North America	Europe and the European Union
Canada	Austria
Mexico	Belgium
United States	Bulgaria
	Croatia
South America	Cyprus
Argentina	Czech Republic
Brazil	Denmark
Peru	Estonia
Chile	Finland
Asia	France
China	Germany
India	Greece
Japan	Hungary
Saudi Arabia	Ireland
South Korea	Italy
Indonesia	Latvia
Malaysia	Lithuania
	Luxemburg
Oceania	Malta
Australia	The Netherlands
New Zealand	Norway
Africa	Poland
Egypt	Portugal
Kenya	Romania
Tanzania	Russia
South Africa	Slovakia
	Slovenia
Antarctica	Spain
	Sweden
The Oceans	Switzerland
	Turkey
	United Kingdom

The environmental agencies of some countries have chosen to form working partnerships with industry with the intent to work more closely together to solve complex environmental issues rather than an adversarial type of approach which much too often results in polarizing points of view. The result of working together saves time, effort, and resources and, in the long run, has resulted in some counties being better at environmental protection than the United States.

Surprising still might be that some countries lack either key environmental protections or the will to enforce the environmental protections they have due to other social or economic challenges or the political realities they face. One overriding difficulty many developing countries face is lack of infrastructure, training, and education to properly implement their own environmental policies and protections. Another overriding factor is climate and climate change which, in some instances, has considerably increased the negative effects of pollution on the environment and humans.

10.6.1 Evaluating Environmental Regulations of the World

We have now completed a review of environmental regulations of more than 50 countries and every continent including a brief assessment of Antarctica and the oceans. With respect to pollution, we have learned that the following points are perhaps the most significant:

- No country, continent, or ocean is pollution free
- Pollution caused by humans is everywhere
- Significant environmental degradation caused by humans has affected every country
- Pollution does not respect country borders
- Pollution will continue to migrate in the air, water, and on land
- The oceans have been treated as the human privy and garbage disposal site of the world

With respect to environmental regulations, the most significant information we have learned include the following:

- Each country has environmental regulations
- No two countries are exactly the same
- Each country has unique attributes whether it be related to its climate, geography, geology, social and economic concerns, or politics which affect the ability of a country to protect human health and the environment that influence many aspects of environmental regulations
- The platform of environmental regulations of each country is similar to that of the United States or the European Union
- Although some countries are doing a better job than others, each country struggles with protecting human health and the environment for various reasons, two of which seem to be universal, that is, a growing population and urban expansion
- Every country has been significantly affected by pollution of the air, water, and land

The largest polluting activity of land worldwide is agriculture. The largest polluter of air worldwide is motor vehicles that include automobiles, diesel trucks, trains, buses, marine vessels, and aircraft. The largest polluter of water worldwide is biological pollutants largely from human waste and from agricultural activities that include pesticides, herbicides, fertilizers, and erosion (UN 2024a).

A disturbing fact concerning pollution is that the final resting places for much of the pollution of the world are the oceans.

10.6.2 Evaluating and Ranking Each Country's Regulatory Effectiveness

To evaluate the effectiveness of each country's environmental regulations, we must examine several factors and assigned integers that appropriately reflect where each country is in the process and to provide an enlightened point of view as to where each country can improve.

This evaluation involves assessing several attributes that we have discussed and highlighted throughout this chapter and include whether each country:

1. Has the appropriate environmental regulations for air, water, and land
2. Has appropriately implemented and enforced environmental regulations
3. Has the appropriate infrastructure to support the implementation and enforcement of environmental regulations
4. Has the economic and financial resources to support continued pollution controls
5. Has other social issues that impede or override environmental concerns such as poverty, corruption, violence, unstable political issues, population, and urban constraints

6. Has sustainability initiatives including pollution prevention, recycling, and waste reduction
7. Has geographical, geological, climate factors, and other natural conditions such as natural disasters that impede environmental performance

To make this evaluation, we must first consider how each country has chosen to protect the air, the water, and the land under the first criterion listed above. Therefore, the regulations that are most important are associated with protecting the air, water, and land and include the following:

- Air quality regulations from both stationary and mobile sources
- Water quality regulations for effluent discharges from industrial sources and public-owned treatment works (POTW) and septic tank discharges
- Drinking water quality regulations
- Solid and hazardous waste regulations relating to characterization, generation, transport, and disposal
- Soil, water, and sediment cleanup regulations including investigation and polluters pay principles
- Regulating different land-use designations such as residential, commercial, industrial, and agricultural

The remaining criteria (those listed as 2 through 7 explained above) are straightforward qualitative evaluations. Table 10.6 presents the assigned values for each criterion for each country and are ranked from most effective to least effective. Each criterion in Table 10.6 corresponds to the list of 1 through 7 described above. Each criterion is assigned an integer of 0 through 4 that corresponds to the following qualitative assessment category:

- A value of zero (0) indicates that country lacks appropriate controls
- A value of one (1) indicates that the country has poor controls
- A value of two (2) indicates that the country has adequate or average controls
- A value of three (3) indicates that the country has good controls

For comparison purposes, the per capita gross domestic product (GDP) is listed for each country (United States Central Intelligence Agency 2024) and a ratio of GDP to the total score of each country is presented in the last column (World Bank 2024). As shown by the analysis of Table 10.6, there should be very little surprises when each country is ranked. The countries with the most room for improvement include Turkey, Egypt, Malaysia, Russia, Saudi Arabia, and India. These countries have a host of challenges that include lack of regulations or enforcement of existing regulations, lack of infrastructure, financial hardship, population pressures, and may have to deal with corruption issues.

The countries in the middle that include Mexico, Indonesia, Brazil, Argentina, South Africa, Kenya, Tanzania, Peru, China, and Chile are developing nations that are struggling to learn and enforce their environmental regulations, have population pressures, lack infrastructure, and are struggling to deal with cleanup of existing sites of contamination.

The countries that rank the highest include Japan, United States, Canada, Australia, New Zealand, European Union, Switzerland, Norway, and South Korea, which are all developed nations with financial resources to address environmental issues but still struggle with issues such as population and urbanization, as well as air and water quality, but have all made significant progress compared to other nations.

Examining the final column in Table 10.6 provides perhaps the most significant information. The countries that rank the highest have a ratio of GDP to total score of less than 2.5.

The countries that have the most room for improvement all have a ratio greater than 10, except for Saudi Arabia. Saudi Arabia ranks poorly for environmental controls but high for per capita GDP,

TABLE 10.6
Country Comparison of Environmental Regulation Effectiveness

Country	Criteria 1	2	3	4	5	6	7	Total Score	GDP Rank	GDP/Score
Turkey	1	1	0	0	0	0	2	4	69	17.25
Egypt	1	1	1	0	0	1	1	5	105	21
Malaysia	1	1	0	1	1	0	1	5	64	16
Russia	1	1	1	1	1	0	1	6	65	10.8
Saudi Arabia	1	1	1	1	1	0	1	6	20	3.3
India	1	1	0	1	1	1	1	6	132	22
Mexico	1	1	2	1	1	1	1	8	82	10.25
Indonesia	2	1	1	2	1	1	1	9	108	12
Brazil	1	1	1	2	1	1	2	9	95	10.5
Argentina	1	1	1	2	1	1	2	9	95	10.5
South Africa	2	1	2	1	1	2	1	10	103	10.3
Kenya	2	2	1	1	1	1	2	10	154	15.4
Tanzania	2	2	1	1	1	1	2	10	156	15.6
Peru	1	1	1	2	2	1	2	10	104	10.4
China	2	1	2	3	2	1	2	13	94	7.2
Chile	2	2	2	2	3	1	1	13	74	5.7
Japan	3	3	3	3	2	2	1	17	38	2.2
United States	2	3	3	3	2	3	2	18	19	1.05
Canada	2	3	3	3	3	2	2	18	32	1.7
Australia	2	3	3	3	3	2	2	18	27	1.5
New Zealand	3	3	3	3	3	2	2	19	44	2.3
European Union	3	3	3	3	3	3	2	20	43	2.1
Switzerland	3	3	3	3	3	3	2	20	17	0.85
Norway	3	3	3	3	3	3	2	20	12	0.6
South Korea	3	3	3	3	2	3	3	20	42	2.1

which indicates that Saudi Arabia has financial resources to improve its environmental controls but chooses to focus its financial resources elsewhere. Egypt and India are the only countries evaluated that have a total score to GDP ratio greater than 20.

China and Chile are unique when examining their ranking and numerical scores. China and Chile rank in between the low performance and high performance. The fact that China has been improving its environmental controls indicates that the country is transitioning from low to high. Chile appears to be currently more stagnant.

10.6.3 Environmental Challenges of Each Country

Some of the challenges each country faces, which have significantly influenced the score of each country as described in this chapter, include the following:

- Turkey is currently facing financial and economic hardship along with a lack of robust environmental standards and enforcement.
- Egypt is experiencing political unrest, a growing population and urbanization, lack of infrastructure, and water shortages.
- Malaysia lacks infrastructure, environmental enforcement, financial resources, and rapid population growth.

- Russia lacks infrastructure, environmental enforcement, financial resources, political framework, and struggles with corruption.
- Saudi Arabia lacks robust regulations and enforcement, some lack of infrastructure but does have financial resources.
- India has large population and poverty pressures, lacks infrastructure, and enforcement of regulations.
- Mexico lacks infrastructure, especially with water treatment, and struggles with corruption and lack of complete regulations (i.e. stormwater) and enforcement struggles.
- Indonesia has pressure from a large population, newer regulations, deforestation, and lack of infrastructure.
- Brazil has population pressure, especially in urban areas, lacks infrastructure, lack of progress in cleaning up contaminated sites, water treatment, deforestation, and corruption.
- Argentina lacks infrastructure which places pressure on enforcement, financial resources, water treatment, and cleanup of contaminated sites.
- South Africa has a degree of political unrest and financial pressures, climate change, and water issues.
- Kenya lacks financial and infrastructure resources.
- Tanzania lacks financial and infrastructure resources.
- Peru has lack of infrastructure to deal with water contamination issues, especially from mining and lacks financial resources.
- China lacks enforcement measures and has population challenges and infrastructure to deal with pollution control measures, especially in the more undeveloped western regions.
- Chile has urban population challenges and water issues in the desert regions of the northern portion of the country.
- Japan has geographical and geologic challenges (i.e. island nation that are prone to earthquakes), population pressure, and lack of landfill space.
- United States faces lack of regulations of agricultural areas, significant chemical use of pesticides and herbicides, erosion, deforestation, and pollution from septic systems, and invasive species.
- Canada lacks some stormwater controls, and faces habitat destruction from mining, deforestation, and petroleum development.
- Australia faces future water issues, population pressure, and habitat destruction, especially the Great Barrier Reef.
- New Zealand faces water quality challenges, especially from agriculture and livestock sources.
- The European Union faces continued challenges from cleaning up historical sites of contamination and population pressure but has done well with regulations and sustainability measures.
- Switzerland has erosion challenges and pollution from agriculture and livestock.
- Norway faces climate change challenges and pollution from agriculture and livestock.
- South Korea faces population pressures and air quality challenges, especially in its capital city of Seoul.

A final point to consider is the objectivity of the evaluation conducted through the selection of countries included in this chapter. As stated at the beginning of this chapter, the countries selected include 70% of the human population and 75% of the land area. The countries not selected include countries such as Somalia, Algeria, Libya, Sudan, Democratic Republic of Congo, Angola, Zimbabwe, Botswana, Mozambique, Congo, Syria, Iran, Pakistan, Afghanistan, Venezuela, and many former Soviet-bloc countries. These countries were not evaluated because of the more severe issues they are experiencing, such as social, economic, poverty, sickness, and violence challenges, which has degraded the natural environment and directly threaten human health. Therefore, when considering environmental degradation and threats to human health and the environment from pollution, the countries examined in this chapter tend to be biased toward those with better environmental performance.

10.7 SUMMARY AND CONCLUSION

Antarctica and the oceans have and will play a significant role concerning measuring future sustainability efforts and survival of our species and many other species on Earth. We have learned thus far that Antarctica and the oceans have become polluted from direct discharge and from migration of contaminates through the air, water, and land. We have also learned that pollution migrates following fundamental laws of physics, and this is evident by examining the impacts observed in Antarctica and the Oceans. We have also learned that pollution does not respect country borders.

The oceans were the origin of life on Earth and remain the prime source of nutrients and food that sustains life on our planet but have become polluted by our own actions. If not for any other reason than our own survival, we must be diligent at assessing the risks posed by pollutants, enact appropriate protections, and act on those protections to prevent further degradation on a global scale.

To summarize this chapter, we must now accept the reality that humans have adversely impacted the entire Earth and that our efforts to improve our environment since the enactment of environmental regulations in the United States and worldwide have simply not been enough. Many Earth scientists are now convinced that out of this fact, we have now moved the needle of geologic time into a new proposed period called the **Anthropocene** (Smithsonian 2021; Zalasiewicz et al. 2018; International Union of Geological Sciences (IUGS) 2016). This is significant because the definition of a Geologic Age is a time period that affects the entire Earth and will be recorded in Earth history and cannot be reversed.

REFERENCES

Antarctic and Southern Ocean Coalition. 2024. *Antarctic Environmental Protection*. www.asocorg/campaign/protecting-antarctica. (Accessed April 16, 2024).

Baum, R., Luh, J., and Bartman, J. 2013. Sanitation: A Global Estimate of Sewerage Connections without Treatment and the Resulting Impact. *Journal of Environmental Science and Technology*. Vol. 47, no. 4, pp. 1994–2000.

Coble, C. R., Murray, E. G., and D. R. Rice. 1987. *Earth Science*. Prentice-Hall Publishers. Englewood Cliffs, New Jersey. 502pp.

International Union of Geological Sciences (IUGS). 2016. *The Anthropocene Epoch: Adding Humans to the Chart of Geologic Time*. International Geological Congress. Cape Town, South Africa. www.35igc.org/verso/5/scientific-programme. (Accessed April 16, 2024).

Lijzen, J. P., Baars, A. J., Otte, P. F., Rikken, M. G., Swartjes, F. A., Verbruggen, E. M., and van Wezel, A. P. 2001. *Technical Evaluation of the Intervention Values for Soil/Sediment and Groundwater*. Institute of Public Health and the Environment. The Netherlands. 147pp.

McConnell, J. R., Maselli, O. J., Sigl. M., Vallelongo, P., Neumann, T. Anschutz, H., Bales, R. C., Curran, M. A., Edwards, R., Das, S. B., Kipfstuhl, S., Layman, L., and Thomas, E. R. 2014. Antarctic-Wide Array of High-Resolution Ice Core Records Reveals Pervasive Lead Pollution Began in 1889 and Persist Today. *Scientific Reports*. Vol. 4, p. 5848.

National Aeronautics and Space Administration (NASA). 2014. *Lead Pollution Beats Explorers to the South Pole*. https://ntrs.nasa.gov/citations/0150001454. (Accessed April 16, 2024).

National Aeronautics and Space Administration (NASA). 2024. *Map of Antarctica*. https://landsat.gsfc.nasa.gov/article/most-accurate-map-of-antarctica-landmass-made-from-landsat-data. (Accessed April 16, 2024).

National Oceanic and Atmospheric Administration (NOAA). 2017. *DSCOVR: Deep Space Climate Observatory*. www.nesdis.noaa.gov/current-satillite-missions/currently-flying/discovr-deep-space-climate-bersatory. (Accessed April 16, 2024).

National Oceanic and Atmospheric Administration (NOAA). 2024a. *Koppen–Geiger Climate Changes*. https://sos.noaa.gov/datasets/koppen-geiger-climate-changes. (Accessed April 12, 2024).

National Oceanic and Atmospheric Administration (NOAA). 2024b. *The Highs and Lows of Climate*. www.climate.gov/news-features/logs/beyond-data/highs-and-lows-of-climate. (Accessed April 13, 2024).

National Oceanic and Atmospheric Administration (NOAA) 2024c. *Ocean Pollution*. www.noaa.gov/resource-collections/ocean-pollution. (Accessed April 16, 2024).

Natural Resources Defense Council (NRDC). 2024. *Ocean Pollution: The Dirty Facts*. https://nrdc.org/stories/ocean-pollution-dirty-facts. (Accessed April 24, 2024).

Netherlands Ministry of Infrastructure and the Environment. 2015. *Dutch Pollutant Standards*. www.scribd.com/document/346050279/Dutch-pollutant-standards t or at www.sanaterre.com/guidelines/dutch.html. (Accessed April 16, 2024).

Norwegian Polar Institute. 2024. *Regulations for Activities in Antarctica*. www.nplar.no/en/regulations/the-antarctic. (Accessed April 16, 2024).

Smithsonian. 2021. *The Age of Humans: Living in the Anthropocene*. Smithsonian Statement on Climate Change. www.si.edu/newsdesk/releases/smithsonian-statement-climate-change. (Accessed April 16, 2024).

United Nations (UN). 2024a. *Human Development Report*. United Nations Development Programme. https://hdr.undp.org. (Accessed April 11, 2024).

United Nations (UN). 2024b. *Global Drinking Water Quality Index and Development and Sensitivity Analysis Report*. UNEP Water Programme Office. Burlington, ON, Canada. 58pp.

United Nations (UN). 2024c. *World Air Pollution Status*. United Nations News Center. New York, NY. www.unep.org/interactive/air-pollution-note. (Accessed April 13, 2024).

United Nations (UN). 2024d. *Great Barrier Reef: UNESO World Heritage Site*. https://news.un.org/en/story/2021/06/1094512. (Accessed April 13, 2024).

United Nations (UN). 2024e. *How Is Climate Change Impacting the World's Ocean*. www.un.org/en/climatechange/science/climate-issues/ocean-impacts. (Accessed April 25, 2024).

United Nations (UN). 2024f. *Oceans and the Law of the Sea*. www.un.org/en/sections/issues-depth/oceans-and-law-sea. (Accessed April 16, 2024).

United States Central Intelligence Agency. 2024. *The World Fact Book*. www.cia.gov/library/publications/the-world-factbook.gos/NA.html. (Accessed April 16, 2024).

United States Department of State. 2024. *United States Department of State Air Quality Monitoring Program*. www.stateair.net/web/post/1/5.html. (Accessed March 31, 2021).

United States Environmental Protection Agency (USEPA) 2024a. *Climate Change Indicators: Ocean Heat*. www.epa.gov/climate-indicators/cliamte-change-indicators-ocean-heat. (Accessed April 25, 2024).

United States Environmental Protection Agency (USEPA). 2024b. *Laws that Protect Our Oceans*. www.epa.gov/beach-tech/laws-protect-our-oceans. (Accessed April 16, 2024).

World Bank. 2024. *Global Economic Prospects: The Turning of the Tide?* World Bank Corp. Washington, DC. 184pp.

World Health Organization (WHO). 2013. *Water Quality and Health Strategy 2013–2020*. www.who.int/water_sanitation_health/dwq/en/. (Accessed April 13, 2024).

World Health Organization (WHO). 2024a. *Ambient Air Pollution Database May 2014*. www.who.org/ambient-air-database. (Accessed April 13, 2024).

World Health Organization (WHO). 2024b. *Guidelines for Drinking-Water Quality*. WHO. Geneva, Switzerland. 516pp.

World Health Organization (WHO). 2024c. *Progress on Drinking Water and Sanitation*. WHO. New York, NY. 90pp.

World Health Organization (WHO). 2024d. *Nine Out of Ten People Breathe Unhealthy Air*. WHO. Geneva, Switzerland. www.who.int/news-room/detail/02-02-2018. (Accessed April 13, 2024).

World Health Organization (WHO). 2024e. *World Health Organization Global Urban Ambient Air Pollution Database*. www.who.int/phe/health_topics/outdoorair/database.htm. (Accessed April 13, 2024).

World Health Organization (WHO). 2024f. *Air Pollution in China: Sources and Amounts*. www.who.org/china-air-pollution. (Accessed April 24, 2024).

World Health Organization (WHO) and United States Environmental Protection Agency (USEPA). 2012. *Animal Waste, Water Quality and Human Health*. IWA Publishing. London, United Kingdom. 489pp.

Zalasiewicz, J. Waters, C., Summerhayes, C., and Williams, M. 2018. The Anthropocene. *Geology Today*. Vol. 34, no. 5, pp. 182–187.

Part III

Risk Mitigation Methods and Sustainability Measures

11 Risk Impact Mitigation Methods and Sustainability

11.1 INTRODUCTION

This chapter is the beginning of the last section of this book and will focus on introducing the concept of sustainability and its connection to geology. The first section of this book focused on the geological and natural environment and the inherent risks associated with an ever-changing and dynamic Earth. The second section examined the many ways human activities have negatively impacted Earth. Now we turn to the last section, which will combine the geological environment with the human-developed world to discover what human actions are necessary to reduce or eliminate the harm that humans inflict on nature and what actions are necessary to protect humans from geologic risks.

We know that we have little to no control over preventing a geologic or environmental hazard from occurring, but we do have control over minimizing or preventing anthropogenic-induced hazards from inflicting harm on us and the natural world. A cartoon that was created by Walt Kelly during the environmental movement perhaps stated anthropogenic-induced issues best, **"We have met the enemy, and he is us"** (Ohio State University 2024).

The first section of this chapter will focus on what mitigation actions are necessary to lower the risks posed by geologic hazards, environmental hazards, and anthropogenic-induced hazards. The second section of this chapter is dedicated to introducing sustainability.

11.2 Risk Impact Mitigation Methods

The risk posed by a hazard is never zero. Humans do not have the means to prevent an earthquake from occurring or preventing a volcanic eruption, hurricane, or tornado. Therefore, the focus is on reducing the impacts to humans through mitigation methods. Risk mitigation methods focus on three key elements. The first element focuses on actions that can be undertaken before the risk of the hazard occurs. The second key element focuses on survival. The third key element focuses on actions taken after the hazard occurs. Risk mitigation focuses on understanding the mechanisms, assessing the potential impacts, identifying areas of highest risk, and designing effective strategies to reduce or eliminate the risks.

One commonality that is recommended in most every type of hazard mitigation is geologic mapping. This follows a logical scientific and geologic principle discussed in Chapter 2 and is known as the Principle of Uniformity. It simply states what has occurred in geologic history will likely occur in the future. For example, if geologic mapping near a volcano identifies evidence that

DOI: 10.1201/9781003463153-14

several lahars have occurred at a certain location, the probability that another lahar will occur at the same location is high.

As stated above, humans do not have the means to prevent a geologic or environmental hazard from occurring. However, humans can undertake measures to minimize the risk posed by geologic and environmental hazards. It is a different story when it comes to anthropogenic-induced hazards. Humans can prevent many or all anthropogenic-induced hazards by changing our behavior.

11.2.1 Risk Mitigation Methods for Geologic Hazards

As we explored and described in the first section of this book, geologic risks or hazards, such as earthquakes, volcanic eruptions, landslides, and coastal hazards, pose significant threats to life, property, and the environment. The impacts of geologic hazards are greatly amplified when they occur in densely populated areas. Geologic hazards have been responsible for millions of human casualties and billions of dollars in economic damage worldwide. As an example, the earthquake in Haiti in 2010 resulted in an estimated 230,000 deaths and cost an estimated $8 billion (United States Geological Survey (USGS) 2024a).

Mitigation strategies are crucial to minimize the devastating effects of natural disasters caused by geologic hazards. Geologic hazard mitigation focuses on understanding the mechanisms, assessing the potential impacts, and designing effective strategies to reduce or eliminate the risks. This includes developing early warning systems, land-use planning, and engineering solutions such as constructing earthquake-resistant buildings. The success of these mitigation strategies depends largely on the incorporation of scientific knowledge into policymaking and the active participation of communities at risk (USGS 2024b).

11.2.1.1 Geologic Hazard Mitigation Steps

This section focuses on living with geologic hazards and describes what actions can be taken to minimize the effects once the geologic hazard is realized. Mitigation actions fall into three categories which include actions that are conducted before the disaster, during the disaster, and after the disaster. Most of the recommended mitigation steps apply to most hazards.

Mitigation strategies for geologic hazards should be conducted in a step-by-step process that includes the following (United Nations 2024a):

1. Collecting and Analyzing Date
2. Risk Assessment and History
3. Analyzing the Range of Geological Hazard Risk
4. Creating the Conceptual Site Model or Risk Map
5. Identifying Areas of Highest Risk
6. Developing an Early Warning System
7. Monitoring
8. Communicating Alerts and Warnings
9. Emergency Planning and Preparedness
10. Establishing an Evacuation Plan
11. Education and Awareness
12. Testing and Training
13. Emergency Supplies
14. Critical Response Effectiveness Review

Each of the mitigation strategies listed above is described in greater detail below.

11.2.1.2 Collecting and Analyzing Data

As with any scientific investigation, collecting and analyzing data is the most important step because future decisions are based on collecting enough data and obtaining the correct data. This information is typically a combination of several types of data that may include chemical and physical properties through direct sampling, seismic data, ground deformation, gas emission, hydrological changes, and most importantly, geologic mapping.

To ensure the integrity of the data, proper adherence to quality control and quality assurance methods are conducted. Two principles that represent the umbrella under which scientific processes should operate are (1) promoting a culture of integrity and (2) using evidence-based interpretation. From these two principles, 10 best practices are recommended to develop and maintain scientific integrity and include the following (National Center for Biotechnology Information 2024):

1. Universal training in scientific methods
2. Maintain strong oversight
3. Reproducibility of data
4. Develop and implement educational tools
5. Establish transparency in methodology
6. Strengthen the peer review process
7. Do not dismiss unanticipated findings
8. Seek consistency across scientific disciplines
9. Design rigorous and comprehensive evaluation criteria
10. Reward the highest levels of integrity

For geological and environmental investigations in urban areas, the following are examples of evaluating the scientific quality of data collecting with consideration of the 10 best practices listed above (United States Environmental Protection Agency (USEPA) 2024a):

- *Initial demonstration of capability:* Demonstrate at a high level that the investigation team is fully qualified and prepared.
- *Method detection limit studies:* Evaluate whether the detection limits required can be achieved given the sample characteristics and potential interferences caused by sample contamination or sample dilution requirements.
- *Blind performance evaluation samples:* Including blind samples within the data set to evaluate method integrity and reproducibility.
- *Trip blanks:* Insert trip blanks within sample containers to evaluate any negative influence on sample integrity from the point of collection to the testing location.
- *Equipment blanks:* Prepare equipment blanks to evaluate the potential for cross-contamination introduction from sampling equipment
- *Instrument blanks:* Testing samples through analytical instruments to ensure there is no carry-over between samples
- *Laboratory method blanks:* Method blanks are used to evaluate the status and cleanliness of analytical instruments.
- *Matrix spikes:* Matrix spikes are samples with a known analyte and known concentration and are used to evaluate the accuracy and precision of analytical instrumentation.
- *Laboratory duplicates:* Running duplicate samples provides data on reproducibility and instrument accuracy.
- *Sample duplicates:* Sample duplicates are used to evaluate sample variation. Sample duplicates are important in evaluating outliers in data sets and also provide valuable information and support confirmation of analyte distribution concentrations.

When collecting geologic and environmental samples in the field, items to constantly consider include the following (USEPA 2024a):

- *Precision:* Test–retest reliability
- *Accuracy:* Comparison of results to consensus value
- *Representativeness:* Does the sample represent the population of samples?
- *Completeness:* Data of acceptable quality and quantity
- *Comparability:* Does the data compare to other data sets?

11.2.1.3 Risk Assessment and History

In assessing and mapping risks, the initial step includes analyzing the risk of the hazard. This involves evaluating available historical data on the risk being investigated to define the investigation area and obtain rough data on the frequency of occurrence and potential severity of impacts. This step influences the decision on the level of effort necessary to adequately address risk and establish boundary conditions.

11.2.1.4 Analyzing the Range of Geological Hazard Risk

The analysis of geological and environmental hazards/risks involves a systematic evaluation of the potential harm or loss that could occur. Therefore, this step focuses on identifying ecological and anthropogenic locations of highest risks and vulnerability to negative impacts.

11.2.1.5 Creating the Conceptual Site Model or Risk Map and Modeling

Building upon the data set collective to this point, the creation of a preliminary conceptual site model is prepared to assess data gaps where additional data is necessary and begin to test the usefulness and integrity of the data set. This step typically includes information collected from historical sources, geographic information system (GIS) technology, surveys, satellite imagery, and other information. Figure 11.1 shows a map of Mount Rainier and the surrounding area that is a varying degree from highest (red) to lowest (yellow) risk of a lahar. Figure 11.2 is a photograph of an urban area in the shadow of Mount Rainier and within the lahar risk zone. Sometimes this information is not necessary if the geologic risk is close enough to a point of exposure as evidenced from Figures 11.3 and 11.4. Figure 11.3 shows the volcano Mount Etna in close proximity to over a million residents in Sicily and the city of Catania. Figure 11.4 shows a home that was engulfed by a lava flow from a 2002 eruption of Mount Etna.

The concept of a conceptual site model, risk, or vulnerability map is important enough that we will dedicate an entire chapter to demonstrate the methodology and significance of geologic hazard sensitivity maps.

Computer modeling is frequently used to assist in preparedness and response. The United Nations (2024a) states that using modeling as a qualitative or quantitative tool to evaluate the nature and extent of potential disaster scenarios provides valuable data on potential response and prevention on elements such as location, intensity, frequency, and probability of occurrence.

11.2.1.6 Identifying Areas of Highest Risk

Through the development of the conceptual site model, identifying the areas of highest risks is the logical next step. Once the preliminary process of identifying areas of highest risk is completed, additional data collection usually is conducted to refine and define the area(s) of highest concern. The identification of areas of highest risk may require the implementation of future land-use restrictions to ensure that certain types of land development do not occur in areas of highest risk.

One example is allowing the construction of a residential community within a zone of high risk of a geologic hazard. Figure 11.2 is a photograph of a residential community in the shadow of

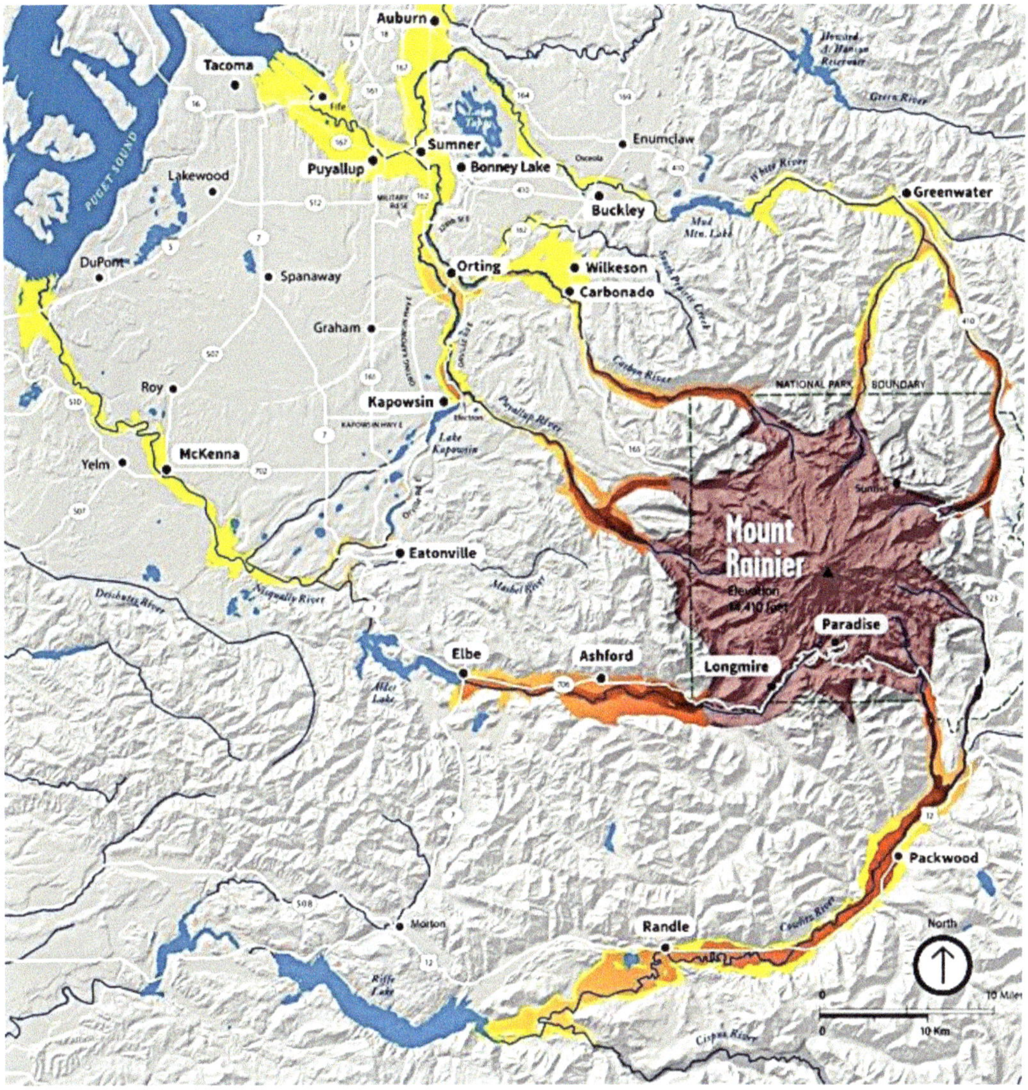

FIGURE 11.1 Example lahar hazard map of Mount Rainier in Washington. (Pierce County, Washington. 2024. *Mount Rainier Hazard Maps*. www.piercecountywa.gov/3800/hazard-maps. (Accessed April 20, 2024).)

Mountain Rainier which is within the high-risk lahar zone identified in Figure 11.1. The lahar risk at the location shown in Figure 11.2 may not be as high as it would be closer to the volcano, but it is a location of highest concern when considering that a residential area is located within a boundary of a geologic hazard.

11.2.1.7 Developing an Early Warning System

Once areas of highest risk are identified, an early warning system can be developed. The concept of an early warning system is to mitigate harm from the risk. A crucial step in this process is to develop a reliable and efficient system capable of actually detecting early warning signs. An effective early warning system incorporates four key elements: risk knowledge, monitoring, communication, and response. High-risk areas for geologic vulnerability maps in urban areas will be discussed in greater

FIGURE 11.2 Residential community built within the historical lahar zone of Mount Rainier. (From United States Geological Survey (USGS). 2024c. *Mount Rainier Geologic Hazards*. www.usgs.gov/Geologic-hazards/mount-rainier-laher-zone. (Accessed April 21, 2024).)

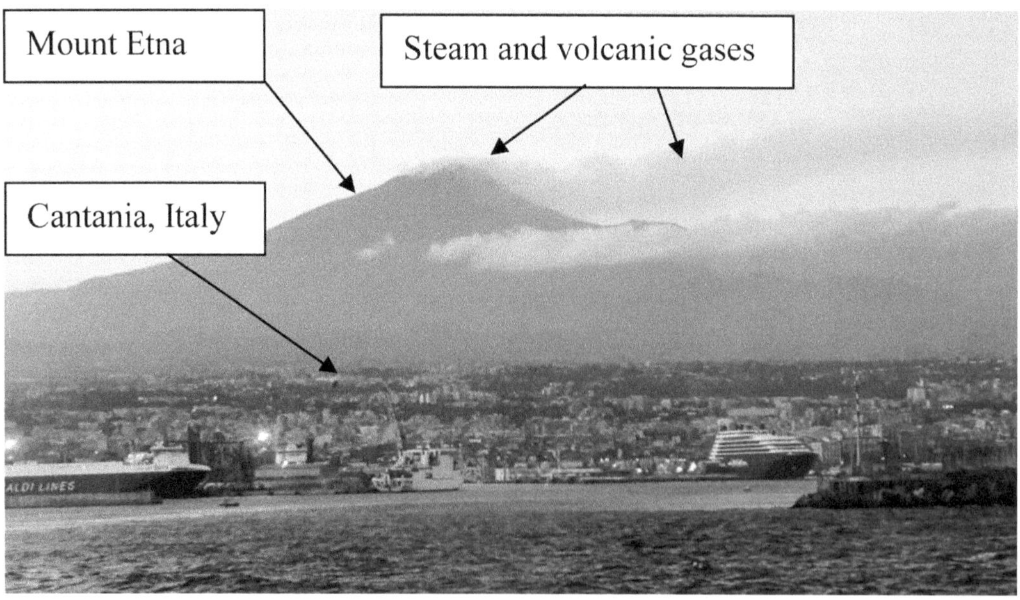

FIGURE 11.3 Photograph showing the Sicilian City of Cantania with Mount Etna in close proximity in the background. (Photograph by Daniel T. Rogers.)

FIGURE 11.4 Photograph of a house engulfed from a lava flow from an eruption of Mount Etna in 2002. (Photograph by Daniel T. Rogers.)

detail in Chapter 13. Vulnerability maps in urban areas can play an important role in urban planning where these maps can serve as an early warning system.

11.2.1.8 Monitoring

Monitoring is a key element in the development of an early warning system. Accurate and timely monitoring can make the difference between life and death in areas prone to earthquakes, volcanic activity, landslides, and geological and environmental risks.

11.2.1.9 Communicating Alerts and Warnings

This step ensures that important information reaches those who are at the highest risk quickly and efficiently. The timeliness is crucial so that the recipients are afforded enough time to react appropriately. Therefore, early warning systems should be clear, concise, and easily understood. In addition, the early warning system should be delivered over multiple methods such are multiple media outlets, social media, mobile applications, public service announcements, warning sirens, through local law enforcement locations, fire and emergency response agencies, and local community groups, if applicable.

11.2.1.10 Emergency Planning and Preparedness

The development of a comprehensive emergency response plan in advance is important so that the response is organized and appropriate. An emergency plan is sometimes called a site health and safety plan in environment-related activities. These plans generally include evacuation routes, identifying exclusion zones, potential hazard identification, protective clothing requirements, communication, and appropriate response actions. Regular practice is necessary to ensure that some specialized equipment is maintained and that efforts are coordinated.

11.2.1.11 Establishing an Evacuation Plan

An evacuation plan often is a critical element in the emergency planning stage. Unfortunately, all too often evacuation routes get clogged and render themselves ineffective during a time of need. Therefore, early warning, multiple routes of evacuation, and monitoring evacuation routes should be conducted to ensure the safe evacuation of at-risk areas.

11.2.1.12 Education and Awareness

Continuous education and awareness are key to the success of responding to risks appropriately and effectively. Continuous education is recommended because strategies may change over time if new information is gathered that alters the risks for certain areas. In addition, new equipment and technology upgrades will require additional training.

11.2.1.13 Testing and Training

To have an effective response to a geological or environmental risk, training and testing are important to ensure a successful response. Response actions and methods may change through time along with the development of new technologies. Therefore, continued education and training on appropriate procedures focused on improving warning time, lowering the response time, and providing improved care must be the prime objective.

11.2.1.14 Storing Emergency Supplies

An integral part of surviving a large and sudden occurrence of a disaster from a geologic risk is preparedness. What is generally recommended is to prepare to be without basic services for 3 to 7 days. Recommended supplies include non-perishable food, water, medical supplies, warm clothing, survival gear, and essential documents (including identification, insurance policies, and medical records). A minimum of one gallon of water per person is recommended.

11.2.1.15 Critical Response Effectiveness Review

As a final step in the process, conducting a critical review afterward to evaluate the effectiveness of the response and also identifying areas for improvement are important so that continuous improvement is achieved.

11.2.1.16 Risk Impact Mitigation Actions for Geologic Risks

Mitigation alternatives and actions to address geologic risks that include earthquakes, volcanic eruptions, tsunamis, landslides, sinkholes and subsidence, and coastal hazards are addressed in this section. This discussion will focus on engineering controls and other actions that can be taken to minimize the impacts of geologic hazards.

11.2.1.16.1 Earthquake Impact Mitigation

In the United States, congress passed the Earthquake Hazards Reduction Act in 1977. The purpose of enacting this legislation was to reduce the risks to life and property from future earthquakes in the United States. By passing the Act, congress recognized that earthquake-related losses could be reduced through improved design and construction methods and practices, land-use controls and redevelopment, prediction techniques and early warning systems, coordinated emergency preparedness plans, and public education and involvement programs.

Four federal agencies are responsible for working cooperatively to implement the objectives of the Act. The agencies involved include the Federal Emergency Management Agency (FEMA), National Institute of Standards and Technology (NIST), National Science Foundation (NSF), and the USGS (National Earthquake Hazards Reduction Program (NEHRP) 2024).

Mitigating the hazards associated with earthquakes represents a unique challenge in coping with uncertainty. While earthquakes cannot be prevented, the consequences of earthquakes in terms of loss of lives and destruction of property and communities can be substantially reduced. Strategies for managing earthquake hazards vary from county to county.

At least 36 nations are exposed to high earthquake risks, making mitigation of earthquake risks a global challenge. Earthquake mitigation strategies are actions that can reduce the impact of earthquakes on people and property. For the individual or family, the following actions can reduce the potential for injury and increase the survivability of an earthquake (USGS 2024b):

- Check for hazards in your home such as bookshelves, tall furniture, water heaters, and other items that could fall and cause damage
- Identify safe places in each room
- Locate safe places outdoors
- Ensure that all family members participate in discussion and training
- Teach children how to call 9-1-1
- Have survival supplies prepared and located in an area that can be easily accessed
- Develop a family emergency communications plan
- Ask an out-of-state-relative or friend to serve as a family contact
- If indoors, take cover under heavy furniture or against an inside wall and stay inside
- If outdoors, move into the open, away from buildings, streetlights, power lines, and trees
- If in a moving vehicle, stop quickly, stay in the vehicle, move to a clear area away from buildings, trees, overpasses, and power lines
- Be prepared for aftershocks
- Help injured or trapped persons and give first aid if necessary
- Stay out of damaged buildings

Earthquake mitigation methods focus on many aspects to reduce the risk of an earthquake. A major difficulty with earthquake mitigation strategies is that there is no real-time warning system because earthquake science does not fully understand when an earthquake will occur. Therefore, the mitigation strategies largely rely on geologic mapping and dating of historical earthquakes to assess risks. Earthquake mitigation strategies are the following (FEMA 2024):

- Mapping the geology in urban areas to identify those locations or areas that are at highest risk of earthquakes and earthquake damage
- Revising building codes so that structures are more resistant to damage from an earthquake
- Conduct building inspections to ensure that the integrity and safety of buildings are maintained
- Protect critical facilities and infrastructure such as hospitals, emergency response facilities, pipelines, natural gas lines, electrical lines, sewers, water lines, bridges, major highways, bracing elevators and generators, installing shutoff valves at strategic locations such as across fault lines, and other actions
- Strengthening and retrofitting non-reinforced masonry buildings
- Retrofitting building veneers to prevent failure
- Building an earthquake-safe room in some buildings
- Installing window film to prevent injuries from shattered glass
- Anchoring rooftop-mounted equipment
- Reinforce chimneys greater than 2 m above the roof line
- Increase earthquake risk awareness
- Conduct outreach activities to builders, architects, engineers, and inspectors
- Provide information on structural and non-structural retrofitting

- Protect against fire following an earthquake
- Protect against looting following an earthquake
- Remove debris and obstacles from critical roadways after an earthquake
- Repair communications, if necessary, after an earthquake

11.2.1.16.2 Tsunami Impact Mitigation

Tsunamis are strongly associated with earthquakes, especially if the epicenter of an earthquake is located offshore. In many cases, the earthquake can be thousands of kilometers away but can still create a deadly tsunami. The following are mitigation strategies for tsunamis (FEMA 2024):

- Mapping the geology in suspect areas to assess the tsunami risk
- Establishing a monitoring and early alert system
- Managing development in tsunami hazard areas
- Protecting against fire risks following a tsunami
- Building tsunami shelters
- Protecting building and infrastructure
- Increasing public awareness
- Installing tsunami warning sirens

11.2.1.16.3 Volcanic Impact Eruption Mitigation

Volcanic eruptions have the potential to cause several types of life-threatening risks that include lava flows, pyroclastic clouds, lahars, landslides, explosions, ash clouds, volcanic bombs, volcanic gases, fallen trees, and other hazards.

According to the USGS (2024d), loss of life from volcanic hazards has been greatly reduced by early alert systems from monitoring volcano behavior such as earthquakes beneath the volcano and monitoring volcano topography (bulging or subsiding slopes), gas emissions, and other volcano activities. Since 1991, the most significant loss of life from volcanoes has been secondary lahars (USGS 2024d). The following are volcano eruption mitigation strategies (USGS 2024d):

- Geologic mapping of potential risk zones from each type of risk. Figures 11.1 and 11.2 present an example of the lahar risks from Mount Rainier. Figures 11.3 and 11.4 indicate the risk area from a volcanic eruption from Mount Etna in Sicily.
- Establishing a monitoring and early alert system
- Establishing evaluation routes
- Limit development in risk zones such as lahar occurrence
- Increase public awareness
- Install volcano hazard sirens
- Mitigation strategies for residents living near super volcanos should focus on early warning and, if an eruption is likely, evacuation a significant distance, perhaps hundreds of kilometers, should be strongly considered.

11.2.1.16.4 Landslide Impact Mitigation

As described in Chapter 5, a landslide is the movement of a mass of rock, debris, or earth down a slope under the force of gravity. Slopes greater than 10 degrees are more likely to slide, as are slopes where the height from the top of the slope to the bottom or toe is greater than 40 feet (see Figure 11.5). Slopes are more likely to slide if vegetation cover is absent or low and water content is high. Potential impacts include environmental disturbance, property and infrastructure damage, and injuries and fatalities (FEMA 2024). The recommended strategies for landslide mitigation include the following (FEMA 2024):

FIGURE 11.5 Photograph showing a high-risk area from landslide activity in Kotor, Montenegro. (Photograph by Daniel Rogers.)

- Geologic mapping of potential risk zones of landslides
- Manage development in identified landslide hazard areas
- Use engineering controls to prevent impacts to roadways through monitoring, installing stabilization measures, planting vegetation, draining excess water buildup, establishing setback requirements, and installing catch-fall netting
- Remove structures and buildings in high-risk locations

11.2.1.16.5 Sinkholes and Subsidence Impact Mitigation

As described in Chapter 5, subsidence is the gradual settling or sudden sinking of the Earth's surface due to the subsurface movement of Earth materials. The degree of subsidence ranges from a broad lowering to a collapse of the land surface. Most causes of subsidence are from anthropogenic sources such as excessive pumping of groundwater, aquifer system compaction, drainage or organic soils, underground mining, and thawing permafrost. The recommended strategies for subsidence and sinkhole mitigation include the following (FEMA 2024):

- Geologic mapping of potential risk zones of subsidence and sinkholes
- Managing development in high-risk locations
- Considering subsidence in building design
- Monitoring the ground surface and subsidence risk factors
- Removing structures and buildings that are damaged or in high-risk areas

11.2.1.16.6 Coastal Hazard Impact Mitigation

Coastal hazards are varied but usually include water. Coastal hazards include slope failure, landslides, flooding, sea-level rise, beach erosion, and tsunamis (which have been previously discussed). The recommended strategies for coastal hazard mitigation include the following (USGS 2024e):

- Geologic mapping of potential risk zones of coastal hazards
- Improve setback requirements for buildings and structures
- Improve flood control measures
- Improve beach topography and landscaping

- Plant vegetation
- Restore or construct wetlands or estuaries
- Construct dikes, levees, berms, sea walls, and pumps

11.2.1.16.7 Erosion Impact Mitigation

Erosion is the wearing area of land, such as loss of riverbank, beach, shoreline, or dune material. Short-term erosion typically results from periodic natural events, such as flooding, hurricanes, storm surge, and windstorms. Erosion also occurs as a result of human activities, such as from wildfires caused by human activity, deforestation, land development, agricultural practices (FEMA 2024). Erosion mitigation strategies include the following (FEMA 2024):

- Geologic mapping of areas prone to erosion
- Manage development in erosion hazard zones
- Require site and building design standards to minimize erosion, such as modifying building design, installing wetlands, constructing specialized landscapes that minimize erosion, planting vegetation, installing rain gardens, and minimizing impervious surfaces
- Stabilizing erosion hazard areas
- Increase awareness

11.2.1.16.8 Aeolian Processes Mitigation

Aeolian process mitigation generally refers to dust, which is defined as particles suspended in the air and originating from a natural source, such as a dust storm in desert regions. Mitigation of dust is generally not possible. Therefore, mitigation then must focus on engineering controls to limit exposure and minimize potential impacts. Some recommended methods of risk mitigation include staying inside, avoiding driving during a dust storm, and wearing a mask and goggles if going outdoors is required (FEMA 2024).

11.2.2 Risk Impact Mitigation Methods for Environmental Hazards

Environmental hazards as defined in this book generally refer to extreme atmospheric events. Solar radiation and wildfires are also placed within the environmental hazard data set. Wildfires are often associated with drought and high winds and solar radiation uses the atmosphere as a pathway for exposure.

A key mitigation factor for environmental hazard avoidance is forecasting and predictions. Warning alerts are the best mitigation factor to minimize loss of life from an environmental hazard. Damage inflicted on the environment may not be avoided, but preparedness and an alert warning system have decreased the loss of life and have limited the damage from environmental hazards.

11.2.2.1 Environmental Hazards Connection to Geology

Geology through tectonic activity determines the growth of mountain ranges, volcanic activity, and the position of the continents. These actions have profound effects on global ocean and atmospheric circulation patterns, and therefore greatly influence weather and climate patterns. In general terms, the influences of plate tectonics include the following (NOAA 2024):

- A decrease in coastal areas and the amount of dry land increases when the continents come together
- More ice sheets form, sea level falls, and temperatures decrease when continents move toward the poles
- Mountain ranges affect atmospheric circulation patterns globally and create orographic precipitation patterns regionally

- Changes in the rate of tectonic activity influence volcanic activity, which influences the amount of solar radiation at the Earth's surface
- Sea-level rise increases when tectonic activity increases seafloor spreading
- Increased formation of limestone and coal locks carbon in rocks and limits the amount of atmospheric heating by limiting greenhouse gases available to be emitted into the atmosphere

11.2.2.2 Types of Environmental Hazards

Environmental hazards as addressed in this book generally refer to extreme atmospheric events and include the following (FEMA 2024):

- Drought
- Hurricanes
- Tornadoes
- Extreme precipitation events
- Extreme heat
- Extreme cold
- Flooding
- Severe storms
- Severe winter weather
- Lightning
- Solar radiation
- Hail
- High winds
- Wildfires
- High-tide floods
- Marine heat waves
- Derecho's

11.2.2.3 Mitigation Methods for Environmental Hazards

Mitigation methods to address environmental hazards impacts include the following (FEMA 2024):

- Construct maps that identify area of highest risk and potential impacts
- Monitor the occurrence of extreme atmospheric events and modify hazard maps as necessary
- Monitor water supplies
- Consider water conservation measures
- Prevent or minimize overgrazing
- Enhance landscaping
- Reduce urban heat island effects by planting more trees, installing vegetated roofs, and by using cool roofing products
- Limit urban development in hazard areas
- Incorporate mitigation impact measures into building codes
- Improve stormwater management planning
- Adopt policies to reduce stormwater runoff
- Elevate to retrofit structures and utilities
- Protect infrastructure
- Protect critical facilities, such as hospitals and emergency response structures
- Consider the construction of flood and sea walls, dams, dikes, levees, and other engineering measures
- Restore natural flood mitigation measures such as wetlands and estuaries
- Increase awareness and education

- Prepare emergency and medical supplies
- Protect critical facilities and equipment from lightning
- Conduct lightning awareness programs
- Protect power lines from wind and storm damage potential
- Improve land-use planning and regulations from severe atmospheric events including wildfires
- Encourage the construction of safe rooms
- Encourage fire-resistant construction techniques and materials
- Create defensible space around structures for wildfires and other severe atmospheric events

11.2.3 Risk Impact Mitigation Methods for Anthropogenic-Induced Hazards

In this section, we highlight risk impact mitigation methods from anthropogenic-induced hazards. Anthropogenic-induced hazards that have impacted the natural environment include the following (Rogers 2023):

- Habitat destruction
- Natural resource exploitation
- Water cycle disruption
- Deforestation
- Overfishing
- Overhunting and poaching
- Invasive species
- Species extinction
- Light pollution
- Noise pollution
- Ocean pollution
- Agricultural pollution
- Agricultural land use
- Nuclear radiation
- Water pollution
- Land pollution
- Air pollution
- Erosion
- Climate change

11.2.3.1 Geology and Connections to the Urban Environment

The urban environment often magnifies and increases the complexity of natural hazards due to large increases in population density. In addition, and perhaps most importantly, urban locations create their own risks and exacerbate geologic risks through the introduction of pollution and other activities. Another additional layer of complexity is that most large urban areas of the world are located near water, which, in many instances, is one of the primary transport mechanisms of pollution. Water often transports chemical contaminants from areas of little or no risk to areas of high risk and potentially exposes large populations to unsafe conditions.

Environmental and geologic risk mitigation methods in urban areas must first start with geologic mapping but at a finer scale than most other geologic maps. The reason is simple: this is where we live and immediately beneath our feet is the contact between civilization and the natural world. The resolution of geologic mapping in urban areas is often down to the centimeter or finer scale in some instances. Geologic maps thus form the foundation for ecosystem management of urban areas (USGS 1998). The natural environment and the humans within it acquire an imposed organization resulting from the geologic framework at any given location. Geologic maps provide us with knowledge of that framework. Specifically, a geologic map facilitates the basic means for evaluating resources, hazards, and environmental data relative to human activities (Thomas 2004). Key elements in creating effective geologic maps in urban areas are evaluating the sensitivity of the geologic environment to the risks identified, evaluating where exposure to those risks may occur, and the adaptive capacity of the urban area (Rogers 2020).

Due to the importance of geologic mapping as a first step in identifying locations where risk is elevated, especially in urban areas, we dedicate Chapter 13 to urban geologic mapping. The importance and value of mapping the geology in urban areas is obvious in one sense because that is where we live and spend most of our lives. The difficulty in mapping the geology of urban areas is that there are added anthropogenic complexities that make the process of mapping time consuming and more challenging.

Mitigation methods for anthropogenic-induced hazards in urban areas typically are very similar to other mitigation methods. Some of the methods that may require added complexities or factors to consider include high-resolution geologic mapping to support a conceptual site model in contaminant fate and transport evaluation, hydrogeological mapping, and geologic sensitivity mapping (Rogers 2020).

An added element that typically does not occur in natural areas is the presence of contamination. Although contamination does occur in natural areas, contaminants are much more common in urban areas and occur in the air, water, and soil from numerous sources. This requires an additional risk evaluation and potential remedial measures if there is an unacceptable risk of exposure. Exposure to contaminants from anthropogenic sources is presented in the following sections.

11.2.3.2 Pollution Risk Mitigation Methods

As highlighted in Chapters 8 and 10, anthropogenic-induced hazards from the release of chemical contaminants into the air, water, and soil are responsible for millions of human deaths each year (United Nations 2024a). However, chemical contaminants are a subset of pollution. Chemical contaminants have affected the entire Earth and include contaminants that are in the air, water, soil, and sediments of Earth. Other forms of pollution that will be addressed in this section include ocean pollution, light pollution, noise pollution, and agricultural pollution. Specific methods of risk reduction techniques are also addressed in the sustainability section discussed later in this chapter.

11.2.3.2.1 Toxicological Risk Assessment

In most cases, the first step in addressing chemical contaminants is to conduct a toxicological risk assessment that assesses the risks associated with contaminants released into the environment. The following sections summarize methods that are employed to conduct a toxicological risk assessment.

11.2.3.2.1.1 Toxicity of Chemical Contaminants

Toxicity or potency is the degree to which a chemical or substance is able to inflict damage to an exposed organism (USEPA 1989a). Note that toxicity does not equal risk. The difference between toxicity and risk is based primarily on the length of exposure to a chemical or substance, and whether the dosage received from this exposure is enough to cause harm. Toxic substances surround us and are present at almost all locations on Earth. However, as stated above, there is only risk if we are exposed to a substance or chemical long enough and at a high-enough dose to cause harm. Toxicity does not assess risk – this determination is reserved for a risk assessment.

11.2.3.2.1.2 Conducting a Toxicological Risk Assessment

A toxicological risk assessment is an evaluation that combines the toxicity of a contaminant with possible exposure pathways. A risk assessment is therefore a measure of the nature and magnitude of health risks and ecological receptors from chemical contaminants that may be present in the environment (USEPA 2024b). There are three basic types of toxic categories:

1. Chemical or substance, including inorganic and organic substances such as acids and bases, flammable liquids, metals, etc.
2. Biological, including bacteria and viruses
3. Physical, including sound and vibration, heat and cold, light, radiation, etc.

Toxicity is measured by the effects on a whole organism, an individual organ, a tissue, or even a cell. Populations are most often used to measure toxicity as any one individual typically may have a different level of response to a toxin at a certain dose or concentration. The most common measure of chemical or substance toxicity is termed the LD_{50}; defined as the concentration or dose that is lethal to 50% of the population being tested (USEPA 1989a). When direct data are not available, the LD_{50} is estimated by comparing the substance to other similar chemicals and organisms.

Another important measure of toxicity is the lowest-observed-adverse-effect level (LOAEL) or threshold effect value. The LOAEL is the lowest tested dose of a chemical or substance causing a harmful or adverse health effect (Agency for Toxic Substances and Disease Registry (ATSDR) 2024). Factors influencing the LOAEL include the following:

- The chemical's solubility in body fluids
- The particle size and state of the chemical
- Route of exposure
- Residence time of the chemical in the body
- Individual susceptibility

Units of dose are expressed as the mass of chemical per unit mass of body weight in milligrams per kilogram (mg/kg). These units are employed to evaluate the relative toxicities between animals of different species and sizes. To develop the most accurate human exposure limits, extensive animal studies are initially used to establish extrapolated human dosage limits for a specified chemical, and these estimates are refined by human health studies conducted on individuals known to have been exposed to the same chemical. The Occupational, Safety, and Health Administration (OSHA) has established a list of exposure limits for over 600 individual chemicals and refers to exposure limits as permissible exposure limits (PELs) (American Conference of Governmental Industrial Hygienists (ACGIH) 2024). Most chemicals or substances have the potential to exhibit some adverse health effect on humans or other organisms given a certain set of circumstances. Adverse health effect is defined as a change in body function or cell structure potentially leading to disease or health problems (ATSDR 2024). The effect that a specific chemical or substance may exhibit on a living organism is dependent upon the following factors (USEPA 1989a, 2005):

- Nature of the chemical or substance
- Concentration
- Route of exposure
- Length of time of exposure
- Individual susceptibility

Chemicals and substances enter the human body through three routes of exposure: inhalation, ingestion, and dermal adsorption (USEPA 1989a). When exposure occurs, chemicals may produce one or more of these symptoms:

- Tissue irritation
- Dizziness
- Vision loss
- Anxiety
- Eye irritation
- Bleeding
- Hearing loss
- Narcosis
- Rash
- Hair loss
- Nausea
- Headache

• Vomiting	• Diarrhea	• Pain
• Fever	• Tremors	• Difficulty breathing
• Psychotic behavior	• Euphoria	• Cancer
• Death	• Burning sensations	• Others

Human response to exposure to a chemical is described as acute or chronic. An acute response is generally characterized as a single high dose with rapid onset and disappearance of symptoms. Chronic response involves a stimulus lingering for a period of time (ATSDR 2024). Exposure to a contaminant may not trigger an immediate response. Instead, there is a latency period – a duration of time without observable effects. Certain chemicals or substances, such as asbestos and forms of mercury, have a latency period of up to 10 to 20 years (USEPA 2005). Exposure to some chemicals or substances may result in the development of cancer, and any chemical, substance, radionuclide, or radiation contributing to the development or propagation of cancer is termed a **carcinogen** (United States Department of Health and Human Services (USDHH) 2005). Cancer exists when cells in the body become abnormal and grow or multiply out of control (ATSDR 2024). There is also often some selectivity involved with chemical exposure, as some chemicals may target certain organs, parts of the body, or reproduction. For example, may have an adverse effect on a developing fetus, mutagens may induce genetic changes that could affect future generations through reproduction, and hepatotoxins are chemicals posing a risk of liver damage (ATSDR 2024). teratogens Carcinogens are grouped into five general categories (ACGIH 2024):

- Group A1: Confirmed human carcinogen
- Group A2: Suspected human carcinogen
- Group A3: Confirmed animal carcinogen with unknown relevance to humans
- Group A4: Not classified as a human carcinogen
- Group A5: Not suspected as a human carcinogen

It is much more difficult to evaluate the toxicity of a mixture of contaminants compared to a single chemical compound or substance. This is because the interaction of a mixture of contaminants may produce enhanced or diminished effects. Common mixtures include gasoline, often containing more than 250 individual chemical compounds, industrial wastes, or a malfunctioning sewage treatment plant, which may result in discharging many chemical and biological contaminants (USEPA 2024c).

11.2.3.2.2 Chemical Contaminant Risk Mitigation Methods

Cleaning up the environment once it becomes contaminated is expensive and time-consuming. A sad and disturbing fact is that sometimes when the environment becomes contaminated, any amount of financial resources or technology will not restore the environment to pre-contamination levels (Rogers 2018). Examples include the Exxon Valdez incident in Alaska, Deep Water Horizon incident in the Gulf of Mexico, numerous sites where groundwater becomes contaminated with chlorinated volatile organic hydrocarbons (CVOCs) or hexavalent chromium, and polychlorinated biphenyls (PCBs) in sediments of large rivers such as the Hudson River in New York and the Willamette River in Oregon. Therefore, in many of these instances, the contamination must be managed through other means so that the spread of contamination is halted or minimized.

Under most scenarios, contaminants released into the environment require clean-up. The clean-up, termed **remediation**, meaning remedy, varies widely and depends on many different factors. However, as stated above and as we shall discover, typical remediation does not result in returning the affected area(s) or media to pre-contamination levels. Remediation of contamination is typically conducted to specific targets for each chemical of concern and is based on ecological and human risk. Remediation goals can be site-specific or to generic standards established by the appropriate regulatory agency.

USEPA estimates that more than 100 billion dollars is spent each year investigating and remediating contaminated sites just in the United States (United States Government Accounting Office (USGAO) 2024; USEPA 2024a). Estimates on the number of sites remediated or under remediation total over 1 million sites. In addition, there are an estimated 500,000 to 1 million abandoned industrial facilities or brownfield sites that are waiting to be investigated. This is because a cost estimate to investigate and remediate brownfield sites has not been conducted. Most of these abandoned industrial sites are located in urban areas of the United States. The majority are located in the 27 urban areas listed in Table 2.4 in Chapter 2. These sites contain some degree of contamination and will require some degree of assessing risk or remediation before they can be redeveloped. Thus, it is likely that the environmental costs to investigate and remediate sites of environmental contamination in the United States will exceed the latest available USEPA estimate of $250 billion (USGAO 2024; USEPA 2024a).

The following is a list of some of the more common remedial technologies for soil, groundwater, surface water, sediment, and air. As with any successful remedial strategy, controlling the source of contamination is critical for achieving the objective. If controlling or eliminating the source is not achieved, any remedial strategy will fail because re-contamination will occur.

11.2.3.2.2.1 Remediation Overview The objective of any remediation project is to prevent, remove, treat, change, destroy, or transform the potentially harmful contaminants from the media or medias of concern so that the risk posed by the presence of the contamination has been effectively eliminated or reduced to an acceptable level (USEPA 2005). As we discussed in Chapter 4, there are several different studies that are conducted at a site of environmental contamination that outline and assist the navigation process through the required procedures and scientific studies that are routinely conducted so that the selected remedy or combination of remedies is successful at achieving the desired goal or objective. However, even with these many checks and balances, many remedial projects are unsuccessful and must be either revised or abandoned in favor of a different technological approach.

The most common reason for unsuccessful remediation to occur is because the site has not been properly characterized. Not identifying a source of contamination, underestimating contaminant mass, not identifying a complete list of contaminants of concern, incomplete sampling plan, and improper sampling techniques and methods all can lead to remediation being unsuccessful and may lead to increased cost, time, and effort to achieve the remedial objective. Other remediation failures include technology misapplication and design, environmental changes such as a change in groundwater flow direction, sudden changes in local climatic conditions such as a drought or flood, changes in geochemistry, and additional releases.

Some of the factors that control or influence the degree to which remediation must be conducted depend upon the following factors:

- Contaminants or contaminants released
- Concentration
- Nature and extent
- Media impacted (e.g., air, surface water, groundwater, soil, sediment)
- Geology
- Hydrogeology
- Hydrology
- Location
- Potential receptors
- Land use
- Future land use
- Climate

- Time necessary to achieve objective(s)
- Results of risk assessment and feasibility study
- Cost

Driven by the need for more effective and less costly clean-up costs, new remedial technologies are constantly being developed (USEPA 2007). Over the last few decades, several remedial technologies have been developed to remediate contaminated soil and groundwater. In fact, USEPA (2007) has estimated that there are approximately 2,000 remedial technologies for environmental contamination. New technologies were developed due to several factors that included (1) the ineffectiveness of early remedial methods, (2) the excessive cost of early methods, and (3) regulatory goals that required remediation to standard-based levels that in many instances were too pristine pre-contaminant or background conditions. In the mid-1990s a transition occurred from the standard-based remediation clean-up goals toward risk-based clean-up goals (American Society for Testing Materials (ASTM) 1995).

Remediating a site to risk-based criteria involves calculating site-specific clean-up levels that are based on the risks posed by the presence of specific contaminants and involve evaluating (1) the toxicity and nature of each contaminant, (2) the site's geological and hydrogeological setting, (3) fate and transport mechanisms, (4) future land use, and (5) analysis of potential receptors. In most cases, applying risk-based clean-up goals raised the concentration required to achieve site closure for a specific contaminant at a particular site, which, in turn, lowered the cost and time required to remediate a site (USEPA 2002).

11.2.3.2.2.2 Soil Remediation Techniques Soil remediation usually involves one or more of the following (Rogers 2020):

- Excavation
- Bioremediation
- Monitored natural attenuation
- Mechanical soil aeration
- Capping
- Land-use restrictions
- Soil vapor extraction
- Phytoremediation
- Soil washing
- In situ thermal desorption
- Ex situ thermal desorption or incineration
- Electrokinetics
- Solidification/stabilization
- Fracturing
- Vitrification
- Chemical oxidation
- Chemical dehalogenation

11.2.3.2.2.3 Groundwater Remediation Techniques Groundwater remediation usually involves one or more of the following (Rogers 2020):

- Pump and treat
- In situ soil flushing
- Air sparging
- Permeable reactive barriers

- Injection of biological agents
- Injection of chemical agents
- Monitored natural attenuation
- Institutional controls
- Multi-phase extraction

11.2.3.2.2.4 Sediment Remediation Techniques Sediment remediation usually involves one or more of the following (Rogers 2020):

- Source control
- Excavation
- Dredging
- Bioremediation
- Capping
- Natural attenuation
- Institutional control

11.2.3.2.2.5 Surface Water Remediation Techniques Surface water remediation usually involves one or more of the following (Rogers 2020):

- Source control
- Engineered wetlands
- Direct removal
- Natural attenuation
- Institutional controls

11.2.3.2.2.6 Air Remediation Techniques Remediation of the air is almost exclusively conducted at the source through the installation of air pollution control equipment (Rogers 2020). An example of source control to improve air quality is the catalytic converter which is required on automobiles to reduce the amount of volatile organic compound emissions into the atmosphere.

11.2.3.3 The Cost of Repairing Nature

Cleaning up the environment once it becomes contaminated is expensive and time-consuming. A sad and disturbing fact is that sometimes when the environment becomes contaminated, any amount of financial resources or technology will not restore the environment to pre-contamination levels (Rogers 2018). Examples include the Exxon Valdez incident in Alaska, Deep Water Horizon incident in the Gulf of Mexico, numerous sites where groundwater becomes contaminated with CVOCs or hexavalent chromium, and PCBs in sediments of large rivers such as the Hudson River in New York and the Willamette River in Oregon. Therefore, in many of these instances, the contamination must be managed through other means so that its spread is halted or minimized. An additional lesson, which perhaps is even more important, is that prevention is key to creating a sustainable balance with nature given that there are over 8 billion humans that inhabit Earth.

Under most scenarios, contaminants released into the environment require clean-up. The clean-up, termed remediation, meaning remedy, varies widely and depends on many different factors. However, as stated above and as we shall discover, typical remediation does not result in returning the affected area(s) or media to pre-contamination levels. Remediation of contamination is typically conducted to specific targets for each chemical of concern and is based on ecological and human risk.

Remediation goals can be site-specific or to generic standards established by the appropriate regulatory agency. USEPA (2024c) estimates that more than 100 billion dollars is spent each year investigating and remediating contaminated sites just in the United States. Estimates on the number of sites remediated or under remediation total over 1 million sites. In addition, there are an estimated 500,000 to 1 million abandoned industrial facilities or brownfield sites that are waiting to be investigated. This is because a cost estimate to investigate and remediate brownfield sites has not been conducted. Most of these abandoned industrial sites are located in urban areas of the United States. The majority are located in the 27 urban areas. These sites have some degree of contamination and will require some degree of assessing risk or remediation before they can be redeveloped. Thus, it is likely that the environmental costs to investigate and remediate sites of environmental contamination in the United States will exceed the latest available USEPA estimate of $250 billion (Rogers 2020).

Remediation takes time, patience, and scientific understanding. Some of the factors that control whether remediation is conducted and how it is conducted and subsequently the cost include the following (Rogers et al. 2006, 2008, 2009):

- *Geology:* Different types of geology can influence the extent of contamination and methods of remediation.
- *Hydrogeology:* The presence of groundwater often plays a significant role in assessing risk, migration pathways, and technology(s) selection.
- *Contaminant type:* The type(s) of contaminants influence migration pathways, mobility, risk, and clean-up targets.
- *Nature and extent:* Nature and extent of contamination impact receptor analysis and potential routes of exposure and clean-up targets.
- *Contaminant mass and concentration:* The mass and concentration of contaminants and whether there is the presence of free phase influences remedial technology(s) selection.
- *Media(s) to be remediated* (e.g., soil, groundwater, surface water, or sediment): Some media are more difficult and costly to remediate than others.
- *Land use:* Residential land typically requires the most stringent clean-up targets.
- *Clean up targets:* Contaminant type, location, and potential receptors and exposure pathways have the most significant influence over establishing clean-up targets.
- *Technology(s) selected:* Some technologies may be more costly than others (i.e., pumping and treating groundwater vs. natural attenuation).
- *Time:* Time may influence cost, especially if the release involves actual human exposure or migration to a sensitive ecological receptor that requires immediate remedial action.

To evaluate (1) whether certain contaminants cost more to remediate than others and (2) what influences geology has on remediation, let us examine remediation that has been conducted at sites of environmental contamination. A total of 370 sites where investigations and remediation had been conducted were compiled and evaluated (Rogers et al. 2006, 2009; Rogers 2018). A few characteristics of each site included: the following

- Of the 370 total sites, 268 were located in the United States and 102 were located in other countries.
- Sites were located in 36 different states and included: AL, AK, AR, CA, CO, DE, FL, GA, IL, IN, IA, KS, KY, MA, MD, MI, MN, MO, MS, NC, NH, NJ, NV, NY, OH, OK, OR, RI, SC, TN. TX, UT, VA, WA, WI, WV.
- International sites were located in 14 countries and included: Australia, Belgium, Brazil, Canada, China, England, France, Germany, Italy, Malaysia, Mexico, Russia, South Africa, and Ukraine.

- Property size ranged from 1.8 acres to more than 300 acres
- All were located in urban areas
- Many of the facilities operated for more than 100 years
- Operations at most facilities included heavy manufacturing

The data from each site included the following (Rogers et al. 2006, 2009; Rogers 2018):

• Contaminant discover facts	• Topography
• Estimated date of release	• Hydrological setting
• Release circumstances	• Fate and transport analysis
• Release operation location	• Analysis of potential receptors
• Source of release	• Remedial technology selected
• Contaminants(s) of concern	• Contaminant mass remediated
• Media(s) impacted	• Contaminant mass left in place
• Media(s) remediated	• Time frame from to remediation
• Surface and subsurface geology	• Projected future cost to closure
• Cost of investigation	• Projected future time to closure
• Cost of Remediation	• Regulatory involvement

Contaminants remediated included:

• Arsenic	• Chlordane
• Chrome VI	• light nonaqueous phase liquid (LNAPL) volatile organic compounds (VOCs)
• Lead	• dense nonaqueous phase liquid (DNAPL) VOCs
• Mercury	• polycyclic aromatic hydrocarbons (PAHs)
• Perchlorate	• PCB's

Several contaminants were detected but were not at high-enough concentrations to be the focus or target of a remedial action. Some of these contaminants also naturally occur. However, concentrations detected exceeded background concentrations and were, therefore, considered anthropogenic. These contaminants included the following:

• Antimony	• Manganese	• Phenols
• Barium	• Nickel	• Phthalates
• Beryllium	• Selenium	• Other SVOCs
• Cadmium	• Silver	• Asbestos
• Copper	• Zinc	

Of the 370 sites evaluated:

- 259 or 70% had soil contamination only.
- 111 or 30% had soil and groundwater contamination.
- 24 or 6.5% had significant free products.
- All the sites were located on unconsolidated sediments that consisted of glacial, fluvial, lacustrine, alluvial, or marine sediments that were composed of gravel, sand, silt, clay, or mixtures of these materials.

Table 11.1 is a summary of the frequency of detection of contaminants, the frequency of when remediation was required, and the percent of total cost. Table 11.2 presents a breakdown of sites by

TABLE 11.1
Summary of Frequency of Detection, Remediation, and Cost

Contaminant	Frequency of Detection	Frequency of Remediation	Percent of Total Cost
PAHs	85%	22.6%	1.5%
LNAPL VOCs	61%	25.9%	42.7%
DNAPL VOCs	31.4%	72.1%	39.4%
Lead	22%	20.0%	1.2%
PCBs	8%	67.0%	3.0%
Chrome VI	9%	80.5%	9.7%
Mercury	3%	82.5%	1.8%
Perchlorate	0.6%	100%	0.1%
Arsenic	5.2%	67%	0.4%
Chlordane	1.2%	100%	0.1%
Total			100%

TABLE 11.2
Site Breakdown by Contaminant Type and Cost

Contaminant	Total Number of Sites	Number of Groundwater Sites	Number of Free Product Sites	Cost (Millions $)
PAHs	103	9	7	11.2
LNAPL VOCs	90	37	17	260.9
DNAPL VOC's	104	65	1	290.2
Lead	26	0	0	7.5
PCBs	19	0	0	28.9
Chrome VI	13	9	0	59.5
Mercury	8	0	0	11.3
Perchlorate	3	2	0	1.1
Arsenic	2	1	0	4.8
Chlordane	2	0	0	0.2
Total	370	113	25	675.6

media that required remediation by contaminant types. Table 11.3 shows the frequency of detection in the soil of the different contaminant groups evaluated and the frequency of when remediation by contaminant type was required. Table 11.4 shows the frequency of detection in groundwater of the different contaminant groups evaluated and the frequency of when remediation by contaminant types was required. Table 11.5 shows a breakdown of the percent of total cost by media remediated. Table 11.6 shows a breakdown of the cost to remediate a kilogram of contaminant by soil type.

Examination of Tables 11.1 through 11.6 reveals that the following:

- PAHs were the group of compounds most often detected but were remediated just one-third of the time and accounted for only 1% of the total cost. This indicates that while PAHs were often detected, remediation was not always required, and when it was required, the cost was less.
- LNAPL VOCs were the group of compounds detected second most often. Remediation of LNAPL VOCs was required approximately as often as PAHs but at a much higher cost. However, over 90% of the cost to remediate LNAPL VOCs was associated with free products.

TABLE 11.3
Breakdown of Soil Remediation

Soil Remediation Technology	Frequency	Frequency of Remediation
Excavation	323	94.5%
Risk assessment	123	35.9%
Institutional controls	108	30.2%
SVE	55	16.2%
Capping	67	19.6%
Thermal treatment	5	0.9%
In situ chemical treatment	5	0.9%

TABLE 11.4
Breakdown of Groundwater Remediation

Soil Remediation Technology	Frequency	Frequency of Remediation
Risk assessment	287	84.1%
Institutional controls	248	72.4%
In situ chemical treatment	161	47.1%
Pump and treat	137	40.2%
Monitored Natural Attenuation (MNA) or Natural Attenuation (NA)	121	35.6%
Air sparging	11	1.8%
Thermal treatment	3	0.5%

TABLE 11.5
Total Cost and Mass Remediated by Media

Remediated Media or Phase	% of Total Cost	% of Total Mass Recovered
Soil	23.4%	17.6%
Groundwater	42.5%	2.1%
Free product	34.1%	80.3%
Total	100%	100%

- Excavation of soil was the preferred remedial option, especially when the volume of soil was less than 5,000 tons and there was no excavation impediment.
- For VOCs, Soil Vapor Extraction (SVE) was the preferred remedial option for course-grained soils with a large, impacted soil volume contaminated or with excavation impediments.
- A combination of alternatives including risk assessment, institutional controls, capping, and thermal treatment were preferred alternatives for large volumes of fine-grained soil or contaminated soil with excavation impediments.
- A combination of alternatives is most often selected when groundwater is impacted, especially with Cr^{6+} or DNAPL.
- Groundwater pump and treat is not currently a preferred remedial method for groundwater unless there is a free product or an incompatible overlapping multi-compound contaminant plume (e.g., heavy metals and DNAPL).

TABLE 11.6
Cost of Remediating a Kilogram of Contaminant

Contaminant	Geology	Remedial Cost ($/kg)
Chrome VI	Clay	1,000
	Silty Clay	1,500
	Sandy and Silty Clay	1,500
	Sand	81,333
DNAPL	Clay	698
	Silty Clay	1,276
	Sandy and Silty Clay	3,019
	Sand	68,363
LNAPL	Clay	311
	Silty Clay	489
	Sandy and Silty Clay	765
	Sand	567
PCBs	Clay	4,500
	Silty Clay	4,800
	Sandy and Silty Clay	5,600
	Sand	5,900
PAHs	Clay	942
	Silty Clay	817
	Sandy and Silty Clay	309
	Sand	27
Lead	Clay	62
	Silty Clay	218
	Sandy and Silty Clay	535
	Sand	462

- In situ chemical treatment is currently the preferred remedial method for Cr^{6+} and DNAPL-contaminated groundwater (e.g., ZVI, calcium polysulfide, ferrous sulfate).
- Cost-effectiveness in remediating groundwater has improved but still remains high.
- Remediating groundwater increased cost by an average of sixfold compared to soil.
- Remediating contaminated groundwater with Cr^{6+} or DNAPL increased cost more than 10-fold compared to soil.
- The average time to investigate and remediate a site with only soil was 2 years.
- The average time to investigate and remediate a site with groundwater contamination was 13 years.
- Groundwater which accounted for 42.5% of the total cost but recovered only 2.1% of total contaminant mass is by far the least efficient compared to free product and soil

The cost of remediating chrome VI and DNAPL VOCs in geologically vulnerable areas results in a cost of remediating a kilogram of contaminant that is greater than $81,000 and $68,000, respectively. The cost of remediating chrome VI and DNAPL VOCs in the least geologically vulnerable areas is more than 50 times less. Table 11.5 also highlights the differential and cost inefficiency of remediating groundwater. The total cost of remediating groundwater at the sites evaluated totaled $260 million or 42.5% of the total cost yet only recovered or remediated 2.1% of the total contaminant mass. This cost disparity highlights the difficulty in cost-effective remediation of groundwater once it has become impacted to such an extent that active remediation is required.

The cost to remediate a kilogram of PCBs is the third highest for the contaminants evaluated. This elevated cost was due to two factors; (1) PCB remediation is heavily regulated and requires intensive investigation and review and (2) PCBs are often detected in sediments. When PCBs are detected in sediments of a river or stream that supports fish and other aquatic life, the expenses for remediation are usually very high due to the expensive nature of cleaning up river and stream sediments with minimal environmental impact.

The remaining types of contaminants evaluated including LNAPL VOCs, lead, and PAHs are not influenced by geology to the extent that geology influences the remedial cost for chrome VI and DNAPL VOCs. As discussed above and in Chapter 8, these contaminants do not exhibit a sufficiently high mobility or persistence which results in a lower Contaminant Risk Factor - Air (CRF_{GW}). Therefore, these contaminants are more cost-effective to remediate on a per-kilogram basis.

11.2.3.4 Habitat Destruction Risk Mitigation Methods

Habitat destruction has been the focus of many scientific studies which have concluded that destroying habitat is a leading cause of species endangerment and extinction. In fact, the current list of endangered species surpassed 1 million worldwide according to the International Union for the Conservation of Nature (IUCN 2024). Much has been documented on the destruction of habitat throughout the world. Almost every country in the world can point to an anthropogenic activity that has resulted in significant harm to the environment through habitat destruction (Rogers 2019a, 2020).

At the global level, restoring habitat to protect from species extinction is of the highest priority. There are some actions that can provide hope in preventing species extinction, which include the following (Tallamy 2019):

- Creating and expanding National Parks throughout the world
- Establishing wildlife corridors that connect National Parks so that unconnected National Parks do not become genetic islands
- Expand World Heritage Sites: The United Nations already has a program in place that designates areas as World Heritage Sites. Currently, there are well over 1,000 World Heritage Sites worldwide (United Nations 2024b). The purpose of World Heritage Sites is to appreciate and learn about our legacy from the past, what we live with today, and what we pass on to future generations. Our cultural and natural heritage are both irreplaceable sources of life and inspiration (United Nations 2024a).

Reducing habitat destruction by connecting National Parks and World Heritage Sites and creating wildlife corridors can also improve many other detrimental anthropogenic-induced activities, including overfishing and poaching, deforestation, pollution in all its forms, land use, erosion, and climate change.

11.2.3.5 Climate Change Mitigation Methods

Risk impacts induced by climate change include increased severity of atmospheric events, sea-level rise, increased temperatures, severe drought, extreme atmospheric precipitation, increased intensity and occurrence of wildfires, and other hazards. Most current hazard maps must be revised to include potential climate change impacts. One of the many challenges in revising and updating hazard maps is that much is unknown as climate change is relatively recent, dynamic, and complex. Not all the facts are known and will require constant research and evaluation to adequately address each risk. This presents some level of difficulty in establishing specific mitigation efforts and engineering controls to adapt and survive climate change impacts. However, many impacts are already being felt throughout the world and reductions in greenhouse gas emissions have not begun in earnest. Therefore, mitigation efforts must focus on climate change adaptations and reducing greenhouse gas emissions.

Preparing for climate change is challenging because climate impacts will vary at each location. Each geographic area and community will need to prepare actions that are prioritized for each location (Snover 2007). The steps for preparing for climate change in any local or regional community include the following recommendations (Snover 2007):

- Collecting data on local climate and historical impacts within the study area
- Evaluating how climate will change in a particular area
- Projecting climate impacts into the future
- Identifying areas of greatest concern
- Establishing community outreach and educational opportunities
- Conducting a sensitivity analysis periodically
- Evaluating adaptive capacity
- Establishing short-term and long-term goals
- Measuring progress
- Updating plans as new information becomes available

At this point, humans have no choice but to start preparing and adapting to climate change. If preparing for climate change is delayed, the economic and ecological impacts will only get worse (Snover 2007).

11.2.3.6 Deforestation Mitigation Methods

Deforestation is one of the primary causes of habitat loss on Earth. Deforestation is a leading cause of reduction in species abundance, endangering species, and species extinction. The following presents a list of action items that the individual and local community can undertake to prevent ocean pollution (Rogers 2023):

- Plant trees
- Consider planting native vegetation rather than grass
- Do not purchase products that contain palm oil
- Do not burn wood excessively
- Consider downsizing your home
- Utilize reusable shopping bags
- Reduce meat consumption
- Print documents only when required
- Use recycled paper
- Choose to purchase items with minimal packaging material (remember if you don't consume or use it, it becomes a waste)
- Recycle plastic, paper, glass, and aluminum (contact for municipality for additional information and recycling opportunities)
- Receive and pay bills online
- Call companies that participate in junk mail advertisements
- Contact your local and state-elected officials and encourage them to support sustainable legislation

11.2.3.7 Agriculture Mitigation Methods

According to the United Nations (2024a), agriculture is the largest land use on Earth and now occupies 40% of the land surface on the planet. In addition, agriculture is the largest pollution source on Earth and is also the largest single factor responsible for species population decline. In addition, agriculture affects ocean health in many ways including excessive sedimentation, pollution from multiple sources, and types of pollution, including fertilizers, pesticides, and herbicides. Agriculture also contributes significantly to climate change and ocean acidification (United Nations 2024b).

Some examples of risk mitigation methods to reduce negative agriculture impacts include the following (Natural Resource Defense Fund (NRDC) 2024):

- Planting cover crops to maintain the soil and reduce erosion
- Reduce or eliminate tilling
- Use organic farm practices
- Reduce or eliminate pesticide and herbicide use
- Composting
- Crop rotation
- Ecological aquaculture
- Strategic crop planting to reduce pesticide use
- Eat less meat

11.2.3.8 Ocean Mitigation Methods

Addressing ocean pollution is a monumental task. Currently, the biggest threat to the oceans is generally considered plastic pollution. Historically, the oceans were used as the dumping ground and human privy for centuries (Rogers 2023). With plastic production increasing, low levels of recycling, poor waste management, and between 4 and 12 million metric tons of plastic dumped into the oceans yearly, the outlook is not anticipated to improve (Rogers 2023). Plastic pollution impacts sea turtles, whales, seabirds, fish, coral reefs, and countless other marine species and habitats. Scientists estimate that more than half of the world's sea turtles and nearly every seabird on Earth have plastic in their bodies (Ocean Society 2024). Plastic is just one example of how humans have mistreated the oceans. Even in the United States, dumping garbage and disposal of toxic waste in the ocean was acceptable up until the late 1980s (Rogers 2020).

Cleaning up the oceans will require a global effort. Mitigating the anthropogenic effects on the oceans will not be solved anytime soon but must be addressed with tenacity and commitment for decades to come. A short list of action items that the individual and local community can undertake to prevent ocean pollution include the following (Rogers 2023):

- Become more educated
- Donate to charitable organizations that focus on cleaning the ocean of pollution
- Contact your local and state-elected officials and encourage them to support sustainable legislation
- Support organizations addressing plastic pollution
- Participate in beach and river clean-up activities
- Spread the word
- Recycle properly
- Avoid products with micro-plastics
- Reduce use of single-use plastic products

11.2.3.9 Nuclear Radiation Mitigation Methods

Approximately one-third of the US population is located within 50 miles of a nuclear power plant (Nuclear Regulatory Commission 2023). Protecting yourself from nuclear radiation is much the same as protecting yourself from solar radiation.

Those protective and mitigation methods include eliminating exposure if possible, distancing yourself from the source, and shielding yourself. Shielding yourself from nuclear radiation can include a barrier with a high lead content. If that is not available, a thick layer of concrete and water can increase protection (USEPA 2023a). In a large-scale radiological release, such as a nuclear power plant accident or terrorist attack, recommended actions include getting inside a building or structure, staying inside, and following the advice of emergency responders (USEPA 2023a, b) While inside a structure, close all windows and doors, take a shower or wipe exposed body parts, and drink bolted water and eat food from sealed containers.

Clean-up of historical nuclear radiation contamination presents unique challenges because of the nature of nuclear contamination. In general terms, management and confinement of the nuclear sources is the primary objective along with proper protection of those conducting remedial actions. Significant sites in the United States where nuclear radiation remediation is being conducted include Georgia, Tennessee, Washington, and Nevada (USEPA 2023b).

11.2.3.10 Summary of Mitigation Methods

In general, mitigation efforts begin by understanding the risks posed by a hazard and that usually involves creating a geological map that identifies the locations of highest risk. The next step is to develop a plan for an early warning system for forecasting future potential events. Geological hazards have a long way to go in forecasting when an earthquake or volcanic eruption will occur and at what magnitude.

Forecasting environmental hazards such as hurricanes has progressed far better in recent decades. Anthropogenic-induced hazards are least understood and, therefore, the most difficult to forecast because of a general lack of understanding of the science behind the interactions of humans and the natural world. This brings us to the last section of this chapter, which introduces the concept of sustainability.

11.3 INTRODUCTION TO SUSTAINABILITY

At this point, we now turn our attention to sustainability. We will discuss the finer points and history of sustainability and then discuss what individuals can do to create a more sustainable environment in their immediate vicinity. In addition, we will outline how each person can get involved in establishing environmental and sustainable policies in their communities. We will focus on three overriding sustainability needs that are of the highest priority and that individuals and local communities can play a significant role. Those sustainability issues include (1) reducing or eliminating chemical contaminants, (2) reducing energy and climate change impacts, and (3) addressing habitat destruction and species decline.

11.3.1 Defining Sustainability

USEPA defines sustainability as creating and maintaining conditions under which humans can exist in productive harmony to support present and future generations (USEPA 2024d, 2024e). Note that the definition states "productive harmony." One may ask, in productive harmony with what? The answer is with the environment. Therefore, to achieve some level of sustainability, we must understand the environment and also understand the aspects of how humans impact the environment. Sustainability is then the outcome of analyzing the aspects of human activity with the natural environment and developing engineering methods to either minimize or eliminate harmful impacts.

Sustainability is based on a simple principle that states that (USEPA 2024d):

> Everything we need for our survival and well-being depends, either directly or indirectly, on our natural environment

The United Nations defines sustainability as (United Nations 2024b, 2024c):

> development that meets the needs of the present without compromising the ability of future generations to meet their own needs

Sustainability had its origin in 1969 in the incorporation of the United States National Environmental Policy Act (NEPA) (USEPA 2024d, 2024e):

> to foster and promote the general welfare, to create and maintain conditions under which humans and nature can exist in productive harmony and fulfill the social, economic, and other requirements of our present and future

The term "sustainable development" was first used by the United Nations in 1987 (United Nations 2024c). In practice and during the course of developing and actually realizing a sustainable balance, the meaning of sustainability is actually a more complex concept and has three core pillars. The three pillars of sustainability are Environmental, Social, and Economic (United Nations 2024c).

These three pillars are the framework to evaluate sustainability for organizations, products services, and technologies (United Nations 2024c). The three pillars mean that sustainable development can be achieved only when environmental protection, social equity, and economic profitability coexist without one area or pillar taking over any of the others. The three pillars should be viewed as interacting with each other at the same level. The three pillars are also referred to as People, Profit, and Planet.

Figure 11.6 presents both interpretations of the three-pillar concept.

The environmental pillar involves regulations, laws, and other tools used to deal with environmental issues such as the management of land, freshwater, oceans, forests, air, natural resources, and wildlife. Environmental management involves the use of environmental science and conservation biology to manage at a high level the allocation of resources such as land, water, and air. Environmental management also involves direct management of the environment through ecosystem restoration and improvement such as planting trees and forest preservation. Another element of environmental management is the direct management of human activities such as managing and reducing pollution, utilizing clean energy, and building energy-efficient buildings (United Nations 2024c).

The social pillar refers to initiatives, public policies planning, and regulations supporting social issues. These actions include addressing poverty, social justice, peace, promoting diversity, quality of living, access to healthcare, education, community development, cultural sustainability and heritage, and some aspects of religion. The social pillar is the least defined and understood pillar compared to environmental and economic. However, the social pillar influences all human activities and is strongly connected to the economic and ecological dimensions of sustainability. Peace, security, and human rights are some of the most important components of social sustainability

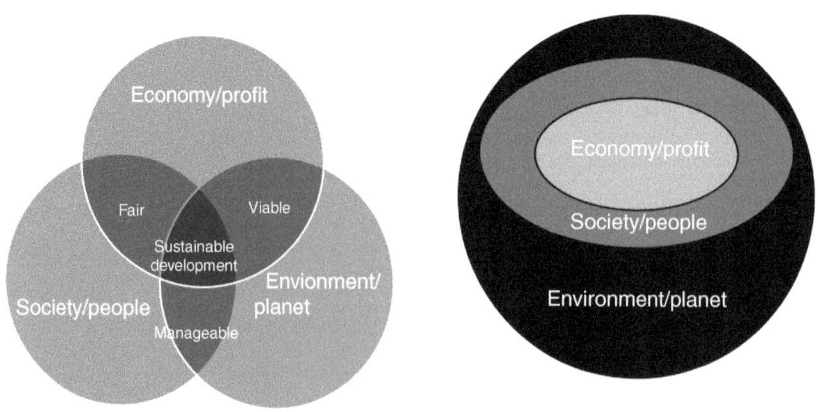

FIGURE 11.6 Two diagram variations depicting the relationships between the three pillars of sustainability.

because wars, crime, and unethical behavior are not only wasting resources, but they adversely impact the environment (United Nations 2024c).

The economic pillar of sustainability is essential for the existence of commerce. A sustainable business should include evaluating and encouraging a more balanced culture, where social and environmental factors are taken into consideration together with profitability and economic sustainability. Unfortunately, often the limited resources available in the environment are not factored into sustainability. This is a significant issue when attempting to achieve long-term economic sustainability when the natural resource is completely depleted (United Nations 2024c).

11.3.2 Effectiveness of Environmental Protection

The United States has made significant progress in restoring the environment and improving sustainability since the USEPA was formed in 1970. Through the enactment of the Clean Air Act (CAA), Clean Water Act (CWA), Resource Conservation and Recovery Act (RCRA), and the Comprehensive Environmental Response and Liability Act (CERLA) in the 1970s to 1980, the general air quality has improved, water quality has improved, and thousands of legacy sites have been cleaned up. An amazing effort and achievement, especially when considering that the population of the United States has doubled during this time period (Rogers 2019b).

Even when considering all that has been accomplished over the last 50 years, it clearly has not been enough because the natural environment continues to decline and more and more species are threatened by a host of human-induced activities worldwide. We have come to know through many different scientific studies that our planet's habitability continues to decline. This is evidenced through polluting the air, water, and land, which is evidenced through climate change, species reduction, and extinction, ocean pollution, decline in coral reef health, pollution in Antarctica, urban expansion and land use, agricultural practices, forest management, and still many other activities.

It is clear that action at the global level must be undertaken to understand the impacts humans have made on our planet and act together and diligently to limit the negative effects and restore ecosystems where possible. These actions must be undertaken before it is too late not just for the survival of other species but for the survival of our own species as well. The first step is to focus efforts on reducing GHG emissions and lowering the many threats that a changing climate may inflict on all life on Earth.

11.3.3 Creating a More Sustainable Environment

Sustainability is complex and will require significant effort at the global, national, local, and individual levels. The rest of this chapter will be dedicated to what actions are necessary at the local and individual level. Many believe that local- and individual-level actions are most important because, in essence, it will take cooperation and effort from every individual on Earth to achieve a global community that is sustainable.

Sustainability involves much more than addressing and reducing climate change impacts. As Chapter 9 discusses, many scientists believe that species extinction is of greater concern at this moment. Challenges that must be addressed are impacts first discussed in Chapter 6 and include the following (Rogers 2023):

- Climate change
- Species extinction
- Decline in species abundance
- Invasive species
- Pollution and contamination of the air, land, and water
- Ocean pollution

- Deforestation
- Agriculture practices
- Erosion
- Habitat destruction
- Land development

11.3.3.1 Mitigating Risks through Prevention Techniques

Addressing pollution and contamination on the individual and local levels should not be considered a monumental task. Individuals can have a significant impact on improving the environment by practicing what is known as ESPM methods. ESPM stands for elimination, substitution, prevention, and minimization (ESPM). ESPM is an acronym for pollution prevention principles. Pollution prevention can be implemented with a stepwise evaluation process that proceeds from the most preventative measure to the least preventative measure.

Practicing ESPM reduces anthropogenic impacts of ocean pollution, noise pollution, light pollution, pollution from agricultural practices, and other sources of air pollution, water pollution, and land pollution. Practicing ESPM methods can also reduce climate change, species extinction and abundance, invasive species, deforestation, erosion, land development, and habitat destruction impacts.

This process is referred to as ESPM, and consists of these steps (Rogers 2023):

- **E**limination: Not using potentially harmful chemicals
- **S**ubstitution: Using a less potentially harmful chemical instead of a harmful one.
- **P**revention: Using engineering controls and other measures to minimize the potential for a release; employed if eliminating or substituting a potentially harmful chemical is not possible.
- **M**inimization: Reducing usage of harmful chemicals or a reduction in generated wastes through process changes, recycling, or other methods

11.3.3.1.1 Elimination

The most aggressive form of pollution prevention is elimination of hazardous chemical use. Where possible, the elimination of hazardous substance use is the preferred pollution prevention method because it is the easiest to manage and has the greatest benefit to the environment. If there is no hazardous chemical use, there is minimal risk. In most cases, total elimination of hazardous chemical use is usually not possible, so the elimination of chemicals should focus on contaminants with high contaminant or pollution risk factors (CRFs). For example, chromium VI and DNAPL VOCs have high CRFs for groundwater, and PAHs, chlordane, PCBs, and mercury have high CRFs in soil (Rogers 2023).

11.3.3.1.2 Substitution

The next most aggressive form of pollution prevention is substitution. Substitution involves using alternative chemicals with the goal of greatly reducing risk. For instance, if a facility uses DNAPL VOCs for cleaning, an effective substitute may be citrus-based cleaners. The Solvent Alternatives Guide provides options and guidance for evaluating available chemical substitutes for common solvents (USEPA 2024f).

Other alternatives exist for chemical substitution. For instance, mercury-containing devices such as switches, thermometers, and monometers can be substituted by digital devices. Liquid transformers containing PCBs can be substituted with dry transformers or with transformers not containing any detectable concentration of PCBs. Other examples of substitution include the following:

- Using paints without VOCs and certain heavy metals
- Using biodegradable oils
- Using paraffin as a lubricant instead of oil

11.3.3.1.3 Prevention

Prevention of contamination entering the environment can be accomplished through engineering controls. Engineering control methods include release prevention, release detection, release containment, and release cleanup immediately after a spill.

The following examples (denoted by bullets) highlight some ways activities and operations can be modified to minimize the potential for the release of hazardous substances to the environment (Rogers 2023):

- Evaluating and eliminating potential points of release. Sumps, pits, trenches, floor drains, and chemical storage and usage areas are common points for hazardous substance release. Conducting an inventory of these locations and locations of chemicals present at a facility will assist in identifying areas where releases may occur.
- Modifying liquid waste storage areas. Storage methods for liquid wastes should include multiple and redundant engineered systems to prevent a release and contain a release if one does occur. These engineered systems typically include:
    - Storing liquids wastes inside and under a roof
    - Coating the floor with epoxy
    - Providing secondary containment if a release were to occur. In this example. Drip pans are located beneath each outer container.
    - Locating the liquid waste storage area at a location without floors drains, sumps, trenches, and pits.
    - Using redundant storage containment. Sealed drums containing liquid wastes are located inside the outer containers pictured in this example.
    - Properly labeling the contents and potential hazards.
- Sealing sumps, pits, trenches, floor drains, and liquid storage areas with an impervious surface. In most instances, leaks will occur from structures composed of concrete storing or conveying liquid wastes. Leaks can occur from seams, cracks, or, in some cases, from the direct migration of liquid through the concrete surface itself. These surfaces should be sealed to prevent a release to the subsurface.
- Providing secondary containment. Engineering redundant systems can effectively prevent an uncontrolled release to the environment.
- Spill containment and cleanup. Despite the existence of engineering controls, accidental releases do occur. Therefore, proper response is necessary to prevent the uncontrolled release of a hazardous substance and to protect human health and the environment. Spill stations outfitted with an assortment of tools, containers, personal protective gear, and instructions can ensure small spills of liquids not presenting an immediate threat to human health are addressed quickly and safely. These stations should be located near hazardous substances.
- Education and training. Education and training are critical for the prevention, response, and cleanup of spills. These efforts can also prevent the response to a spill or accidental release of a hazardous substance when evacuation and immediate notification are the necessary courses of action, and qualified emergency personnel are required on site.

11.3.3.1.4 Minimization

Waste minimization involves using less hazardous substances through conservation efforts or process changes. The result is a reduction in the amount of waste requiring disposal. In some instances, minimization also includes recycling. For instance, reducing the discharge or generation of liquids through process changes, recycling of water, or the evaporation/recycling of liquid or solid waste can greatly lower the volumes and mass of waste streams.

Recycling of metals, plastic, glass, wood products, and many other materials can greatly minimize the amount of solid waste generated and disposed of in a landfill. Often, these activities result in significant cost savings. The future in this arena is promising, as the discovery of new options for the beneficial reuse of waste materials parallels the appearance of new waste sources. Minimization is also achieved through awareness, tracking, and accountability.

Energy reduction strategies evaluate the energy types consumed, and their usage rates by location over time. In addition, energy loss and potential recovery/recycling may also come into play.

Purchasing habits can greatly assist minimization efforts. When supplies are purchased in bulk containers, the number of containers requiring disposal is reduced. It is also possible to realize reductions in energy usage and air emissions through production efficiencies. For instance, switching to more efficient gas turbines in certain production processes may save energy, while conducting energy-intensive activities at night when energy demand is low can lower air emissions. At off-peak times, power companies are more likely to substitute a non-fossil fuel such as hydroelectric power for coal.

Product packaging and transportation methods are also important areas where many improvements can be made from an environmental perspective. For instance, using packaging made from biodegradable material instead of plastic can have a significant positive impact on the environment. In transportation, the use of rail or barges to transport products consumes much less fossil fuel and reduces air emissions.

The following presents a list of action items that the individual and local community can undertake for pollution prevention (Rogers 2023):

- Do not consume water from plastic bottles
- Recycle plastic, paper, glass, and aluminum (contact for municipality for additional information and recycling opportunities)
- Properly dispose of electronic equipment (contact your local municipality for more information concerning other items such as light bulbs, batteries, mercury switches)
- Properly dispose of other household items including cleaners, solvents, paints, and other items (contact your local municipality for a comprehensive local list)
- Utilize reusable shopping bags
- End the use of plastic bags and plastic bottles
- Choose to purchase items with minimal packaging material (remember if you don't consume or use it, it becomes a waste)
- Print documents only when required
- Use recycled paper
- Drink more tap water
- Don't drink from disposable cups
- Limit dry cleaning clothes and other items
- Call companies that participate in junk mail advertisements
- Participate in volunteer outdoor clean-up efforts in your community
- Purchase bleach-free paper products
- Collect and donate packing peanuts to a local shipping store
- Purchase rechargeable batteries
- Recycle motor oil, ink cartridges, and tires (contact your local municipality for more information)

11.3.3.2 Addressing Habitat Destruction and Species Extinction

Creating a living landscape with indigenous species instead of a grass lawn along with other actions will create more habitat, increase biodiversity, and improve ecosystem productivity (Darke and Tallamy 2019). Many ecologists, entomologists, wildlife biologists, and other scientists have a term for grass lawns; they refer to them as "The Green Desert." This is largely because the grass used for lawns in the United States is not indigenous. It arrived in the United States from Europe in the early period of colonization of North America. Indigenous insect species have not adapted to these grasses and generally do not consider them a food source, so they tend to ignore them. If the insects ignore them, then so do the birds that feed off the insects. This begins a cycle of abandonment and ecosystem decline (Darke and Tallamy 2019).

Restoring habitat for ecosystem diversity requires many considerations to be examined to construct the richest habitat possible. A way to think of building a healthy diverse habitat is to think in layers. The layers include the canopy, understory trees, shrubs, herbaceous plants, the ground layer, the dynamic edge, wet edges, wetlands, and below-ground considerations. The next step after an individual living landscape is created is to invite neighbors to construct their own living landscapes and connect them. This creates wildlife corridors and will attract a multitude of birds and other mammals (Tallamy 2019; Darke and Tallamy 2019). The following presents a list of actions that the individual and local community can undertake to increase species' habitat and improve biodiversity, and ecosystem productivity (Rogers 2023):

- Establish plant and wildlife corridors on individual homesites and coordinate with neighbors to establish larger and connected corridors
- Plant a variety of native trees and understory plants to increase food and shelter options for insects, birds, and other wildlife
- Plant milkweed
- Compost yard waste and food waste
- Limit or eliminate pesticide and herbicide application and use
- Limit fertilizer use
- Consider collecting rainwater from roofs to water plants and vegetation during dry periods
- Join a local garden club
- Plant a vegetable garden
- Plant additional flower gardens or other native species
- Consider installing rain barrels on gutter downspouts for watering gardens and plants
- Purchase organically grown and eco-friendly food
- Become more ecologically educated
- Plant a butterfly garden
- Plant indigenous or native species in gardens
- Plant indigenous or native species rather than grass
- Leaf litter and understory plant and vegetative matter should remain in place during late fall and winter to improve the number and health of native insect species the following spring.

11.3.3.3 Reducing Energy and Climate Change Impacts

Local and individual actions that can be undertaken to reduce energy and climate change impacts include the following (Rogers 2023):

- Bike to work and to school
- Walk
- Carpool
- Consider living closer to work
- Drive a more fuel-efficient automobile, hybrid, or electric automobile
- Use mass transit
- Consider an electric lawn mower and snow blower
- Install solar house panels
- Purchase green electricity, if possible, from your utility company
- Lower water heater temperature
- Dry clothes by hanging them outside
- Upgrade appliances with improved energy efficiency
- Connect your outside lights to a timer

- Install a heat blanket on the hot water heater and lower its temperature by 10% to 20%
- Use low-flow shower heads
- Take shorter showers
- Wash clothes in cold water
- Purchase "Energy Star" products
- Plant deciduous trees in locations around your home to provide shade in the summer and light into your home during the winter
- Turn your computer off rather than using sleep mode or screen saver
- Conserve energy – turn off appliances and lights when you leave a room
- Keep woodstoves and fireplaces well maintained. Consider replacing old wood-burning stoves with an EPA-certified model.
- Keep vehicle tires at recommended air pressure
- Attempt to limit or avoid drive-thru lines
- Get regular engine tune-ups and regular maintenance checks
- Request to work from home, if possible
- Do full loads of laundry
- Run the dishwasher only when full
- Turn down the heat or turn air conditioning when not at home
- Lower the thermostat during winter and especially at night and increase the thermostat during the summer
- Unplug electrical devices such as toasters, printers, chargers, computers, and other items when not in use
- Improve the insulation of your home

The following presents a list of action items that the individual and local community can undertake to prevent ocean pollution (Rogers 2023):

- Donate to charitable organizations that focus on cleaning the ocean of pollution
- Reduce use of single-use plastic products
- Support legislation to curb plastic production and waste
- Recycle properly
- Participate in beach and river clean-up activities
- Avoid products with micro-plastics
- Spread the word
- Support organizations addressing plastic pollution
- Recycle plastic, paper, glass, and aluminum (contact for municipality for additional information and recycling opportunities)
- Utilize reusable shopping bags
- Become more educated
- Contact your local and state-elected officials and encourage them to support sustainable legislation
- Avoid purchasing food in individual serving sizes
- Avoid purchasing items that are considered single use
- Plant native trees

11.4 SUMMARY AND CONCLUSION

The individual and local actions outlined in this chapter are necessary. Success will require cooperation not often experienced in our short history. We must act individually and as a community. We must also endeavor to fully understand the impacts humans have inflicted on the natural world and diligently pursue focused efforts to limit the negative effects and restore ecosystems where possible. These actions must be undertaken before it is too late not just for the survival of other species but for the survival of our own species as well.

REFERENCES

Agency for Toxic Substances and Disease Registry (ATSDR). 2024. *ATSDR Glossary of Terms. Center for Disease Control.* http://atsdr.cdc.gov/glossary.html. (Accessed May 10, 2024).

American Conference of Governmental Industrial Hygienists (ACGIH). 2024. *Guide to Occupational Exposure Values.* ACGIH. Cincinnati, OH. 498pp.

American Society for Testing Materials (ASTM). 1995. *Standard Guide for Risk-Based Corrective Action.* ASTM Publication E1739-95. ASTM. West Philadelphia, PA.

Darke, R., and Tallamy, D. 2019. *The Living Landscape.* Timber Press. Portland, OR. 392pp.

Federal Emergency Management Agency (FEMA). 2024. *Mitigation Ideas: A Resource for Reducing Risk to Natural Hazards.* www.fema.gov/default/files/2020-06/fema-mitigation-ideas. (Accessed May 10, 2024).

International Union on the Conservation of Nature (IUCN). 2024. *Endangered Species.* www.iucn.org/endangered-species. (Accessed April 8, 2024).

National Center for Biotechnology Information. 2024. *Scientific Integrity Principles and Best Practices.* www.ncbi.nih.gov/pms/articles/scientific-integrity-best-practices. (Accessed May 10, 2024).

National Earthquake Hazards Reduction Program (NEHRP). 2024. *Earthquake Mitigation Overview and Background.* www.nist.gov/materials-and-structural-systems.national-earthquake-hazards-reduction-program-office. (Accessed May 10, 2024).

National Oceanic and Atmospheric Administration. 2024. *Geology and Climate Extremes.* www.gldl.noaa.gov/xtremes. (Accessed May 8, 2024).

Natural Resource Defense Fund (NRDC). 2024. *Regenerative Farming 101.* www.nrdc.org/regenerative-agriculture-101. (Accessed May 9, 2024).

Nuclear Regulatory Commission. 2023. *Radiation Dispersion Maps and Information.* www.nrcgov/radiatio-dispersion-maps. (Accessed June 24, 2024).

Ocean Society. 2024. *The Ocean Plastic Pollution Problem.* www.oceansociety.org/resources/ways-to-reduce-ocean-plastic. (Accessed May 10, 2024).

Ohio State University. 2024. *University Libraries. We Have Met the Enemy and He Is Us.* www.library.osu.edu/sitewe=have-met-the-enemy-and-he-is-us. (Accessed May 19, 2024).

Pierce County, Washington. 2024. *Mount Rainier Hazard Maps.* www.piercecountywa.gov/3800/hazard-maps. (Accessed April 20, 2024).

Rogers, D. T. 2018. Derivation of a Comprehensive Environmental Risk Model for Urban Groundwater Protection. *International Association of Hydrogeologists Congress.* Vol. 1, p. 879.

Rogers, D. T. 2019a. Key Policy Changes Required to Reduce Environmental Degradation of the United States and Beyond. In: *International Association of Hydrogeologists 46th Congress.* Vol. 1. Malaga, Spain. p. 529. International Association of Hydrogeologist, London, UK.

Rogers, D. T. 2019b. *Compliance and Sustainability: Global Challenges and Perspectives.* CRC Press. Boca Raton, FL. 547pp.

Rogers, D. T. 2020. *Urban Watersheds: Geology, Contamination, Environmental Regulations, and Sustainability.* CRC Press. Boca Raton, FL. 606pp.

Rogers, D. T. 2023. *Environmental Compliance Handbook: Sustainability and Future Environmental Regulations.* Vol. 4. CRC Press. Boca Raton, FL. 559pp.

Rogers, D. T. Kaufman, M. M., and Murray, K. S. 2006. Improving Environmental Risk Management through Historical Impact Assessments. *Journal of Air and Waste Management.* Vol. 56, pp. 816–823.

Rogers, D. T., Kaufman, M. M., and Murray, K. S., 2008. *Empirical Analysis of Contaminant Risk at Brownfield Sites.* United States Environmental Protection Agency National Brownfield's Conference Proceedings. USEPA. Detroit, MI.

Rogers, D. T., Kaufman, M. M., and Murray, K. S. 2009. *An Analysis of Remedial Technology Effectiveness at Brownfield Sites.* Association of American Geographers Annual Meeting. Abstracts with Programs. Las Vegas, NV. American Association of Geographers, Washington, D. C.

Snover, A. K., Binder, W., Lopez, J., Willmont, E., Kay, J., Howell, D., and Simmonds, J. 2007. Preparing for Climate Change: A Guidebook for Local, Regional, and State Governments. Association with and Published by ICLEI – Local Government for Sustainability. ICLEI. Oakland, CA. 186pp.

Tallamy, D. W. 2019. *Nature's Best Hope*. Timber Press. Portland, OR. 254pp.

Thomas, W. A. 2004. *Meeting the Challenges with Geologic Maps*. American Geological Institute (AGI). Alexandria, VA.

United Nations. 2024a. *Computer Modeling for Disaster Readiness*. www.un.org/disaster-preparedness. (Accessed May 10, 2024).

United Nations. 2024b. *Human Development Report*. United Nations Development Programme. https://hdr.undp.org. (Accessed April 11, 2024).

United Nations. 2024c. *What Is Sustainable Development*. www.un.org/sustainable-development. (Accessed May 10, 2024).

United States Department of Health and Human Services (USDHH). 2005. *Report on Carcinogens*, 11th Edition. Public Health Service, National Toxicology Program. Washington, DC.

United States Environmental Protection Agency (USEPA). 1989a. *Risk Assessment Guidance for Superfund. Volume I. Human Health Evaluation Manual (Part A)*. EPA/540/1-89/002. Office of Emergency and Remedial Response. Washington, DC.

United States Environmental Protection Agency (USEPA). 2002. *A Citizen's Guide to Cleanup Methods*. EPA 542-F-01-007. USEPA. Washington, DC.

United States Environmental Protection Agency (USEPA). 2005. *Guidelines for Carcinogenic Risk Assessment*. EPA/630/P-03/001F. USEPA. Washington, DC.

United States Environmental Protection Agency (USEPA). 2007. *Treatment Technologies for Site Cleanup: Annual Status Report*. 12th Edition. EPA-542-R-07-012. Office of Solid Waste and Emergency Response, USEPA. Washington, DC.

United States Environmental Protection Agency (USEPA). 2023a. *Protecting Yourself from Radiation*. www.epa.gov/radiation/protecting-yourself-radiation. (Accessed June 25, 2024).

United States Environmental Protection Agency (USEPA). 2023b. *Radiological Environmental Remediation*. www.epa.gov./radiation/cleanup-of-radiation-at-superfund-sites. (Accessed June 25, 2025).

United States Environmental Protection Agency (USEPA). 2024a. *QA/QC and Method Performance Considerations*. www.epa.gov/qaqcmethodperformance. (Accessed May 10, 2024).

United States Environmental Protection Agency (USEPA). 2024b. *Risk Assessment: What Is It?* www.epa.gov/risk_.html. (Accessed May 10, 2024).

United States Environmental Protection Agency (USEPA). 2024c. *EPA's Report on the Environment (ROE)*. www.epa.gov/report-environment. (Accessed May 10, 2024).

United States Environmental Protection Agency (USEPA). 2024d. *What Is Sustainability?* https://epa.gov/sustainability. (Accessed May 10, 2024).

United States Environmental Protection Agency (USEPA). 2024e. *National Environmental Policy Act*. www.epa.gov/NEPA. (Accessed May 10, 2024).

United States Environmental Protection Agency (USEPA). 2024f. *SAGE 2.1: Solvent Alternatives Guide*. www.epa.gov/solvent-alternatives-guide. (Accessed May 10, 2024).

United States Geological Survey (USGS). 1998. *Forum on Geologic Mapping Applications in the Washington–Baltimore Urban Area*. United States Geological Survey Circular 1148. USGS. Washington, DC.

United States Geological Survey (USGS). 2024a. *Deadliest Earthquakes*. www.usgs.gov/deadliest-earthquakes. (Accessed May 10, 2024).

United States Geological Survey (USGS). 2024b. *Earthquake Mitigation Strategies*. www.usgs.gov/earthquake-mitigation-strategies. (Accessed May 10, 2024).

United States Geological Survey (USGS). 2024c. *Mount Rainier Geologic Hazards*. www.usgs.gov/Geologic-hazards/mount-rainier-laher-zone. (Accessed April 21, 2024).

United States Geological Survey (USGS). 2024d. *Volcanic Eruption Mitigation Strategies*. www.usgs.gov/volcanic-eruption-mitigation-strategies. (Accessed May 10, 2024).

United States Geological Survey (USGS). 2024e. *Coastal Hazard Mitigation Strategies*. www.usgs.gov/coastal-hazard-mitigation-strategies. (Accessed May 10, 2024).

United States Government Accounting Office (USGAO). 2024. *Cost of Cleaning Up the Environment*. www.gao.gov/products/gao-22-104662. (Accessed May 10, 2024).

12 Geologic Investigations in Urban Areas

12.1 INTRODUCTION

This chapter presents science-based procedures used to evaluate a specific land parcel or larger area for environmental contamination and assess its environmental risk to humans and the environment. We will present specific methods available to environmental professionals for collecting samples from different environmental media that include soil, groundwater, sediment, surface water, and air. First, we define and present some basic concepts necessary for understanding and framing geologic investigations for environmental purposes.

The **environment** is a broad term encompassing all living and non-living things on the Earth or some region thereof (United Nations 2024). This definition identifies the natural world existing within the atmosphere, hydrosphere, lithosphere, and biosphere and also includes the built or developed world (anthrosphere). **Environmental risk** is defined as the probability of an event resulting in an adverse impact upon the environment or humans (Fletcher and Paleologos 2000). Environmental risk increases when a completed human pathway occurs, or a sensitive ecological system is impacted (United States Environmental Protection Agency (USEPA) 2002).

Conducting an **environmental assessment** contributes to our understanding of human impacts on Earth and is considered the first step in evaluating and ensuring the long-term viability of Earth as a habitable planet. The basic definition of an environmental assessment is to evaluate the environmental condition and potential environmental risk of a property, parcel, or site. For hundreds of years, the practice of conducting assessments to determine adverse environmental impacts was never a significant priority. Since the last quarter of the 20th century, however, significant amounts of human and financial resources have been mobilized to evaluate the actual or potential environmental damage incurred from human activities of the past, present, or potential impacts from future activities.

Environmental investigations are generally conducted when there is the likelihood certain environmental contaminants exist at a specific property or site having the potential to cause material harm to human health or the environment historically, presently, or in the foreseeable future. Some subjectivity is inherent in the decision to proceed with an environmental investigation at any given property. This uncertainty occurs because most properties contain, store, or at some time have used substances posing potential harm to human health and/or the environment. For instance, most households store and use contaminants, including the following (Rogers 2020):

- Cleaners (ammonia)
- Solvents (acetone, mineral spirits)
- Gasoline for lawn maintenance, snow removal, and other purposes

DOI: 10.1201/9781003463153-15

- Pesticides (e.g., permethrin, used for flea and tick control)
- Herbicides (e.g., glyphosate)
- Lawn fertilizers (organophosphates)
- Some paints (volatile organic compounds (VOCs))
- Oil
- Grease

In addition, some household products contain substances posing contamination risks to the environment if discarded improperly. Some of the most common or recognizable products include the following:

- Computer equipment (e.g., lead, cadmium, mercury)
- Televisions (e.g., lead, cadmium, mercury)
- Some electrical equipment (e.g., lead, copper)
- Most batteries (e.g., cadmium, lead)
- Some building products (e.g., formaldehyde and benzene)
- Wood with certain applied preservatives or coatings (e.g., arsenic)

The purpose for conducting any environmental investigation is to evaluate whether there is an unacceptable environmental risk to a specific site or property being studied (Jain et al. 1993).

Because we live in the midst of so many harmful chemicals, some well-informed and educated interpretation is necessary to evaluate the actual risk potential when environmental contamination is identified at a site. Martin (2003) notes the triggering mechanisms for conducting an environmental investigation at a property occur when:

- A property is involved in a real estate transaction
- Contamination of some kind is released or suspected to have been released, such as a tanker spill
- A property is developed or redeveloped
- A change in land use is proposed
- A lending institution is involved with a property and requires an environmental investigation before committing any financial resources
- Something is observed at a property indicating contamination may be present

Subsurface environmental investigations often begin at the ground surface – the intersection of air, land, water, and human activity – and frequently involve hazardous materials and remnants of contamination from the past. These characteristics of subsurface investigations impart a distinct set of rigorous requirements (Rogers 2020):

- Investigations focus on evaluating the potential existence of hazardous substances. Therefore, the investigation must be conducted with great care to minimize the potential risks of chemical exposure and other site hazards to investigators.
- The results of investigations may be used in a court of law, so the expertise level of the investigator must be unimpeachable, and any method used – including those for site investigation, quality control, and documentation – must be expertly performed. Interpretations and conclusions must rely upon a firm foundation of science.
- Evaluating the possible presence of chemicals in the environment often requires searching for specific compounds at extremely low concentrations routinely in the parts per billion range. A part per billion is equivalent to 1 second per 31 years, an analogy that demonstrates why these low quantities of evidence require great care to ensure the integrity of the investigation.

- Well-defined activities characterize subsurface environmental investigations, including scientific inquiry, historical review, data collection, interpretation and assessment, and, in some cases, remedial action to lower or eliminate any environmental risk considered unacceptable. During these activities, special emphasis is placed on evaluating the presence or potential likelihood of uncontrolled releases or migrations of contaminants.
- Once discovered, the key to defining the nature and extent of contamination is through a detailed understanding of the site geology and hydrogeology. Geological and hydrogeological expertise and experience are crucial to achieving a successful outcome for the site/area being investigated.

Subsurface environmental investigations exist as a subset of a larger set of environmental investigation types. Throughout the chapter, when we refer to an environmental investigation, the reference is to a subsurface environmental investigation, unless otherwise noted.

12.2 TYPES AND DESCRIPTION OF SUBSURFACE ENVIRONMENTAL INVESTIGATIONS, STUDIES, PLANS, OR REPORTS

Environmental investigations are conducted for many different purposes and objectives but generally follow a stepwise, logic- and progressive-based approach where the results of one investigation are used to evaluate the need and scope of subsequent investigations if necessary. Most environmental investigations originate from the Resource Conservation and Recovery Act of 1976 (USEPA 1976) and the Comprehensive Environmental Response and Compensation and Liability Act of 1980 (USEPA 1980). The most common types of environmental investigations include the following (Rogers 2020):

- Phase I Environmental Site Assessment
- Phase II Investigation
- Geophysical Investigation
- Remedial Investigation
- Feasibility Study
- Ecological and Human Health Risk Assessment
- Natural Resource Damage Assessment (NRDA)
- Environmental Impact Statement (EIS)
- Remedial Action Plan (RAP)
- Confirmatory Investigation

Each type of investigation is discussed in greater detail in the following sections.

12.2.1 Phase I Environmental Site Assessment

The **Phase I Environmental Site Assessment** (ESA) is typically the first environmental investigation conducted at a specific property. The Phase I ESA is conducted with the objective of qualitatively evaluating the environmental condition and potential environmental risk of a property or site. A **property** is defined here as a parcel of land with a specific and unique legal description. A **site** is defined here as a parcel of land including more than one property or easement and typically refers to an area of contamination potentially affecting more than one property. As the first environmental investigation, the Phase I ESA is often regarded as the most important activity because all subsequent decisions concerning the property are, in part, based on the results of the Phase I ESA (Rogers 1992). Therefore, great care, scrutiny, scientific inquiry, and objectivity should be exercised while conducting the Phase I ESA.

Standards for conducting Phase I ESAs were published by the American Society for Testing Materials (ASTM) in 1993 and were revised in 1997, 2000, and 2005 (ASTM 2005a). On November 1, 2006, the USEPA established federal standards for conducting Phase I ESAs termed "Standards and Practices for All Appropriate Inquiries" (AAI) under CERCLA, commonly known as Superfund (USEPA 2005). Under CERCLA, United States courts have ruled that a buyer, lessor, or lender may be held responsible for remediation of hazardous substance residues, even if a prior owner caused the contamination. However, performance of a Phase I ESA may create a safe harbor or protection from liability, known as the "Innocent Landowner Defense," for new purchasers or lenders. Therefore, Phase I ESAs have become the standard type of environmental investigation employed when initially investigating a property.

According to USEPA (2005) requirements, an environmental professional, such as a geologist or environmental scientist, must conduct the Phase I ESA. General requirements for conducting a Phase I ESA include the following:

- Extensive review of current and historical written records, operations, and reports
- Extensive site inspection
- Interviews with knowledgeable and key onsite personnel
- Assessment of potential environmental risks from offsite properties
- Data gap or data failure analysis

A Phase I ESA is typically a non-invasive assessment, performed without sampling or analysis. In some instances, limited sampling may be conducted on a case-by-case basis, if in the professional judgment of the person conducting the assessment – or due to other requests or mitigating factors – sampling is justified. In most cases, collecting and analyzing samples is usually deferred to the Phase II investigation, but if it does occur, the sampling conducted during a Phase I ESA may include the following:

- Sediment
- Drinking water
- Surface water
- Groundwater
- Waste material
- Soil
- Lead-based paint
- Suspected asbestos-containing materials
- Mold
- Radon

The environmental professional conducting the Phase I ESA must perform extensive research and review all available written records. These researching activities apply not only to the property in question but also to the surrounding properties within a radius of up to 1 mile of the investigated property or site. The research/review process typically encompasses the following items and actions (Rogers 2020):

- Title search and environmental liens
- Historical operations documents
- Historical chemical ordering documents
- Historical photographs
- Historical aerial photographs
- Engineering reports and diagrams

- Previous environmental reports
- Previous accident reports indicating whether a spill or release of a hazardous substance occurred
- Material safety data sheets (MSDS)
- Historical and current environmental compliance documents
- Hazardous substance inventories and mass balance evaluations
- Fire insurance maps
- Soil Conservation Service soil maps and reports
- United States Geological Survey (USGS) topographic maps
- USGS investigation reports
- USGS geologic maps
- Environmental agency records
- Native American Tribal records
- Local governmental records, including sewer and utility departments
- Building permits
- Constructing diagrams and blueprints

The site inspection consists of a walk-through of the property or site. Items to evaluate and document during the site inspection include (modified from Rogers 1992):

- Interviews with knowledgeable and key facility personnel
- Stressed vegetation
- Areas absent of vegetation
- Stained soil or pavement
- Aboveground storage tanks (ASTs)
- Underground storage tanks (USTs)
- Chemical storage areas
- Back doors
- The "back 40," meaning the rear of a property
- Signage
- Storage sheds
- Refuse storage and containers
- Special labeling
- Evidence of fill or mounding
- Recent excavations or land disturbance
- Depressions in land surface
- General topography
- Wetlands
- Mold
- Animal droppings
- Insects
- Pits
- Trenches
- Floor and roof drains
- Broken concrete
- Areas not inspected or inaccessible
- Weather conditions
- Recent precipitation events
- Nearest water body
- Onsite and offsite drainage patterns

- Onsite or offsite evidence of wells or borings
- Potential asbestos-containing materials
- Potential septic systems
- Utilities
- Onsite or offsite dumping
- Offsite inspection
- Soil type(s)
- Potential contaminant migration pathways
- Potential ecological and human receptor pathways

Once the environmental professional has completed the data collection and site inspection portion of the Phase I ESA, an evaluation of whether there is evidence of an existing release, a past release, or a material threat of a release of any hazardous substance or petroleum is made. If a product has been released and made its way into structures on the property or into the ground, groundwater, or surface water of the property, then this situation is termed a **Recognized Environmental Condition (REC)**. If a REC is discovered, further investigation will likely be recommended to evaluate its potential significance. Many sites have more than one REC, and many of these RECs may require further evaluation. An example of a REC with significant amounts of oil-stained soil, stressed, and dead vegetation is presented in Figure 12.1.

RECs are intended to exclude *de minimis* conditions generally not presenting a threat to human health or the environment and typically would not be the subject of an enforcement action if brought to the attention of appropriate governmental agencies. However, there may be impacts encountered whose perceived severity falls between a *de minimis* condition and a REC. In these situations, the term **environmental concern** is applied. Items listed as environmental concerns become RECs if left unattended or lead to a release.

Figure 12.2 is a photograph of a paved parking space with a residual amount of what appears to be a petroleum product discharged from an automobile. In the opinion of the environmental professional who conducted a Phase I ESA at this property, this condition was not evaluated to be a REC but was characterized as *de minimis*.

Historical aerial photographs are effective sources of information and often help with environmental investigations. In urban areas especially, aerial photographs from several sources are readily available. These sources include the following:

- Private local companies specializing in aerial photography
- Private national companies specializing in aerial photography

FIGURE 12.1 Recognized environmental condition. (Photograph by Daniel T. Rogers.)

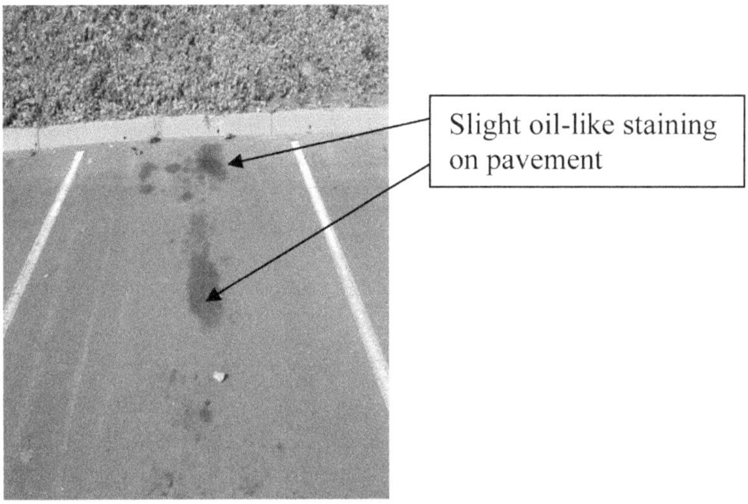

FIGURE 12.2 Example of a de minimis release. (Photograph by Daniel T. Rogers.)

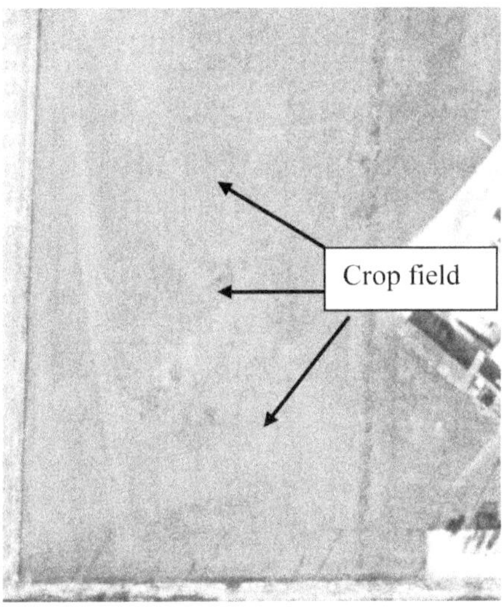

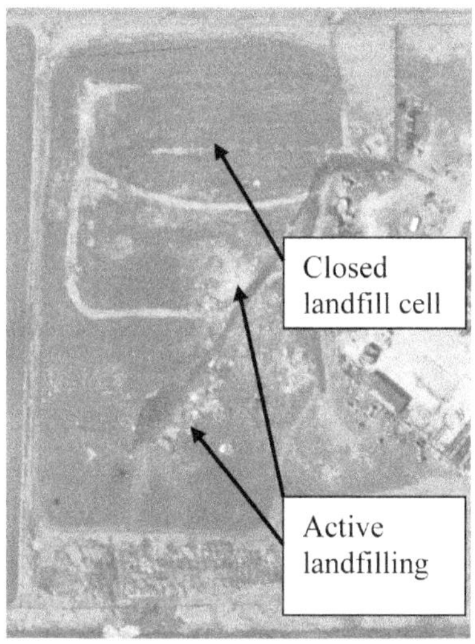

FIGURE 12.3 (A) Aerial photograph showing farmland. (B) Earlier aerial photograph showing landfilling activities.

- Local and state historical societies
- Local and state agencies
- Utility companies
- Local companies
- Federal agencies, such as USEPA, USGS, Soil Conservation Service, National Forest Service, Bureau of Land Management, and National Park Service.

Figure 12.3 demonstrates how historical aerial photographs can help identify *RECs*. On the left, Figure 12.3A shows a particular property as farmland with no identifiable REC. Analysis of an earlier aerial photograph taken a few years earlier of the same property (Figure 12.3B to the right) indicates the property was used as a landfill.

12.2.2 Health and Safety

If the result of the Phase I ESA indicates further investigation is needed, a health and safety plan must be prepared before the Phase II investigation can begin (Occupational Safety and Health Administration (OSHA) 1989, as revised). This plan can only be prepared by a qualified professional because onsite Phase II investigations are conducted at locations containing physical and chemical hazards. The potential for exposure to these hazards requires specialized health and safety training for personnel going into the field – an absolute necessity for anybody doing this work – and especially critical in urban settings with their additional array of hazards.

The items below represent the minimum level of health and safety planning required before fieldwork begins (Rogers 2020):

- List of activities, or scope of work to be conducted
- List of emergency contact names, titles, and contact information
- Identification of potential risks excluding chemical exposure including:
  - Potential physical hazards
  - Confined space entry
  - Overhead hazards
  - Traffic hazards
  - Weather
  - Other natural and anthropogenic hazards
- List of chemicals that may be present
- MSDS for every chemical that may be encountered during onsite activities
- Review medical procedures for each chemical that may be encountered during onsite activities to ensure that proper safety equipment is onsite and readily available
- Map showing nearest medical treatment facility with directions
- List of appropriate personal protective clothing
- List of other clothing requirements such as hard hats, steel-toed shoes, gloves, reflective safety vests, and hearing protection
- Review safety procedures for non-routine activities
- Review buddy system requirements, if appropriate
- Review emergency hand signals, if required
- Check safety equipment to ensure that it has been properly decontaminated and is in working order
- Contingency plans in case an incident occurs
- Update the health and safety plan as appropriate as new information is obtained

Specific health and safety equipment may be required when conducting Phase II investigations. The list below also contains quality control and quality assurance considerations:

- Hard hats
- Steel-toed boots
- Specialized gloves
- Hearing protection
- Safety glasses
- Reflective vests, if appropriate
- Face protection, when appropriate
- Boot covers, when appropriate
- Vapor monitors and detectors for compounds or conditions, including:
  - Explosive conditions
  - Oxygen levels

- Hydrogen sulfide
- Carbon monoxide
- VOCs
- Chlorine gas
- Temperature
- Other potential hazards, as appropriate

Depending on the specific physical and chemical hazards encountered, increased levels of health and safety procedures and personal protective equipment may be required to ensure the health and safety of personnel conducting the investigation. The highest level of protection would include a specialized suit and breathing apparatus supplying purified air along with several layers of specialized gloves and boots. In many cases, personal protective equipment is promptly removed and disposed of properly after field activities to prevent the inadvertent spread of any contamination beyond the impacted area.

12.2.3 Demolition

In many instances, evaluating options for demolition of onsite structures should be made before starting an anticipated robust subsurface investigation. The advantages to conducting demolition activities include the following (Rogers 2020):

- Lowering the potential for trespass and injury
- Increases potential for discovery of sources of potential contamination, such as historical floor drains, sumps, pits, tranches, tanks, etc.
- Access to the subsurface by not having any obstructions that could impede evaluating the nature and extent of contamination or interim remedial activities, if necessary
- An improved Conceptual Site Model (CSM)
- Removing and proper disposal of potential residual wastes not previously discovered
- Lowers maintenance costs

In many instances, demolition of onsite structures provides the investigation team with direct access to the developmental and operational history of the site. This provides the investigation team with valuable information as to the storage, use, and waste disposal practices of hazardous substances at the site, which improves the reliability of the Conceptual Site Model, which we will discuss in greater detail later in the chapter.

12.2.4 Phase II Environmental Site Assessment

When a Phase I ESA has been completed and there are one or more RECs identified, a Phase II investigation will likely be required to confirm the existence of the REC(s). The scope of Phase II investigations varies widely and largely depends on the RECs identified in the Phase I ESA. Under normal circumstances, a Phase II investigation will involve the collection and analysis of samples collected from the area identified as a REC in the Phase I ESA. These samples are collected and analyzed to evaluate the specific contaminants present and if they exceed federal- or state-mandated concentrations for protection of human health and the environment.

Phase II investigations are designed to use the scientific data gathered from each step to evaluate the need for or to plot the direction of the subsequent step(s). The USEPA has published guidance for conducting investigations under the CERCLA (1988) and under the Resource Conservation and Recovery Act (RCRA 1986). In addition, many states have published guidelines for conducting investigations; e.g., New Jersey (2005), Ohio (1995, 2006), and California (1995).

The investigative process follows three steps consisting of find, define, and refine. Each step in the sequence is described in greater detail below (Rogers 2016a):

- **Find**: This type of investigation has the dual purpose of finding and identifying specific contaminants and is accomplished by collecting and analyzing biased or "worst-case" samples. Examples of "worst-case" samples are those nearest the source of the release, directly beneath an outfall, appearing the most heavily stained, exhibiting the strongest odor, or those registering the highest readings on field screening equipment.
- **Define**: After contamination is found, the purpose of the define-type of investigation is to evaluate the nature and extent of contamination. Define-type investigations may consist of several phases requiring a greatly expanded scope compared to the find-type of investigation. The resulting scope depends upon the following factors:
  - Federal- or state-specific requirements for each contaminant of concern
  - The media being evaluated (e.g., soil, groundwater, sediment, surface water)
  - The pathways being considered (e.g., ingestion, inhalation, dermal contact)
  - Background concentrations of heavy metals such as lead or arsenic

The extent is derived laterally and vertically by these two equivalencies: (1) to the concentration equal to or less than the lowest applicable cleanup criteria for non-naturally occurring compounds and (2) to background concentrations for those compounds naturally occurring in the environment, such as heavy metals.

- **Refine**: The purpose of the "refine" type of investigation is to gather additional information about the contamination and determine if an unacceptable risk to human health and the environment exists and requires remedial action. Refine-type investigations include collecting and analyzing samples from locations of highest risk for specialized analytical parameters to evaluate certain remedial options or to gather more specific data on the nature and extent of contamination.

Depending on several co-dependent factors, many phases of investigation may be necessary to fully understand the nature and extent of contamination detected at any one site. The three main factors that influence the depth of scientific inquiry required encompass three factors that include the following:

1. The nature of the release(s). This includes information that may be included in the Phase I ESA but may require further evaluation based on Phase II results. In basic terms, scientific information contained in the Phase I should generally match information in the Phase II. If not, the **Conceptual Site Model** must be revised. A CSM is a model that tells the science story concerning all aspects of a release of hazardous substance(s) into the environment. Items that the investigation team needs to consider are questions that are listed below. If the investigation team suspects that the CSM that was developed does not match the scientific evidence collected, the CSM must be revised, and further scientific data will likely be required until the CSM and the scientific data are in general agreement. The questions to ask include (Rogers and Welty 2017; Rogers 2016a):
   a. Was the source(s) of the contamination properly identified?
   b. Were other potential points of contaminant release(s) ruled out?
   c. Was the estimate of contaminant mass within reason?
   d. Was the time frame of contaminant release(s) reasonable?
   e. Did the investigation team properly rule out other potential contaminants?
   f. Did the investigation team evaluate all potential contaminant pathways and media?

g. Was the extent of contamination defined, including the vertical extent?

2. The nature of the contaminant(s) released. Evaluating the nature of the contaminants involves understanding the physical chemistry of each contaminant and how each behaves in differing geologic environments.
3. The nature of the geological environment. This involves where and how to evaluate the nature and extent of contamination in the specific geologic environment to which the release occurred.

As we shall discover in future chapters, there are certain instances where more intense investigation techniques should be conducted. This occurs where certain types of contaminants are released into a geologic environment that favors the spread of contaminants quickly and has the potential to transport the contaminants to sensitive locations where humans or other life forms are put at a higher exposure risk. Under these circumstances, high-resolution sampling may be justified to understand contaminant migration pathways. High-resolution sampling involves the collection and analyses of significant additional data points and using several different investigative techniques and methods to fully characterize the nature and extent of contamination (Rogers 2016a, 2016b).

12.2.5 Geophysical Investigation

There are occasions when direct sampling under a traditional Phase II investigation approach does not adequately characterize a specific site. Examples may include areas not accessible to sampling equipment, extremely large sites, or locations where buried objects are of great concern. In these cases, surface geophysical techniques may be used to fill in data gaps. As geophysical investigations provide supplemental data, they rarely are the only type of environmental investigation conducted at a site. There are a wide range of geophysical techniques to accommodate different objectives, investigative requirements, and site limitations, and these techniques offer certain advantages, including the following:

- Increasing the accuracy and area of coverage of subsurface investigations
- Increasing data density
- Collecting data on subsurface geology and hydrogeology at locations inaccessible to other investigative techniques, or areas considered dangerous to investigate using other methods
- Decreasing the cost and time necessary to characterize certain sites, and they are generally non-invasive

Geologists conducting environmental investigations apply geophysical techniques for the following reasons:

- Collect geologic and hydrogeologic data
- Detect and map contaminant plumes
- Detect and map subsurface features such as buried drums, utilities, tanks, trenches, landfills, buried wastes, or other buried objects
- Evaluate soil properties
- Evaluate anthropogenic subsurface structures

Geophysical investigations include the following techniques and benefits:

- Electrical resistivity. Electrical resistivity measures the apparent resistivity averaged over a volume of material (ASTM 1999a, 2000a). Electrical resistivity measurements can be used to evaluate the depth and thickness of geologic units, map saltwater intrusion and contaminant plumes, locate buried wastes, and evaluate lateral changes in geologic units.

- Electromagnetics. Electromagnetic methods measure the conductivity of subsurface materials and are frequently used to detect buried metal objects (ASTM 1999b).
- Gravity survey. Gravity or microgravity surveys measure changes in subsurface density (ASTM 1999b). Natural variations in the density of subsurface materials such as soil and rock types, cavities, buried channels, faults, fractures, and joints can be detected and evaluated using gravity surveys.
- Ground-penetrating radar (GPR). GPR uses high-frequency electromagnetic waves to evaluate subsurface strata (ASTM 2005b). GPR is one of the more popular geophysical techniques used in environmental investigations because it is easy to use and economical and employable in a variety of challenging accessibility situations. GPR can determine objects having a dielectric contrast with surrounding areas, i.e., buried drums, disturbed soil, etc.
- Borehole geophysics. Borehole geophysics uses instruments to measure and record different properties outside of a well or borehole as the instrument is lowered down the borehole (Keys 1990; ASTM 2005c). Borehole geophysics is used to evaluate well construction, lithology, permeability and porosity, and water quality.
- Seismic refraction and reflection. Seismic refraction and reflection measurements involve the measurement of seismic waves traveling through the subsurface (ASTM 2000b). Stratigraphy, structure, and material properties can be evaluated using seismic methods.

12.2.6 Ecological and Human Health Risk Assessment

After the investigation phase has been completed and the nature and extent of contamination have been defined, an environmental **risk assessment** is conducted. A common objective for conducting a risk assessment is to evaluate whether there is an unacceptable risk posed by the presence of contamination at a given site for the future intended use of the property (USEPA 2002). Intended land use usually fits into the three broad categories of industrial, commercial, and residential. In general, if the intended land use is industrial or commercial, higher concentrations of contaminants are allowed if certain site conditions are satisfied to minimize the exposure potential to the contamination.

An environmental risk assessment follows scientific procedures for evaluating the risk to humans and the environment posed by the contamination present at a specific site. Three basic components are used to evaluate risk: (1) toxicity or potency of contaminants present, (2) exposure pathways, and (3) receptors (California Environmental Protection Agency 1996; Oregon Department of Environmental Quality (ODEQ) 1998; USEPA 2002).

Toxicity. Toxic substances, or toxicants, are health threats because of the resultant effects on biotic receptors when exposure to these toxicants occurs. The following chemical hazards found commonly at hazardous waste sites may be classified as toxic: chemicals fatal to humans in low doses or to over 50% of test animals at certain concentrations; hazardous substances: chemicals that can cause harm because they are flammable, explosive, strongly acidic or basic, or can induce allergic reactions; carcinogens: chemicals along with radiation and viruses that can cause or promote the growth of malignant tumors or cancer; mutagens: chemicals, including ionizing radiation, and viruses that causes mutations – inheritable changes in DNA molecules in the genes found in chromosomes; teratogens: chemicals, ionizing radiation and viruses that cause birth defects. Toxicants include a wide variety of substances such as metals, most commonly Sb, As, Ba, Be, Cd, Cr, Co, Cu, Hg, Mn, Ni, Pb, and Zn, asbestos, nitrate, and nitrite or organic chemicals. The toxicities of organic chemicals are difficult to fully assess. We need to take into account not only the exposure level but also the duration of exposure (Yong 2001).

Exposure pathways fall into three broad categories: inhalation, dermal adsorption, and ingestion (USEPA 1989). Any chemical can affect the human body, with the outcome depending on several factors:

- The nature and toxicity of the chemical itself
- The concentration of chemical
- The distribution of the chemical at a particular site
- The chemical's route of entry into the human body (inhalation, dermal adsorption, or ingestion)
- The length of the exposure
- Individual human susceptibility

Receptors implies biotic receptors but can also pertain to the physical land environment, including surface water and groundwater. Consequently, the nature and extent of the threat of a particular pollutant will not only depend on the nature and distribution of the pollutant but also the target that is threatened.

Risk assessment is accomplished using a four-step process:

1. Hazard Identification – answers this question: Does exposure to a chemical or agent cause an increase in the incidence of an adverse health effect (e.g., cancer or birth defects)?
2. Dose–Response Assessment – quantitative characterization of the relationship between the dose of a chemical or agent and the incidence of an adverse health effect.
3. Exposure Assessment – evaluation of the intensity, frequency, duration, and routes of exposure to the chemical or agent.
4. Risk Characterization – estimation of the potential incidence of a health effect, calculated by obtaining information from the dose–response assessment along with information from the exposure assessment.

The outcome of a risk assessment usually produces a hazard quotient (HQ), which is the result of evaluating each chemical of concern for potential carcinogenic risk and chronic health risk. USEPA (2002) currently has an acceptable carcinogenic risk of 1 in a 100,000 (1×10^{-5}) cancer incidence, with values under 1 being an acceptable non-carcinogenic chronic health risk. However, many risk assessments evaluate cumulative risk. For instance, there may be ten different chemicals present, and one chemical alone may not result in an unacceptable risk, but exposure to more than one may result in an unacceptable exposure. In most circumstances this is an acceptable methodology, especially if contaminants overlap.

USEPA and most states have published guidelines for conducting ecological and human risk assessments, and these address whether a cumulative risk evaluation is warranted. A review of federal, local, and state requirements is recommended before conducting an ecological or human health risk assessment.

12.2.7 Feasibility Study

A feasibility study is conducted after the nature and extent of contamination have been defined. The timing of this study corresponds with the risk assessment if contamination is present at sufficient levels to justify the possibility of remedial action. Selecting the most appropriate remedial technology or technologies – if more than one is necessary – is a key objective of the feasibility study. Technology selection can be evaluated within this framework: (1) the nature and extent of contamination at a specific site, (2) the future intended land use or uses, (3) the results of the risk assessment, and (4) other sites' conditions, if present (USEPA 1988; New York Department of Environmental Conservation 2009).

A feasibility study evaluates many different methods and technologies to reduce contaminant risk. Some cases may not involve a specific technology but include alternatives such as institutional controls and deed restrictions. The purposes of institutional controls are to restrict or prevent access to areas posing exposure risk or to reduce the risk of accidentally spreading any of the remaining onsite contamination. Examples of institutional controls include the following:

- Vapor barriers
- Pavement
- Maintaining building structures
- Drainage control
- Access restrictions
- Fencing
- Constructing other physical barriers.

Common examples of deed restrictions include special handling of materials if they are disturbed and restricting the use of surface water or groundwater.

As part of a feasibility study, a bench scale and/or pilot level test is conducted using a particular remedial technology to evaluate its effectiveness as applied to real site conditions, contaminants, and onsite geologic and hydrogeologic conditions. A bench scale test involves collecting samples of contaminated material from a site and conducting tests in the laboratory under controlled conditions. The samples are then used to assess the efficacy of the particular technology, and, if successful, additional engineering and design data for the pilot test are collected. A pilot test is larger in scale than a bench scale test and is conducted *in situ* (at the site of the contamination). Results from the pilot test are used for final engineering design, dosing (if used), and evaluating the time required to complete remediation. Information obtained from the bench scale and pilot tests assists in selecting and designing a remedial technology.

12.2.8 Remedial Investigation

Remedial investigations represent the actions taken at a site of environmental concern based on a combination of the Phase I ESA, Phase II investigation, Risk assessment, and Feasibility assessment outcomes. Remedial investigations generate large amounts of geological and hydrogeological information about the area and often the region being investigated but often require 2 or more years to complete and may cost several million dollars. These time and cost factors dictate that these investigations are conducted only when it is generally accepted the site poses an obvious and significant risk to human health and the environment. Combining the investigation and evaluation process can shorten the time required to investigate a site and save money, provided the necessary amount of environmental risk is removed to protect human health (USEPA 1988).

12.2.9 Natural Resource Damage Assessment

A NRDA is a study that evaluates the damage or injuries to the environment (USEPA 2009). The term **injury** refers to an actual adverse impact to, or loss of, a natural resource. Damage refers to the monetary cost of restoration or replacement of the natural resource. The USEPA (2009) defines a natural resource as:

- Land
- Air
- Water
- Fish
- Biota
- Wildlife
- Groundwater

- Drinking water supplies
- Other potentially identifiable natural resources

NRDAs are usually conducted after the extent of contamination has been defined and before any final remedy is implemented. Typically, an NRDA is conducted at large sites where there is clear evidence a human-induced environmental impact significantly degraded the ecological and natural resources of a specific region.

12.2.10 Environmental Impact Statement

An EIS is conducted to identify and evaluate the positive and negative biophysical, social, and other environmental effects that a proposed development action may have on the environment (Glasson et al. 2005). An EIS includes:

1. Describing the proposed action or activity to be conducted
2. Describing the extent to which the proposed action or activity will potentially affect the environment
3. Analyzing each potential environmental impact
4. Presenting a range of development alternatives for the public to consider regarding the potential environmental impacts of the proposed action. Included in this range of alternatives is a "no action" option, meaning the project should not be undertaken.

If the EIS process concludes the proposed action has a high potential to cause significant environmental degradation, other options may be required to reduce the environmental impact. The process thus does not prohibit the proposed action from causing any harm to the environment; it merely requires disclosure and understanding of the potential impacts prior to taking any action (Glasson et al. 2005).

12.2.11 Remedial Action Plan

A RAP describes the actions implemented to lower the risk posed by the presence of contamination at a specific site to an acceptable level. RAPs include the following (USEPA 1988; New Hampshire Department of Environmental Services 2008):

- A summary of each investigation completed
- A list of the contaminants of concern (COC)
- Cleanup objectives and goals for each COC
- The technology selected to reduce concentrations of each COC
- Methodology and justification for technology selection
- A timeframe to complete each activity
- A plan outlining methods to confirm the remedial action has achieved the desired result and has met the cleanup standard
- Contingency plan in case the selected remedial action or actions do not achieve the desired result

A RAP includes detailed engineering and planning drawings to precisely depict the methods and systems proposed for lowering the contaminant concentrations at a particular site to an acceptable level.

12.2.12 Completion Evaluation Study

A Completion Evaluation Study is conducted after the site is remediated. The purpose of the Completion Evaluation Study is to evaluate the effectiveness of the remediation and to confirm the desired result of permanently reducing contaminant concentrations to acceptable levels has been achieved. Samples of previously affected media, including soil, groundwater, sediment, surface water, and air, are collected to help confirm the remediation was successful. To avoid bias, requirements for the Completion Evaluation Study are usually outlined in the RAP. In addition, many individual states have general guidelines for determining when remediation has been completed. A Completion Evaluation Study may require several years to finish, especially if groundwater is a media of concern and has undergone an active remedial action. The requirements for a Completion Evaluation Study include the following (Rogers 2020):

- Confirmatory sample locations
- Number of samples
- Media to be sampled
- Frequency of sampling
- Quality control and quality assurance requirements
- Analysis and analytical procedures
- Statistical methodology, if statistical methods are used
- Cleanup criteria for each COC
- Contingency plan in case one or more samples indicate remediation has not been completed

12.2.13 Summary of Environmental Investigations

Table 12.1 lists each type of environmental investigation and its major purpose. The Phase II Investigation, Remedial Investigation, and the Geophysical Investigation typically generate the most geological and hydrogeological information, and, in some instances, generate a large volume of detailed information. In addition, the Phase I ESA often has valuable information useful for geological purposes, especially if a Phase II or other investigation is subsequently conducted.

TABLE 12.1
Types and Purposes of Environmental Investigations

Investigation Type	Purpose
Phase I ESA	Qualitative review of a specific site to evaluate if there is any scientific basis indicating that the site poses any environmental risk
Phase II	Quantitative study that finds, defines, and refines the nature and extent of contamination
Geophysical investigation	Used to supplement and add geologic and hydrogeologic information to Phase II investigations; helps identify and locate buried objects, utilities, or other buried structures
Risk assessment	Evaluate whether the presence of contamination poses an unacceptable risk to human health and the environment
Feasibility study	Evaluate potential remedial alternatives with the objective of selecting the most appropriate technology(s) to lower risk to an acceptable level
Remedial investigation	Defines the nature and extent of contamination at large sites likely posing significant risks to human health and the environment
NRDA	Assess damages to natural resources
EIS	Evaluate potential impacts of future development on natural resources and the environment
RAP	Plan outlining what actions will be conducted to lower risk to an acceptable level
Completion study	Quantitative investigation conducted to confirm the remediation at a site is sufficient and complete

12.3 COMMON ENVIRONMENTAL SAMPLING METHODS

Sampled items during an environmental investigation typically include soil, groundwater, sediment, surface water, or air. The primary objectives when collecting any sample during an environmental investigation are: (1) collecting a representative sample and (2) ensuring the integrity of samples collected by taking measures not to contaminate or cross-contaminate the samples or altering any of the original properties of the sample (USEPA 1988).

With these objectives in mind, acceptable sampling techniques have been developed by the USEPA. In addition, most states have published protocols for the collection of samples for environmental purposes. An overview of common sampling methods for soil, groundwater, sediment, and surface water is described in the following sections. In addition to sample collection requirements and guidelines, transportation or shipment of samples may be subject to Department of Transportation (DOT) rules described in Title 49 of the Code of Federal Regulation (CFR), Parts 171 through 179. Therefore, applicable local, state, and federal guidelines should be followed when conducting an environmental investigation at a specific site.

12.3.1 Soil Sampling and Description

Soil sampling is usually initiated during the Phase II investigation and involves collecting surface and subsurface soil samples from soil borings drilled within and perhaps adjacent to RECs identified in the Phase I ESA. The number and depth of the borings are dependent upon the source and nature of contamination suspected to have been released. Soil samples are collected in a wide variety of methods, including the following (Rogers 2020):

- Taking surface samples using a stainless steel trowel or another similar device
- Using manual hand augers to depths of approximately 5 to 10 feet
- Hydraulically driving probes called a Geoprobe™, Salina, Kansas, USA, to depths of 20 feet
- Excavating test pits using a backhoe or a similar type of equipment to depths approaching 20 feet
- Boring with hollow-stem augers to depths greater than 50 feet
- Using roto-sonic boring to depths of 100 feet or more
- Using a cone penetrometer (CPT) for depths greater than 100 feet

12.3.1.1 Surface Sampling

Many investigations begin simply by collecting surface soil samples in each area of concern. When collecting an initial sample, selecting a "worst-case" sample location is generally preferred so that the presence and magnitude of COC can be identified analytically. Surface soil samples are generally collected using a stainless steel trowel or another appropriate device. Wearing appropriate health and safety clothing and adhering to sampling protocol should always be maintained.

12.3.1.2 Excavating Test Pits

When possible, excavating test pits using a backhoe is a preferred method of collecting and characterizing shallow subsurface geology. Test pits expose much more of the subsurface than other methods and give the geologist greater opportunity to observe and gather geologic information. An added advantage with test pits is the ability to select optimal locations for collecting soil samples for analysis. Recording observations in the field in real time is a critical and important step in the data collection process (Rogers 2020).

While test pits may be the preferred method of characterizing subsurface geology and providing optimal sample collection points, excavating test pits in an urban setting is uncommon. In many cases, test pits cannot be excavated because they disturb much larger areas than drilling a soil boring using a Geoprobe™ or other similar method. In addition, many sampling locations are located near utilities, buildings, or beneath paved areas, making test pits very difficult, dangerous, or impossible to excavate.

12.3.1.3 Drilling Using a Hand Auger

Using a hand auger to characterize subsurface geology has severe limitations. The twisting required to advance the auger deeper into the subsurface causes the loss of most, if not all, depositional structures and features in the recovered samples. For this reason, a hand auger is used to characterize soil or sediment type and is not used to interpret most complex geological features or structures. A hand auger may also be used to drill the first few feet to explore and avoid buried utilities before a drill rig completes the rest of the boring.

12.3.1.4 Mechanical Drilling Methods

For drilling in unconsolidated materials, preferred mechanical drilling methods include a Geoprobe™, hollow-stem auger, or roto-sonic methods. Characterizing subsurface geology from soil borings obtained by mechanical methods is much more difficult than test pits, because these methods create a long and very narrow sample (four to ten feet long, and only two or four inches in diameter). These dimensions of sample size increase the difficulty of the analysis. In addition:

1. The outer edges of soil samples collected from a mechanical method may be smeared from the sample collection method; and,
2. The sample probe must be re-inserted into the borehole after each sample is collected, and this action may lead to cross-contamination or result in a gap in the depositional history of the area. Additional care and scrutiny must be undertaken when characterizing subsurface geologic features from samples collected using these methods.

Using a Geoprobe™ for soil sampling is the preferred technology in states like Michigan. A Geoprobe™ is typically faster than other methods, can penetrate hard surfaces such as pavement and concrete, and can be used inside buildings and other locations with limited access and space. Some Geoprobe™ drill rigs are capable of directional drilling, providing advantages in some unique situations. Soil samples collected using a Geoprobe™ are typically two inches in diameter and either two or four feet long. The sample is collected inside a plastic sleeve inserted into the bottom of a metal rod hydraulically pushed and vibrated into the ground. There are drawbacks to the use of Geoprobe™ devices, however. First, great care must be taken not to "push" contaminated soil downward to lower levels, resulting in misidentification of the vertical extent of contamination. For this reason, Geoprobe™ devices are not allowed in California when conducting remedial investigations. Second, the usefulness of a Geoprobe™ is limited to the vertical extent of its capability, generally to within 20 to 30 feet. However, under either favorable conditions or specially outfitted Geoprobes™, depths of greater than 75 feet have been achieved. Finally, if a soil boring is to be converted to a semi-permanent monitoring well, a hollow-stem truck-mounted drill rig would need to be brought onto the site, resulting in additional costs.

Soil samples collected from a hollow-stem auger drilling rig are typically 2 inches in diameter and are 2 feet long. The sample is collected inside a plastic sleeve inserted into a tool called a split spoon. A split spoon is a geotechnical soil sampling tool comprised of two halves of a hollow metal tube. The tube is threaded on each end where end caps are screwed onto the split spoon to hold it in place while the sample is collected. The split spoon is lowered down the borehole to the desired sampling depth through the center of the hollow-stem auger and then is pounded into the ground using a 140-pound hammer. The number of hammer blows required to drive the split spoon into the ground two feet is recorded to provide geotechnical information concerning soil type and resistance.

FIGURE 12.4 Photograph showing a Geoprobe. (Photograph by Daniel T. Rogers.)

FIGURE 12.5 Photograph showing a hollow-stem auger drilling rig. (Photograph by Daniel T. Rogers.)

Figure 12.4 is a photograph of a Geoprobe™, and Figure 12.5 is a photograph of a hollow-stem auger drilling rig. Figure 12.6 shows a 2-inch diameter by 4-foot-long soil sample collected using a Geoprobe™ being inspected by a geologist. The plastic sleeve has been sliced in half longitudinally to provide proper access to the sample for inspection.

The roto-sonic method is similar to the geoprobe method of advancing a boring into unconsolidated subsurface deposits. One advantage of the roto-sonic method when compared to the Geoprobe is its ability to reach greater depths with a larger diameter borehole. This capability enables a larger volume of samples to be collected and also results in improved monitoring well installation success rates. Roto-sonic also produces less waste and has a faster drilling rate than the hollow-stem auger.

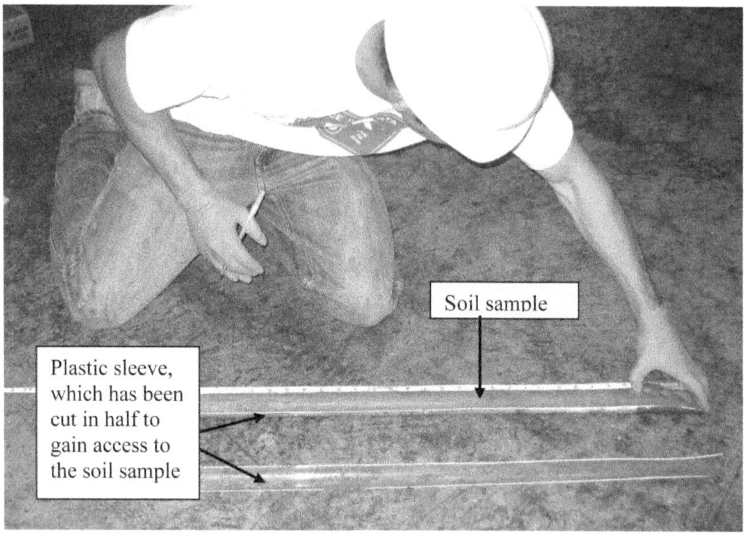

FIGURE 12.6 Photograph showing a 3-foot-long soil sample collected from a Geoprobe. Note that the person in the photo is wearing protective gloves. (Photograph by Daniel T. Rogers.)

Other drilling methods include direct mud rotary, directional drilling, solid flight and bucket augers, jetting, direct air rotary, cable tool, and rotary diamond. These methods are used when drilling through solid rock and not preferred for unconsolidated materials.

12.3.1.5 High-Resolution Sampling

There are instances where more intense investigation techniques should be conducted that involve the collection of perhaps an order of magnitude of more environmental samples and using additional sampling techniques to more fully understand the subsurface and contaminant distribution and extent. This occurs where certain types of contaminants are released into a geologic environment that favors the spread of contaminants quickly and has the potential to transport the contaminants to sensitive locations where humans or other life forms are put at a higher exposure risk. Under these circumstances, high-resolution sampling may be justified to better understand contaminant migration pathways. High-resolution sampling involves the collection and analyses of significantly more data points and using several different investigative techniques and methods to fully characterize the nature and extent of contamination (Rogers 2016a, 2016b). High-resolution sampling should be considered when one or more of the following factors has been satisfied (Rogers 2016a, 2016b; Welty et al. 2010) including evaluating the following factors:

- Types of contaminant present. In general, when the site being investigated has known releases of chlorinated volatile organic compounds (CVOCs), also commonly known as chlorinated solvents, or hexavalent chromium, especially from multiple sources.
- Subsurface geology. If subsurface geology is such that contaminant migration is not impeded through either natural conditions or from anthropogenic pathways.
- The site is located near sensitive or complex ecological exposure pathways.
- The site is located near human routes of exposure.

High-resolution sampling essentially provides a higher magnification of existing subsurface conditions that enhance the interpretation of the fate and migration of contaminants being evaluated (see Figure 12.7). The advantages of conducting high-resolution sampling are many, some include the following:

FIGURE 12.7 Difference in resolution using high-resolution investigation techniques and more tradition techniques.

- More accurate characterization
- Greatly improved and more accurate CSM
- Ensures that remediation and risk reduction measures are properly located and sized
- Lowers risk to human health and the environment by a greater understanding of natural conditions

Using high-resolution techniques must also include enhanced data organization and analysis techniques, which may include electron data management and analysis and three-dimensional representations of the data and computer modeling.

12.3.1.6 Sample Containers

Soil samples using any of the methods described above are transferred into appropriate sample containers following their collection. Most containers consist of glass having a Teflon inner lid designed to minimize the loss of vapors and are filled to capacity. Specification of sample containers, and the methods, protocols, and equipment required for sample transfer, vary with the specific contaminants being evaluated. In addition, sampling procedures and methodology differ from state to state for certain analytes. To avoid confusion, a work plan identifying the appropriate sampling procedures should be prepared prior to the initiation of sampling activities (Rogers 2020).

12.3.1.7 Documentation

Proper description of each soil sample by a qualified scientist is crucial. A field logbook houses this information.

Information recorded in a field logbook when each soil boring is being advanced should include the following:

- Drilling method
- Size of auger or drill bit
- Drilling or advance rates
- Drilling or advance difficulties
- Equipment failures
- Drilling difficulties
- Weather conditions

After extraction, each soil sample is inspected and information related to the site's geology and soil properties is recorded in the field log book, and shown below:

- Soil or sediment type (i.e., sand, clay, etc.) using the Unified Soil Classification System
- Grain size
- Grain shape
- Grain sorting
- Grain composition
- Consistency
- Plasticity
- Stratigraphy
- Observable depositional features and bedding
- Presence of fill or anthropogenic materials or substances
- Color using the Munsell Color Chart (developed by Munsell (1905))
- Color changes
- Degree of saturation
- Field screening results
- Odors
- Location of sample collection for laboratory analysis, if any

Table 12.2 lists the grain size dimensions used to differentiate between clay, silt, and sand. Figure 12.8 is a diagram that shows the relative size difference between a particle of sand, silt, and clay. Figure 12.9 is a soil textural chart used in the field as a guide to identify and describe soils on a consistent basis (USDA 2019a). Figure 12.10 describes the Unified Soil Classification System and Chart (Virginia Department of Transportation 2003).

Consistency refers to the relative ease a sediment can be deformed. There are various methods for determining consistency, but in the field, consistency can be estimated using the methodology described in Table 12.3. To determine consistency using Table 12.3, a sample of the soil or sediment is necessary; the recommended sample size is approximately 5 to 10 cm^3. Moisture content should be estimated in the field using the criteria listed in Table 12.4.

Plasticity is the property of soil or rock allowing it to be deformed beyond the point of recovery without cracking or exhibiting appreciable change in volume – its "plastic limit." Criteria for estimating plasticity of soil samples in the field are presented in Table 12.5 (OEPA 2006) (Figure 12.11).

TABLE 12.2
Soil Classification by Grain Size

Soil Type	Diameter(range in millimeters or sieve size)
Very coarse sand	1.0 to 2.0
Coarse sand	0.5 to 1.0
Medium sand	0.25 to 0.50
Fine sand	0.125 to 0.25
Very fine sand	0.0625 to 0.125
Silt	0.0039 to 0.0625
Clay	Less than 0.0039

Source: Modified from New Jersey Department of Environmental Protection. 2005. *Field Sampling Procedures Manual*. Trenton, NJ.

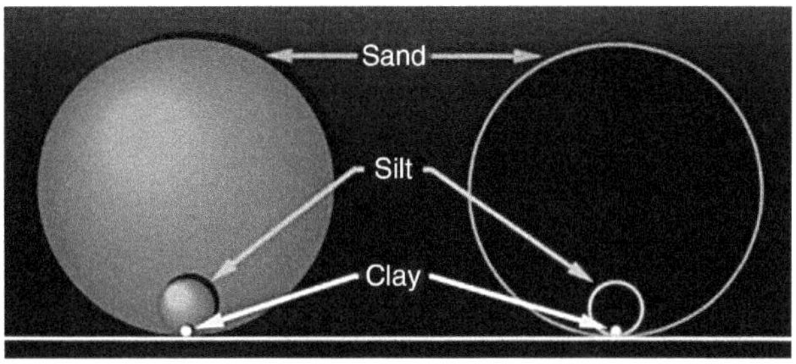

FIGURE 12.8 Relative size difference between a sand, silt, and clay particle. (From United States Department of Agriculture. 2019a. *From the Surface Down. An Introduction to Soil Surveys for Agronomic Use*. Washington, DC. https://directives.nrcs.usda.gov/sites/default/foles2/1712940824.pdf. (Accessed February 6, 2025).)

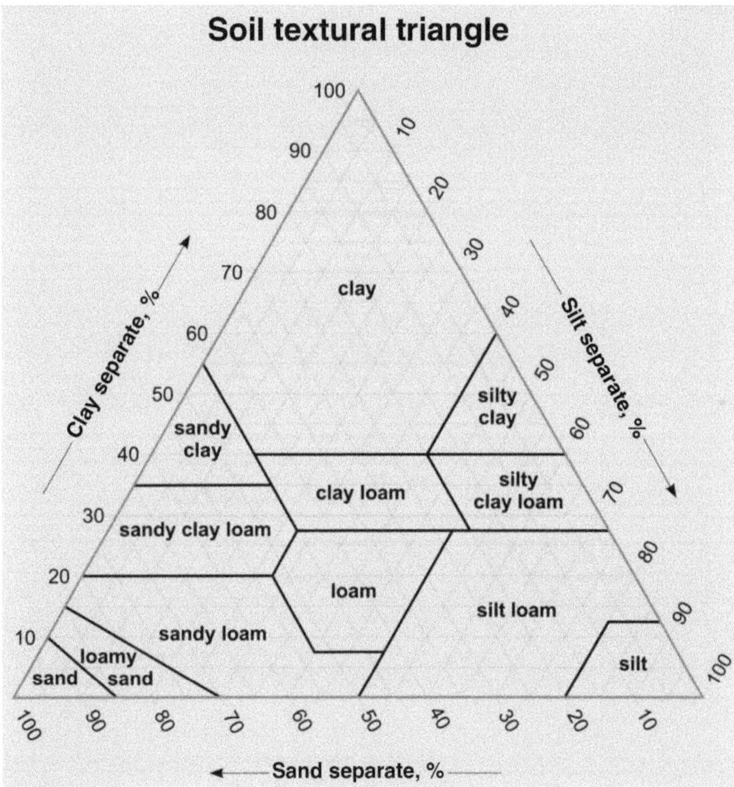

FIGURE 12.9 Soil textural chart. (From United States Department of Agriculture. 2019b. *Guide to Texture by Feel*. http://soils.gov/education/resources/lessons/texture. (Accessed May 11, 2024).)

Additional physical parameters may be used to characterize subsurface solid materials, but these cannot be reliably measured in the field and most often require determination at a laboratory. These additional parameters include the following (Rogers 2020):

- Bulk density
- Cation exchange capacity
- Organic carbon content
- pH

UNIFIED SOIL CLASSIFICATION AND SYMBOL CHART		
COARSE-GRAINED SOILS (more than 50% of material is larger than No. 200 sieve size.)		
GRAVELS More than 50% of coarse fraction larger than No. 4 sieve size	Clean Gravels (Less than 5% fines)	
	GW	Well-graded gravels, gravel-sand mixtures, little or no fines
	GP	Poorly-graded gravels, gravel-sand mixtures, little or no fines
	Gravels with fines (More than 12% fines)	
	GM	Silty gravels, gravel-sand-silt mixtures
	GC	Clayey gravels, gravel-sand-clay mixtures
SANDS 50% or more of coarse fraction smaller than No. 4 sieve size	Clean Sands (Less than 5% fines)	
	SW	Well-graded sands, gravelly sands, little or no fines
	SP	Poorly graded sands, gravelly sands, little or no fines
	Sands with fines (More than 12% fines)	
	SM	Silty sands, sand-silt mixtures
	SC	Clayey sands, sand-clay mixtures
FINE-GRAINED SOILS (50% or more of material is smaller than No. 200 sieve size.)		
SILTS AND CLAYS Liquid limit less than 50%	ML	Inorganic silts and very fine sands, rock flour, silty of clayey fine sands or clayey silts with slight plasticity
	CL	Inorganic clays of low to medium plasticity, gravelly clays, sandy clays, silty clays, lean clays
	OL	Organic silts and organic silty clays of low plasticity
SILTS AND CLAYS Liquid limit 50% or greater	MH	Inorganic silts, micaceous or diatomaceous fine sandy or silty soils, elastic silts
	CH	Inorganic clays of high plasticity, fat clays
	OH	Organic clays of medium to high plasticity, organic silts
HIGHLY ORGANIC SOILS	PT	Peat and other highly organic soils

FIGURE 12.10 Unified soil classification system. (Oklahoma Department of Transportation. 2021. *Geotechnical Investigation Guidance Document*. ODOT Geotechnical Specifications 25Aug21.pdf. oklahoma.gov. (Accessed September 25, 2024).)

- Specific gravity
- Mineral content

12.3.2 Groundwater Sampling Methods

If groundwater is encountered during the drilling of soil borings or excavating test pits, then the environmental investigation must also evaluate the potential for groundwater contamination.

TABLE 12.3
Criteria for Describing Consistency

Description	Criteria
Very soft	Thumb will penetrate easily, typically more than 1 inch
Soft	Thumb will penetrate approximately 1 inch
Firm	Thumb will indent soil approximately 0.25 inches
Hard	Thumb will not indent soil, but readily indented with thumbnail
Very hard	Thumbnail will not indent soil

Source: Ohio Environmental Protection Agency. 2006. *Technical Guidance Ground Water Investigations*. Columbus, OH.

TABLE 12.4
Criteria for Describing Soil Moisture

Description	Criteria
Dry	Absence of observable moisture, dry to the touch
Moist	Damp, but no visible water
Wet	Water is observed when sample is squeezed
Saturated	Free water present without squeezing the sample

Source: Ohio Environmental Protection Agency. 2006. *Technical Guidance Ground Water Investigations*. Columbus, OH.

TABLE 12.5
Criteria for Describing Plasticity

Description	Criteria
Non-plastic	A sample cannot be rolled to an approximate diameter of ½ inch
Low plasticity	A sample can be rolled, but is lumpy and difficult and cannot be rolled when sample is drier than the plastic limit
Medium plasticity	A sample is easily rolled and not much time is required to reach the plastic limit
High plasticity	A sample is very easily rolled and it takes considerable time to reach the plastic limit

Source: Ohio Environmental Protection Agency. 2006. *Technical Guidance Ground Water Investigations*. Columbus, OH.

Sampling groundwater usually requires the installation of a monitoring well, but this is not always the case. Occasionally, a water sample is collected directly from the borehole or excavation pit, or a temporary monitoring well is installed and removed after a groundwater sample is collected.

12.3.2.1 Direct Groundwater Sample Collection

A groundwater sample collected directly from an open borehole is typically called a grab sample. Groundwater samples collected from a temporary monitoring well may also be called grab samples depending on the construction of the temporary well, how long the well had been in place before sampling, and how the well was developed. In general, groundwater samples collected from open boreholes or temporary monitoring wells installed without great care are used as a rough screening of groundwater

SOIL BORING LOG AND MONITORING WELL CONSTRUCTION DIAGRAM

Boring / Well Number: **MW-6**	Facility Name: **ABC Facility**	Facility Street Address: **123 West Avenue - Somewhere, Iowa**
Boring Depth (ft) X Diameter (in): **15.0 x 6**		Drilling Method: **HSA**
Well Contractor Name: **BEST Consultants** Registration Number: **C98765**		Logged by: **A. Bore**

Ground Surface Elevation (ASL): **1234.55**	Top of Casing Elevation (ASL): **1235.43**

Date: **1/2/02** Start Time: **8:00 am**	Date: **1/3/02** End Time: **12:30 pm**	UST Number: **9901234**	LUST Number: **9ZYXW00**

Depth (feet)	Well Construction Details	Blow Count if applicable	Sample No.	Sample Type*	PID / FID Reading	USCS	Rock Formations, Soil, Color and Classifications, Observations (moisture, odor, etc.) First column for USCS
0–5	Cement grout	20	1	SS	0	CL	Brown Sandy Clay, medium stiff, moist (Fill)
5–10	Bentonite pellets; Sand filter; Slotted PVC casing	25	2	SS	50	SM	Light Brown to Gray Silty Sand to Clayey Sand, medium dense to dense, moist to wet
10–15		30	3	SS	100	ML	Dark Brown to Black Clayey Silt, stiff to very stiff, saturated
15	Bentonite bottom seal	100/6"	4	SS	ND		
							Bottom of MW-6 @ 15 feet

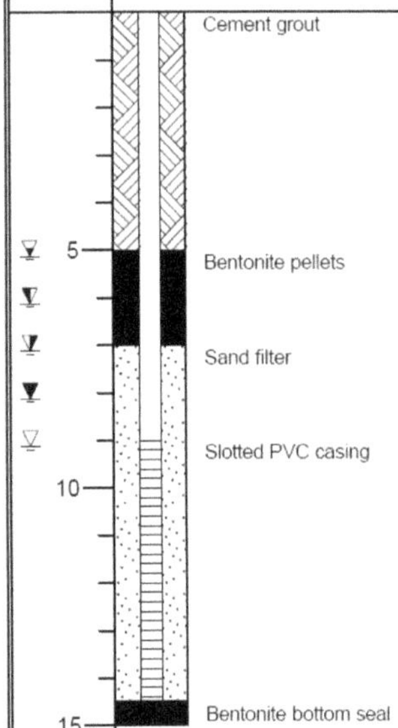

* SS (split spoon) HS (hollow stem auger)

Observations	Date:	1/3/02	1/3/02	1/3/02	1/3/02	1/3/02
Water Levels (ASL)	Level:	1,225.43	1,226.43	1,227.43	1,228.43	1,229.43
Static Water Level Symbol	Time:	8:00 am	9:00 am	10:00 am	11:00 am	12:30 pm

DNR FORM 542-1392

FIGURE 12.11 Sample boring log. (From Iowa Department of Natural Resources. 2002. *Sample Boring Log DNR Form 542-1392*. Iowa Geological Survey. Des Moines, IA.)

to evaluate the need for a more permanent monitoring well. Collecting and analyzing groundwater samples from an open borehole or temporary monitoring well are not always recommended because the analytical results may not be representative of the actual groundwater in the aquifer.

Figure 12.12A–C shows the process of collecting a groundwater grab sample from a borehole. Figure 12.12A shows a stainless steel well screen, which is placed inside the lead drill rod and is lowered into the borehole. Once the desired depth is reached, the outer rod is pulled up to expose the screen. A polyvinyl chloride (PVC) flexible hose is inserted into the inside of the drill rods, as shown in Figure 12.12B, and groundwater is pumped from the subsurface using a pump. Groundwater is then pumped into a drum, as depicted in Figure 12.12B. After the pumping continues for several minutes, a groundwater sample is collected in an appropriate container (Figure 12.12C).

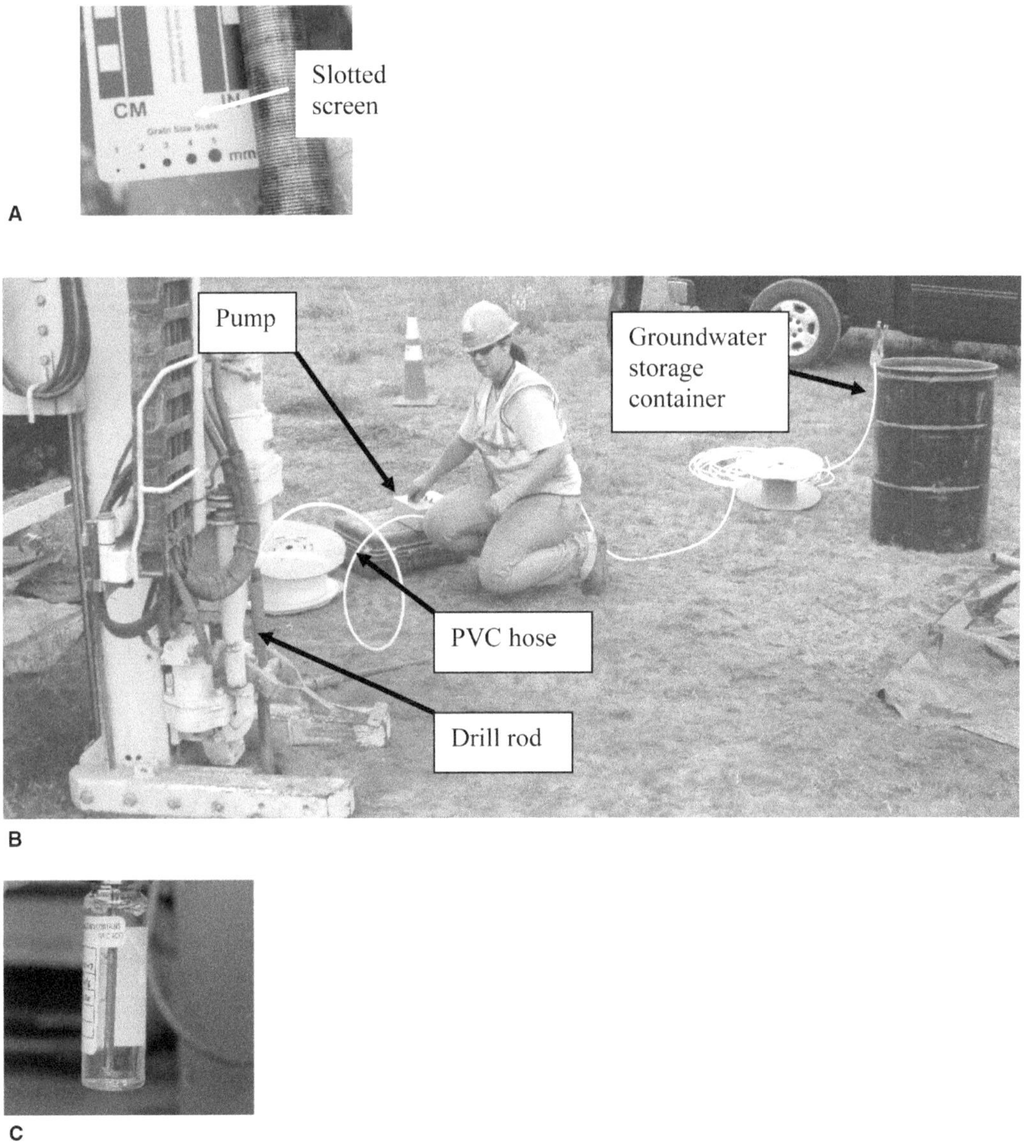

FIGURE 12.12 (A) Stainless steel well screen; (B) pumping groundwater; (C) groundwater sample in an appropriate 40 mL glass sampling container(Photographs by Daniel T. Rogers.)

12.3.2.2 Installing a Monitoring Well

Collecting a representative groundwater sample requires the installation of a monitoring well. The objectives of installing a monitoring well include the following (Rogers 2020):

- Collecting a representative sample of groundwater for analytical testing to evaluate groundwater quality
- Evaluating the direction of groundwater flow
- Estimating the hydraulic conductivity
- Evaluating specific yield
- Monitoring seasonal fluctuations in flow direction
- Future monitoring of groundwater quality

In many cases, monitoring wells should not be used for evaluating the nature and extent of groundwater contamination. Monitoring wells should be used to monitor groundwater, not to define the nature and extent of contamination. Evaluating the nature and extent of contamination in complex contaminant and geological strata should be done using high-resolution techniques in most instances. After defining the nature and extent of contamination, monitoring wells should then be considered to monitor groundwater quality at strategic locations based on the results of the define stage of the investigation. Installing monitoring wells too early in an investigation or using monitoring wells to define the extent of contamination should be avoided. Monitoring wells are tailormade for the opportunity to monitor groundwater quality through time. However, unless the monitoring well is installed at the precise and optimal location within a contaminant plume, it will not provide the investigation team with the necessary scientific data required to evaluate the nature and extent of contamination. To prevent this scenario, high-resolution aquifer profiling should be employed.

As depicted in Figure 12.13, monitoring wells have the advantage of high precision but have the potential for low accuracy, whereas high-resolution aquifer profiling may perhaps be less precise but often has much better accuracy.

Groundwater samples are typically analyzed for specific chemical compounds having detection limits in the microgram per liter (μg/L) or parts per billion (ppb) range. These detection limits mandate a representative sample not only because of these low detection limits – cleanup standards are also very low for certain chemical compounds.

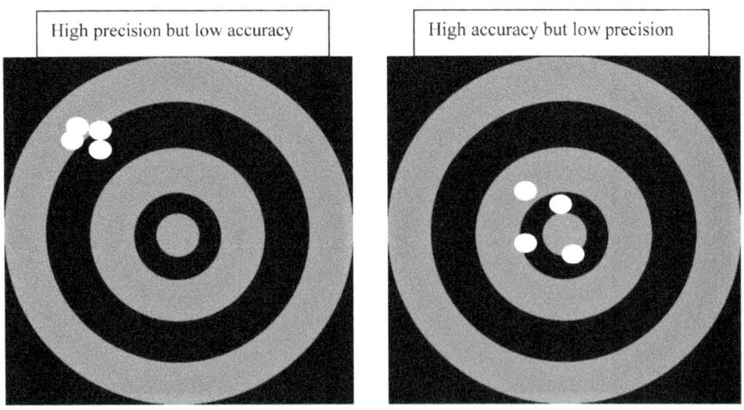

FIGURE 12.13 Comparing data precision and accuracy between monitoring wells and high-resolution aquifer profiling.

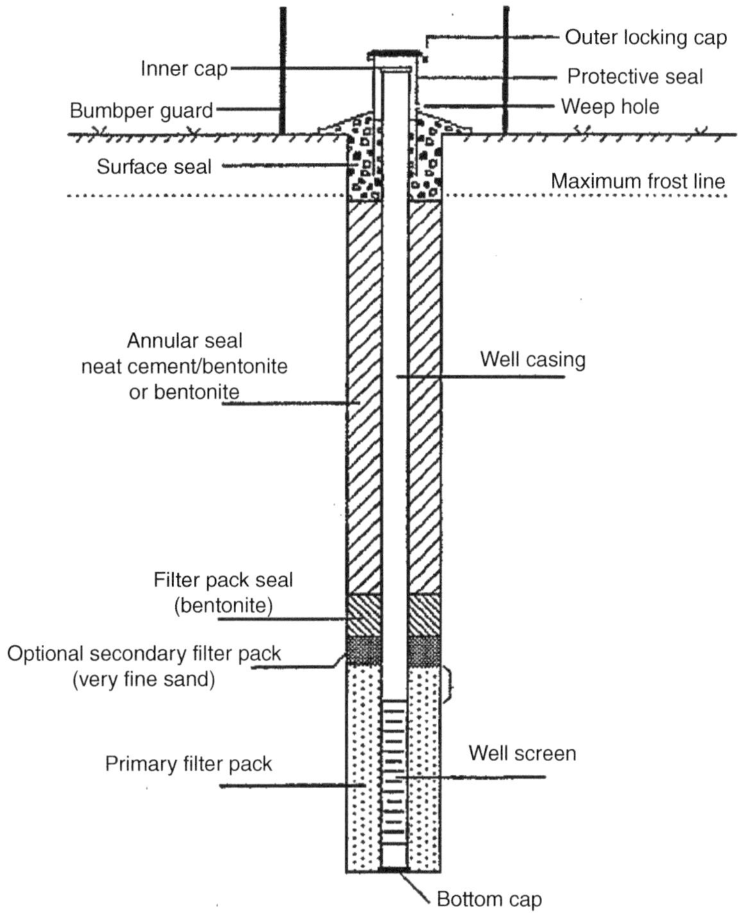

FIGURE 12.14 Monitoring well construction diagram. (From Ohio Environmental Protection Agency. 1995. *Technical Guidance Manual for Hydrogeologic Investigations and Ground Water Monitoring*. Division of Drinking and Ground Waters. Columbus, OH.)

Monitoring wells are typically constructed of PVC but may also be constructed of stainless steel (ss), Teflon, or steel. Figure 12.14 has a construction diagram of a typical monitoring well.

The **casing** in a monitoring well is the hollow tube placed inside the borehole and provides access to the subsurface. The **screen** at the bottom of the well allows water to seep into the well from the aquifer. The slits have a very small diameter, so the water enters the well but not the filter pack around the outside of the well screen. A filter pack is placed around the well screen and serves to prevent any fine-grain sediment in the aquifer formation from entering the well. The **filter pack** contains granular sand or gravel of uniform size and is placed between the aquifer formation and the well screen. The filter pack aids in screening out the fine particles within the aquifer mobilized by well pumping, so this device is important in aquifers composed of unconsolidated sediments that tend to be poorly sorted. Removing solid particles before they enter the well keeps the groundwater sample free from analytes of concern that may be sorbed onto sediment particles.

Many states require that the filter pack and the slot size of the well screen be "engineered" based on the textural characteristics of the aquifer material. This is accomplished by performing a sediment size analysis of the aquifer material and creating a grain size distribution curve. The desirable percentage of open slots in a well screen should be at least equal to the porosity of the filter pack.

Although desirable, this is not always achievable, particularly if the aquifer material consists of fine sand. Moreover, the cost of performing a grain size analysis of aquifer material to optimize the filter pack and the well screen is frequently too great when you consider the cost of having the drill rig and crew standing by while this analysis is being performed. As a result, we cannot overstate the importance of contracting an experienced drilling company that has already successfully installed monitoring wells in the area where you are working.

An **annular or bentonite** seal composed of bentonite clay is placed above the filter pack and above the well screen surrounding the casing. This placement of the seal helps to minimize potential infiltration from surface water and other fluids via the outside casing of the well.

Most monitoring well screens are 5 or 10 feet long and 2 inches in diameter. Many local regulatory agencies require well screens to be no more than 5 feet in length. Monitoring wells installed in an unconfined aquifer generally straddle the water table, and these usually have a 10-foot-long screen (if allowed by the regulatory agency) to allow for the seasonal fluctuation of the water table. Half the well screen (5 feet) is installed below the water table, with the other half above the water table. Monitoring wells installed entirely below the water table (termed fully submerged) have well screens 5 feet in length.

Installing more than one monitoring well per location may be required in cases of significant aquifer thickness, those with more than 20 feet of saturated thickness. Multiple monitoring wells at a single location are installed to evaluate the groundwater quality vertically through the length of the aquifer. Figure 12.14 shows a cross section of an aquifer composed of sand with a saturated thickness of approximately 75 feet. In addition to the monitoring wells installed straddling the water table, three monitoring wells were installed as a set at three depths within the aquifer. One set straddles the water table, one set was placed at the approximate mid-point of the aquifer, and the last set was located at the bottom of the aquifer. However, in some cases, even installing three monitoring wells at one location may not be sufficient for gathering enough data to adequately understand the inner dynamics of a contaminant plume. An example of this is where multiple plumes are present. In this scenario, a technique termed **high-resolution aquifer profiling** is used and involves collecting numerous groundwater samples vertically at several locations to develop a detailed three-dimensional depiction of a groundwater plume or plumes (Rogers 2016a, 2016b; Welty et al. 2010). Figure 12.15 presents an example of using high-resolution aquifer profiling to characterize a groundwater contaminant plume. In Figure 12.16, the horizontal distance is 3 km and the vertical distance is 30 m. Approximately 3,000 groundwater samples were collected and analyzed to characterize the contaminant plume.

Requirements for monitoring wells exist and vary state by state and often county to county. The differences occur for items such as obtaining a permit to install, location, depth, seals, certification, length of well screen, diameter of well screen, placement of screen, construction materials, and installation procedures. Before installing a monitoring well or using high-resolution aquifer profiling, state and local requirements should be examined.

12.3.2.3 Groundwater Sampling

Collecting groundwater samples consists of a process involving the following nine steps (modified from Harter 2003):

1. Preparation
2. Inspecting and accessing the monitoring well
3. Measuring the water level
4. Purging the well
5. Letting the well recover
6. Measuring the water level
7. Collecting the water sample

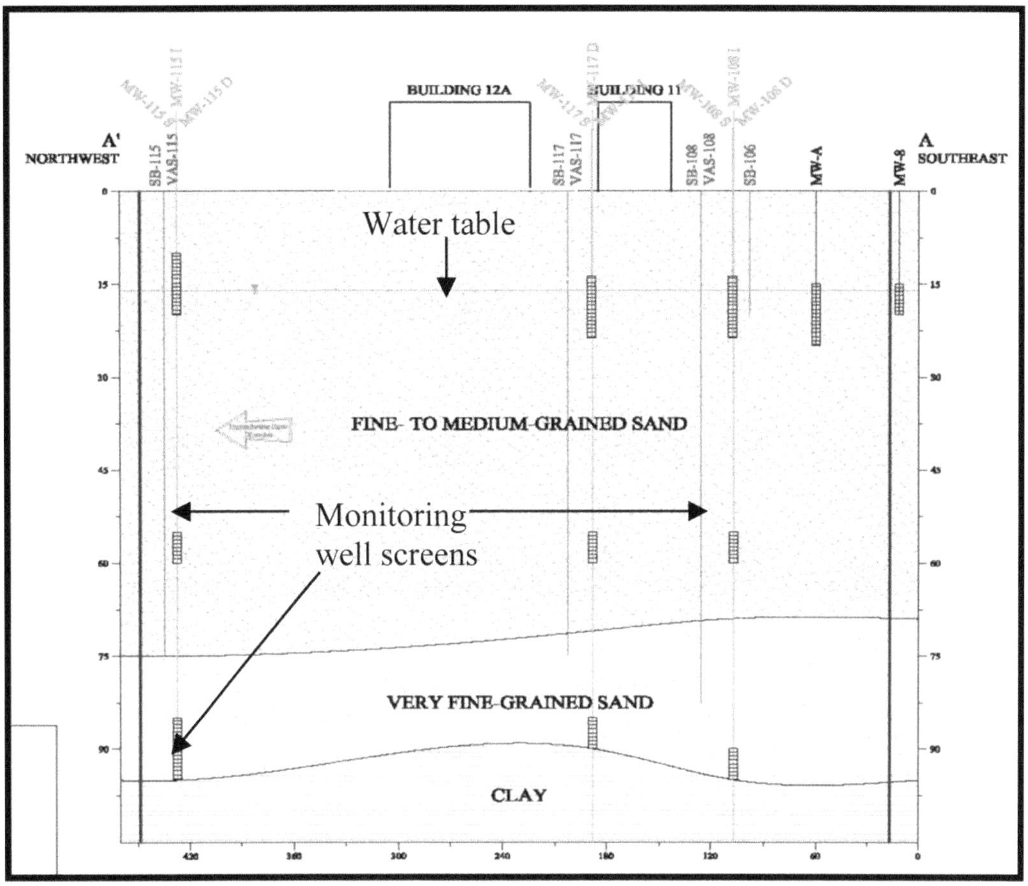

FIGURE 12.15 Cross section showing multiple monitoring wells horizontally and by depth.

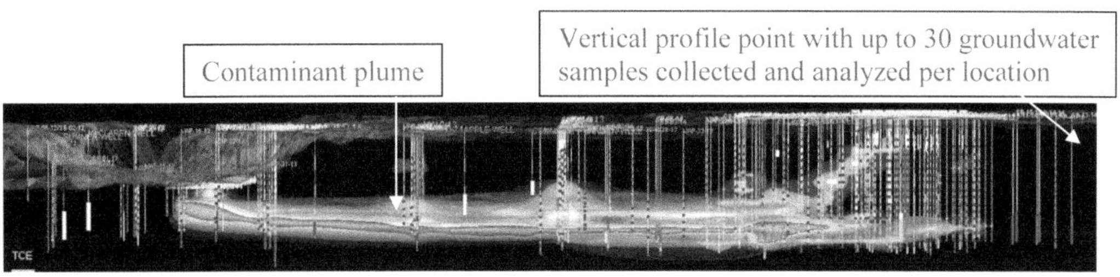

FIGURE 12.16 Cross section demonstrating high-resolution aquifer profiling.

8. Decontamination
9. Securing the well after sampling

The preparation step involves the following:

- Establishing the list of parameters for analysis
- Ensuring the purging and sample collection equipment, including bailers and pumps, are in working order; these items should be new or properly decontaminated

- Verifying the correct type and quantity of gloves, splash protection, and other potentially required health and safety equipment are working properly; these items should be new or properly decontaminated
- Checking the proper function of the water quality measuring equipment; these items should be new or properly decontaminated. Equipment may include temperature, pH, conductivity, dissolved oxygen, reduction–oxidation potential, hydrogen sulfide, photoionization detector (PID), explosivity, and lower explosive limit detection devices or meters
- Ordering appropriate sample containers
- Establishing appropriate protocol for sample labels
- Making sure chain of custody and laboratory log-in sheets have been prepared in advance to the extent possible
- Ensuring the sampling preservation, transport, and packing materials are appropriate, including coolers, ice, sample preservation chemicals, filters, and any specialized laboratory instructions

Figures 12.17A and B show accessing a monitoring well and lowering a Teflon bailer into a well. To obtain a measurement of the depth to water, a water level indicator is typically used. A water level indicator consists of a graduated tape measure attached to an electronic sounding device at the tip that emits an audio alarm when water is encountered.

The depth to groundwater using the water level indicator should be measured from the top of the monitoring well casing to the water table surface. Usually a reference mark is placed at some common point on the top of the monitoring well casing, because accurate measurements to within 1/100 of a foot are routinely required when measuring the depth to groundwater. This mark should be consulted every time a water level measurement reading is conducted to ensure consistency and accuracy because the top of the monitoring well casing is not always level.

The elevation of the top of the casing of the monitoring well must also be measured accurately to determine the direction of groundwater flow. The geographic position and elevation at the top of the casing of each monitoring well should be measured by a licensed surveyor.

If the monitoring wells will be used for any length of time, usually more than 1 year, consideration should be given to checking the elevation of each monitoring well to ensure the elevation of the casing has not changed. The elevation can change from natural causes, such as frost heave, and from anthropogenic causes, such as damage caused by snow removal or other equipment accidentally running into the monitoring well.

A B

FIGURE 12.17 (A) Monitoring well cover; (B) lowering a Teflon bailer down into a monitoring well. (Photographs by Daniel T. Rogers.)

FIGURE 12.18 Teflon bailer containing groundwater. (Photograph by Daniel T. Rogers.)

Figure 12.18 is a photograph of a Teflon bailer containing groundwater with relatively high turbidity (cloudy appearance) extracted from the monitoring well located near the bottom of the picture. Teflon bailers are routinely used for collecting groundwater samples. The bailer shown in Figure 12.18 has a check valve at the bottom that allows water to fill the bailer when groundwater is encountered as the bailer is slowly lowered into the well. After the check valve is engaged upon lifting of the bailer, water is prevented from escaping.

Two types of high-resolution aquifer profiling well points are depicted in Figure 12.19. The desired well point can be attached to the bottom of the drilling rod and then advanced to the desired vertical depth in the aquifer to be sampled. Water enters the well point through openings pictured. Groundwater can then be pumped to the surface by inserting Teflon tubing through the center of the drilling rods as depicted in Figure 12.12 A–C.

12.3.3 Sediment Sampling

Sediment sampling is conducted with several types of specialized equipment. One instrument commonly used is called a **Ponar dredge**. A Ponar dredge is equipped with a set of jaws and is deployed in the open position while being lowered through the water column until it becomes embedded into the bottom sediment. Once the instrument is pulled upward with minimal force to avoid pulling it off the bottom, the jaws are engaged and come together as the instrument is raised. This sequence of

FIGURE 12.19 Two types of high-resolution aquifer profiling well points.

events traps sediment in the instrument bucket. These types of samplers allow the collection of small or large samples depending on the size of the bucket (USEPA 2007).

Another popular sediment sampler is called a tube or grab sampler; it consists of a hollow tube typically constructed of stainless steel, PVC, or Teflon (Capri et al. 2005). These types of samplers are intended for collecting shallow sediment samples within water columns of less than 5 to 10 feet. Proper function of these devices depends upon the presence of a strong water current or other associated factors that may cause the sampler to drift and not collect an appropriate sample size or sample the wrong location.

12.3.4 Surface Water Sampling

Sampling surface water uses many of the same procedures as groundwater sampling. The first step in sampling surface water is evaluating the morphology and hydrology of the surface water body, so appropriate sampling location(s) can be identified. Collecting an appropriate and representative sample of surface water also depends upon the type of surface water body, e.g., a lagoon, stream, lake, river, or spring (USEPA 1994).

Factors affecting sampling location include the following:

- Chemical compounds of concern
- Depth of water body
- Flow rate
- Size
- Whether there is a specific location of concern
- Topography and composition of the bottom of the water body

Each surface water sample location should be marked and surveyed in case a future sampling event is required. Placements of stakes, flags, or buoys are common ways to mark surface water sampling locations. The elevation of a surface water body is also measured during every sampling event to assess the current hydrologic conditions, e.g., drought or flood. Placing a benchmark on a stationary object, such as the bottom or side of a bridge, the edge of a paved roadway, or a building corner usually works.

In complex surface water settings, measuring water quality data such as temperature, pH, and dissolved oxygen helps to know if any stratification exists either horizontally or vertically within the water column. Monitoring these factors improves the chances you will select appropriate sampling locations and an effective number of samples.

FIGURE 12.20 Collecting a surface water sample using the dip method. (Photograph by Daniel T. Rogers.)

Surface water samples can be collected using a glass tube, pumps, bailers, or obtained directly using a sample container. Specialized surface water instruments include the Kemmerer Bottle, Bacon Bomb Sampler, or Dip Sampler. These types of instruments are used from a boat or a bridge and are lowered into the water column and can collect a water sample from any depth. The exception here is the Dip Sampler, which only collects samples from the surface and is essentially a sample container dipped into the water. Dip samplers are useful where access is limited (see Figure 12.20).

12.3.5 Air Sampling

Conducting subsurface investigations sometimes requires sampling air. Some contaminants migrate through subsurface soil and groundwater and may come into contact with subsurface structures such as basements. When this event occurs, certain contaminants have the potential to accumulate in these subsurface void spaces and contaminate the air. Additionally, contaminants are released directly into the air and may become deposited on the ground or water surfaces.

Under certain conditions, air sampling is justified when conducting a subsurface environmental investigation for three purposes (USEPA 1998):

1. Evaluating whether the source of soil or water contamination originated from air deposition
2. Protecting the health and safety of people during an investigation
3. As part of conducting a risk assessment.

Sampling for contaminants in the air falls into two broad categories: contaminants in the vapor or gas phase and contaminants as particulates or sorbed to particulates (USEPA 1991). The objectives of air sampling include the assessment of exposure for health reasons and for normal or ambient air quality. Exposure monitoring usually is performed with indoor air, and ambient air monitoring is typically done with outdoor air.

12.3.5.1 Indoor Air Sampling

Before conducting any air sampling, a site inspection should be completed – especially if the sampling will be conducted indoors. The inspection seeks to identify two conditions: anything affecting or interfering with the proposed testing and obvious sources of air contaminants.

The site inspection should include the following (New York Department of Health 2005):

- Type of structure
- Physical conditions
- Odors such as solvents, mold, sweet, sour
- Airflow and ventilation engineering design of the building, structure, or residence
- An analysis of potential contaminant sources including
  - Chemicals used
  - Chemical storage areas
  - Chemical use areas
  - Chemical disposal methods
  - Chemical disposal areas
- Weather conditions

Conducting a chemical inventory is recommended prior to any indoor sampling. This inventory will greatly assist in providing valuable information concerning the identification of the COC and whether air sampling is necessary. MSDSs for each chemical at the site should be reviewed.

At least 24 hours prior to conducting any indoor air sampling, the measures listed below should be avoided as they may skew the sample results (New York Department of Health 2005):

- Opening windows
- Operating ventilation fans
- Smoking
- Painting
- Using wood stoves or fireplaces
- Kerosene heaters
- Operating or storing automobiles in a structure connected to the location where indoor sampling will be conducted (i.e., attached garage)
- Cleaning, waxing, or polishing furniture or other surfaces with petroleum- or oil-based products
- Operating any lawn maintenance equipment, including lawn mowers or other mechanical instruments, using liquid fuels
- Using cosmetics, e.g., hairspray, deodorant, nail polish, nail polish removers, perfume, and cologne
- Applying pesticides and herbicides

Air samples should be collected from an adequate number of locations to understand the likely sources of potential chemical exposure and to evaluate their potential exposure to occupants at various locations. In private residences, air samples should be collected from all floors including the basement, and outdoors.

For comparison purposes, indoor samples collected from commercial or industrial settings should be collected during normal operating hours and again during times when the facility or location is not in operation and occupied. To ensure a robust sample – one that is representative of the conditions being tested and minimizes other influences – samples should be collected for at least 1 hour and not more than 8 hours.

Sample collection techniques vary depending on the analytical method(s) being used and must conform with the specific sample collection method being utilized. Some methods also specify that duplicate samples must be collected. Indoor air sampling is either conducted on individuals or at stationary locations. Sampling conducted on individuals is either passive or by a sampler attached to a pump that pumps air through a filter or sampling media. This process captures potential contaminants and stores them until analysis can be conducted. To be as representative as possible, indoor air samples are usually collected within the breathing zone between 3 and 6 feet above the floor.

12.3.5.2 Outdoor Air Sampling

Outdoor air sampling for particulates is conducted using high-volume air sampling equipment (USEPA 1998). The ambient air sampling device shown pumps air from all directions across a 7-inch by 9-inch exposed filter inside the shelter at flow rates ranging from 39 to 60 cubic feet of air per minute. The sizes of particles collected using the sampler range from 0.3 μm to as large as 25 to 50 μm in aerodynamic diameter and are dependent upon wind direction and speed (USEPA 2008). Filters used for collecting outdoor air sampling for particulates are composed of either glass or cellulose fibers. Glass fiber samples have been extensively used, and cellulose fiber filters are commonly used whe n sampling for heavy metals. One drawback to cellulose fibers is their potential to increase the potential for adsorption of water and enhance artifact formation of nitrates and sulfates on the filter. The filters are kept in a clean and sterile environment prior to sampling at a constant temperature and a relative humidity of approximately 3% and are precisely weighed (USEPA 1998). After the sample is collected, several parameters must be recorded, including the following:

- Starting time and date
- Ending time and date
- Air flow rate
- Temperature ranges during sampling
- Relative humidity difference during sampling
- Summary of conditions that may affect results (construction activities in the area, etc.)
- Barometric pressure at start and end

12.4 SUMMARY AND CONCLUSION

Environmental subsurface investigations are very detailed scientific investigations. These investigations collect enormous amounts of geologic and hydrogeologic information and significant amounts of other information. All of this information allows the professionals conducting the investigation to determine the existence of contamination at a site or property and if the level of any contamination present requires a remedial action because it presents an unacceptable risk to human health or the environment.

Environmental investigations must be conducted with a high level of scientific care for several reasons: (1) because contaminants requiring a remedial action may be present at very low levels, and precautions must be taken to minimize the chances of any cross-contamination; (2) because the investigation and potential remediation must be conducted in such a manner to enable their admission in a court of law; and

(3) to protect the investigators from any exposure to contaminants while collecting and handling samples taken during field activities.

Over the last 25 years, environmental subsurface investigations have been the most common type of environmental investigation conducted. In urban areas, they have provided extensive geologic and hydrogeologic information of the highest quality. Therefore, environmental subsurface investigations can form the cornerstone of mapping the geology and hydrogeology of an urban watershed.

The next chapter uses geologic and hydrogeologic information obtained from urban environmental investigations along with other sources of geologic information and fieldwork to map the geology of an urban watershed. This chapter provides the insight and tools necessary to take geologic information from numerous subsurface investigation sites within an urban watershed and then generate a geologic map capable of telling the story of the near-surface depositional history of that same watershed.

REFERENCES

American Society for Testing Materials (ASTM). 1999a. *Standard Guide for Using the Direct Current Resistivity Method for Subsurface Investigation. D6431-99.* ASTM. West Philadelphia, PA.

American Society for Testing Materials (ASTM). 1999b. *Standard Guide for Using the Gravity Method for Subsurface Investigation. D6430-99.* ASTM. West Philadelphia, PA.

American Society for Testing Materials (ASTM). 2000a. *Standard Guide for Selection of Methods for Assessing Ground Water or Aquifer Sensitivity and Vulnerability.* ASTM. West Philadelphia, PA.

American Society for Testing Materials (ASTM). 2000b. *Standard Guide for Using the Seismic Refraction Method for Subsurface Investigation. D5777-00.* ASTM. West Philadelphia, PA.

American Society for Testing Materials (ASTM). 2005a. *Standard Practice for Environmental Site Assessments: Phase I Environmental Site Assessment Process. E1527-05.* ASTM. West Philadelphia, PA .

American Society for Testing Materials (ASTM). 2005b. *Standard Guide for Using the Surface Ground Penetrating Radar Method for Subsurface Investigation. D6432-99.* ASTM. West Philadelphia, PA.

American Society for Testing Materials (ASTM). 2005c. *Standard Guide for Planning and Conducting Borehole Geophysical Logging.* D5753-05. ASTM. West Philadelphia, PA.

California Environmental Protection Agency. 1995. *Guidelines for Hydrogeologic Characterization of Hazardous Substance Release Sites. Volume 1: Field Investigation Manual.* CalEPA, Sacramento, CA.

California Environmental Protection Agency. 1996. *Guidance for Ecological Risk Assessment at Hazardous Waste Sites and Permitted Facilities.* Department of Toxic Substances and Control. Sacramento, CA.

Capri, J., Schumacher, B. A., Wanning, S., Smith, E., Zimmerman, J., and Vanover, J. D. 2005. *Collecting Undisturbed Surface Sediments.* United States Environmental Protection Agency. EPA/600/R-05/076. Washington, DC.

Fletcher, C. D., and Paleologos, E. K. 2000. *Environmental Risk and Liability Management.* American Institute of Professional Geologists. Westminster, CO.

Glasson, J., Therivel, R., and Chadwick, A. 2005. *An Introduction to Environmental Impact Assessments.* Routledge Publishing. London, England.

Harter, T. 2003. *Groundwater Sampling and Monitoring.* University of California ANR Publication 8085. University of California, Division of Agriculture and Natural Resources. Oakland, CA.

Iowa Department of Natural Resources. 2002. *Sample Boring Log DNR Form 542-1392.* Iowa Geological Survey. Des Moines, IA.

Jain, R. K., Urban, L. V., Stacey, G. G. and Balbach, H. E. 1993. *Environmental Assessment.* McGraw-Hill, Inc. New York, NY. 524p.

Keys, W. S. 1990. *Borehole Geophysics Applied to Ground-Water Investigations.* United States Geological Survey Techniques of Water-Resources Investigation. Book 2. United States Government Printing Office, Washington, D.C.

Martin, S. N. 2003. *Site Assessment and Remediation Handbook.* CRC Press. Boca Raton, FL.

Munsell, A. H. 1905. *A Color Notation.* Geo. H. Ellis and Company Publishers. Boston, MA.

New Hampshire Department of Environmental Services. 2008. *Contaminated Site Management Remedial Action Plan Checklist.* NHDES, Concord, NH.

New Jersey Department of Environmental Protection. 2005. *Field Sampling Procedures Manual.* New Jersey DEP, Trenton, NJ. 574p.

New York Department of Environmental Conservation. 2009. *Remedial Investigation and Feasibility Study Fact Sheet.* www.decny.gov/chemical/8658.html. (Accessed November 4, 2019).

New York Department of Health. 2005. Indoor Air Sampling and Analysis Guidance. www.hwlth.y.gov/environmental/indoor/air.guidance.html. (Accessed February 6, 2025). Occupational Safety and Health Administration (OSHA). 1989. *Hazardous Waste Operations and Emergency Response (HAZWOPER).* Code of Federal Regulations 40 (CFR) 1910.120. Washington, DC.

Ohio Environmental Protection Agency (OEPA). 1995. *Technical Guidance Manual for Hydrogeologic Investigations and Ground Water Monitoring.* Division of Drinking and Ground Waters. Columbus, OH. 255p.

Ohio Environmental Protection Agency (OEPA). 2006. *Technical Guidance Ground Water Investigations.* OEPA, Columbus, OH. 67p.

Oklahoma Department of Transportation. 2021. Geotechnical Specifications. Oklahoma City, Oklahoma. 85p. https://okahoma.gov/content/dam/ok/en/odot/documents/Geotech%20Specifications.pdf. (Accessed February 6, 2025)

Oregon Department of Environmental Quality. 1998. *Guidance for Ecological Risk Assessment: Levels I, II, III and IV.* Oregon Department of Environmental Quality. Portland, OR.

Rogers, D. T. 1992. *The Importance of Site Observation and Follow-up Environmental Site Assessment – A Case Study. Groundwater Management Book 12.* National Groundwater Association. Dublin, OH. pp. 563–572.

Rogers, D. T. 2016a. Next Generation of Urban Hydrogeologic Investigations. *Journal of Italian Geological Society.* Vol 39, no. 1, p. 349.

Rogers, D. T. 2016b. *Scientific Advancements that Improve the Conceptual Site Model in Urban Hydrogeologic Investigations.* 35th International Geological Congress. Paper No. 3019. Cape Town, South Africa. pp. 325–337.

Rogers, D. T. 2020. *Urban Watersheds: Geology, Contamination, Environmental Regulations, and Sustainability.* CRC Press. Boca Raton, FL. 606p.

Rogers, D. T., and Welty, N. 2017. *From Characterization to Closure: A Large TCE Plume.* Michigan Department of Environmental Quality and American Institute of Professional Geologists. Lansing, MI.

United Nations. 2024. United Nations Environment Programme – UNEP. 2024. www.unep.org/who-we-are/about.us. (Accessed May 12, 2024).

United States Department of Agriculture (USDA). 2019a. *From the Surface Down. An Introduction to Soil Surveys for Agronomic Use.* Washington, DC. www.nrcs.usda.gov/conservation-basics/natural-resource-concerns/soil/nation-soil-survey-center . (Accessed February 6, 2025).

United States Department of Agriculture. 2019b. *Guide to Texture by Feel.* www.nrcs.usda.gov/resources/education-and-teaching-materials/soil-education. (Accessed February 6, 2025).

United States Environmental Protection Agency (USEPA). 1976. *Resource Conservation and Recovery Act (RCRA). U.S.C. 6901.* United States Government Printing Office. Washington, DC.

United States Environmental Protection Agency (USEPA). 1980. *Comprehensive, Environmental Response, Compensation and Liability Act U. S. C. 9601-9675.* United States Government Printing Office. Washington, DC.

United States Environmental Protection Agency (USEPA). 1986. *Guidance for Conducting Subsurface Investigations at RCRA Sites.* Office of RCRA Corrective Actions. OSWER No. 9902.3-2A. United States Printing Office. Washington DC. 259p.

United States Environmental Protection Agency (USEPA). 1988. *Guidance for Conducting Remedial Investigations and Feasibility Studies Under CERCLA.* Office of Emergency and Remedial Response. EPA 540/G-89-004. United States Printing Office. Washington, DC. 173p.

United State Environmental Protection Agency (USEPA). 1989. *Risk Assessment Guidance for Superfund. Volume 1: Human Health Evaluation Manual (Part A).* United States Environmental Protection Agency, Office of Emergency and Remedial Response. OSWER Directive 9285.7 Ola. Washington, DC.

United States Environmental Protection Agency (USEPA). 1991. *Evaluating Exposures to Toxic Air Pollutants.* United States Environmental Protection Agency EPA Document 450/3-90-23. Washington, DC.

United States Environmental Protection Agency (USEPA). 1994. *Standard Operating Procedure for Surface Water Sampling.* S.O.P. 2013. Washington, DC.

United States Environmental Protection Agency (USEPA). 1998. *Guidance for Using Continuous Air Monitors.* United States Environmental Protection Agency Office of Air Quality Planning and Standards. Washington, DC.

United States Environmental Protection Agency (USEPA). 2002. *Risk Assessment Guidance for Superfund; EPA-540/R-02/002.* United States Government Printing Office. Washington, DC.

United States Environmental Protection Agency (USEPA). 2005. *Standards and Practices for All Appropriate Inquiries. 40 Code of Federal Regulation (CFR), Part 312.* United States Government Printing Office. Washington, DC.

United States Environmental Protection Agency (USEPA). 2007. *Sediment Sampling.* USEPA Region 4, Science and Ecosystem Support Division. Athens, GA.

United States Environmental Protection Agency (USEPA). 2008. *List of Designated Reference and Equivalent Methods for Air Sampling*. United States Environmental Protection Agency National Exposure Research Laboratory. Washington, DC.

United States Environmental Protection Agency (USEPA). 2009. *Natural Resource Damage Assessment*. www.epa.gov/superfund/programs/nrd/nrda2.htm#pagetop. (Accessed November 4, 2019).

Virginia Department of Transportation, 2003. Unified Soil Classification System. www.occ.state.ok.us/Divisions/PST/Forms/Technical%20Forms/Orbca%20Guidance/unified%20soil%classification.pdf. (Accessed February 6, 2025).

Welty, N., Rogers, D. T., and Quinnan, J. 2010. *High Resolution Profiling of Aquifer Permeability and Contaminant Mass*. Battelle Chlorinated Remediation Conference. Monterey, CA.

Yong, R. N. 2002. *Geoenvironmental Engineering, Contaminated Soils, Pollution Fate, and Mitigation*. CRC Press. Boca Raton, FL.

13 Geologic Sensitivity Maps of Urban Areas

13.1 INTRODUCTION

This chapter explains the process of interpreting and integrating information from environmental investigations. We combine this information with field mapping techniques to construct a geologic map of an urban area. The methods presented in this chapter provide the basic building block for understanding the geology and hydrogeology beneath urban regions.

Geologic maps tell a story of the history of a particular area, and their intended users are geoscience professionals and the general public. **Urban geologic maps** present many new challenges to geologists and are typically complex to construct. One of the key challenges is to create an easily understandable and usable geological map for members of the non-scientific community. This should not come as a surprise to the geoscience community, as it has long been one of the major goals to construct geologic maps with the ability to clearly communicate scientific information to a diverse audience. John Wesley Powell, who as director of the United States Geological Survey (USGS), stated in 1888, "maps are designed not so much for the specialist as for the people, who justly look to the official geologist for a classification, nomenclature, and system of convention so simple and expressive as to render his work immediately available" (Powell 1888).

Effectively communicating complex information is joined by other unique challenges when developing urban geologic maps. As many geologists studying the geology of urban areas know, access and disturbance of the natural landscape present some of the greatest problems when trying to develop a map. The geologist must be clever, tenacious, and opportunistic because a large geological map base covering urban areas does not exist. Therefore, the geologist has limited historical information to begin the process of understanding the geology of many urban areas. Only recently has interest in geologic mapping of urban areas become a focus in the United States (USGS 1998, 2024a; Great Lakes Geologic Mapping Coalition 2024).

The unconsolidated near-surface geology of urban regions has been largely ignored because urban areas are not in the spotlight like areas with valuable natural resources, or risk zones such as those prone to earthquakes or volcanoes. Despite this lack of high-profile attention, characterizing, understanding, and evaluating the geology at urban locations has been quietly conducted at tens of thousands of sites throughout the United States, and these efforts collectively represent a massive information database of fundamental geologic knowledge.

The complexity associated with urban geologic maps is directly related to the characteristics of urban environment. The vast majority of large urban areas in the United States are located on unconsolidated sedimentary deposits. Most of these areas are less than 2 million years old – young by geological standards – but complex in nature by virtue of them having natural- and anthropogenic-induced unconformities as a common occurrence. These urban areas are also

DOI: 10.1201/9781003463153-16

proximate to large amounts of surface water and groundwater, and this water plays a fundamental role in functioning as the prime transport mechanism for contaminants. This combination of complex sedimentary deposits and flowing water tells why the shallow geologic environment beneath urban areas must be understood at a level of detail at least equal to, or even greater than other areas of the Earth because these places are where we live in and form the tightest bonds between nature and society.

Geologic maps thus form the foundation for ecosystem management (USGS 1998). The natural environment and the humans within it acquire an imposed organization resulting from the geologic framework at any given location. Geologic maps provide us with knowledge of that framework. Specifically, a geologic map facilitates the basic means for evaluating resources, hazards, and environmental data relative to human activities (Thomas 2004). The arrangement, thickness, and composition of the sediment layers beneath our feet have a profound influence on where cities are located, how buildings are constructed, where roads are built, and how contaminants behave and affect people. Constructing a geologic map of a region or a specific site is critical and should be the first step in evaluating the development or re-development potential of an urban region or site (USGS 1998; Berg et al. 1999; Berg 2002). It is therefore not an exaggeration to say that sustainable urban environments cannot be achieved without accurate and usable geological maps (Rogers 2015a, 2015b).

13.2 GEOLOGIC MAPPING IN AN URBAN ENVIRONMENT

Mapping the geology of urban areas increases our understanding of their fundamental geology and identifies the types and locations of anthropogenic impact to consider during the sustainable development and re-development of city cores and outlying areas. Standards exist for geologic mapping (USGS 2024b); they always require that extensive fieldwork be done when constructing a geologic map of a desired area (Maley 2005). However, many urban locations are not accessible, often making it imperative to gather and evaluate as much reliable historical data and information as possible from multiple sources. The next nine sections present a sequence beginning with the collection of relevant information and ending with the preparation of the final urban geologic map.

13.2.1 Information and Mapping Challenges

Mapping the geology of an urban region presents many challenges to geologists and can be more difficult than mapping other regions (USGS 1998; Farrand 1988; Rogers 2020). Some of these challenges include the following:

- Maintaining safety: Conducting field activities in urban areas poses many safety concerns, including physical hazards from equipment, machinery, automobiles, and other mechanical equipment. In addition, there may be other hazards such as heat stress and related hazards imposed by weather conditions and the requirement for wearing specialized personal protective clothing to prevent exposure to contaminants.
- Inaccessibility due to legal aspects of trespass and private property. Many locations of interest are not on public property and can only be accessed by receiving permission from private property owners.
- Areas covered with buildings and other anthropogenic structures obscure original features and eliminate the possibility of physical inspection.
- Disturbed areas such as parks, landfills, or other landscaped areas may cover, destroy, or remove original deposition features.
- Fieldwork – such as excavating test pits in a historic park – may significantly disturb an area and create the need for expensive measures to restore the land to an acceptable standard.

- Influencing and altering the natural environmental balance through large construction projects affects the subsurface environment. For example, the construction of a dam may raise the regional groundwater levels or flood areas with water before a sufficiently detailed geologic study can be conducted.
- Vegetative cover in urban areas occupies large amounts of surface area, especially in regions with humid climates. This extensive vegetative cover in urban areas contributes to the difficulty in observing natural deposits and also increases the potential for disturbing the natural deposits through either physical or chemical means.
- Introduction of chemicals and anthropogenic sources of water may destroy, disturb, or dissolve original depositional features and deposits. Sources of these external change agents include:
  - Stormwater runoff
  - Leaking sewers
  - Landfills
  - Old dumps
  - Underground tanks
  - Underground utilities
  - Pipelines
  - Septic systems
  - Wastewater treatment plant discharges
  - Industrial sites
  - Fertilizer application
  - Pesticide and herbicide application
  - Vehicle exhaust deposition
  - Old cemeteries
  - Golf courses
  - Fire-fighting efforts
  - Tanker spills
  - Brownfield sites
  - Abandoned buildings
  - Power plant particulate matter and water discharges (nuclear and coal)
  - Acidic precipitation

13.2.2 Information Sources, Integration, and Evaluation of Usability

To counter the challenges faced by mapping the geology of urban regions, geologists have to be more creative and diligent in pursuing the information necessary to adequately understand the geology of the near-surface urban environment.

There are several sources of information in urban regions useful for geological purposes, which in some cases may be more extensive than what is available for mapping undeveloped regions (Rogers 2014). Often, to be successful in gathering information on the geology of urban areas, the geologist must become a forensic geologist, because the near-surface unconsolidated deposits in many urban areas of the United States have experienced significant disturbance. Being a forensic geologist means knowing and evaluating whether the information collected is reliable and valid, and whether it should be included in the final urban geologic map.

Mapping the geology of an urban region begins with a literature review of existing data and publications using the following sources as a guide:

- Current and historical maps:
  - Geologic maps
  - Planning maps

    - Historical highway maps
    - Fire protection maps (often referred to as "Sanborn maps")
    - Topographic maps (7.5-minute scale is best)
    - Trail maps
    - Historical interest maps
    - Park and recreational maps
    - Maps prepared for site planning purposes; these often contain accurate locations of existing underground infrastructure and above-ground water management structures, such as detention basins
    - Others
- United States Geological Survey studies and reports
- State Geological Survey studies and reports
- State and local water well records
- Army Corps of Engineers reports and studies
- Oil and gas records, studies, and reports
- Department of transportation studies and reports
- Building construction reports and permits
- Geotechnical reports and studies
- State and local environmental investigation reports and studies at sites of contamination
- State and local historical societies studies; publications and photographs
- State and local universities and colleges
- Sources of historical aerial photographs
- Previously published scientific literature

You may think historical highway maps cannot provide valuable information to help understand the geology of a particular area or region – but this is simply not true. Old and even new highway maps provide valuable information on mapping the geology of urban areas. A few examples include:

- *Names of highways and roads:* The names of highways and roads sometimes provide insight into a region's geology. For instance, a highway named "Ridge Road" creates a potential association between topography and geology. Others performing this task include Muddy Road, Sandy Flats, Rocky Road, Sandy Hill Road, Wash Road, Winding Road, Coastal Highway, Beach Highway, Marsh Creek Road, Peat Bog Road, etc.
- *Location of cemeteries:* Cemeteries typically provide information on a region's geology because cemeteries during the 19th and early 20th centuries were usually located in areas of well-drained soil such as sand or gravel. This siting prevented the occasional casket from popping out of the ground after burial; an event possible in more poorly drained soil such as clay.
- *Location of parks and recreational areas:* Parks and recreational areas are often found in areas with a scenic or educational theme related to the geology of a region.
- *Historical markers and historical sites:* These locations often exist over or near significant natural disasters, battlefields, and trails.
- *Locations of historical settlements and buildings:* Before urbanization occurred many historical settlements used foundation materials of native stone collected from nearby locations and are useful when assessing a region's geology.

These sources of information must be carefully inspected for integrity and usefulness. Perhaps the information was not initially collected for geological purposes and may not have been collected by a qualified geologist. What if you are consulting historical geological maps? Unfortunately, most historical mapping of urban regions was produced in two dimensions and did not map the subsurface geology to a depth of more than a few meters. Relying upon a two-dimensional map is problematic because effective and useful urban geologic maps produced for environmental purposes must represent three dimensions. This spatial requirement means these maps are much more complex and require significantly more information (Barnes 1993). Some forensic methods described throughout this chapter can help evaluate the accuracy and ultimate usability of questionable historical geologic information (Rogers 2014, 2015b).

Just when things are starting to sound bleak, we get some relief in the form of geologic and hydrogeologic information obtained from environmental investigations. These activities can supply much of the baseline information to evaluate whether questionable geologic information is usable and also supply the necessary geologic information to enable detailed three-dimensional mapping. In many urban areas, environmental investigations have been conducted at thousands of locations. A large percentage of these environmental investigations yield very detailed geologic and hydrogeologic information collected by qualified and certified professionals. This information is readily accessible through Federal, State, and local environmental regulatory agencies. For example, over 3,000 reports of the subsurface geologic investigations performed in the Rouge River watershed are available for review in one state agency, the Michigan Department of Environmental Quality (MDEQ 2008). It is therefore quite possible the best sources of geologic and hydrogeologic information are within easy reach.

Figure 13.1 shows the location of hundreds of sites where detailed and large-scale environmental investigations have been conducted within the Rouge River watershed (Rogers 1996). In terms of

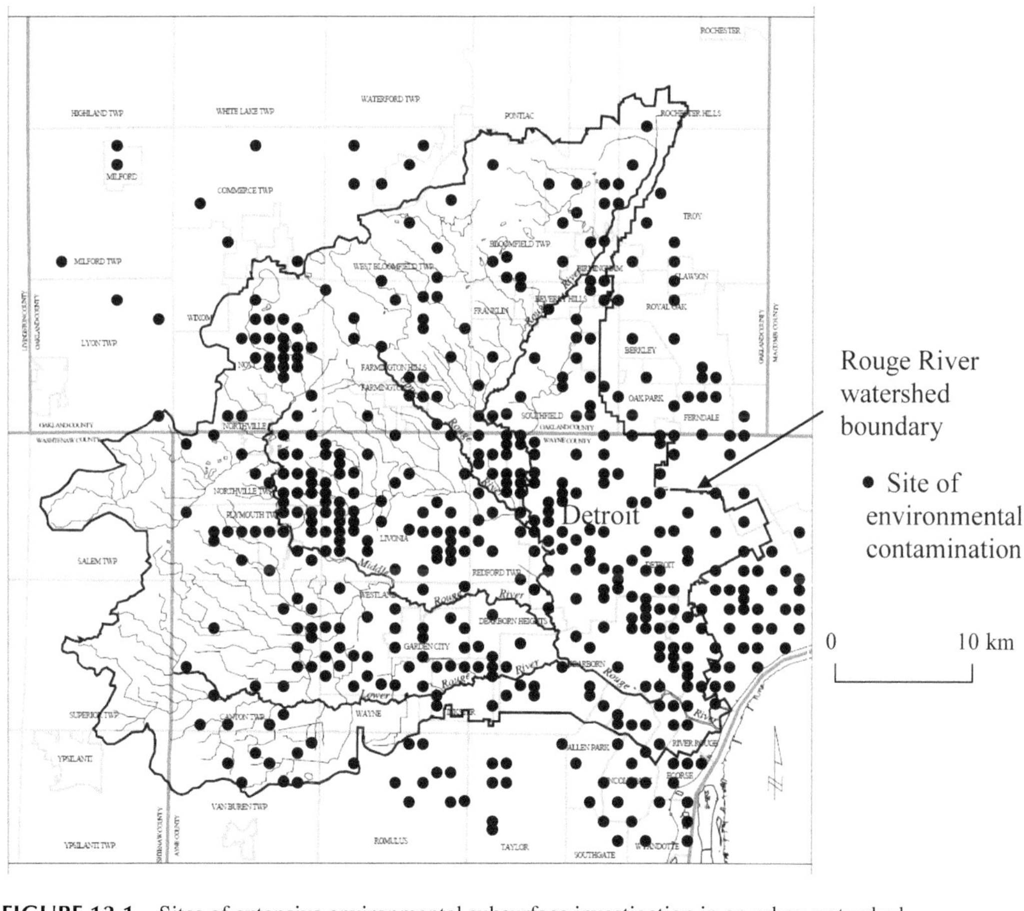

FIGURE 13.1 Sites of extensive environmental subsurface investigation in an urban watershed.

the geologic and hydrogeologic data collected, each site had a minimum of 20 to sometimes more than 300 individual soil borings drilled and up to 80 monitoring wells installed. Investigative depths reached as deep as 30 to 50 m beneath the ground surface.

Environmental investigations in urban areas usually do not extend beyond 15 m beneath the surface. However, unconsolidated sediments in many urban areas extend to depths greater than 15 m and may even approach 300 m or beyond as in Salt Lake City, Los Angeles, Phoenix, and Las Vegas, to name a few (Stokes 1989; Allen 2003; Halka 1983; Page et al. 2005; Bilodeau et al. 2007). The evaluation of data from these investigations can be supplemented and perhaps confirmed by the information supplied from well drillers and well logs.

Water supply and exploratory wells are routinely drilled to depths exceeding 30 m beneath the surface and some may extend to over 150 m. Drillers installing water supply wells are often required to register the well with the appropriate local, county, or State officials and submit a written well log. Water supply well logs often contain valuable and abundant sources of geologic and hydrogeologic information. However, as stated earlier, care should be taken before accepting the accuracy of data from a water supply well or other similar sources providing geologic or hydrogeologic data because they may not have been conducted by a qualified professional. Therefore, some forensics are required. Cross-referencing the geologic information obtained during an environmental investigation with a water supply well log that was installed very close to the investigation site can provide information on the accuracy and ultimately acceptability of the water supply well log. For instance, assume a water supply well was historically installed to a depth of 100 m on the same property where an environmental subsurface investigation was conducted to a depth of 15 m. If the water well log correctly interpreted the geology in the upper 15 m, a reasonable conclusion is that the water well driller correctly interpreted the geology at depths greater than 15 m.

There is another point with respect to the installation of water wells worth mentioning here from a forensic point of view. Historically, it was not uncommon for a qualified geologist to accompany water supply well drillers when drilling wells because this was the only source of affordable geologic information. On many occasions, the geologist's name would appear on the water supply well log (Sherzer 1913; Mozola 1954,1969). There is some benefit then to consulting with geologists in the area who were historically employed by the United States Geological Survey, or State Geological Survey, or who teach at local colleges and universities. This action will help evaluate and cross-reference the information when reviewing historical water supply well logs.

Geophysical information in various forms is capable of providing valuable geologic information to assist in geologic mapping both horizontally and vertically. Geophysical information is collectible in most urban areas without causing any disturbance and can be extremely useful in interpreting data between boreholes where the geology is known. Geophysical information is most helpful when the areas are inaccessible to other methods of collecting geologic or hydrogeologic information and are within areas of complex geology.

13.2.3 Building the Conceptual Geologic Model

Following the review of existing information, a conceptual geologic model can generally be constructed. The conceptual model is a working draft geologic map of the desired region that identifies areas by their basic geologic processes. For instance, most urban areas are composed of unconsolidated sediments resulting from more than one type of geologic process, for example, fluvial, glacial, alluvial, and marine. The conceptual model simply identifies geologic processes that formed the deposits along with their general location and horizontal and vertical boundaries.

In many cases, enough information has been gathered to allow the initially identified geologic processes to be subdivided into individual geologic units or formations. It is during this step that data gaps and difficulties in interpreting geologic information usually become apparent. These data gaps then form the basis for concentrating the field work activities described in the next section.

Listing every data gap, irregularity, or area of interest is recommended until each has been resolved with adequate certainty. After each data gap, irregularity or area of interest has been addressed, re-interpretation of the conceptual geologic model is recommended to evaluate if any changes are necessary (Welty et al. 2016).

Another on-the-fly decision to make while building the conceptual geologic model is whether anthropogenic deposits such as fill material and natural deposits such as top soil or peat should be included in the mapping process. Fill material is usually noted, but not included on a geologic map unless the aerial extent is significant. Top soil may be noted at specific locations, but is generally not mapped unless the deposit is either: (1) significant in areal and vertical extent; and/or (2) represents peat or other similar material resulting from deposition in a current or former swamp, bog, or wetland. The investigator should pay attention to these deposits during the information-gathering phase of the mapping process and make a determination whether they are sufficiently large in scale compared to the map being developed and contribute to the understanding and history of the area being mapped.

13.2.4 Fieldwork

Information gathered and evaluated and building the conceptual geologic model will assist the geologist in understanding the area to be mapped before extensive field work begins. The geologist must be skilled in understanding sedimentary processes and in identifying the difference between a natural and anthropogenic deposit. This is sometimes difficult even when there is a detailed understanding of the geology of a site or region. An example is the anthropogenic deposition of dredge materials, which may take on the appearance of a natural deposit. Complications such as this in many urban areas may require several rounds of fieldwork, literature search, and historical forensic review before sufficient understanding can be achieved to construct a map. Based on the authors' review of large volumes of historical information and the construction of numerous conceptual geologic models, fieldwork in urban areas is concentrated at the following locations:

- Areas with limited or no information, or areas with gaps or missing information in the existing data set
- Locations with geologic irregularities or complexities
- Areas with extensive anthropogenic disturbance
- Areas of geologic significance such as formation contacts and unconformities
- Areas where geologic processes of formation overlap or are located in close proximity
- Locations requiring confirmation of previous interpretations
- Areas not easily accessible.

Geologic mapping in an urban area often requires the geologist to excavate through potentially disturbed soil profiles to reach material that represents naturally occurring deposits. This fieldwork is often a dangerous activity due in part to broken glass, metal fragments and other debris, underground pipes, and the existence of potentially harmful contamination. There may also be other hazards present, such as high vehicular traffic. Therefore, extreme care and proper planning should be employed, and the geologist should perform the following actions before conducting geologic mapping activities in the field (Rogers 2016a, 2016b):

- Conduct a reconnaissance of each area or location of interest
- Prepare a written scope of work for each desired location, including:
    - Purpose and objective
    - Activities to be conducted
    - Equipment list

    - Methods of investigation
    - Sampling collection procedures
    - Chain of custody (if required)
    - Photographic documentation
- Obtain written permission of private property owner(s) if any type of invasive fieldwork may be conducted
- Obtain written permission from the local municipality or appropriate governmental agency if any type of invasive fieldwork has the potential of being conducted on public property
- Contact local utilities to ensure the proper marking of all utilities in the area where fieldwork is to be conducted
- Prepare an appropriate written health and safety plan before beginning fieldwork

Fieldwork may simply consist of digging a test pit with a shovel or drilling a soil boring by hand using a hand auger. In some cases, test pits may be excavated using a backhoe or other type of excavator. If you are fortunate, many opportunities exist for conducting fieldwork without much effort. Areas under construction often offer the urban geologist ample opportunity to conduct field activities and mapping because the subsurface geology is already exposed for other purposes. Some of these locations or activities include the following:

- Road building
- Subsurface utility work
- Excavation of foundations for new buildings or for repairing existing buildings
- Demolition activities where the footings and foundation of existing buildings are excavated and removed
- Sewer repair
- Stormwater control construction activities

Conducting inspections and fieldwork at construction sites provides the geologist with many opportunities to collect valuable data at a much lower cost, and in a much shorter time. Contact local municipalities to assist with fieldwork planning and to obtain the necessary permissions before construction activities begin. Review the building permits before the start of construction and contact the appropriate Department of Transportation and utility company personnel in areas where fieldwork is to be performed. Then secure the necessary permissions before conducting any activities and employ the appropriate health and safety measures. Figures 13.2, 13.3, and 13.4 document the exposure of the near-surface geology in an urban area and show the opportunities a construction site offers the geologist. Each example represents a good opportunity to gather an abundance of geologic information in a short amount of time, and at no cost. Next to each figure is a stratigraphic column used later in this chapter to assist in building a geologic map and evaluating the recent geologic history of the region.

After examining the depositional layers from excavation pits as shown in Figure 13.4, evaluating and recording the strike and dip of the depositional units is required before a three-dimensional map can be developed (Lahee 1961). **Strike** refers to the attitude or trend of a particular deposit. For instance, if a geologist were mapping a sandy beach deposit from an ancient lake and recording the trend of the deposit at several locations, the geologist would be able to determine the size of the lake. The strike of a deposit, outcrop, or other planar feature is represented on a geologic map as a short straight-line segment oriented parallel to the compass direction of the strike. The **dip** is defined as the angle at which the geologic deposit, feature, or structure is tilted relative to the horizontal plane. The dip is represented on a map as a line segment perpendicularly attached to the strike symbol. For unconsolidated deposits, evaluating strike and dip is much more difficult to measure as the sediments being evaluated are not consolidated and may be loose. However, measuring the strike and dip is still important and may provide valuable information concerning the origin of formation

FIGURE 13.2 Geologic strata exposed at a construction site. (Photograph by Daniel T. Rogers.)

FIGURE 13.3 Construction site with exposed geologic strata. (Photograph by Daniel T. Rogers.)

and whether the unit has experienced any type of deformation since being deposited. Evaluating a beach deposit from a former glacial lake provides an example of natural deformation. Many former beach deposits from former glacial lakes are not horizontal because the strike of many glacial lake deposits dips very slightly away from the former glacial front. This dip provides an indication of the amount of isostatic rebound that has occurred as the glacier retreated and provides a basis for estimating the thickness of the glacier. Measuring the slight angle produced by this process requires high accuracy and precision as many deposits only dip 1 to 2 m/mile at most locations. In addition, measuring and evaluating strike and dip of unconsolidated deposits can also be used to evaluate whether the unit has experienced any anthropogenic disturbance.

FIGURE 13.4 Excavation at a construction site exposing geologic strata. (Photograph by Daniel T. Rogers.)

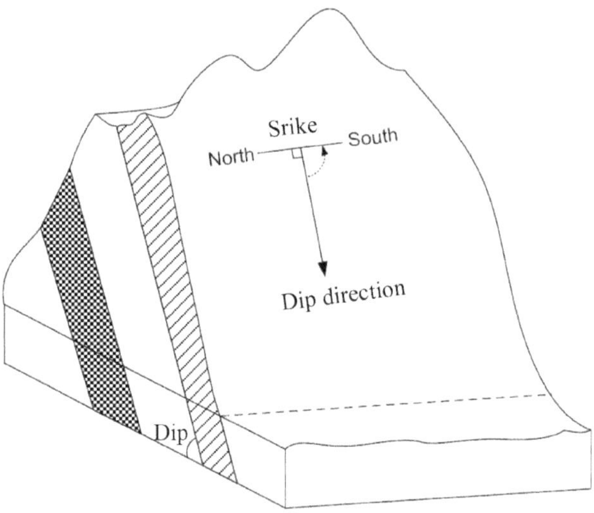

FIGURE 13.5 Concept of strike and dip.

Figure 13.5 shows strike and dip of a set of layered deposits. Note the perpendicular dip direction with respect to the strike direction. Figure 13.6 shows the strike and dip of several geologic units and the corresponding information provided by strike and dip when constructing a three-dimensional representation of the data collected from a site. The numerical values in the figure represent the angle at which the units plunge into the subsurface.

Producing detailed intermediate maps showing the geology of an individual site is a crucial step in the process of constructing a three-dimensional geologic map of an entire region and watershed. Figure 13.7 shows a basic geologic map beneath a site where each number represents a distinct deposit; Figure 13.8 illustrates a planar representation of the individual strata present at a given

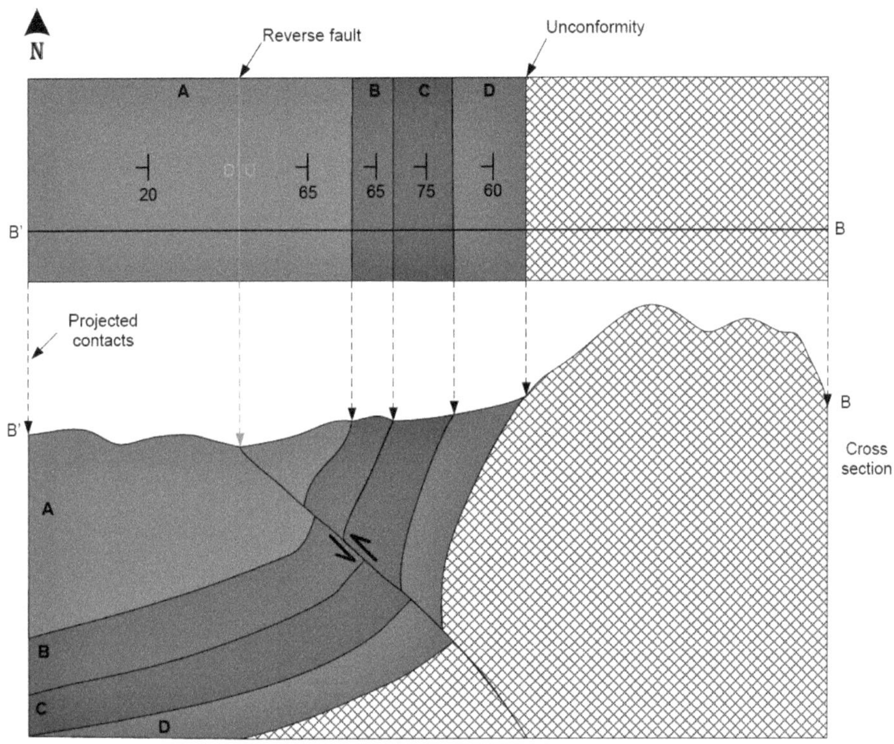

FIGURE 13.6 Concept of strike and dip as depicted on a map.

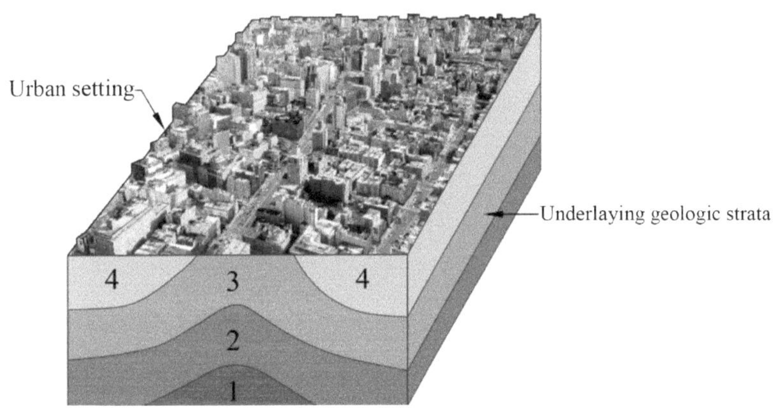

FIGURE 13.7 Urban setting showing underlying geologic units at a site.

region; and Figure 13.9 is a composite map of the underlying geology shown in Figure 13.8 and the individual strata depicted in Figure 13.9.

As stated earlier, geologic mapping in an urban environment is very difficult because it is typically covered, is not accessible, and for the most part is below the surface and cannot be readily observed. However, there are some instances where subsurface geology is readily observed and when this occurs, every opportunity to gather scientific data should not be ignored.

Combining geologic information from many individual sites using facies analysis is the next step in constructing a three-dimensional geologic map for a region.

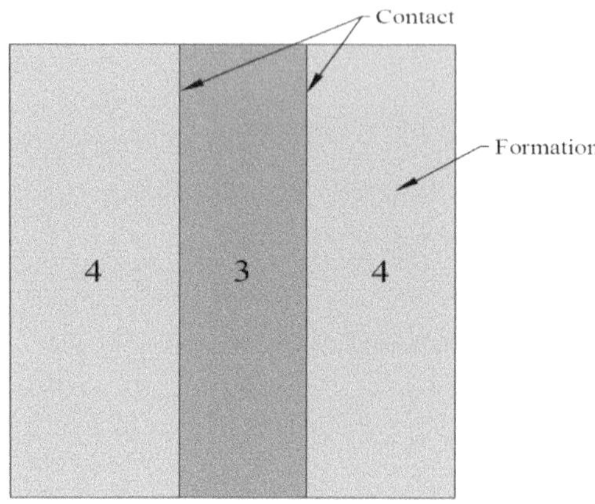

FIGURE 13.8 Geologic site map of area shown in Figure 13.7.

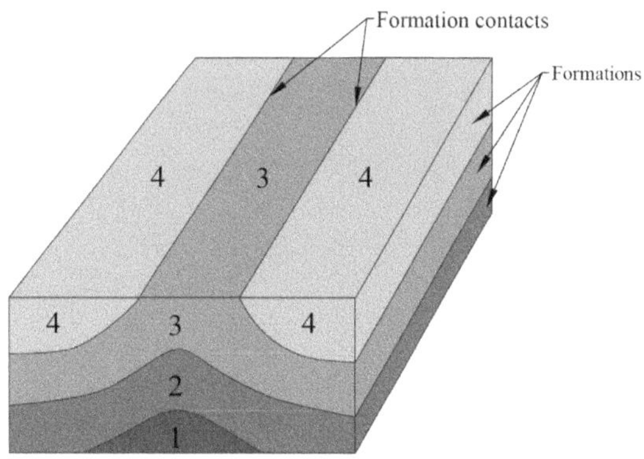

FIGURE 13.9 Composite block map developed from Figures 13.7 and 13.8.

13.2.5 Facies Analysis

Facies are the characteristics of a sedimentary rock mass that reflects its depositional environment. **Facies analysis** or cross-dating is a technique used to help understand sedimentary environments and the processes leading to the origin of their sediments (Anderton 1985). Facies analysis can be performed when two or more **stratigraphic columns** are developed for different locations within a given area. Facies analysis is conducted to:

- Evaluate the completeness of sedimentary deposits at any one location
- Identify any gaps in the depositional record
- Reconstruct the geologic history of a region
- Identify or confirm the historical occurrence of geological processes within a given region
- Reconstruct a succession of events
- Create a vertical succession of depositional record

The type of facies analysis most often employed to reconstruct the geologic history of a particular region is termed **relative dating**. This technique uses the four principles of stratigraphy which include: (1) principle of original horizontality; (2) principle of superposition; (3) principle of cross-cutting relations; and (4) principle of faunal and floral succession. Using these four principles, facies analysis can evaluate and establish the chronology of depositional events within the area being mapped. For example, during the course of reviewing historical literature and conducting fieldwork, it usually becomes apparent that a few locations contain a nearly complete and undisturbed vertical geologic section. These accessible columns essentially have preserved the depositional history so it can be studied in detail and are called **type sections**. Once the chronology of deposition has been firmly established, a stratigraphic column can be constructed for the entire area being mapped (USGS 2007).

Many types of geologic processes are responsible for building the geologic history of the large urban areas within the United States. We will demonstrate the process of facies analysis using the stratigraphic columns from Figures 13.2, 13.3, and 13.4 because they contain a good record of the depositional history of the Rouge watershed. The stratigraphic columns from each site are presented in Figure 13.10.

The distance separating the three sites and represented by each column shown in Figure 13.10 is approximately 1.5 km. Each site is perpendicularly aligned (north to south) to the strike of the geologic units identified at each location (west to east). The lines drawn between each stratigraphic column connect geologic units appearing in more than one stratigraphic column such as the Lacustrine Beach Sand Deposit and the Lacustrine Silty Clay Deposit. The units of Fill Material and the Beach Dune Sand Deposit exhibit converging lines because they are not present in the other stratigraphic columns within the study area.

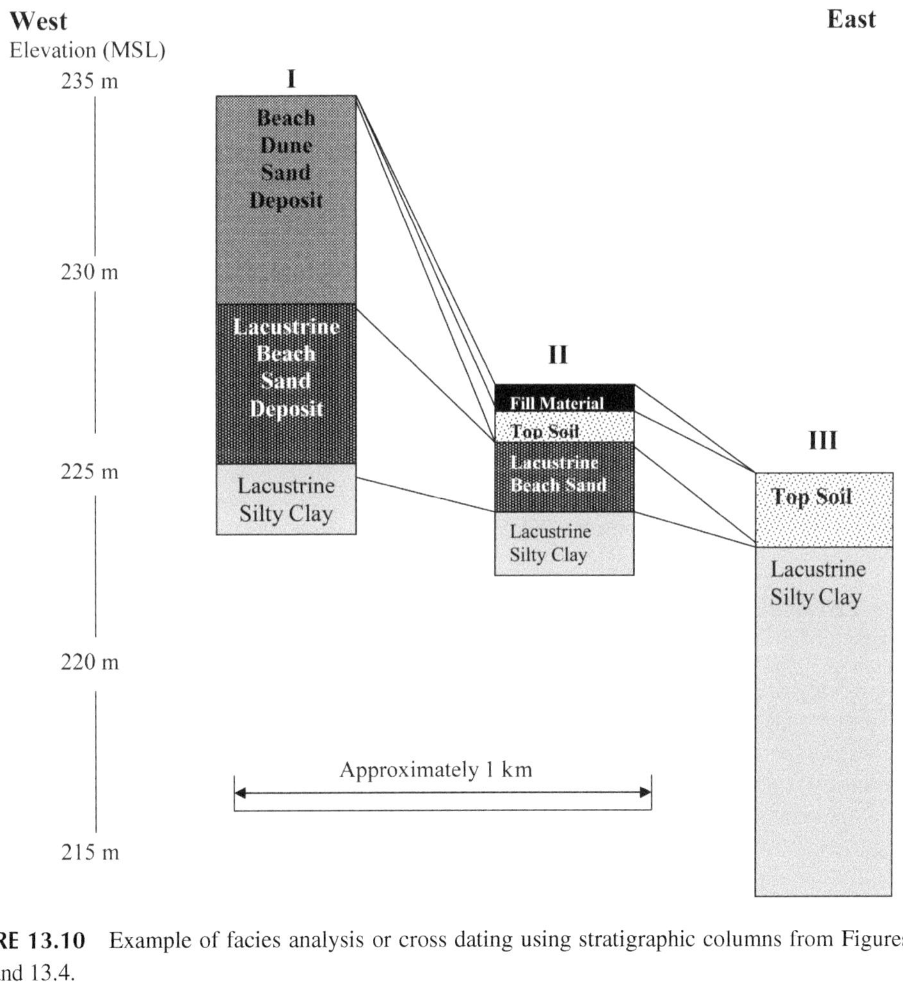

FIGURE 13.10 Example of facies analysis or cross dating using stratigraphic columns from Figures 13.2, 13.3, and 13.4.

A simple analysis of the facies represented by each depositional unit in Figure 13.10 indicates each site was once part of a former lake, with the beach and associated dune located toward the west and the deeper water. The small amount of dip present between columns I and II (a few meters per kilometer) is also characteristic of these glacial lacustrine features. The presence of a lake between columns I and III is further confirmed by the absence of the sand and dune deposits in column III toward the east, and the continued presence of the lacustrine silty clay deposit.

It should be noted that the presence of sedimentary deposits that dip in this example represents the slope or contour of the original land surface and does not represent a structural cause, such as faulting, for the dipping deposits. However, some isostatic adjustment has occurred as the materials were deposited and is evident from an elevation rise in the beach elevation of approximately 1 m every 2 km toward the northeast and is likely a result of the historical ice unloading of the region.

Additional analysis of facies also suggests the lake was once larger because a deposit of lacustrine silty clay is present beneath the beach sand deposit in Stratigraphic Column I. This suggests the lake was getting smaller and in a regressive stage before it disappeared. The absence of top soil in Stratigraphic Column I and the presence of fill material in Stratigraphic Column II also indicates anthropogenic disturbance. Additional analysis of facies also suggests the lake was once larger because a deposit of lacustrine silty clay is present beneath the beach sand deposit in Stratigraphic Column I. This suggests the lake was getting smaller and in a regressive stage before it disappeared. The absence of top soil in Stratigraphic Column I and the presence of fill material in Stratigraphic Column II also indicates anthropogenic disturbance.

Do not get too comfortable – facies analysis is not always this straightforward. Gradational or interpolated boundaries between geologic units require separate and careful treatment. Gradational boundaries occur both laterally and vertically and are fairly common in sedimentary deposits. Gradational boundaries should be identified and explained in the legend and accompanying text according to the following guidelines:

- Nature of the gradational boundary
- Cause of the gradational boundary
- Extent of boundary

A sufficient facies analysis contains numerous stratigraphic columns with sufficient areal coverage to enable an understanding of the region's depositional history. This information allows for the construction of cross sections. Cross sections display the geologic deposits without any breaks or gaps between the stratigraphic sections, and help facilitate the construction of a three-dimensional geological map (USGS 2002). Some degree of interpretation is usually necessary to construct a cross section, and these finer points are explained in greater detail in the next section.

13.2.6 Developing and Constructing Cross Sections

A cross section is similar to facies analysis, but cross sections add levels of complexity and additional information. A cross section represents a view of the subsurface as though it was cut open and viewed from the side. Geologic cross sections are intended to be interpretive and usually constructed to demonstrate the arrangement, relationship, structural composition, and depositional history of geologic units displayed in the cross section. General guidelines for constructing a geologic cross section include the following:

- Choosing appropriate horizontal and vertical scales
- Properly locating control points or points along the cross section where stratigraphic sequences are known
- Ensuring the legend incorporates and explains each geologic material and feature

- Using appropriate symbols to identify each geologic unit
- Using appropriate orientations and landmark features
- Including vertical and horizontal scales
- Including a statement of vertical exaggeration.

When constructing a cross section, the amount of geologic interpretation between stratigraphic sections depends upon the:

- Number of stratigraphic sections
- Distribution of stratigraphic sections
- Number of data gaps requiring interpretation
- Complexity of each stratigraphic section.

Selecting the line or slice direction of a cross section is perhaps the most important operation when constructing a cross section. Usually, several cross sections are constructed to create a more complete picture of the vertical dimension and depositional sequence. Generally, the cross section containing the most significant information is one constructed as a perpendicular slice to the strike or trend of the geologic units being evaluated. These types of cross sections are most effective because more geologic units are usually represented and the relationship, structure, and depositional history are more easily revealed and understood. This is especially relevant to urban watershed restoration efforts because most of the deposits existing below urban areas are sedimentary in nature.

An example of a cross section constructed perpendicular to the strike is shown for the Rouge River watershed in southeastern Michigan in Figure 13.11 (Rogers 1996). This cross section includes several types of sedimentary deposits of glacial, lacustrine, and fluvial origin commonly found in most urban areas of the United States. Figure 13.12 is a three-dimensional representation of several intersecting cross sections (USGS 2024a). Cross sections of this type aid in interpreting the depositional history of the area being mapped.

13.2.7 Creating the Base Map

After the review of existing data, conducting field work, facies analysis, and constructing cross sections, a base map can be prepared. The steps required to prepare a base map include: (1) using appropriate nomenclature; (2) selecting appropriate symbols; (3) selecting the scale for the map; and (4) determining what type of map is most appropriate (USGS 2024a).

Use of recognized and acceptable nomenclature and symbols ensures the geologic map being prepared maintains consistency and minimizes ambiguity. The standard nomenclature for geologic maps in North America is The North American Stratigraphic Code (American Association of Petroleum Geologists 2005; USGS 2024b). This code is used for naming, describing, establishing, redefining, and abandoning geologic units. For symbols, the standard is the Federal Geographic Data Committee Digital Cartographic Standard for Geologic Map Symbolization (USGS 2006).

Items such as line symbols, point symbols, and colors and patterns used to portray different features on a geologic map are covered by this standard. Additional information, methodology, and context within the international mapping community is the International Cartography Association (ICA). The goal of ICA is to promote the disciplines and professions of cartography and standardization in an international context (ICA 2024).

The first step in preparing a geologic map is to start small, and then add adjacent locations to build the entire map. One way to accomplish this process is, to begin with a single geologic unit that is relatively simple or easy to identify in the field and then add adjacent geologic units both horizontally and vertically through the entire unconsolidated strata until bedrock is encountered. For reference, the applicable 7.5-minute series USGS Quadrangle map ("quad") is recommended to

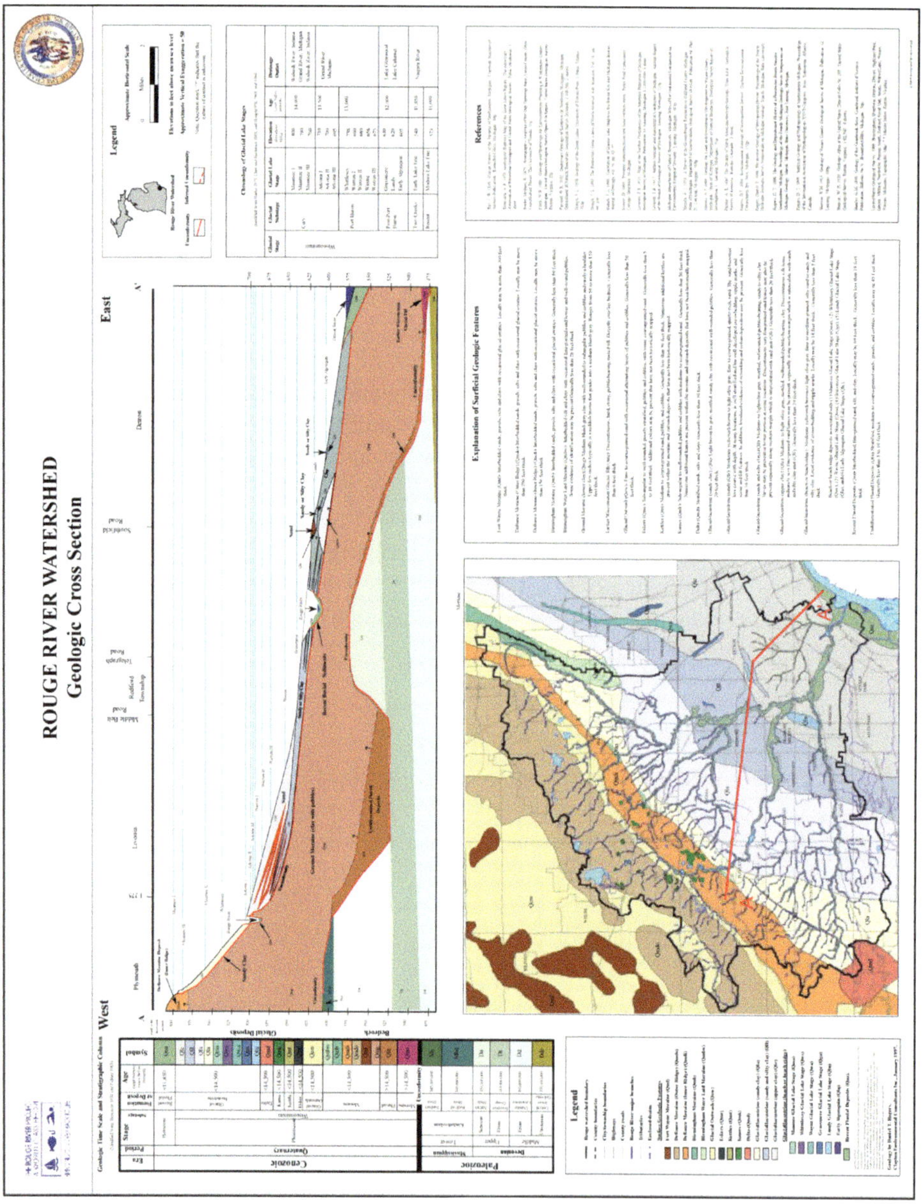

FIGURE 13.11 Detailed cross section of Rouge River watershed in southeastern Michigan. (From Rogers, D. T. 1997a. *Surficial Geologic Map of the Rouge River Watershed in Southeastern Michigan*. Wayne County, MI. Map Scale 1:62,500. 2 Sheets.)

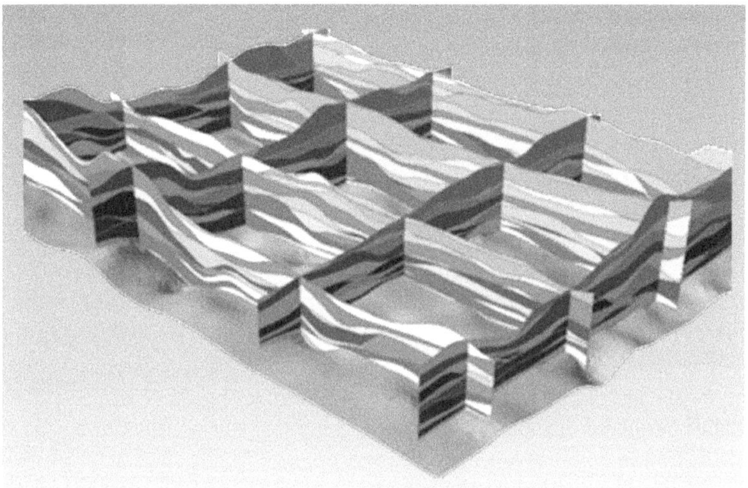

FIGURE 13.12 Several interconnecting cross sections. (From United States Geological Survey. 2024a. *National Cooperative Geologic Mapping Program*. http://ncgmp.usgs.gov. (Accessed May 12, 2024).)

construct the field base map. The 7.5-minute series quad is a 1:24,000 scale map – a scale suitable for many environmental applications (USGS 2024c).

These maps also have valuable information to greatly assist in mapping any given area, including the following:

- Each map is very accurate and reliable.
- Most urban areas in the United States have readily available 7.5-minute series Quadrangle maps.
- Elevation contours are present to assist with the identification of geologic units and help establish formation boundaries.
- Many land features of interest to geologists are identified.
- Anthropogenic features such as developed areas, roads, and parks are identified.
- Natural features such as lakes, rivers, streams, forested areas, and points of interest are identified.
- Each map is well labeled.
- Coordinates of longitude and latitude are present on each map.
- Magnetic declination is included on each map.
- The location of springs and wells are located on many maps.
- Information can be easily transferred to other mapping technologies such as Geographic Information Systems (GIS) and three-dimensional maps
- Many maps are available electronically so areas with complex geology or areas requiring finer resolution can be mapped using larger scales as appropriate

13.2.8 Data Analysis

One of several methods used to evaluate whether sufficient information has been obtained to construct a geologic map is called a rarefaction analysis. A rarefaction analysis involves constructing a model, testing the data's reliability and sensitivity through the collection of additional data, and then evaluating if the original assumptions of the model have changed. The result indicates the adequacy of the model developed for the area being mapped (Birks and Line 1992; Siegel 2006). If the assumptions have not changed when the additional data are collected, it is a reasonable indication enough data were collected. A flattening curve shown indicates the data set is approaching an asymptotic condition (its maximum level), so the collection of additional data is probably not necessary. Most geologic maps have areas where some professional interpretation has been conducted, as new information becomes available, maps are usually updated. Therefore, the purpose of rarefaction or other types of statistical analysis is to evaluate major depositional sequences.

13.2.9 Final Draft Map Preparation

Over 100 years ago John Wesley Powell stated that "maps are designed not so much for the specialist as for the people". He elaborates more on this topic (and we paraphrase here) by saying the system of convention selected should be simple and expressive, so the concepts and processes depicted can be easily understood. Conveying geologic information with the simplest of terms is difficult and challenging because the information is often very complex. Do not despair – there are several methods available to allow an effective presentation of the information to intended users. Selecting an appropriate method or combination of methods depends on:

1. the level of complexity, detail, and scale of the information and area being mapped;
2. the intended users, such as scientific or academic, land-use planners, or the general public; and
3. cost.

Some common methodologies used to prepare the map include the following:

1. Preparing a series of maps showing the regional development and evolution of the landscape of a particular region. These maps tend to lack great detail because the scale is generally very large. The purpose of these maps is to convey major concepts of regional significance.
2. Preparing several maps beginning with the deposition of the bottom or oldest layer of sediment and then continuing with the display of distinctive intervals of equal time or significant geologic events of deposition. These maps have advantages because they can be very detailed and can focus on significant geological events.
3. Adding geological processes that formed the sedimentary deposits such as stream or river succession, lake elevations, or glacial advances or retreats. This type of map combines the areal distribution of geological deposits with the methods of their formation. Although this type of map is very effective at conveying processes with geological deposits, it may lack the desired detail depending on the scale of the map.
4. Producing maps in three dimensions. Constructing maps in three dimensions generally requires more time and effort, but has the benefit of providing better data gap analysis and integrating landscapes into the geologic map.
5. Using geographic information systems to combine geological maps with other digital data layers such as roads, buildings, parks, utilities, drainage patterns, proposed developments, and population is helpful. A composite map representing the overlay of some or all of these layers can produce new information about the study area.

The most common methods are #3, #4, and #5, or some combination of each.

13.2.9.1 Succession of Maps

A succession of maps is one method used to decrease the time necessary to understand the geological processes responsible for the development of sedimentary deposits in urban areas. This technique presents complex processes gradually using several maps that increasingly add complexities and depositional processes. At a regional scale, building layers showing different geological

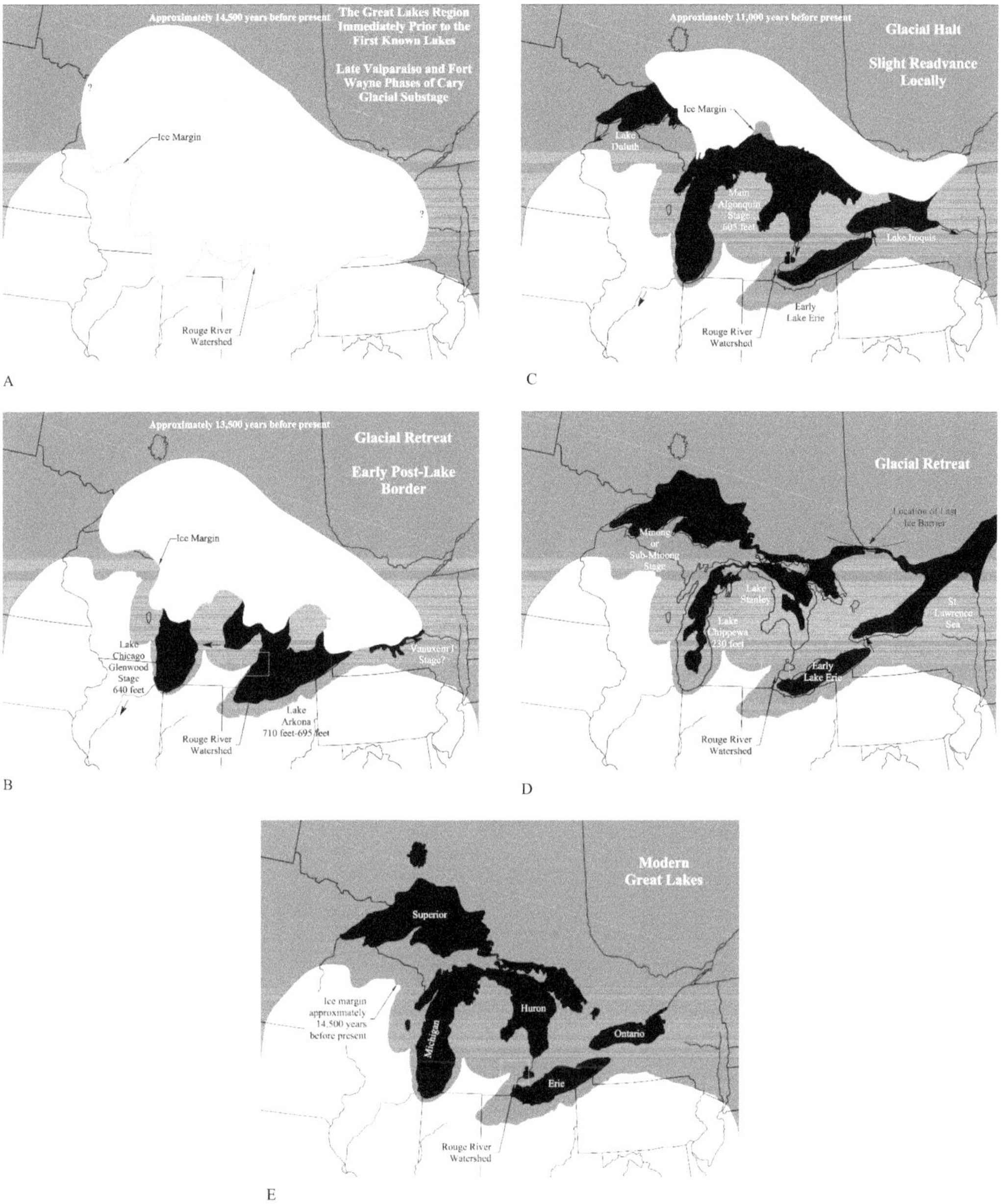

FIGURE 13.13 (A) Great Lakes region approximately 14,500 years ago. (B) Approximately 12,000 years before the present. (C) Approximately 11,000 years before the present. (D) Approximately 9,000 years before the present. (E) Modern time. (Modified from Hough, J. L. 1958. *Geology of the Great Lakes*. University of Illinois Press. Champaign–Urbana, IL; Hough, J. L. 1963. The Prehistoric Great Lakes of North America. *American Scientist*. Vol. 51, pp. 84–109.)

processes occurring at specified intervals and/or depicting the sequences forming deposits can simplify the communication of geologic information. These types of maps can also incorporate three-dimensional and Geographic Information Systems (GIS) formats to present the information. An example of a succession of regional maps of the Great Lakes region is presented in Figure 13.13A

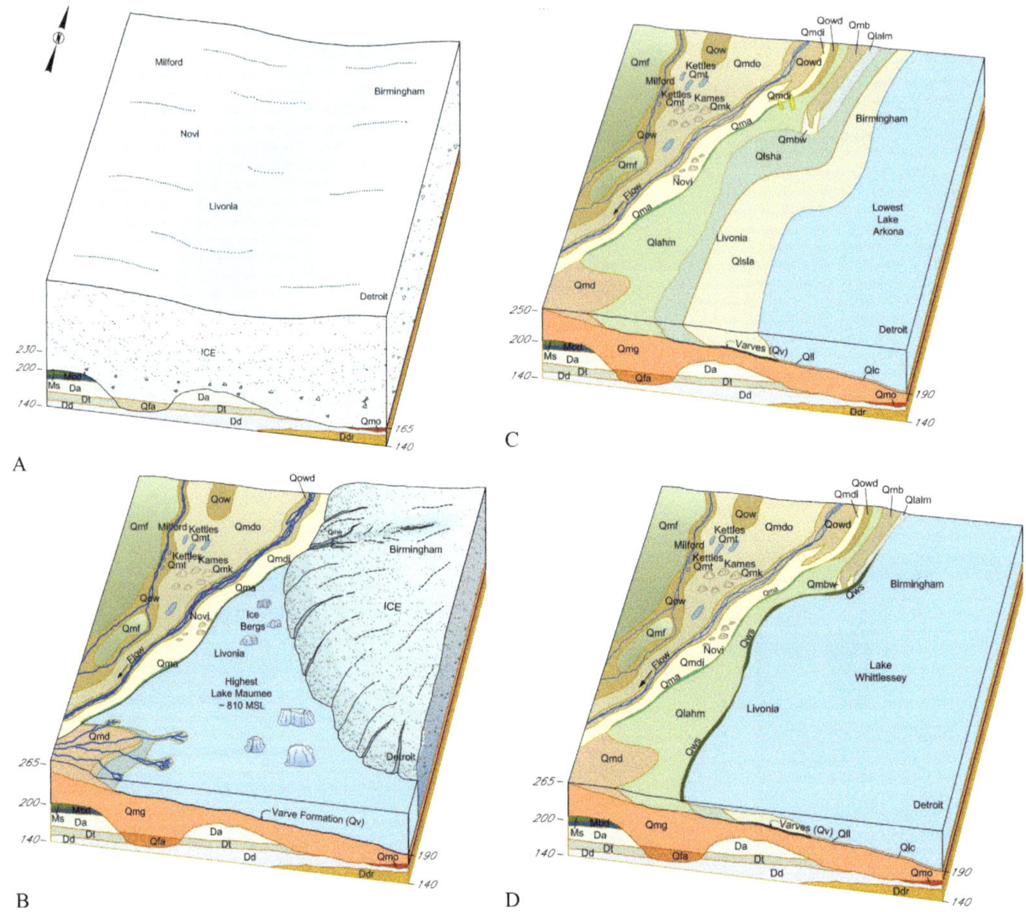

FIGURE 13.14 (A) Approximately 14,500 years before the present. (B) Approximately 13,500 years before the present. (C) Approximately 13,000 years before the present. (D) Approximately 12,900 years before the present. (Modified from Rogers, D. T. 1997. *Surficial Geologic Map of the Rouge River Watershed in Southeastern Michigan*. Wayne County, MI. Map Scale 1:62,500. 2 Sheets.)

through 13.13E. This map succession shows the regional development and evolution of the landscape for a particular region at a specific time – the Great Lakes during the retreat of the last glacial advance of the late Pleistocene. These maps are general in nature and lack detail because their scale is very large.

The purpose of these maps is to: (1) convey the concept of regional geological significance, that is, the glacial origin of unconsolidated sediments; and (2) provide a reference point for the next set of more detailed maps presented as the Figure sequence 13.14A through 13.14E. Starting at the general regional scale and moving to the more detailed watershed scale where more information of a specific nature can be included, this illustrates the educational capability of this method. The map succession shown in Figures 13.14A through 13.14D provides an example of how the use of progressive complexity at the smaller scale of an urban watershed can depict the processes forming the sedimentary deposits. For example, this sequence shows the glacial retreat and the associated sedimentary deposits in three dimensions. Figure 13.15 represents the current geology of the watershed in an enlarged format with a complete legend describing the stratigraphic column.

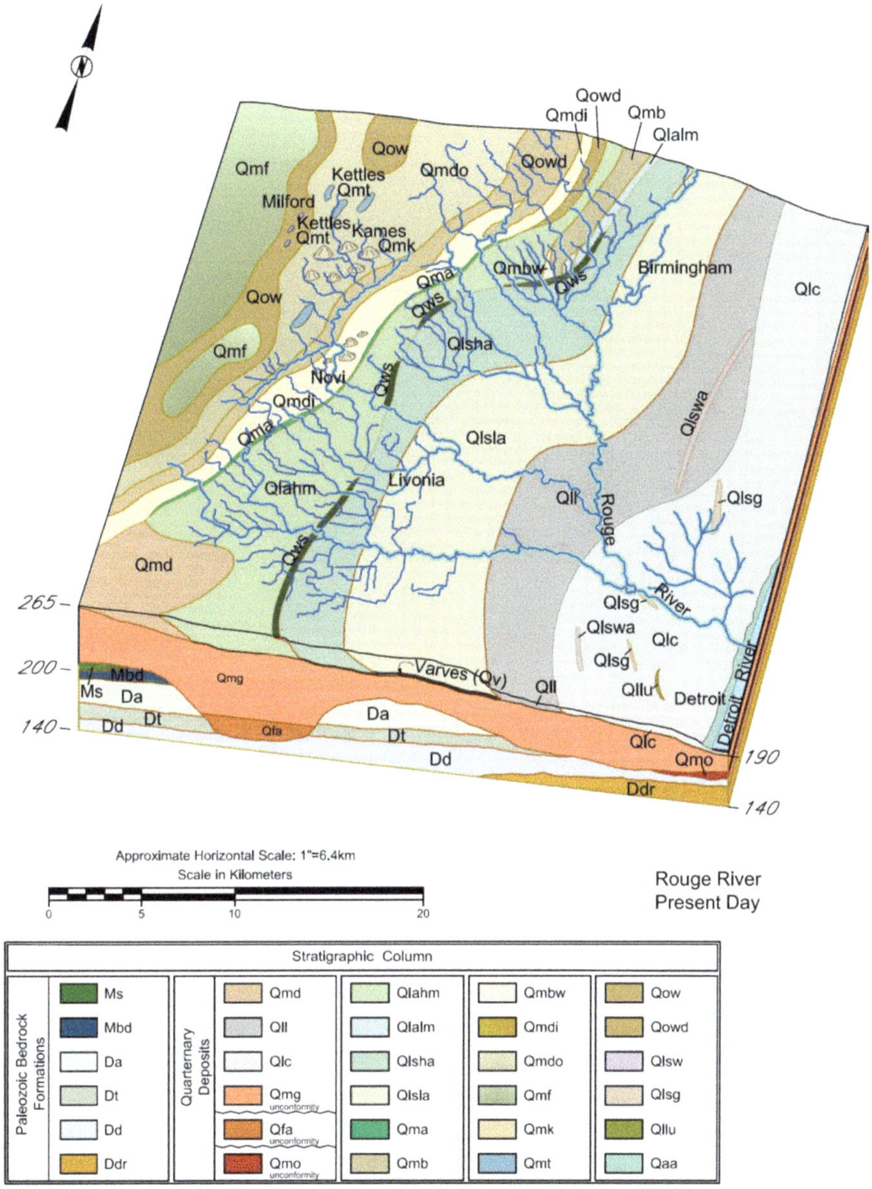

FIGURE 13.15 Enlarged and annotated current geologic map of the Rouge River watershed in southeastern Michigan. (From Rogers, D. T. Kaufman, M., and Murray, K. S. 2008. Quaternary Geology of the Rouge River Watershed in Southeastern Michigan, USA. In: *33rd World Geological Congress.* Abstracts with Programs. Oslo, Norway.)

13.2.9.2 Three-Dimensional Maps

In addition to map sequences showing increasing amounts of information, three-dimensional digital techniques are also available to develop geologic maps (USGS 2005).

13.2.9.3 Maps Using Geographic Information System

GIS are useful for displaying and integrating many sources of information, including geological maps of urban areas (Chang 2007). One advantage of using GIS is its popularity in urban areas, and

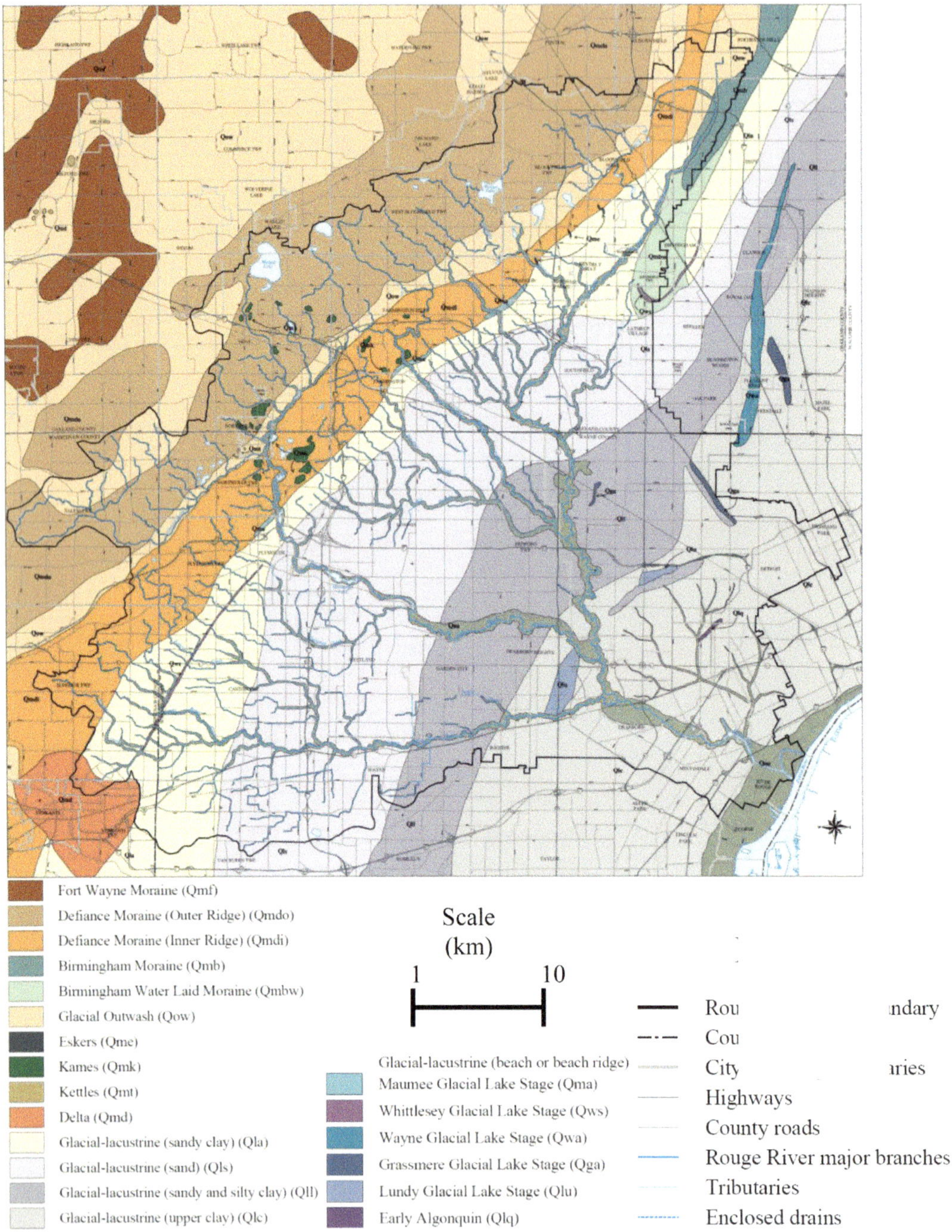

FIGURE 13.16 Geologic map of the Rouge River watershed produced using GIS. (From Rogers, D. T. 1997a. *Surficial Geologic Map of the Rouge River Watershed in Southeastern Michigan*. Wayne County, MI. Map Scale 1:62,500. 2 Sheets.)

most major urban areas in the United States have significant GIS capabilities. As the systems already exist, integrating geological information with other types of readily available layers of digital data provides a distinct cost advantage. A common operation of a GIS is to combine specific digital data layers of similar geographic extent to produce new maps containing the input layers' spatial and descriptive attributes.

The use of multiple layers is recommended when constructing the base map using geologic data of similar types of formation. Subdividing the input of information into layers also assists with the editing process and may also be used to develop multiple maps. An example of two-dimensional geologic map constructed using GIS of the Rouge River watershed is shown in Figure 13.16 (Rogers 1997a).

Figure 13.17 is the stratigraphic column for the Rouge watershed. Stratigraphic columns are essentially a list of all geological formations depicted on a map and are typically listed in sequential order by age. Stratigraphic columns are a necessary part of constructing a geologic map because they list each deposit. Many of the hydrogeologic principles demonstrated through the mapping of the Rouge watershed in southeastern Michigan are applicable to many other areas in the United States and the world. Some of these principles include the following:

1. Sediments are composed of geologic materials that have originated from several different sedimentary processes including fluvial, lacustrine, and glacial; this is very typical of most watersheds in the United States
2. The drainage pattern within the watershed is reflective of the geology. Specifically, the drainage pattern changes density once the sand unit is encountered
3. The dendritic drainage pattern of the Rouge watershed is similar to most inland urban watersheds lying over sedimentary rock anywhere in the world (even those not glaciated); coastal zones may be different because of regional tectonic influences.
4. The trend or strike of the glacial and lacustrine deposits is generally from the southwest to the northeast and follows the retreating ice as the glaciers melted.
5. The types of deposits change, sometimes significantly, over short distances. This is evident from the map as the deposits vary from glacial to lacustrine to fluvial.

The advantages of a three-dimensional map compared to a two-dimensional map become apparent by comparing Figure 13.15 – a simplified three-dimensional map of the Rouge River watershed – with the two-dimensional geologic map of the Rouge River shown in Figure 13.16. These advantages include showing the thickness of the sediment layers; showing formations, such as the Lower Clay Unit, which is not even represented in the two-dimensional map; and showing the dip, if any, of the geologic units.

13.3 CREATING A GEOLOGIC SENSITIVITY MAP

The concept of geologic sensitivity of the subsurface to contamination originated in France during the 1960s and was introduced into the scientific literature by Albinet and Marget (1970). Since then, the concept has evolved to include both a distinction between and a combination of sensitivity and risk assessment. Groundwater sensitivity is currently interpreted as a function of the natural properties of the overlying soil or sediments of the unsaturated zone, aquifer properties (e.g., effective porosity and recharge area), and aquifer material (Foster and Hirata 1988; Robins et al. 1994; Rogers et al. 2007). Geologic sensitivity mapping can be divided into two groups: subjective rating methods and statistical and process-based methods (Focazio et al. 2001). The subjective rating methods are characterized by numerical scales representing low to high vulnerabilities. Typically, the results are applied to large areas and used for policy and management objectives. By contrast, the statistical process-based methods produce finite values, such as areas exceeding specific water quality values.

Era	Period	Epoch	Stage		Age (approximate years before present)	Symbol	Unit Name
Cenozoic	Quaternary	Holocene	Wisconsinan	Fluvial	<10,000		Qaa - Recent Rouge River Fluvial Deposits
		Pleistocene		Glacial lucustrine	<14,500		Qllu - Glacial Lacustrine Beach Deposit Lake Lundy
							Qlsg - Glacial Lacustrine Beach Deposit Lake Grassmere
							Qlswa - Glacial Lacustrine Beach Deposit Lake Wayne
							Qlc - Glacial Lacustrine Clay
							Qll - Glacial Lacustrine Silty Clay
							Qws - Glacial Lacustrine Beach Ridge Lake Whittlessey
							Qlsla - Glacial Lacustrine Sand, Lowest Lake Arkona
							Qlsha - Glacial Lacustrine Sand, Highest Lake Arkona
							Qv - Varves
							Qlahm - Glacial Lacustrine Sandy Clay Maumee Lake Deposit
							Qma - Glacial Lacustrine Beach Highest Lake Maumee
				Moraine	<14,500		Qmbw - Birmingham Waterlaid Moraine
							Qmb - Birmingham Moraine
				Delta	<14,500		Qmd - Delta
				Outwash	<14,500		Qowd - Outwash Defiance
				Esker	<14,500		Qme - Esker
				Moraine	<14,500		Qmdi - Inner Defiance Moraine
							Qmdo - Outer Defiance Moraine
				Glacial Outwash	<14,500		Qow - Outwash Fort Wayne
				Kame	<14,500		Qmk - Kame
				Kettle	<14,500		Qmt - Kettle
				Moraine	<14,500		Qmf - Fort Wayne Moraine
							Qmg - Ground Moraine / Lodgement Till
					unconformity		
				Fluvial	22,000		Qfa - Fluvial (Interglacial) >22,000 ybp
					unconformity		
				Moraine	22,000		Qmo - Pre-Late Wisconsinan Glacial Till Depost >22,000 ybp
					unconformity		
Paleozoic	Mississipian	Lower	Kinderhookian	Sunbury Shale	345,000,000		Ms - Sunbury Shale
				Bedford Shale	345,000,000		Mbd - Bedford Shale
	Devonian	Upper	Senecan	Antim Shale	350,000,000		Da - Antrim Shale
			Erian	Traverse Group	355,000,000		Dt - Traverse Group
		Middle	Erian	Dundee Limestone	360,000,000		Dd - Dundee Limestone
			Ultsterian	Detroit River Dolomite	370,000,000		Ddr - Detroit River Formation

FIGURE 13.17 Stratigraphic column of Rouge River watershed in southeastern Michigan. (From Rogers, D. T. 1997a. *Surficial Geologic Map of the Rouge River Watershed in Southeastern Michigan*. Wayne County, MI. Map Scale 1:62,500. 2 Sheets.)

With these methods, the results are usually not applied to large areas due to data gaps and variable geology. In addition, the results are generally obtained under more detailed site-specific assessments and used for purely scientific purposes (Focazio 2001). In practice, subjective rating methods are preferred for conducting sensitivity assessments on a watershed scale (Murray and Rogers 1999a).

The concept of geologic sensitivity relies on the assessment and representation of various hydrogeologic parameters such as vadose zone characteristics (e.g., thickness and infiltration capacity), depth to water, and amount of recharge (Zaporozec and Eaton 1996; Eaton and Zaporozec 1997). The utility of this concept, however, becomes more important when the geologic data are supplemented with environmental, economic, and political insight gained through past environmental clean-up efforts (Foster et al. 1994; Loague et al. 1998).

Successful development of geologic sensitivity maps can be difficult to achieve in areas experiencing rapid growth. Urbanization and the artificial infrastructure it produces (e.g., sewers and detention ponds) can have a profound influence on regional hydrogeology (Vuono and Hallenbeck 1995; Zaporozec and Eaton 1996; Kibel 1998). Basic processes affecting surface water and groundwater are modified, including surface water drainage patterns and velocities, evaporation rates, infiltration, and aquifer recharge (Burn et al. 2007; Fresca 2007; Howard et al. 2007; Mohorlok et al. 2007). The difficulties in sensitivity map development are also compounded by the differences in the amount and type of geologic and hydrogeologic information available in urban areas and rural settings. For these reasons, a uniform assessment of data while conducting sensitivity studies in urbanizing areas is difficult to achieve.

Geologic sensitivity mapping provides a starting point for quantifying anticipated environmental risk at a particular site and also highlights locations where additional information is required. In lieu of specific site information, this risk assessment can serve, if necessary, as a proxy for anticipated future clean-up costs. Additionally, this method can be used by other interested parties during the recycling of industrial sites to estimate the liability of sites (Stiber et al. 1995; Murray and Rogers 1999b). With respect to water resource allocation, as surface and groundwater interact, mapping groundwater also provides valuable information concerning their respective distributions (Rogers and Murray 1997; Pierce et al. 2007).

It is often impossible or prohibitively expensive to clean up polluted groundwater effectively. This is why a better approach for ensuring groundwater quality is to map both actual groundwater pollution and groundwater potentially sensitive or vulnerable to pollution, and then use this information in land use planning and management (Zaporozec and Eaton 1996; Rogers 2020). The information gathered through the process of geologic sensitivity assessment would allow decision-makers to assess the current and future environmental risks associated with any particular site as long as it was contained within the area mapped.

13.3.1 Geologic Sensitivity Criterion and Methods

Murray and Rogers (1999a, 1999b) and Kaufman et al. (2003, 2005) have developed a method for geologic sensitivity mapping using a modified DRASTIC model (Aller et al. 1987). This method contains a subjective numerical rating system that uses different weighting coefficients for various geologic and hydrogeologic parameters of concern and incorporates potential receptors and political data into the model. Listed below are those geologic and hydrogeologic factors, along with the political, ecological, and environmental factors that are collected during environmental subsurface investigations at known or suspected sites of contamination (Rogers 1992, 1996, 2002; Murray and Rogers 1999a;Kaufman et al. 2005):

- Soil or sediment type, composition, color, texture, thickness, and soil moisture
- Stratigraphy of geologic units
- Horizontal and vertical extent of geologic units

- Variation within geologic units (heterogeneity and anisotropism)
- Type of primary and secondary features
- Presence, extent, and structure of unconformities
- Presence, depth, and relative abundance of groundwater
- Groundwater flow direction
- Areas of groundwater recharge
- Areas of groundwater discharge
- Groundwater–surface water interaction features
- Anthropogenic features including:
    - Storm sewer placement relative to groundwater elevations
    - Utility corridors
    - Building footings and other subsurface structures
    - Landfills
    - Artificial recharge basins
    - Surface water confinement features
    - Dams
    - Stormwater retention or detention basins
    - Large, paved areas
    - Artificial surface drainage pattern alteration
    - Roads and road cuts
- Potential receptors or points of potential exposure:
    - Water supply wells
    - Irrigation wells
    - Surface water bodies such as lakes, streams, rivers, swamps, bogs, springs, groundwater seeps, etc.
    - Buildings and building foundations
    - Parks, schools, playgrounds, day care facilities, retirement communities, hospitals, long-term care facilities, etc.
- Potential sources of contamination:
    - State and Federal lists of environmental contamination
    - Hazardous waste facilities
    - Abandoned dumps
    - Brownfield sites
    - Historical industrial manufacturing sites
    - Electrical generating facilities
    - Gasoline service stations
    - Dry cleaning facilities
    - Refineries
    - Other known or identified sources

These factors provide a framework for the evaluation of the geologic sensitivity within any urban area using a sensitivity matrix and scoring system as shown in Table 13.1 (Rogers 2020). A sequential description of the sensitivity matrix parameters presented in Table 13.1 is further explained herein below using the following parameters:

- *Depth to groundwater:* The closer to the surface groundwater is encountered, the higher the geologic sensitivity because contaminants have a shorter distance to migrate vertically before encountering saturated conditions.

  Depth to groundwater varies widely across the United States depending on geology, climate, and anthropogenic influence. For example, groundwater is routinely present at depths very near the surface along the east coast and southeast; from a meter to a few meters in the

TABLE 13.1
Geologic Sensitivity Matrix and Scoring System

Parameter Identification Number	Parameter Description	Rating Strength
1	**Depth to Groundwater**	
	Less than 3 m below the ground surface	10
	3 to 10 m	5
	Greater than 10 m	1
2	**Composition, areal extent, and thickness of soil units in the unsaturated zone**	
	Thick and extensive sequence of sand and gravel	10
	Interbedded sands and clay deposits	5
	Thick and extensive sequence of clay	1
3	**Composition, areal extent, and thickness of saturated zone**	
	Thick and extensive sequence of sand and gravel	10
	Interbedded sands and clay deposits	5
	Thick and extensive sequence of clay	1
4	**Occurrence and relative abundance of groundwater**	
	25% or less likelihood before encountering an aquiclude	10
	25% to 74% likelihood	5
	Greater than 75% likelihood	1
5	**Area of groundwater recharge**	
	Significant area of recharge	10
	Moderate area of recharge	5
	Not a significant area of recharge	1
6	**Areas of groundwater discharge**	
	Significant area of recharge	10
	Moderate area of recharge	5
	Not a significant area of recharge	1
7	**Travel time and distance to point of potential exposure**	
	Less than 10 years	10
	10 to 25 years	5
	Greater than 25 years	1
8	**Source of potable water**	
	Current source of potable water	10
	Potential source of potable water	5
	Not a potential source of potable water	1

Source: Rogers, D. T. 2020. *Urban Watersheds: Geology, Contamination, Environmental Regulations, and Sustainability*. 2nd Edition. CRC Press. Boca Raton, FL. 606pp.

Midwest; to perhaps several hundred meters in the southwest. With respect to anthropogenic influence, the landscape in urban areas is characterized by extensive modifications. The natural hydrology is altered by impervious surfaces, the construction of stormwater retention basins, and the use of more permeable areas for artificial recharge or groundwater storage. In addition, leaks from sanitary sewers, storm sewers, potable water lines, and surface watering (especially golf courses) may profoundly impact the depth of groundwater and the direction of flow locally or over larger areas.

- *Composition, areal extent, and thickness of soil units in the unsaturated zone:* The composition of soil or sediment above the water table is important to evaluate because it helps to determine whether the soil or sediments above the water table will impede the vertical migration of contaminants through the soil or sediment column. For instance, clay soils are fine-grained and generally impede or slow migration, whereas sand and gravel deposits tend to facilitate contaminant migration. Interbedding of geologic units is also considered.

Other significant factors include the primary and secondary porosity and the potential for unsaturated soils located above aquifers to be subject to anthropogenic disturbance. For instance, a clay layer that is 6 m thick may appear to significantly impede the vertical migration of contaminants. However, if portions are excavated, or if the integrity of the clay has been compromised by vertical fractures, root fragments, or anthropogenic activities, these events increase the potential for contaminant migration through the overlying clay into groundwater (Murray et al. 1997).

Scoring this value therefore involves a consideration of multiple factors. With the 10-point scoring system, sands and gravels typically receive 10 points. Clay deposits of adequate integrity, thickness (generally more than 3 to 6 m thick), and areal extent receive a score of 3 or less.

- *Composition, areal extent, and thickness of saturated zone:* Composition, areal extent, and thickness of aquifer materials are important factors because they influence the rate and direction of groundwater flow. The risk may be higher in a scenario where the saturated thickness is 25 m and the material is composed of sand as opposed to a saturated thickness of less than a meter composed of silt. The areal extent and thickness of the aquifer must also be evaluated to establish whether multiple aquifers exist and are hydraulically connected. For instance, using the example above, if the meter-foot thick saturated silt layer was hydraulically connected to the much thicker saturated sand layer; this would change the risk evaluation.

  In an urban environment, perched saturated zones often result from anthropogenic influences. Construction activities such as roads, building foundations, and utility corridors are backfilled with porous materials, and then become more porous than natural soils and sediments. Over time, these porous backfilled materials become saturated with water. Therefore, investigative activities must be very detailed in nature to evaluate the significance of potential anthropogenic influences on the hydrology of the area being investigated and mapped.

  Scoring this parameter is similar to scoring the composition of the unsaturated zone. In practice, however, the factors determining the score are the areal extent of the saturated zone and whether the aquifer is hydraulically connected to additional or larger aquifers. If a saturated zone is encountered and evaluated to be perched or discontinuous and therefore isolated, the scoring value of 1 may be assigned. However, if the saturated thickness is 25 m and lies within a mappable geologic formation, a scoring value of 10 may be assigned.

- *Occurrence and relative abundance of groundwater:* This parameter is similar to the depth of groundwater but focuses on whether groundwater is present at relatively shallow depths, and of a sufficient quantity to sustain a rate of withdrawal. A key aspect of this measure is whether any groundwater is encountered before an aquiclude. The abundance of groundwater relates to the transmissivity of the formation and is not intended to focus on whether groundwater is encountered.

  With this parameter, there is high-scoring variability depending on the type of materials in the deposit. Consider an extensive deposit composed of saturated sand and gravel. This deposit would tend to have a much higher transmissivity value and would likely receive a high sensitivity score of 8 or more. When this deposit is compared to a similarly extensive and saturated silty clay deposit with a low relative transmissivity, the silty clay deposit would receive a score of 5 or less.

- *Area of groundwater recharge:* This parameter focuses on the source of groundwater. If the area being evaluated is a source of groundwater recharge, its risk will be higher than an area not considered a significant source of groundwater discharge.

Special care should be taken in urban areas to evaluate the anthropogenic sources of groundwater recharge. Significant anthropogenic influences include stormwater infiltration, sanitary sewer leakage, and water supply leakage. Other anthropogenic impacts to consider are detention basins and wetland modification.

- *Areas of groundwater discharge:* This parameter requires an understanding of groundwater flow at different geographic scales, including specific sites, off-site areas of larger extent, and the entire watershed. The migration and final discharge locations for groundwater include an understanding of the relationship between groundwater and surface water and permit an evaluation of the potential for interconnecting aquifer systems and discharge to deeper aquifers. It is also essential to understand the geology of a region and the potential influence of unconformities associated with the depositional units within the watershed.
- *Travel time and distance to the point of potential exposure:* In general, the longer the time and distance required for a contaminant to reach its potential point of exposure, the lower the risk. This is true for most contaminants that degrade naturally in the environment. Evaluating anthropogenic influence with this parameter is also important. For instance, if groundwater is used as a source of potable water and pumped from the ground using extraction wells, the residence time of potential contaminants in the aquifer may be greatly reduced.
- *Source of potable water:* If the aquifer encountered beneath a particular site is used as a source of potable water or is connected to a source of potable water, the risk increases. Evaluating this parameter must also include other potential uses besides the provision of drinking water. Further information may be necessary to identify those uses and accurately score this parameter.

13.3.2 Creating a Geologic Sensitivity Map

Because of its diversity, a good example of building a geologic sensitivity map is provided by the Rouge River watershed located in southeastern Michigan. The key features making the Rouge River watershed an ideal watershed to study include the following:

- It is a heavily urbanized area.
- It has varied geology and hydrogeology.
- There are different stream patterns and densities throughout.
- The population is approximately 1.5 million people.
- The urban area is within the Rouge River watershed, which has been studied extensively.
- There is an abundance of geologic and hydrogeologic data.
- It has a long history of industrial output.
- It has significant environmental impacts.
- It has geology similar to that of many other cities in the world.

The Rouge River watershed has over 120 miles of streams, tributaries, lakes, and ponds and encompasses an area of approximately 450 square miles (Murray and Rogers 1999a). The fan-shaped watershed includes all or part of 47 municipalities in three counties. Over 66% of the watershed has been developed, and in the year 2000, 99% of the watershed's population lived within the US Census-defined urbanized areas – making it the watershed with the highest population density in the eastern United States (Kaufman et al. 2003).

The near-surface geology of the study area is dominated by glacial deposits, glacial lacustrine deposits, and the recent fluvial deposits from the Rouge River itself. The glacial deposits are generally greater than 75 m in thickness, the glacial lacustrine deposits are rarely more than 10 m thick, and the recent fluvial deposits are generally less than 3 m thick (Farrand 1982,1988; Rogers 1996,

1997a, 1997b). Distinctive geologic units identified and mapped within the watershed include the following:

- Four different surface moraine units that include the Fort Wayne Moraine, the Outer and Inner Defiance Moraine, and the Birmingham Moraine
- Glacial outwash deposits
- Several beach deposits composed of sand from historical glacial lakes
- Silt and clay deposits from historical glacial lakes
- Recent fluvial deposits from the Rouge River
- A ground moraine or lodgment till composed of clay underlies the entire watershed beneath the near-surface deposits described above.
- A sand and gravel deposit is located beneath the ground moraine in the center portion of the watershed and represents a fluvial deposit from a large river that was present before the last glacial advance some 22,000 years before the present.
- Weathered bedrock is encountered beneath the ground moraine at other locations.
- Bedrock composed of shale, limestone, and sandstone of Paleozoic age is present beneath the unconsolidated units.

The geologic map, stratigraphic column, and geologic cross section for the metropolitan Detroit area are included in Figures 13.11, 13.16, and 13.17, respectively.

Anthropogenic effects within the study area have been numerous and significant. Surface water and groundwater within the watershed have been degraded. At one time, the Rouge River was ranked as one of the most toxic sites in Michigan (MDEQ 2008). The watershed is also the focus of ongoing intense scientific study and restoration. It has been identified as an area of concern by the International Joint Commission (Hartig and Zarull 1991) and cited as a significant source of pollution to the lower Great Lakes (Murray and Bona 1993). Rouge River sediments have shown a significant presence of heavy metals and polychlorinated biphenyls (PCBs) (MDEQ 2008; Murray 1996; Murray et al. 1999a, 1999b).

Shallow groundwater within the Rouge River discharges to surface water, and accounts for most of the baseflow (Rogers and Murray 1997). Average annual precipitation amounts and the geology of the watershed determine this flow pattern. Climatically, the Rouge watershed is situated within the humid microthermal zone of the Midwestern United States, meaning the Rouge River is effluent and is fed by groundwater entering as baseflow. With respect to its geology, the Lower Clay Unit (ground moraine or lodgment till) that completely underlies the watershed is very thick (ranging from 10 to more than 60 m), has a very low hydraulic conductivity (less than 1×10^{-8} cm/sec), and does not show signs of unconformities or features suggesting any significant secondary porosity (i.e., vertical fractures or root fragments). Therefore, the Lower Clay Unit is a very effective aquiclude. This type of formation is not uncommon. Most other areas in North America glaciated during the Pleistocene have similar ground moraine or lodgment till deposits of similarly low hydraulic conductivity (Keller et al. 1989). This is a significant finding because any contamination that does not degrade and reaches groundwater will eventually migrate and discharge to the Rouge River. From there, the contamination subsequently discharges into the lower Great Lakes (Rogers and Murray 1997; Rogers 1997b).

Using this geologic, hydrogeologic, and anthropogenic-impact information, a sensitivity matrix was developed and is presented in Table 13.2. The geologic units with the highest geological sensitivity (represented by the total score column) were the Outwash Unit, Recent Fluvial deposits associated with the Rouge River, and the Sand Unit of glacial lacustrine beach origin. The next group of units with moderate risk were the four surface moraine units. The geological units with the lowest risk were a sandy clay unit, a silty clay unit, and an upper clay unit, all associated with the glacial lacustrine deposits. The Outwash, Recent Fluvial, and Sand Units were evaluated to

TABLE 13.2
Rouge River Geologic Sensitivity Scoring

	Parameters and Sensitivity Scoring								Total	
Geologic Unit	**1**	**2**	**3**	**4**	**5**	**6**	**7**	**8**	**Score**	**Rank**
Outwash Unit	10	9	9	10	10	10	10	10	79	1
Recent Fluvial	10	6	6	8	10	10	10	10	70	2
Main Sand Unit	10	9	9	10	10	10	6	1	65	3
Other Sand Units	10	8	8	10	10	10	6	1	63	4
Moraine Unit 1	5	7	7	8	8	10	5	10	60	5
Moraine Unit 2	5	6	6	7	7	10	5	10	56	6
Moraine Unit 3	5	5	5	7	7	10	5	10	54	7
Moraine Unit 4	5	4	4	6	6	10	5	10	50	8
Sandy Clay Unit	4	3	3	4	1	1	3	1	20	9
Sandy and Silty Clay Unit	4	2	2	4	1	1	3	1	18	10
Upper Clay Unit	3	1	1	3	1	1	2	1	13	11
Lower Clay Unit (Ground Moraine Unit)	1	1	1	1	1	1	1	1	8	12

Source: Rogers, D. T. 2020. *Urban Watersheds: Geology, Contamination, Environmental Regulations, and Sustainability*. 2nd Edition. CRC Press. Boca Raton, FL. 606pp.

be the most geologically vulnerable geologic units in the watershed because they are generally composed of coarse-grained sediments, have an abundance of groundwater, and account for the majority of the base flow of surface water to the Rouge River. The four moraine units were of moderate risk because they are composed of finer-grained deposits, have less abundance of groundwater, and do not serve to recharge groundwater or discharge to surface water as significantly as the Outwash, Recent Fluvial, or Sand Units. The low-sensitivity units are the Sandy Clay Unit, Sandy and Silty Clay Unit, Upper Clay Unit, and the Lower Clay Unit.

These geological units generally do not have an abundance of groundwater, are very fine-grained units, and generally impede the migration of groundwater. Using the surface geological map of the Rouge River watershed (Rogers 1997a) as a base map, and the information contributing to the sensitivity scoring, Figure 13.18 was developed as the geologic sensitivity map for the Rouge River watershed (Rogers 1997c).

13.3.3 Geologic Sensitivity Case Study

As noted by Foster et al. (1994) and Loague et al. (1998), the significance of geologic sensitivity is not appreciated until it can be put into environmental, economic, or political perspective by actually cleaning up sites of environmental contamination. Therefore, to evaluate whether certain geologic units are vulnerable to contamination, a comparison between specific sites located in low geologic sensitivity areas to sites located in high geologic sensitivity areas must be conducted. If valid, the geologic sensitivity mapping should confirm that the sites situated above high-sensitivity locations pose a greater risk of exposure than sites located above low-sensitivity locations.

For this analysis, a site of low sensitivity located in a geological environment predominantly composed of clay sediments ("Site 1") is compared to a high-sensitivity site located in a geological environment predominantly composed of sand ("Site 2"). Both sites are located in the Rouge River watershed and are separated by only 12 km. However, Site 1 is significantly larger than Site 2 (approximately twice the size), has a much longer heavy industrial operational history

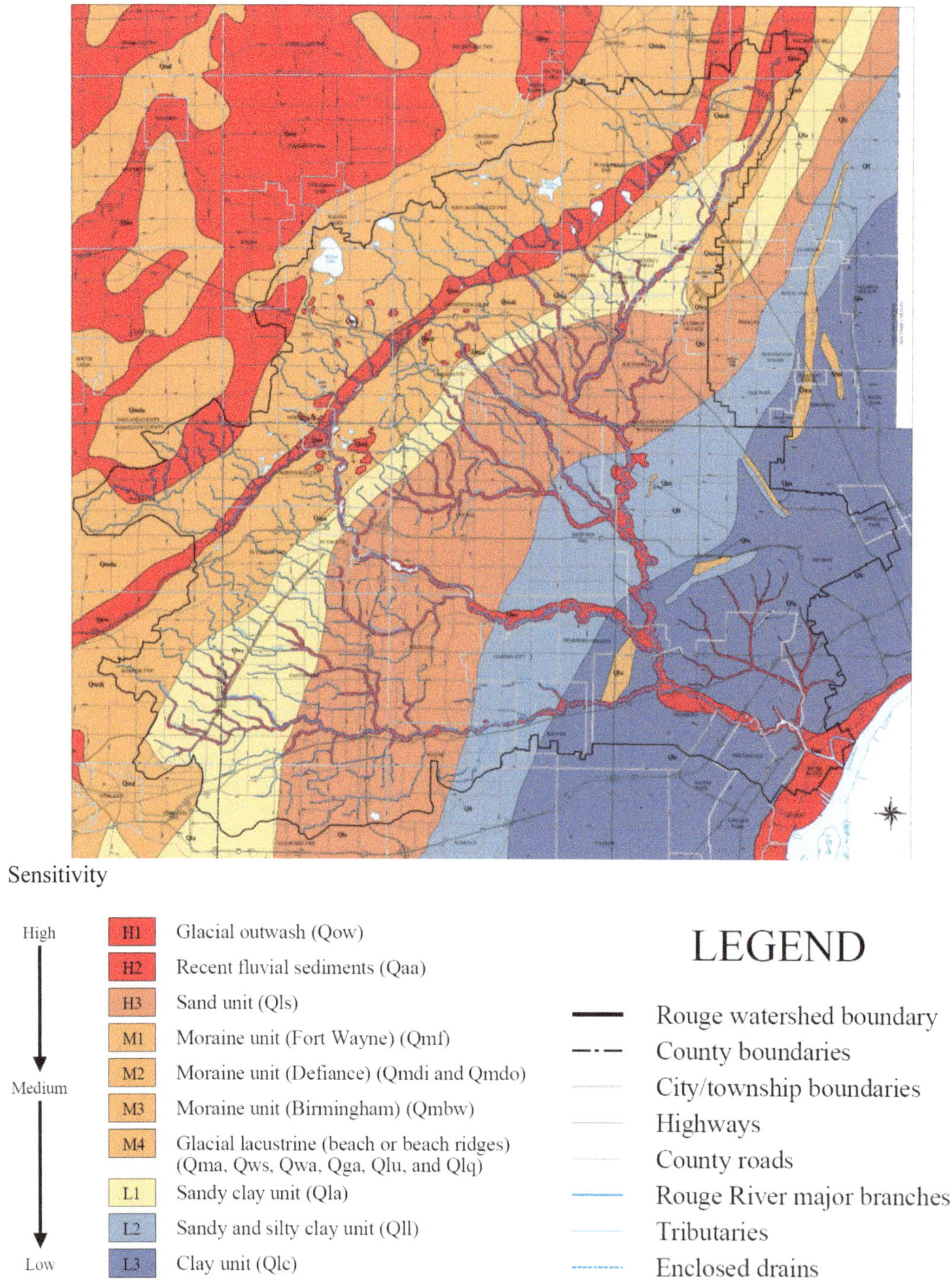

FIGURE 13.18 Geologic sensitivity map: Rouge River watershed. (From Rogers, D. T. 1997c. Geologic Sensitivity Map of the Rouge River in Southeastern, Michigan. Wayne County, MI. 1:62,500. 1 Sheet.)

(operated approximately 40 years longer); and has significantly more contaminant mass and types of contaminants (nearly 10 times the mass and three times as many contaminants) released into the environment. Without considering the geology of each site, it would be logical to assume the environmental risks are higher at Site 1, and the associated clean-up costs would be higher and reflect its contamination history.

13.3.3.1 Site 1 – Low Geologic Sensitivity Site

Site 1 is a former heavy manufacturing facility located on approximately 5 hectares of land and operated for approximately 70 years. The initial Phase I environmental site assessment identified six recognized environmental conditions (RECs). During the next investigational period (Phase II), several subsurface investigations were conducted at the facility and the main sources of contaminant release were identified that required remediation included surface spills, an aboveground storage tank, and spills and leaks of hazardous liquids located in waste storage areas. Other sources or releases were identified during the course of evaluating the site but were not severe enough to warrant further action.

The initial Phase II investigation involved drilling soil borings in areas with the highest likelihood of detecting contamination. This initial subsurface investigation had the objective to begin to characterize the subsurface geology of the site, identify the contaminants that were suspected to have been released, and evaluate the magnitude of contaminants present. This was accomplished by collecting and analyzing "worst-case" samples from each area of suspected contamination. The results confirmed the presence of contamination at five of the six RECs identified during the Phase I environmental site assessment. Those compounds detected were at significant concentrations and required further evaluation. Three additional investigation phases were conducted to define the nature and extent of contamination. During the subsequent phases, a total of 40 additional soil borings were drilled, and six monitoring wells were installed to evaluate whether there was enough groundwater present for analysis and to establish the direction of groundwater flow. In many of the soil borings, more than two soil samples were analyzed to gather data on the vertical extent of contamination. Characterization of the geology at the site was accomplished by drilling soil borings to a depth of 15 m in non-impacted locations. In general, the site was underlain by a clay deposit from a Pleistocene-age glacial lake that occupied the region more than 12,000 years before the present and was encountered immediately beneath the surface to a depth of less than 2 m. Historical geological literature of the region indicates a ground moraine or lodgment till deposit extended to depths of approximately 2 to 60 m beneath the ground. Small quantities of groundwater were encountered primarily at the contact of the upper glacial lake clay deposit and the underlying ground moraine/lodgment till deposit. Groundwater was present in only three of the six monitoring wells and was not present in sufficient quantities to sustain even a low rate of withdrawal and did not recover within 24 hours. Additional important technical and geological information concerning the site included the following:

- Storm sewers in the immediate vicinity did not intersect any of the contamination.
- Surface water drainage was controlled by storm sewers.
- No buried utilities intersected contaminated areas.
- Potable water was supplied by the municipality and the source was more than 10 km away.

The types of contaminants detected at the facility included the following:

- Volatile organic compounds (VOCs) including:
  - Dense nonaqueous phase liquids (DNAPLs) commonly referred to as chlorinated solvents used to degrease and clean metal surfaces

- Light nonaqueous phase liquids (LNAPLs) are used as solvents, paint thinners, and cleaning products, and are common constituents in fuels such as gasoline
- Polynuclear aromatic hydrocarbons (PNAs or PAHs) commonly used as lubricants, motor oils, and cutting fluids
- PCBs used in electrical equipment
- Heavy metals (arsenic, chromium, and lead) commonly used in paints and pigments, batteries, and metal plating

A list of specific chemical compounds, highest concentrations detected, and estimated contaminant mass remediated are listed in Table 13.3.

TABLE 13.3
Contaminant Types, Concentration, Mass Remediated for Site 1

Contaminant	Maximum Concentration (μg/kg)	Estimated Contaminant Mass Remediated (kg)
Volatile Organic Compounds (VOCs)		**80**
DNAPL Compounds		
Tetrachloroethene (PCE)	60,100	
Trichloroethene (TCE)	45,000	
Cis-1,2-dichoroethene	20,000	
Trans-1,2-dichloroethene	3,000	
Methylene chloride	800	
LNAPL Compounds		
Ethyl benzene	10,000	
Xylenes	10,000	
Acetone	280	
Carbon disulfide	200	
Polynuclear Aromatic Hydrocarbons (PNA or PAH)		
Naphthalene	339,000	**115**
Acenaphthalene	18,000	
Fluorene	22,000	
Phenanthrene	280,000	
Fluoranthene	156,000	
Pyrene	13,000	
Benzo(a)anthrcene	4,800	
Benzo(a)pyrene	3,200	
Dibenzo[a.h]anthracene	1,800	
Indeno[1,2,3-cd]pyrene	1,400	
Chrysene	11,000	
Polychlorinated Biphenyls (PCBs)	16,000	**2**
Heavy Metals		
Arsenic	23,000	**23**
Chromium	530,000	
Lead	930,000	

Source: Rogers, D. T. 2020. *Urban Watersheds: Geology, Contamination, Environmental Regulations, and Sustainability*. 2nd Edition. CRC Press. Boca Raton, FL. 606pp.

μg/kg = micrograms per kilogram.

Soil excavation and disposal of the contaminated soils at a licensed landfill was the remedial method of choice for the soil contamination at this site. The overriding considerations for selecting this method were: the contaminated areas were less than a meter deep and were not located beneath any buildings; the activities could be conducted quickly without disturbing ongoing facility operations, and it represented the lowest cost alternative. Approximately 7,000 m^3 – equivalent to 10,000 metric tons of soil – were excavated and transported to a local landfill for disposal. The total cost for investigation and remediation was approximately $400,000. This translates into a remediation cost per kilogram of contaminant of nearly $2,000. After the remediation was verified by the regulatory authority through the collection and analysis of the soil samples taken from each area remediated, closure was granted and a "No Further Action Required" letter was issued for the site. The closure was deemed unrestricted, meaning the site had been remediated to comply with residential land-use requirements. Eighteen months had elapsed since the Phase I environmental site assessment to closure granted by the regulatory authority.

13.3.3.2 Site 2 – High Geologic Sensitivity Site

Site 2 is a former heavy manufacturing facility slightly greater than 2 hectares in size that operated for 30 years. The Phase I environmental site assessment identified four RECs that included two outdoor chemical storage areas, a former underground storage tank, and surface staining near an outdoor shed used to store paint and solvents. Six phases of subsurface investigations were conducted to define the nature and extent of contamination. During these phases, a total of 132 additional soil borings were drilled, with many of the soil borings having multiple samples analyzed to assist in characterizing the vertical extent of contamination and 80 monitoring wells were installed.

The general subsurface geology of the site immediately beneath the surface consisted of a sand deposit originating from a Pleistocene-age glacial lake that occupied the region more than 12,000 years before the present. Specific geology beneath the site consisted of sand from the surface to a depth of nearly 10 m. Beneath this glacial lacustrine beach sand deposit was a ground moraine or lodgment till deposit extending to a depth of at least 25 m. Historical geological literature of the region indicated that the ground moraine or lodgment till deposit extended to depths of approximately 60 m.

During the investigation, multiple groundwater monitoring wells were installed at the same location but were screened at different depths within the saturated zone to evaluate the vertical distribution of contaminants within the aquifer. Several samples of the ground moraine deposit beneath the aquifer were also collected and analyzed for the presence of contamination and for certain hydrologic parameters (such as hydraulic conductivity and grain size analysis) to evaluate whether the ground moraine deposit was an effective aquiclude preventing contaminant migration to deeper aquifers. In addition, three deep soil borings were drilled to a depth of 25 m in non-impacted areas of the site to verify the horizontal distribution and approximate thickness of the ground moraine/lodgment till deposit. Other technical and geologically related information relevant to the site analysis included the following:

- Storm sewers in the immediate vicinity did not intersect any of the contamination
- Surface water drainage was controlled by storm sewers
- No buried utilities intersected contaminated areas
- Potable water was supplied by the municipality and the source was more than 10 km away. However, some local residences used groundwater within the same aquifer for irrigation purposes
- Contaminated groundwater extended beyond the property more than a kilometer.

The types of contaminants detected at the facility included the following:

- Volatile organic compounds (VOCs) including:
  - DNAPLs commonly referred to as chlorinated solvents used to degrease and clean metal surfaces
  - LNAPLs used as solvents, paint thinners, and cleaning products, and are common constituents in fuels such as gasoline
- PCBs used in electrical equipment

The specific chemical compounds with their highest concentrations detected and estimated contaminant mass remediated are listed in Table 13.4.

The highly permeable soil and shallow groundwater depth allowed VOCs to rapidly infiltrate and migrate to groundwater. In addition, the high groundwater seepage velocities resulted in a VOC plume extending more than 1 km to a downgradient spring, and it ultimately discharged into a recreational lake and the Rouge River.

PCBs were not detected in groundwater. Therefore, excavation and disposal of the PCB-contaminated soils at a licensed landfill was the remedial method of choice because the impacted soils were less than a meter in depth, were not located beneath any buildings, could be conducted quickly, did not disturb ongoing facility operations, and were the lowest cost alternative.

Excavation was also the remedial action of choice for soils highly impacted by VOCs. This method was chosen because there was not a substantial volume of impacted soil with VOCs, as the sandy soils at the site had a low capacity for retaining VOCs. As a result, the VOCs tended to migrate downward through the soil column and contaminate groundwater without adsorbing to soil grains.

TABLE 13.4
Contaminant Types, Concentration, Mass Remediated for Site 2 High Sensitivity Site

Contaminant	Maximum Concentration in Soil (μg/kg)	Maximum Concentration in Groundwater (μg/L)	Estimated Contaminant Mass Remediated (kg)
Volatile Organic Compounds (VOCs)			**45**
DNAPL Compounds			
Tetrachloroethene (PCE)	22,000	3,250	
Trichloroethene (TCE)	8,000	2,200	
Cis-1,2-dichoroethene	5,000	1,800	
Trans-1,2-dichloroethene	480	280	
1,1-Dichloroethene	640	220	
1,1,1-Trichloroethane	2,800	2,100	
Vinyl chloride	1,100		
LNAPL Compounds			
Ethyl benzene	32,000	240	
Xylenes	28,000	130	
Polychlorinated Biphenyls (PCBs)	7,000,000	Not detected	**12**

Source: Rogers, D. T. 2020. *Urban Watersheds: Geology, Contamination, Environmental Regulations, and Sustainability*. 2nd Edition. CRC Press. Boca Raton, FL. 606pp.

μg/kg = micrograms per kilogram; μg/L = microgram per liter.

The remedial method chosen for groundwater was air sparging and soil vapor extraction because the contaminants in groundwater were VOC compounds and did not extend past the mid-portion of the saturated zone. Air sparging involves the injection of air beneath the impacted groundwater and then letting the air rise naturally through the saturated zone. As the air migrated upward through the saturated zone it volatilized the contaminants, and the vapors containing the VOCs were removed using a soil vapor extraction system in the vadose zone. The vapors were removed from the air by passing them through a granular activated carbon tank.

The VOCs contaminating groundwater at this site were DNAPL compounds having a specific gravity slightly greater than water. Therefore, when present at sufficient concentrations, DNAPL compounds may sink through the water column and contaminate lower portions of an aquifer. This sinking action did not occur at this site because its geology and hydrogeology – in effect a stratigraphic control, prevented the VOCs from migrating to the bottom of the aquifer. The composition of the aquifer gradually became finer-grained with depth and the hydraulic conductivity decreased proportionately. This reduction of hydraulic conductivity, combined with low contaminant mass in groundwater, resulted in restricting contaminants to the upper portion of the aquifer. With the VOCs restricted to the upper portion of the aquifer, air sparging became the most practical remediation technique.

The time duration between the Phase I environmental site assessment and the receipt of the closure letter from the regulatory authority was approximately 14.5 years. The cost of PCB-contaminated soil, including investigation, was $1.1 million. The cost of hot-spot VOC excavation was $0.1 million. VOC-contaminated groundwater, including investigation, was $6.6 million. The total cost for investigation and remediation of this site was $7.8 million dollars, which translates into a cost of $120,000 per kg of recovered contaminant.

A risk assessment was also conducted for the site and was designed to determine if the residual contamination at the site would pose an ongoing and unacceptable risk to human health and the environment. Locations included in this evaluation were the onsite areas with persistent soil contamination, and the onsite and offsite areas with contaminated surface water and groundwater. The results of the risk assessment indicated that a deed restriction was appropriate for the site property and banned the use of groundwater for any purpose. A cap consisting of asphalt pavement was also required for portions of the property where some soil contamination remained in place, and if any soil were to become disturbed or exposed, another evaluation must be conducted to evaluate the need for further remedial actions. Land use for the property was restricted to industrial use.

13.3.3.3 Site Comparison Analysis

A profound difference between these two examples is seen in the cost per kilogram to remediate the contaminants. Despite the smaller acreage, shorter time of industrial operations, and low contaminant mass released to the environment, the cost to remediate a kilogram of contaminant at the site of high sensitivity (Site 2) was 150 times greater than to remediate a kilogram of contaminant at the low sensitivity site (Site 1). This cost differential is logically due to the geology present at each site.

The high-sensitivity site is located in a geologic area (Sand) that is environmentally sensitive to contamination because the highly permeable soil allows contaminants to infiltrate readily and migrate to groundwater and the contaminant plume was transported at relatively high seepage velocities in the sand aquifer to potentially sensitive receptors that included potential potable water wells and surface water. The costs for the high-sensitivity site would have been much higher had the contaminants migrated along the bottom portion of the aquifer; this is sometimes the case with the type of contaminants present (DNAPLs) as they are denser than water. The decreased hydraulic conductivity within the lower portion of the aquifer prevented this from happening at the higher sensitivity site. In addition, the costs would have been significantly higher had there been a completed human pathway represented by the ingestion of contaminated groundwater. Table 13.5 summarizes

TABLE 13.5
Site Comparison Table

Parameter	Site 1 – Low Sensitivity	Site 2 – High Sensitivity
Predominant geology	Clay	Sand
Presence of shallow groundwater	No	Yes
Size of site	16 acres	8 acres
Length of operation	70 years	30 years
Land use	Heavy industry	Heavy industry
Types of contaminants	VOCs PNAs, PCBs, and heavy metals	VOCs and PCBs
Number of contaminants remediated	27	9
Contaminant mass remediated	500 pounds	130 pounds
Cleanup criteria	Same as Site 2 for overlapping compounds	Same as Site 1 for overlapping compounds
Cleanup cost	$400,000	$7,800,000
Remedial methods (soil)	Excavation	Excavation
Remedial methods (groundwater)	Remediation not required	Air sparging and soil vapor Extraction
Other remedial control measures	None, unrestricted closure	Restricted closure included: • Deed restriction • Institutional controls • Industrial land use only
Timeframe	18 months	14.5 years
Cost per kilogram of contaminant[a]	$2,000	$150,000
Sensitivity ranking using Table 6.2	13	65

Source: Rogers, D. T. 2020. *Urban Watersheds: Geology, Contamination, Environmental Regulations, and Sustainability*. 2nd Edition. CRC Press. Boca Raton, FL. 606pp.

[a] Cleanup costs include costs for investigation and remediation.

the major differences between these two sites of environmental contamination. Please note that the sensitivity map ranking accurately predicted the relative costs of remediation.

13.4 SUMMARY AND CONCLUSION

By combining the methods and concepts presented in this chapter, a detailed geologic map of any urban watershed in the United States can be produced. Geologic mapping in urban areas requires a special skill set that includes (1) a thorough geologic knowledge of near-surface sedimentary environments, (2) the ability to identify the differences between natural and anthropogenic deposits, (3) knowledge of the urban landscape, (4) awareness and expertise in the methods used for obtaining access to sites of interest and information sources, and (5) using a simple and expressive system to display the concepts and processes. The most critical, other than getting the geology right, is the presentation of data. Even the most detailed information database is rendered useless if it cannot be effectively and easily communicated.

The use of multiple maps designed to add geologic knowledge and complexity through time, used alone or coupled with geographic information systems and three-dimensional mapping techniques, also helps simplify the communication of complex information. In addition, these tools can integrate geological information with other useful forms of information for the intended audience.

Through the construction of a sensitivity map, we have combined the knowledge gained to explain the relationship between the natural environment and human influence in urban areas.

As demonstrated in the case study, the arrangement, thickness, and composition of the soil layers beneath our feet have a profound influence on how contaminants behave and how they affect the environment and people.

The realization that certain locations or areas within urban regions are especially sensitive to contamination offers great promise for resolving future environmental issues through the prevention of the spread of contamination (Rogers 2018). Geologic sensitivity analysis of urban regions produces essential information for evaluating the environmental and financial risks associated with development and redevelopment. By minimizing the impact of pollution once a release has occurred, certain geological features may play, if we so choose, a central role in the development and redevelopment of any urban area.

REFERENCES

Albinet, M., and Margat, J. 1970. Cartographie de la Vulnerabilite a la Pollution des Nappes d''neau Souterrians. *Bull. Bur. Rech. Geol. Min. Sect 3 (Fr.)*. Vol. 2, pp. 13–22.

Allen, R. 2003. *The Rocks of the Valley of the Sun*. Phoenix, Arizona. www.Gemland.com. (Accessed November 4, 2019).

Aller, L., Bennett, T., Lehr, J. H., Petty, R. J., and Hackett, G. 1987. *DRASTIC: A Standardized System for Evaluating Ground Water Pollution Potential Using Hydrogeologic Settings*. United States Environmental Protection Agency USEPA-600/2-87-35. USEPA. Ada, OK.

American Association of Petroleum Geologists. 2005. The North American Stratigraphic Code. The North American Commission of Stratigraphic Nomenclature. *American Association of Petroleum Geologists Bulletin*. Vol. 89, no. 11, pp. 1547–1591.

Anderton, R. 1985. *Clastic Facies Models and Facies Analysis*. Vol. 18. Geological Society of London, Special Publication. Geological Society of London. London, England.

Barnes, J. 1993. *Basic Geologic Mapping*. 2nd Edition. Geological Society of London Handbook. John Wiley and Sons. London, England.

Berg, R. C. 2002. Geoenvironmental Mapping for Groundwater Protection in Illinois, USA. In: Bobrowsky, P.T., editor. *Geoenvironmental Mapping; methods, Theory, and Practice*. Balkema Publishing. The Netherlands.

Berg, R. C., Bleuer, N. K., Jones, B. E., Kincare, K. A., Pavey, R. R., and Stone, B. D. 1999. *Mapping the Glacial Geology of the Central Great Lakes Region in Three-Dimensions – A Model for State-Federal Cooperation*. United States Geological Survey Open File Report 99-349. USGS. Washington, D.C.

Bilodeau, W. C. Bilodeau, S. W., Gath E. M., Oborne, M., and Proctor, R. J. 2007. Geology of Los Angeles. *Environmental and Engineering Geoscience Journal*. Vol. 13, no. 2, pp. 99–160.

Birks, H. J., and Line, J. M. 1992. The Use of Rarefraction for Estimating Palynological Richness from Quaternary Pollen-Analytical Data. *Journal of the Holocene*. Vol. 2, pp. 1–10.

Burn, S., Eiswirth, M. Correll, R., Cronin, A., DeSilva, D., Diaper, C., Dillon, P., Mohrlok, U., Morris, B., Rueedi, J., Wolf, L., Vizintin, G., and Vott, U. 2007. Urban Infrastructure and Its Impact on Groundwater Contamination. In: Howard, K. W. F., editor. *Urban Groundwater – Meeting the Challenge*. Taylor & Francis. London, England.

Chang, K. 2007. *Introduction to Geographic Information Systems*. 4th Edition. McGraw Hill. New York, NY.

Eaton, T. T., and Zaporozec, A. 1997. Evaluation of Groundwater Vulnerability in an Urbanizing Area. In: Chilton, J., et al., editors. *Groundwater in the Urban Environment*. Balkema Publishers. Rotterdam, The Netherlands.

Farrand, W. R. 1982. *Quaternary Geology of Southern (and Northern) Michigan*. Michigan Department of Natural Resources, Geological Survey Division. Lansing MI.

Farrand, W. R. 1988. *The Glacial Lakes around Michigan*. Michigan Department of Natural Resources, Geological Survey Division Bulletin Number 4. Lansing, MI.

Focazio, M. J., Reilly, T. E., Rupert, M. G., and Helsel, D. R. 2001. *Assessing Ground-Water Vulnerability to Contamination: Providing Scientifically Defensible Information for Decision Makers*. United States Geological Survey Circular 1224. USGS. Denver, CO.

Foster, S. S. D., and Hirata, R. 1988. *Groundwater Risk Assessment: A Methodology Using Available Data*. CEPIS. Lima, Peru.

Foster, S. S. D., Morris, B. L, and Lawrence, A. R. 1994. Effects of Urbanization on Groundwater Recharge. In: Wilkinson, W. B., editor. *Groundwater Problems in Urban Areas: ICE Conference Proceedings*. London, England.

Fresca, B. 2007. Urban-Enhanced Groundwater Recharge: Review and Case Study of Austin, Texas, USA. In: Howard, K. W. F., editor. *Urban Groundwater – Meeting the Challenge*. Taylor & Francis. London, England.

Great Lakes Geologic Mapping Coalition. 2024. *Geologic Mapping Introduction*. Indiana Geologic Survey. http://igs.indiana.edu/GreatLakesGeology/introdcution.html. (Accessed May 12, 2024).

Halka, C. 1983. *Roadside Geology of Arizona*. Mountain Press Publishing Company. Denver, CO.

Hartig, J. H., and Zarull, M. A., 1991. Methods of Restoring Degraded Areas in the Great Lakes. *Reviews of Environmental Contamination and Toxicology*. Vol. 177, pp. 127–154.

Hough, J. L. 1958. *Geology of the Great Lakes*. University of Illinois Press. Champaign–Urbana, IL.

Hough, J. L. 1963. The Prehistoric Great Lakes of North America. *American Scientist*. vol. 51, pp. 84–109.

Howard, K. W. F., Di Biase, S. Thompson, J. Maier, H., and Van Egmond, J. 2007. Stormwater Infiltration Technologies for Augmenting Groundwater Recharge in Urban Areas. In: Howard, K. W. F., editor. *Urban Groundwater – Meeting the Challenge*. Taylor & Francis. London, England.

International Association of Cartography (ICA). 2024. *Mission, Vision, Objectives*. www.ica.org/mission. (Accessed May 30, 2024).

Kaufman, M. M., Rogers, D. T., and Murray, K. S. 2003. Surface and Subsurface Geologic Risk Factors to Ground Water Affecting Brownfield Redevelopment Potential. *Journal of Environmental Quality*. Vol. 32, pp. 490–499.

Kaufman, M. M., Rogers, D. T., and Murray, K. S. 2005. An Empirical Model for Estimating Remediation Costs at Contaminated Sites. *Journal of Water, Air and Soil Pollution*. Vol. 167, pp. 365–386.

Keller, K. C., Van Der Kamp, G., and Cherry, J. A. 1989. A Multiscale Study of the Permeability of a Thick Clay Till. *Journal of Water Resources Research*. Vol. 25, no. 11, pp. 2299–2317.

Kibel, P. S. 1998. The Urban Nexus: Open Space, Brownfields, and Justice. *Boston College Environmental Affairs Law Review*. Vol. 25, no. 3, pp. 589–618.

Lahee, F. H. 1961. *Manual of Field Geology*. 6th Edition. McGraw Hill. New York, NY.

Loague, K. Corwin, D. L., and Ellsworth, T. R. 1998. The Challenge of Predicting Nonpoint Source Pollution. *Environmental Science and Technology*. Vol. 26, pp. 127–154.

Maley, T. S. 2005. *Field Geology*. Sheridan Books. Ann Arbor, MI.

Michigan Department of Environmental Quality (MDEQ). 2008. *Michigan Sites of Environmental Contamination*. Lansing, MI.

Mohorlok, U., Cata, C., and Bucker-Gittel, M., 2007. Impact on Urban Groundwater by Wastewater Infiltration into Soils. In: Howard, K. W. F., editor. *Urban Groundwater – Meeting the Challenge*. Taylor & Francis. London, England.

Mozola, A. J. 1954. *A Survey of Groundwater Resources of Oakland County Michigan*. Publication No. 48. Michigan Geologic Survey. Lansing, MI.

Mozola, A. J. 1969. *Geology for Land and Groundwater Development in Wayne County, Michigan*. Michigan Geologic Survey. Lansing, MI.

Murray, J. E., and Bona, J. M. 1993. *Rouge River National Wet Weather Demonstration Project, Wayne County*. Wayne County, MI.

Murray, K., Bazzi, C., Ehlert, A., Harris, A., Kopec, M.. Richardson, J., and Sokol, H., 1997. Distribution and Mobility of Lead in Soils at an Outdoor Shooting Range. *Journal of Soil Contamination*. Vol. 6, no. 1, pp. 79–93.

Murray, K. S. 1996. Statistical Comparison of Heavy-Metal Concentrations in River Sediments. *Journal of Environmental Geology*. Vol. 27, pp. 54–58.

Murray, K. S., and Rogers, D. T. 1999a. Groundwater Vulnerability, Brownfield Redevelopment and Land Use Planning. *Journal of Environmental Planning and Management*. Vol. 42, no. 6, pp. 801–810.

Murray, K. S. and Rogers, D. T. 1999b. Evaluation of Groundwater Vulnerability in an Urban Watershed. In: *Proceedings of the 2nd International Congress on Water Resources and Environmental Research*. International Association of Hydrogeologists, London, UK. pp. 877–883.

Page, W. R., Lundstrom, S. C., Harris, A. G., Langenheim, V. E., Workman, J. B., Mahan, S. A., Paces, J. B., Dixon, G. L., Rowley, P. D., Burchfiel, B. C., Bell, J. W., and Smith, E. I. 2005. *Geologic and Geophysical Map of the Las Vegas 30′ × 60′ Quadrangle, Clark and Nye Counties, Nevada, and Inyo*

County, California. United States Geological Survey Pamphlet to Accompany Scientific Investigations Map 2814. USGS. Washington, DC.

Pierce, S. A., Sharp, J. M., and Garcia-Fresca, B. 2007. Evaluating Groundwater Allocation Alternatives in an Urban Setting Using a Geographic Information System Data Model and Economic Valuation Technique. In: Howard, K. W. F., editor. *Urban Groundwater – Meeting the Challenge*. Taylor & Francis. London, England.

Powell, J. W. 1888. *Methods of Geologic Cartography in Use by the United States Geological Survey*. USGS. Washington, DC.

Robins, N., Adams, B., and Foster, S. S. D. 1994. Groundwater Vulnerability Mapping: The British Perspective. *Hydrogeologie*. Vol. 3, pp. 35–42.

Rogers, D. T. 1992. The Importance of Site Observation and Follow-up Environmental Site Assessments – A Case Study. In: *Proceedings of the National Ground Water Association Phase I ESA Conference*. National Groundwater Associate, Dublin, OH. pp. 218–227.

Rogers, D. T. 1996. *Environmental Geology of Metropolitan Detroit*. Clayton Environmental Consultants. Novi, MI. 152pp.

Rogers, D. T. 1997a. *Surface Geological Map of the Rouge River Watershed in Southeastern, Michigan*. Wayne County. MI. 1:62,500. 2 Sheets.

Rogers, D. T. 1997b. *The Influence of Groundwater in Surface Water in Michigan's Rouge River Watershed. Conjunctive Use of Water Resources: Aquifer Storage and Recovery*. American Water Resources Association. pp. 173–180.

Rogers, D. T. 1997c. *Geologic Sensitivity Map of the Rouge River in Southeastern, Michigan*. Wayne County, MI. 1:62,500. 1 Sheet.

Rogers, D. T. 2002. The Development and Significance of a Geologic Sensitivity Map of the Rouge River Watershed in Southeastern Michigan, USA. In: Bobrowsky, P. T., editor. *Geoenvironmental Mapping: Methods, Theory, and Practice*. A. A. Balkema Publishers. The Netherlands. pp. 295–319.

Rogers, D. T. 2014. *Scientists Call for a Renewed Emphasis on Urban Geology*. Vol. 95, no. 47. Earth and Space News. Eos. American Geophysical Union. pp. 431–432.

Rogers D. T. 2015a. Next Generation of Urban Hydrogeologic Investigations. In: 42th *International Association of Hydrogeologists Congress*. Rome, Italy. International Association of Hydrogeologists, London, UK.

Rogers, D. T. 2015b. Urban Geologic Mapping: Methods, Implications, and Benefits. In: *Illinois State Geological Survey Seminar Series*. University of Illinois. Champaign–Urbana, IL.

Rogers, D. T. 2016a. Scientific Advancements that Improve the Conceptual Site Model in Urban Hydrogeologic Investigations. In: *35th International Geological Congress*. Paper No. 3019. International Geologic Congress. Cape Town, South Africa.

Rogers, D. T. 2016b. Next Generation of Urban Hydrogeologic Investigations. *Journal of the Italian Geological Society*. Vol. 39, no. 1, p. 349.

Rogers, D. T. 2018. *Derivation of a Comprehensive Environmental Risk Model for Urban Groundwater Protection*. Vol. 1. International Association of Hydrogeologists Congress. Daejeon, Korea.

Rogers, D. T. 2020. *Urban Watersheds: Geology, Contamination, Environmental Regulations, and Sustainability*. 2nd Edition. CRC Press. Boca Raton, FL. 606pp.

Rogers, D. T., Kaufman, M. M., and Murray, K. S. 2008. Quaternary Geology of the Rouge River Watershed in Southeastern Michigan, USA. In: *33rd World Geological Congress*. World Geological Congress. Oslo, Norway.

Rogers, D. T., and Murray, K. S. 1997. Occurrence of Groundwater in Metropolitan Detroit, Michigan, USA. In: Chilton et al., editors. *Groundwater in the Urban Environment*. Vol. 1. Balkema Publishers. The Netherlands. pp. 155–160.

Rogers, D. T., Murray, K. S., and Kaufman, M. M. 2007. Assessment of Groundwater Contaminant Vulnerability in an Urban Watershed in Southeast Michigan, USA. In: Howard, K. W. F., editor. *Urban Groundwater – Meeting the Challenge*. Taylor & Francis. London, England.

Sherzer, W. H. 1913. *Geology of Wayne County*. Geological Survey of Michigan. Lansing, MI.

Siegel, A. F. 2006. Rarefaction Curves. In: *Encyclopedia of Statistical Sciences*. Springer-Verlag. New York, NY.

Stiber, N. A., Small, M. J., and Fischbeck, P. S. 1995. The Relationship between Historic Industrial Site Use and Environmental Contamination. *Journal of the Air and Waste Management Association*. Vol. 48, no. 9, pp. 809–818.

Stokes, W. L. 1989. *Geology of Utah*. Utah Museum of Natural History Occasional Paper No. 6. Utah Museum of Natural History. Salt Lake City, UT.

Thomas, W. A. 2004. *Meeting the Challenges with Geologic Maps*. American Geological Institute (AGI). Alexandria, VA.

United States Geological Survey (USGS). 1998. *Forum on Geologic Mapping Applications in the Washington–Baltimore Urban Area*. United States Geological Survey Circular 1148. USGS. Washington, DC.

United States Geological Survey (USGS). 2002. *3D Geologic Maps and Visualization: A New Approach to the Geology of the Santa Clara (Silicon) Valley, California*. United States Geological Survey Open-File Report 01-223. USGS. 13pp.

United States Geological Survey (USGS). 2005. *3D Modeling Techniques for Geological and Environmental Visualization and Analysis*. United States Geological Survey Open-File Report 2005-1428. USGS. Washington, DC.

United States Geological Survey (USGS). 2006. *Federal Geographic Data Committee (FGDC) Digital Cartographic Standard for Geologic Map Symbolization*. FGDC Document Number FGDC-STD-013-2006. USGS. Washington, DC.

United States Geological Survey (USGS). 2007. *Divisions of Geologic Time – Major Chronostratigraphic and Geochronologic Units*. United States Geological Survey Fact Sheet 2007-3015. USGS. Washington, DC.

United States Geological Survey (USGS). 2024a. *National Cooperative Geologic Mapping Program*. http://ncgmp.usgs.gov. (Accessed May 12, 2024).

United States Geological Survey (USGS). 2024b. *Geologic Mapping Standards*. National Cooperative Geologic Mapping Program. USGS. http://ncgmp.usgs.gov. (Accessed May 12, 2024).

United States Geological Survey (USGS). 2024c. *Geographic Information Systems*. USGS. Washington, DC. www.usgs.gov/media/files/map-projections-poster. (Accessed May 24, 2024).

Vuono, M., and Hallenbeck, R. P. 1995. Redeveloping Contaminated Properties. *Journal of Risk Management*. Vol. 42, pp. 58–69.

Welty, N., Quinnan, J., Yanites, A., and Rogers, D. T. 2016. A New Way of Investigating Sites: Smart Characterization. In: *Proceedings of the Tenth International Conference on Remediation of Chlorinated Solvents and Recalcitrant Compounds*. Paper 554. Battelle Memorial Institute, Washington, DC.

Zaporozec, A., and Eaton, T. T. 1996. Ground-Water Resource Inventory in Urbanized Areas. In: *Hydrology and Hydrogeology of Urban and Urbanizing Areas*. American Institute of Hydrology (AIH) Annual Meeting Proceedings. American Institute of Hydrology. St. Paul, MN.

14 Improved and New Environmental Regulations

14.1 INTRODUCTION

In this chapter, we address the need for new or improved environmental regulations that more broadly address the pressures exerted on the environment by an expanding human population. This concern is nothing new, after all this was first addressed by congress in 1969 with the passage of the National Environmental Policy Act (NEPA). NEPA expressly stated that the objective was to live in productive harmony with nature and address over population. Although there is much to be proud of when examining the improvements in the United States since the enactment of NEPA more than 50 years ago, it clearly has not been enough. Perhaps the hidden benefit with the robustness of existing environmental regulations in the United States is the platform to address the serious environmental issues facing all humans on Earth is already in place.

Since NEPA was enacted, the achievements have included improving air quality, water quality, and clean-up of sites of environmental contamination, which now number in the tens of thousands across the United States. However, there are several areas where improvements can be made if some existing regulations were modified or expanded and new regulations were developed. A current concern is that sustainability efforts will focus on climate change only and will not address other concerns that are also of high priority, such as species extinction, the oceans, and habitat loss.

The first portion of this chapter addresses those actions that can be enacted to improve the environment before development occurs. The second portion this chapter addresses improvements to existing environmental regulations and the development of new regulations.

14.2 IMPROVING URBAN AREAS

For many decades, nature was categorized as perhaps the enemy, something that needed to be conquered or tamed. The concept was that progress and development would achieve this desired result. Scientists now realize that this may not have been a good approach and that our relationship with nature must improve. The natural setting of our urban areas has a crucial role to play in how contaminants behave and how they affect us when they are released into the environment. All of us walk on a history book in each urban area of the world, but yet most of us are unaware of its significance and the potential impact it may have on all our lives and it is just centimeters beneath our feet.

These soil and sediment layers beneath our feet are the meeting place between human civilization and the natural world. This contact is the proving ground and the point at which contaminants released into the environment begin their destructive journey that has now impacted on all living things on Earth. We must understand our natural world in at least the same scientific detail that we need to understand pollution. The reason we must do this is simple, this is where we live, and nature

DOI: 10.1201/9781003463153-17

and civilization interact with each other. The near-surface geologic environment has become very important in characterizing and remediating sites of environmental contamination and in evaluating risk to the environment and human health. The near-surface geologic environment frequently acts as the migration pathway for contaminants to travel from a specific point of release to a point of ecologic or human exposure. In many cases, this pathway goes undetected because the contaminants migrate beneath the surface of the ground and cannot be observed or detected until it is perhaps too late.

14.2.1 Geologic Sensitivity

We do not have control over the natural environment. Therefore, we must understand the natural environment where our urban areas are located and develop methods to minimize or eliminate the potential harmful effects of contaminants upon human health and the environment. A logical first step to this end is through an *understanding* of urban geology, followed by an *evaluation* of the extent that a given urban area's geology influences the migration of contaminants. Also, as water plays a critical role in assessing a region's vulnerability to contamination, the analyses performed during the evaluation step require an understanding of water occurring at the Earth's surface and beneath (Rogers 2014). These factors control the severity of the damage and are as follows: (1) the geologic and hydrogeologic environment, (2) the physical chemistry of the contaminants and amounts released, and (3) the mechanism in which the release occurs (Rogers 2018, 2020).

Despite the availability of specific methods and procedures, the environmental assessment of many urban areas can become a daunting task. This situation arises because the near-surface geologic deposits in urban areas are poorly understood, difficult to study, complex, have been anthropogenically disturbed, and exhibit high variability over short distances. Therefore, to achieve any level of success in mitigating environmental contamination, it becomes a prerequisite to understand how contaminants migrate in any given urban area which makes it necessary to understand its geology and hydrology (Rogers 2016a, 2016b, 2020, 2021; Rogers et al. 2016).

14.2.2 Physical Chemistry of Pollutants

If it were not for hazardous substances, there would be little need for the majority of environmental regulations. However, that is certainly not the case. There are currently more than 84,000 chemicals in use today and more and more every year. Contaminants are everywhere – in the air, soil, water, inside buildings, and in our homes. Most households contain chemicals that would be considered contaminants if they were released into the environment or disposed of improperly. Therefore, we must understand which chemicals are most harmful and limit their use or eliminate them all together. United States Environmental Protection Agency (USEPA) has not banned any chemicals since 1984, so it is up to industry and the consumer to evaluate the risk and take appropriate action.

In addition, chemicals or substances become pollution when they are released into the environment either inadvertently or improperly – at the wrong place or in the wrong amounts. For example, milk becomes a contaminant when large quantities are released into a stream. Therefore, it is not just what chemicals are used, but preventing releases of any chemicals, especially to sensitive areas (Rogers 2020).

14.2.3 Informed Urban Development

The human population is increasing and requires more and more space to live, grow food, and for build roads. The rapid physical growth of urban areas leads to complex processes of landscape transformation which alters the structure and function of ecosystems. Urban invasion into natural areas decreases productivity, decreases infiltration of water, increases impervious surfaces, increases flooding, increases erosion, increases pollutant load, increases energy demand, and decreases species biodiversity. The size of a single family home in the United States has more than doubled in the last 70 years (United States Department of Housing and Urban Development 2024; United States Census Bureau 2024a).

Many opportunities for smart urban development are available within many cities of the world. Some of the numerous opportunities include the following:

- Less single-family homes and more multifamily units. The average size of a condominium is approximately 130 m^2 (1,200 square feet) versus 300 m^2 (2,700 square feet) for a single-family home. The average land required for a single-family home in the United States is 815 m^2 or 0.25 acres per home. This includes the land required for easements and roads for a typical subdivision (National Association of Home Builders 2024).

Building large skyscrapers that have hundreds to sometimes over a thousand condominium units have enormous sustainability improvements over the single-family home. Take for example a large skyscraper currently being constructed in Chicago in the United States that will have an estimated 1,200 condominium units on a footprint of approximately 5,000 m^2 (2 acres). This represents a 99.5% reduction in occupied land if those 1,200 condominium units were single family homes. There are also significant reductions in the amount of construction materials that would be used if each were a single-family home that includes forest materials, insulation, roofing, paint, metal piping, sewers, stormwater, roads and curbs, and sidewalks. There are also significant long-term energy savings including heat and electrical.

- Reducing the size of housing units. The size of a single-family home in the United States has nearly doubled since 1970 and nearly tripled since 1940 (United States Census Bureau 2024b). This is in contrast with the fact that the children per family in the United States have decreased by more than 60% from 3.76 children per family in 1940 to 1.58 children per family in 2018 (United States Census Bureau 2024b).
- Living closer to work. The average commute to work in the United States is 20 miles one way and takes nearly 30 minutes (United States Census Bureau 2024c).
- Increase mass transit availability. Mass transit is much more efficient and emits much less pollution than an automobile per person per kilometer travel. Estimates are as high as 95% more efficient when traveling by commuter train compared to an average automobile when considering fuel use alone and not taking into consideration of the cost and inefficiencies of building and maintaining highways (United States Department of Energy 2024; National Geographic Society 2024).
- Increase other forms of transport that include using scooters, bicycles, and walking. Many cities in the United States and throughout the world are improving bike-friendly and foot transport routes to increase these forms of transportation by promoting health and enjoyment benefits.
- Integrating more mixed land use into urban areas. The United States has much to learn from other countries when it comes to this urban development technique. Historically, the United States has not taken an integrated approach to land use. Many European and Asian countries, namely South Korea, have embraced integration of farmland, industry, and housing.
- Increasing clean energy production such as solar and wind. Both solar and wind energy have become more efficient and less costly. A key point for solar efficiency is prevention of dust buildup on the solar panels. Wind turbines are available in different engineering configurations that can be applied in small areas and minimize bird deaths (United States Department of Energy 2024).

- Increasing permeable surfaces. To decrease flooding, increasing infiltration of precipitation through several available and simple techniques is widely available and will be discussed later in this chapter.

Currently, examples of sustainable and smart urban development using some or all the items listed above are mainly confined to Europe, Asia, and some parts of the United States. Two impediments that have been slow in becoming a reality in the United States concerning smart urban development is the American love affair with the automobile and achieving the American dream in which owning a single-family home has often been viewed as a degree of measure that defines making that dream come true.

14.2.4 Improving Our Relationship with Nature

In the 19th and 20th centuries, many supported the concept of progress, taming and conquering nature, and treating nature as a commodity to be exploited for profit (Tallamy 2019; Darke and Tallamy 2019).

We have discussed invasive species and how destructive they can be at reducing native populations of most all form of indigenous life. In addition, climate change has already affected many forms of life such as the Bark Beetle's march north and destroying millions of trees in North America (United States Forest Service 2024). Other impacts include the reduction in many bee species and significant reduction on the number of Monarch butterflies, just to name a couple (Michigan State University 2021).

At the national level, there is much that can be conducted, a few include the following (Tallamy 2019; Darke and Tallamy 2019):

- Planting native plant species. One example is planting milkweed, which is a plant species required for the survival of the Monarch butterfly. Native species planting is a trend that is expanding within many state departments of transportation and should be encouraged and funded at higher levels at the national level (United States Fish and Wildlife Service 2024a).
- Reducing biological wastelands. Namely lawns which are symbols of wealth and status but are generally biological deserts.
- Establishing biological and wildlife corridors.
- Improving insect habitat. In most ecological systems, insects form the foundation of ecosystem health and without a healthy insect population, which provide food through all levels of life, ecosystems may collapse
- Limit use of herbicides and pesticides
- Limit use of fertilizers
- Restore and expand wetlands
- Practice regenerative agricultural methods. Regenerative farming is a philosophy of common principles that include re-establishing relationships between humans and land through improving soil health, reducing or eliminating the use of harmful chemicals, growing diverse crops, holistic livestock management, innovative and efficient use of resources, and equitable labor (Natural Resource Defense Council (NRDC) 2024). This type of farming is not new. Indigenous cultures, and even those in the United States in the 19th century, practiced this type of farming before the widespread use of pesticides, herbicides, and extensive use of fertilizers. Current farm and food production in the United States is not designed to prioritize climate, ecosystems water quality, human health, and establishing a healthy and productive relationship with nature (NRDC 2024).

14.2.5 Infrastructure Upgrades

Many residents in older urban areas are at an elevated risk of contaminant exposure because of a deterioration of infrastructure. Before moving forward with future sustainability measures, we must first repair and upgrade existing systems. Specific infrastructure needs include the following:

- Replacing old and deteriorated water supply lines. The incident involving the Flint, Michigan water supply highlighted the presence of lead in water pipes in older urban area of the United States, which places children at risk of exposure to lead (USEPA 2024a, 2024b).
- Upgrading water treatment of drinking water. With the detection of many new and emerging contaminants and with newer and better water treatment technologies, many water treatment plants should be upgraded to protect drinking water and drinking water sources (USEPA 2024c).
- Upgrading wastewater treatment plants. Related to the previous bullet point, if the treatment of drinking water is upgraded, so should its treatment before its discharge back into the environment. This is necessary to remove many of the household chemicals that may be present in wastewater, which may include biological pathogens, heavy metals, volatile organic compounds, and other chemicals (USEPA 2024d).
- Improving municipal landfills. Municipal landfills may contain many types of waste materials, including wastes that would otherwise be considered hazardous. Therefore, additional engineering measures should be undertaken to minimize the potential for a release of contaminants into the environment. In addition, further actions should be undertaken to eliminate or minimize the potential for hazardous contaminants disposed of in municipal landfills (USEPA 2024e).
- Improving erosion controls. Erosion remains a significant issue that impacts rivers, lakes, streams, and the oceans, especially from agricultural areas. These areas contribute pesticides, herbicides, and fertilizers to surface water and ultimately the oceans, which have been well documented (USEPA 2024f).
- Increasing mass transit. Mass transit is significantly more efficient than the automobile. However, the United States lacks sufficient infrastructure, namely rail transport, to make mass transit by rail more attractive and convenient to consumers.

14.3 EXPANDING, IMPROVING, AND NEW ENVIRONMENTAL REGULATIONS

Improving environmental regulation of the United States may seem unnecessary as many believe the environmental regulations of the United States are the most comprehensive, most complex, and most effective. However, there is much that the United States can do to improve their environmental regulations. In addition, and perhaps most importantly, the environmental regulations of the world have been used as a platform for environmental regulations worldwide. In most instances, the rest of the world relies on the United States to lead the way forward with respect to environmental regulations. This is significant because, as we know, contaminants do not respect country borders and if the United States forges ahead in improving its environmental regulations, it is only logical that the rest of the world will follow (Rogers 2019).

14.3.1 Modify Environmental Enforcement Emphasis and Policy

At this point, it is not apparent that a need for modifying USEPA's approach to enforcement is in need of change, especially as it has been seemingly effective up to this point. However, it is time to move beyond a litigation-heavy, punishment-based enforcement model and enter into more of a reform- and repair-based enforcement policy that emphasizes sustainability. In addition, expanding application of environmental regulations to other areas that are either not regulated or lightly regulated is long overdue. For instance, according to the United Nations, agricultural practices are the largest polluter on Earth yet environmental regulations of the United States largely exempt agriculture from being regulated. This is no more evident than when examining pesticide and herbicide use and fertilizer use.

We will learn in the next chapter that some countries are more advanced compared to the United States in how they have chosen to organize and implement their environmental regulations and where they focus internal resources. In addition, the governments of some countries have different attitudes and relationships with industry. The regulatory agencies of some countries have chosen to form working partnerships with industry with the intent to work more closely together to solve complex environmental issues rather than an adversarial approach which much too often results in polarizing points of view and expensive litigation.

The result of working together saves time, effort, and resources and, in the long run, has resulted in the fact that some counties are better at protecting human health and the environment than the United States.

14.3.2 Science-Based Regulation Improvements

We discovered in earlier chapters that contaminants behave differently and are more prone to releases in different geologic, hydrogeologic, geographic, and climatic regions. For instance, regulations specific to geologic risk factors such as earthquakes (e.g., California) or climate change (eastern seaboard and the coast along the Gulf of Mexico) should better reflect those areas where there is the highest risk.

14.3.3 Regulating Agriculture

Agriculture is now the biggest polluter in the United States and the world (United Nations 2016). The United Nations estimates that most land pollution is caused by agricultural activities, such as grazing, use of pesticides, fertilizers, irrigation, plowing, and confined animal facilities (United Nations 2016). Agricultural activities release significant amounts of greenhouse gases (GHGs) from plowing, decaying vegetation, and from livestock. Fugitive emissions and erosion are also significant and release significant amounts or chemicals into the air and the water. In addition, fertilizers are a significant source of pollution to the oceans of the world.

As environmental regulations have been so successful at reducing the harmful effects from industry in the United States, it is now time to reduce harmful effects on the environment from agriculture.

14.3.4 Improving Urban Air Quality

The Clean Air Act has made a major impact at improving the air quality throughout the United States, especially with respect to automobile exhaust, ozone, particulate matter, and the other criteria pollutants. According to USEPA (2018), compared to emissions in 1980, carbon monoxide emissions have been reduced by 72%, lead by 99%, nitrogen oxides by 61%, volatile organic compounds by 54%, particulate matter (PM10) by 61%, and sulfur dioxide by 89%.

Improvements on exhaust from diesel trucks and buses have been slowed and are especially noticeable in urban areas. In addition, other sources of particulate matter are of high concern in urban areas as well and needs improvement. This need is in part the result of an increasing population and that urban areas continue to grow as population centers that will inevitably increase automobile congestion and emissions if not addressed as quickly as possible.

As we shall see in the next chapter, even though the United States is in need of improving urban air, many locations in the rest of the world are far worse when it comes to urban air pollution.

14.3.5 Banning Harmful Chemicals

Let's examine the regulations of the United States for just a moment from an overview perspective. The majority of environmental regulations focus on wastes that are generated and whether a facility has a permit. They are most concerned, and for good reason, on air emissions, water discharge, and solid and hazardous waste generation and disposal. The reasoning is that these are the primary pathways for harmful substances to migrate from a point of generation to a point of human exposure where they may cause harm to us or the environment. In general, the regulations do not restrict what substances can be used at a facility.

TSCA is the set of regulations that has this authority and, to date, TCSA has very few chemicals that have been banned in the United States. In its history, TSCA has banned nine chemical substances and ten others are under current review. Generally, those compounds that are the most persistent, bioaccumulative, and toxic are of the highest priority.

You may have wondered and even asked yourself a question. Why didn't we know about the harm certain chemicals have on humans and the environment before they are widely used. This is a very good question, and it happens to be a mistake our society has made time and time again. It happened with dichlorodiphenyltrichloroethane (DDT), polychlorinated biphenyl (PCBs), asbestos, meythel-tert-butyl-ether (MTBE), and many other compounds that, at first, were even advertised as a miracle chemical (i.e., DDT and PCBs), or a miracle substance (i.e., asbestos). However, as time marched on and these chemicals or substances were widely used, evidence began to surface that these substances had unintentional and unknown side effects that had a negative effect on human health, the environment, or both. Sometimes there is a latency period, sometimes 20 or more years after exposure, that a symptom or negative effect surfaces.

There are over 84,000 chemicals in the United States currently being used. USEPA has been successful in restricting only 9 chemicals in its history of over 40 years since promulgated in 1976 and none since 1984. Those substances currently restricted include the following (Rogers 2020):

- PCBs
- Chlorofluorocarbons (CFCs)
- Dioxin
- Asbestos
- Hexavalent chromium
- Four nitrite compounds that are:
  - Mixed mono and diamides of an organic acid
  - Triethanolanime salt of a substituted organic acid
  - Triethanolanime salt of tricarboxylic acid, or
  - Tricarboxylic acid

In June 2016, TSCA was amended significantly and gave USEPA additional authority and responsibility to proactively evaluate the risks of all chemical substances. USEPA has identified ten chemicals as high priority and are currently conducting risk evaluations. They include the following (Rogers 2020):

• 1,4-Dioxane	• 1-Bromopropane
• Asbestos	• Carbon tetrachloride
• Cyclic aliphatic bromide cluster	• Methylene chloride
• N-methylpyrrolidone	• Pigment violet 29
• Tetrachloroethylene	• Trichloroethylene

Analysis of the above list includes four of substances that belong to the same chemical group called chlorinated volatile organic compounds (CVOCs), also known as chlorinated solvents.

Those include carbon tetrachloride, methylene chloride, tetrachloroethylene, and trichloroethylene. In addition, 1,4-dioxane is sometimes associated with the breakdown of tetrachloroethylene and trichlorethylene, so perhaps there may be as many as five in the CVOC family of chemicals. Therefore, of the ten chemicals USEPA has initially placed in its high priority list, half or more belong or are associated with the same group of compounds, CVOCs, or chlorinated solvents (Rogers 2020).

There is perhaps negative feedback that has occurred by not banning more chemicals in the United States that has resulted in providing a false sense of security concerning chemical use. This sense of false security has developed an attitude of reliance that the USEPA will protect us, after all that is their job, right? Many users may say that if a particular chemical were harmful, it would have been banned. This is where we can improve but have not, because TSCA was not strong enough when enacted and the burden of scientific proof required to ban a chemical is too high and takes too long to establish. This is evident when we examine some of our example sites in this chapter that include Woburn Massachusetts, Toms River New Jersey, and the Hinckley Site in California.

14.3.6 Dealing with Emerging Contaminants

Emerging contaminants are chemicals or substances that have been recently discovered in the environment and, in some instances, have caused harm or have impacted resources to the extent that they can no longer be used for fear of health risks. Emerging contaminants include many compounds including per-fluor-alkal-substances (PFAS), many pharmaceutical compounds, 1,4-dioxane, and others. Many of these emerging compounds are now being detected in the environment and have not been adequately evaluated. This highlights a need for improved prevention measures and the need for improving handling and disposal practices.

14.3.7 Addressing Invasive Species

Some consider invasive species a type of emerging contaminant. For clarity, we treat them separately because there are so many invasive species and their behavior in the environment to so different compared to chemicals and other substances because their mode of migration is through breeding. Invasive species have been present for the last few hundred years, but environmental damage is increasing and in need of further regulatory protections.

14.3.8 Improving Pollution Prevention

The Pollution Prevention Act (PPA) and Oil Spill Prevention Act (OSPA), both of 1990, do not get enough credit in the history of environmental regulations of the United States. The PPA and OSPA were the first real efforts to turn from regulating the environment at the "end of the pipe" to promoting the prevention of pollution from entering the environment. One of the driving factors behind the PPA and OSPA was the realization of how much it cost to clean up the environmental after a release. It was a sobering lesson that just one incident could cause so much harm, cost billions of dollars, and, in some instances, could not be cleaned up to acceptable levels because technologies simply did not exist. It was during this time that USEPA and many others discovered that sometimes you can throw money at a contaminated site or incident and make it go away and then there are those where all the money and technology in the world would not be able fix the problem before enormous harm was a reality. Therefore, out of this realization, USEPA began to focus efforts on pollution prevention, in part because it made common sense, in that it is proactive at environmental protection and lowers response cost much like fire prevention. As we shall see in the next chapter, it was out of these and similar efforts in the European Union that the term "sustainability" took hold.

14.3.9 Improving Urban Land Use

One important item that should be added to pollution prevention efforts is urban land use. There is much that we can accomplish at protecting human health and the environment by not locating high-risk activities in areas that are especially sensitive to pollution. Currently, urban planning for industrial activities is usually located near transportation routes and not near residential areas. This is a good practice. However, there is often a complete disregard for evaluating geologic and hydrogeologic risk factors that can greatly influence the migration of contaminants through air, water, and soil that ultimately lead to human and ecological exposure from these areas. The examples for Woburn Massachusetts, Hinkley Site in California, Toms River, Love Canal, and Times Beach are just a few of the hundreds of examples that highlight this risk but has not been adequately addressed with effective legislation for land-use planning.

A worldwide estimate of bird deaths due to collisions with buildings is 4 billion a year (World Wildlife Fund 2022). Design considerations for bird-safe buildings to reduce bird loss in the United States and world include the following (United States Fish and Wildlife Service 2024b):

- Using bird-friendly glass treatment that is more visible to birds
- Minimizing transparent building corners
- Preventing light pollution and exterior lighting
- Turning off unnecessary lighting
- Avoiding up-lighting and spotlights
- Minimizing rooftop antennas and decorative spires
- Applying window films
- Adding exterior screens
- Installing UV patterns
- Placing decals on windows
- Reducing lighting at night
- Landscape for birds such as planting native vegetation, shelter, and resting places
- Monitoring effectiveness and modifying if necessary

A worldwide estimate of bird deaths from collisions with wind turbines has yet to be estimated. However, wind turbine bird deaths may be higher within raptor populations. Modifying wind turbines to twist instead of moving in a circular pattern has been one suggestion to reduce the amount of bird deaths (Audubon Society 2020).

14.3.10 Regulating Noise Pollution

An additional type of pollution we have not addressed in this chapter is noise pollution. The reason is that noise pollution is largely not regulated in the United States at a federal level. However, as we shall see in the next chapter, noise pollution is a regulated form of pollution in many countries of the world, including the European Union, Mexico, and even China.

14.3.11 Improving Household Waste Regulation

Some local efforts have been initiated at educating the public on the potential dangers of some household wastes that include cleaners, solvents, electrical equipment, waste paints, oil and grease, and many other substances. However, much more effort and incentives are needed. The landfills where much of the waste from households is disposed are not as adequately constructed to prevent releases to groundwater from many of the substances that are located within the majority of municipal landfills. In addition, septic tanks are additional source of pollution to the ground and groundwater and highlight the need for expanding local public owned treatment works (POTWs) so that this untreated waste is not directly discharged into the environment.

14.3.12 Non-point Source Pollution Improvements

As described in the summary of the Clean Water Act, non-point source pollution is in need of regulation, especially from rural and agricultural land to minimize the build-up and discharge of pollutants that include fertilizers, pesticides, herbicides, bacteria, and other pollutants into the waters of the United States and causing massive marine deaths in the Gulf of Mexico and perhaps other coastal locations (e.g., Florida). This will be more difficult to regulate in part because it affects areas that are not within the geographical boundary of the United States. The source of much of this pollution originates from non-point sources located in rural and agricultural land within the interior portions of the country.

Environmental regulations concentrate on urban land which only accounts for 2.5% of developed land in the United States. Other areas that include crop land, range land, and managed forest account for 75% of developed land in the United States and are generally not regulated by the USEPA. USEPA has been successful in improving the environmental quality in urban areas to such a degree that pollution effects from non-point sources, which account for the 75% of other developed land in the United States, are much more noticeable and in need of regulation and protection.

14.3.13 Sustainability Regulations

Sustainability legislation presents the United States an opportunity to move its environmental regulations from reactive to proactive. In addition, the United States is in need of a clear picture of what it means to be sustainable. To many, it is a buzz word lacking specific goals or even a basic plan for achieving sustainability.

The European Union and other countries have established a clear picture and have set goals for sustainability. It is overdue for the United States to formally develop sustainability legislation and set goals at the national level. If there are delays in enacting sustainability measures, it may be too late.

14.3.14 Consumer Products and Packaging Improvements

Plastic has become one of the more important current pollution issues and was the focus of Earth Day 2018 (Earth Day 2024). Plastic has many valuable uses. However, humans have become addicted to single use of disposable plastic. Around the world, 1 million plastic drinking bottles are purchased every minute and up to 5 trillion single-use plastic bags are used worldwide every year. In total, half of all plastic produced is designed to be used only once and then thrown away. Plastic waste is now so common in the natural environment that some scientists have even suggested that plastic could serve as a geological indicator of the Anthropocene Era (United Nations 2024).

From the 1950s through the 1970s, only a small amount of plastic was produced and was not a significant issue. However, by the 1990s plastic waste had more than tripled. In the early 2000s plastic output increased more than in the previous 40 years. As of the end of 2018, more than 300 million metric tons of plastic waste is generated per year, which is equivalent to the weight of the entire human population (United Nations 2024). Although plastic containers and packaging are nearly impossible to regulate by USEPA, there is a clear need to address this issue.

14.3.15 Addressing Public Lands

Areas of public land addressed here are the National Parks which are managed by the National Park Service, National Forests which are managed by the United States Forest Service, and range land which is largely under the management of the Bureau of Land Management. From an administrative viewpoint, each is regulated separately, which sometimes leads to inconsistencies of regulations and environmental protections. Each segment of public land is under stress in some form or another that includes effects from forest harvesting, overgrazing, overuse, erosion, lack of funds, and habitat destruction.

14.3.16 Households

Regulating pollution from households has been a point of contention for many decades because of the perception or realization that regulating household wastes may be interpreted as an invasion of privacy. However, this should not be the case if future regulations focus on those wastes that are discarded and, therefore, would ultimately not remain on private property. The following are a few areas where new regulations can potentially have significant impacts at lowering environment risks:

- Household waste. Municipal landfills primarily receive household waste which can contain wastes that would be considered hazardous and would be more strictly regulated if generated from an industrial source. From a sustainability point of view, it would be preferred that households not dispose of certain materials in their waste, but this sometimes is not the case. Some examples of these wastes include cleaners, solvents, pesticides, herbicides, batteries, gasoline, and others. Therefore, more strict measures are needed to further prevent hazardous wastes entering municipal landfills to minimize costly clean-up measures and protect human health and the environment. USEPA regulates municipal landfills under Resource Conservation and Recovery Act (RCRA) and does require certain construction measures to prevent or minimizes the potential for contaminants to migrate beyond the footprint of the landfill. However, this sometimes is not enough (USEPA 2024e).
- Septic tanks. USEPA estimates that approximately 20% of homes in the United States use septic systems to treat wastewater (USEPA 2024g). When a septic system is not properly maintained, elevated nitrogen and phosphorus can be released and migrate to surface water, groundwater, or both. USEPA estimated that up to 20% of septic systems fail at some point in their operational history and that common causes are age, inappropriate design, and overuse (USEPA 2024g). According to USEPA, the responsibility to properly maintain a septic system is the homeowner (USEPA 2024g).
- Private drinking water wells. Private drinking water wells are not regulated by USEPA. In addition, USEPA does recommend criteria or standards for individual drinking water wells. USEPA estimates that more than 13 households rely on private wells for drinking water in the United States (USEPA 2024h). Permits are generally always required for installation of a private well and local regulations vary widely. From a sustainability point of view, overuse of groundwater resources in the United States, predominantly from agriculture, is significant (United States Geological Survey (USGS) 2024). The effects from overuse of groundwater resources lead to groundwater depletions, which in turn can lead to a reduction of water in streams and lakes, land subsidence, and deterioration of water quality (USGS 2024).
- Home heating oil tanks. A home heating oil tank is essentially what is considered an aboveground storage tank (AST), or if buried, an underground storage tank (UST). USEPA strictly regulates ASTs and USTs under Subtitle I of RCRA. However, home heating oil tanks are generally exempt under RCRA (USEPA 2020). The actual number of home heating oil tanks in the United States is unknown but is likely significant. Many states have published material to assist homeowners with managing heating oil tanks. Releases from heating oil tanks can cause a significant environmental impact as those regulated under RCRA. In fact, USEPA stated in 1989 that tanks represented the largest risk to groundwater than any other contaminant source (USEPA 2020).

- Construction materials. Many building materials historically contained hazardous substances, including asbestos, lead, PCBs, CFCs, and others. These still may be present in older building (USEPA 2024i). In addition, wood preservative chemicals including pentachlorophenol (PCP), chromate arsenicals, and creosote were largely historically used in industrial settings. Recent wood preservatives for residential use include alkaline copper quaternary (ACQ), borates, copper azole, copper naphthenate, copper-HDO, and polymeric betaine (USEPA 2024h).
- Improve landscaping in public areas. Landscaping on public lands, especially along highways, should be improved. These areas represent a large area when considering the thousands of kilometers of interstate highways in the United States. Recent management of these areas is one of many reasons likely responsible for the decline in the Monarch butterfly when milkweed, which the Monarch butterfly needs for reproduction, was removed in favor of other species, including along roadway easements. Other documented influences for the decline include mortality during migration and climate change (Michigan State University 2021).
- Residential landscaping improvements. Minimizing grass lawns have numerous advantages for increasing habitat for native species to saving water by planting native vegetation that include plants, shrubs, bushes, and trees (Tallamy 2019).

14.3.17 Climate Change

The United Nations has played perhaps the most significant role in educating and promoting actions for reducing GHGs throughout the world. The Intergovernmental Panel on Climate Change (IPCC) was established by the United Nations Environment Programme (UNEP) and the World Meteorological Organization in 1988 (IPCC 2021). The IPCC is the United Nations body for assessing the science related to climate change. The IPCC is widely considered the world leader in setting the standards for assessing, mitigating, and reducing GHG emissions (IPCC 2021).

Targets for GHG emission reduction have focused on a 50% reduction by 2030 and to become carbon neutral by 2050. The IPCC must continue to assess the science of climate change and take a lead role of reducing GHG emissions at the global level. However, even though targets have been set, achieving those targets has yet to become clear.

14.3.18 Summary of Expanding Environmental Regulations

The sections highlighted above describe the majority of subject areas where environmental regulations in the United States are either in need of improvement or where new regulations are needed. For reference purposes and to list those areas that have the greatest need for new or improved environmental regulations the following is provided (Rogers 2019):

- Modify environmental enforcement emphasis and policy
- Science-based improvements to environmental regulations to account for regional risks and climate change
- Agriculture
- Pesticides and herbicides
- Fertilizers
- Septic tanks
- Invasive species
- Urban planning and land use
- Residential and household waste
- Urban air
- Further or new restrictions on harmful chemicals, especially those that are persistent and mobile in the environment

- Emerging contaminants
- Noise
- Plastic
- Climate change
- Sustainability

The purpose of environmental regulations is to protect human health and the environment. It makes little sense if human health and the environment are protected from nine out of ten exposure risks when the remaining risk factor causes harm. Therefore, environmental regulations should be as comprehensive and complete as possible. Otherwise, they will fail to achieve their intended objective. This example is realized when considering the lack of regulations of the items listed above. Another item worthy of discussion is that there are political factors and areas where there is overlapping jurisdictions that may have negative outcomes that must be addressed when discussing the need for further environmental regulations in the United States. For instance, the United States Department of Agriculture would almost certainly interject an interest in environmental regulation of farm land. In addition, the Department of Interior would also be interested in sustainability regulations as it pertains to national forests and parks and environmental regulations on rangeland.

14.4 SUMMARY AND CONCLUSION

There is much to be proud of when examining the improvements in the United States compared to 50 years ago that have significantly improved the air quality, water quality, and cleanup of sites of environmental contamination, which now number in the tens of thousands of sites across the United States. However, there are several areas where improvements can be realized if some existing regulations were modified or expanded and if some new regulations were developed. In addition, sustainability actions tend to focus on proactive measures which will take some time to adjust since much of our response to enacting environmental legislation has been reactive in nature. An additional concept that requires attention is our relationship with nature. For many decades, nature was categorized as perhaps the enemy, something that needed to be conquered or tamed. The concept was that progress and development would achieve this desired result. Scientists now realize that this may not have been a good approach and that our relationship with nature must improve. The step toward this goal is research and understanding

REFERENCES

Audubon Society. 2020. *Analysis: Is It Possible to Have Wind Power While Keeping Birds Safe?* www.audubon.org./news/birds-and-wind-turbines. (Accessed June 22, 2024).

Darke, R., and Tallamy, D. 2019. *The Living Landscape*. Timber Press. Portland, Oregon. 392p.

Earth Day. 2024. *Earth Day Network Campaign: End Plastic Pollution*. https://earthday.or/campaigns/plastic-campaign. (Accessed May 5, 2024).

Intergovernmental Panel on Climate Change (IPCC). 2021. *Climate Change 2021: The Physical Science Basis*. IPCC AR6 WGI. New York, NY. 3949p.

Michigan State University. 2021. *Why Is the Eastern Monarch Butterfly Disappearing?* MSUToday. Michigan State University. https://msutoday.msu.edu/Why is the eastern monarch butterfly disappearing?. (Accessed May 5, 2024).

National Association of Home Builders. 2024. *Typical America Subdivisions*. www.nahbclassic.org. (Accessed May 5, 2024).

National Geographic Society. 2024. *Energy Efficiency of Modes of Transportation*. www.nationalgeographic.com/carbon-footprint-transportation-efficiency. (Accessed May 5, 2024).

Natural Resource Defense Council (NRDC). 2024. *Regenerative Agriculture*. www.nrdc.org/resources/regenerative-agriculture. (Accessed May 5, 2024).

Rogers, D. T. 2014. Scientists Call for a Renewed Emphasis on Urban Geology. *American Geophysical Union. Earth and Space News. Eos*. Vol 95, no. 47, pp. 431–432.

Rogers, D. T. 2016a. Next Generation of Urban Hydrogeologic Investigations. *Journal of the Italian Geological Society*. Rome, Italy. Vol. 39, no. 1. p. 349.

Rogers, D. T. 2016b. *Scientific Advancements That Improve the Conceptual Site Model in Urban Hydrogeologic Investigations*. 35th International Geological Congress. Paper No. 3019. CapeTown, South Africa.

Rogers, D. T. 2018. Derivation of a Comprehensive Environmental Risk Model for Urban Groundwater Protection. *International Association of Hydrogeologists Congress*. Vol. 1, p. 879. Daejeon, Korea.

Rogers, D. T. 2019. *Environmental Compliance and Sustainability: Global Challenges and Perspectives*. CRC Press. Boca Raton, FL. 583p.

Rogers, D. T. 2020. *Urban Watersheds: Geology, Contamination, Environmental Regulations, and Sustainability*. CRC Press. Boca Raton, FL. 606p.

Rogers, D. T. 2021. *Empirical Results of a Geological- and Anthropogenic Contaminant-based Environmental Risk Sustainability Model*. Geological Society of America. Paper No. 8325. Portland, Oregon.

Rogers, D. T., Kaufman, M. M., and Murray, K. S. 2016. *Development of an Environmental Risk Management and Sustainability Model Using Chemistry, Geology, and Human Activity, and Response Inputs*. 35th International Geological Congress. Paper No. 2090. Cape Town. South Africa.

Tallamy, D. W. 2019. *Nature's Best Hope*. Timber Press. Portland, Oregon. 254p.

United Nations. 2016. *Human Development Report*. United Nations Development Programme. www.hdr.undr.org/sites/2017. (Accessed May 5, 2024).

United Nations. 2024. *The Facts on Plastic Pollution*. www.unenvironment.org/beat-plastic-pollution. (Accessed May 5, 2024).

United States Census Bureau. 2024a. *Census Bureau Average Population Per Household*. www.census.gov/population/socdemo/hh-fam/tabHH-6.pdf. (Accessed May 5, 2024).

United States Census Bureau. 2024b. *Median and Average Square Feet of floor Area in New Single-family Homes*. www.census.gov/const/c25ann/sftotalmedavsqft.pdf. (Accessed May 5, 2024).

United States Census Bureau. 2024c. *Average One-Way Commuting Time and Distance*. www.census.gov/library/visualization/interactive/travel-time.html. (Accessed May 5, 2024).

United States Department of Energy. 2024. *Energy Efficiency and Renewable Energy*. Alternative Fuels Data Center. www.afdc.energy.gov.data. (Accessed May 5, 2024).

United States Department of Housing and Urban Development. 2024. *Median and Average Square Feet of Floor Area in New Single-family Homes*. United States Census Bureau. www.census.gov/const/c25ann/sftotalmedavsqft.pdf. (Accessed May 5, 2024).

United States Environmental Protection Agency (USEPA). 2018. *Air Quality – 2018 National Summary*. www.epa.gov/air-trends/air-quality-national-summary. (Accessed May 5, 2024).

United States Environmental Protection Agency (USEPA). 2020. *Summary of the Resource Conservation and Recovery Act*. www.epa.gov/rcra/summary (Accessed May 5, 2024).

United States Environmental Protection Agency (USEPA). 2024a. *Flint Drinking Water Crisis*. US EPA. www.epa.gov/Flint Drinking Water Response. (Accessed May 5, 2024).

United States Environmental Protection Agency (USEPA). 2024b. *Reducing Lead in Drinking Water*. US EPA. www.epa.gov/Leaders in Reducing Lead in Drinking Water. (Accessed May 5, 2024).

United States Environmental Protection Agency (USEPA). 2024c. *Drinking Water Treatment Technologies*. US EPA. www.epa.gov/Drinking Water Technologies. (Accessed May 5, 2024).

United States Environmental Protection Agency (USEPA). 2024d. *Emerging Technologies for Wastewater Treatment*. Emerging Technologies for Wastewater Treatment and In-Plant Wet Weather Management (epa.gov). www.epa.gov/. (Accessed May 5, 2024).

United States Environmental Protection Agency (USEPA). 2024e. *Municipal Solid Waste Landfills*. US EPA. www.epa.gov/Municipal Solid Waste Landfills | (Accessed May 5, 2024).

United States Environmental Protection Agency (USEPA). 2024f. *Climate Adaptation and Erosion and Sedimentation*. US EPA. www.epa.gov/Climate Adaptation and Erosion & Sedimentation. (Accessed May 5, 2024).

United States Environmental Protection Agency (USEPA). 2024g. *The Sources and Solutions: Septic Tanks*. US EPA. www.epa.gov/The Sources and Solutions: Wastewater. (Accessed May 5, 2024).

United States Environmental Protection Agency (USEPA). 2024h. *Private Drinking Water Wells*. US EPA. www.epa.gov/Private Drinking Water Wells. (Accessed May 5, 2024).

United States Environmental Protection Agency (USEPA). 2024i. *Overview of Wood Preservative Chemicals*. US EPA. www.epa.gov/Overview of Wood Preservative Chemicals. (Accessed May 5, 2024).

United States Fish and Wildlife Service. 2024a. *Accessing the Status of the Monarch Butterfly*. fws.gov. (Accessed May 5, 2024)

United States Fish and Wildlife Service. 2024b. *Buildings and Birds*. www.fws.gov/story/threats-birds-collisions-buildings-glass. (Accessed June 21. 2024).

United States Forest Service. 2024. *Bark Beetles*. www.fs.usda.gov/internet/fse_documents/fseprd563458. (Accessed April 2, 2024).

United States Geological Survey (USGS). 2024. *Groundwater Decline and Depletion*. www.usgs.gov/Groundwater Decline and Depletion (usgs.gov). (Accessed May 5, 2024).

World Wildlife Fund. 2022. *Living Planet Report – Building a Nature Positive Society*. In: Almond, R. A., Grooten, M., Bignoli, J., and Petersen, D., editors. Chapter 1 – The Global Double Emergency. WWF. Gland, Switzerland. 114p.

15 Global Sustainability Actions

15.1 INTRODUCTION

In this chapter, we focus on sustainability efforts at the global level. The United States is not isolated and pollution along with many other environmental issues has no boundaries. We are an interacting global community. Those concerns that affect the environmental health of one country or region can eventually affect us all. Therefore, we must work together as one community if we are to repair the harm we have inflicted on our planet with a sense of urgency and intention of living with nature rather than against it.

We also know through many thousands of scientific studies that our planet's habitability is in decline. This is evidenced through our expanding civilization and polluting the air, water, and land, which can be directly observed through climate change, species reduction and extinction, ocean pollution, decline in coral reef health, pollution in Antarctica, urban expansion and land use, agricultural practices, forest management, and many other activities. We must act as a global community to more fully understand the impacts humans have made on our planet and act together and diligently through focused efforts to limit the negative effects and restore ecosystems where possible. These actions must be undertaken before it is too late not just for the survival of other species but for the survival of our own species as well.

This chapter is dedicated to understanding, implementing, and improving sustainability at the global level. In addition, we address sustainability methods that can be applied in different settings and locations from urban to agricultural to wilderness areas on land, the oceans, and the atmosphere. Many items listed at the global level also apply to actions required at the national and even the individual and local level but have a different scope and focus. Therefore, there may appear to be duplication but there are differences in scale and application that will be discussed in this chapter.

15.2 GLOBAL CHALLENGES THAT SUSTAINABILITY MUST OVERCOME

Challenges exist between countries concerning sustainability and include differences in power and responsibility, environmental, economic concerns, population, cultural, and religious differences that must be addressed before sustainability can proceed on a path to become an effective reality on a global level (United Nations 2024a). However, the recent report by the Intergovernmental Panel on Climate Change (IPCC 2021), which clearly stated the dire need for immediate action to reduce greenhouse gas (GHG) emissions may create an opportunity for the countries of the world to set aside differences and inequalities and unite in a common effort to overcome the potential cataclysmic effects that climate change may inflict on all living species on Earth. These are monumental

DOI: 10.1201/9781003463153-18

challenges and humans are now facing perhaps the greatest threat to our species. Ironically, that threat has largely been caused by humans. This is nothing new, early in the environmental movement in 1971, an Earth Day poster read "We have met the Enemy and He Is Us" (Ohio State University 2024). This is similar to the United States Advisory message that stated People Started Pollution, People Can Stop It." This challenge is also a huge opportunity for international cooperation and innovation.

Religious conviction and population challenges are generally not openly discussed and likely represent the most challenging to overcome to an extent is meaningful, especially in Western developed countries. This is likely because these two challenges will be interpreted as invading or challenging personal freedoms.

15.2.1 Power and Responsibility

Points of contention are realized between developing and developed countries. It is of concern for developed nations because they are perceived to be imposing their new-found sustainable values upon developing countries. Developed nations have already greatly benefited from the exploitation of environmental resources, whereas developing nations have not had the chance or enough time to exploit their natural resources and feel resentment toward developed nations. A significant focus in the sustainable development movement has been on freeing people in developing nations from the bonds of poverty and starvation. Freeing nations from poverty and starvation would mean consuming enormously more energy and resources to raise the standard of living of these individuals which number in the billions. One of the most important arguments of sustainability states that the rights provided by governments, such as the right to vote and freedom of speech, are not of much use unless the society provides its citizens with an opportunity of for education, access to food, and employment (United Nations 2024a).

15.2.2 Economic

This brings us to perhaps one of the two most problematic hurdles in achieving a sustainable world and that is sustainability does not yet appear to be compatible with a capitalistic economic system (United Nations 2024a). An economic system of capitalism relies on an ever-growing economy. How can humans achieve sustainability with a growing population, increasing energy needs, lifting vast human populations (e.g. more than 2 billion people) in undeveloped nation's standard of living, and achieve a sustainable world where we all live in harmony with our environment. This all must take place in a fragile biosphere called Earth. So far, our economic system continues to grow and gather momentum and as it does it threatens to destabilize the global ecological balance in many significant ways. How can humans modify a capitalistic system that has ecological destructive outcomes and achieve a sustainable planet?

There are specific issues, such as reversing much of the damage to the ozone layer, which have been successful and lend hope and a call to action that not all is lost and that balance can be restored. After all, humans are an adaptable species, and we are just recently putting our collective intellectual and cooperative skills to work to confront this challenge which we must win.

15.2.3 Religion

Scientists and theologians have written eloquently about their awe and wonder at the history of the universe and life on Earth. Attempts to pit science and religion against each other create controversy where none needs to exist especially when attempting to address negative anthropogenic impacts on Earth caused by humans (National Academies of Sciences, Engineering, and Medicine 2024).

In May 2015, Pope Francis of the Roman Catholic Church released an Environmental Encyclical from the Vatican that addressed climate change, global warming, pollution, economics, and population control. The Environmental Encyclical as it is now referred to states that (Pope Francis 2015):

> humans must forcefully reject the notion that we were given dominion over the Earth and absolute dominion over other creatures

This is yet another example of hope and resolve that we must do what must be done and that we all have a part to play as Pope Francis further states that (Pope Francis 2015):

> there is an urgent need for a radical change in the conduct of humanity

15.2.4 Globalization and Governance

Globalization of economies has developed connections between varied secular cultures and political ideologies through cooperative commerce and trade. In many ways, this has brought the world closer together and has increased understanding and respect between widely varying cultures. Globalization through commerce and trade has also identified some challenges that include poverty, fairness, war, public health, and other differences. To overcome differences so that improved consistency and fairness in trade and governance can be better established will require effort, guidance, and financial assistance from a global community. Additionally, climate change and other environmental impairments have become more apparent and costly, especially in developing nations. As a result of these challenges, the concept of Earth System Law was born (Cadman et al. 2022).

Law has historically provided stability, certainty, and predictability in the ordering of social relations, predominantly between humans and mainly within the same country or jurisdiction. However, in recent decades, the world has changed dramatically. First, economies went global, and secondly, and perhaps more importantly, how humans treat the Earth has come under scrutiny. This involves a belief system that recognizes that Earth has inherent rights to the environment including flora, fauna, and other elements of the biosphere and solid Earth. This concept challenges a long-held legal belief that the law only applies to humans. However, many are reassessing the belief that Earth, as a system, has intrinsic rights. This raises many questions and challenges that must be overcome as humans move into the future in attempting to create a balance between different cultures and our relationship and treatment of our only home planet. This will require input from each country and different fields of study and professions including the physical and life sciences, political science, anthropology, economics, law, ethics, sociology, psychology, corporations, citizens, and governments (Cadman et al. 2022).

15.3 UNITED NATIONS GLOBAL SUSTAINABILITY GOALS

The United Nations has framed sustainability goals termed "Global Goals," which include the following (United Nations 2024b):

• No Poverty	• Zero Hunger	• Good Health
• Well-Being	• Quality Education	• Gender Equality
• Clean Water	• Effective Sanitation	• Affordable Energy
• Clean Energy	• Economic Growth	• Satisfactory Employment
• Industry	• Innovation	• Infrastructure
• Reduced Inequalities	• Sustainable Cities	• Responsible Consumption
• Climate Action	• Life Below Water	• Life on Land
• Peace and Justice	• Partnerships	• Life in the Air

Many of the goals listed above may sound unrealistic. However, the United States has made significant progress in curbing environmental degradation and increasing sustainability since the United States Environmental Protection Agency (USEPA) was first formed nearly 50 years ago. This has been achieved through improving air quality, water quality, and land impact through the Clean Air Act (CAA), Clean Water Act (CWA), Resource Conservation and Recovery Act (RCRA), Comprehensive Environmental Response, Compensation and Liability Act (CERCLA), and a multitude of other environmental regulations (USEPA 2024a). In addition, actions by the United Nations, the World Bank, private nonprofit organizations, thousands of businesses, and many other nations have had positive results.

We have covered several subjects within this book that now can be pulled together into a cohesive framework that can be used to implement a comprehensive sustainability program at any level with the end goal of environmental stewardship. The pursuit of sustainability involves many subject areas and levels. In this chapter, we will address actions required at the global level. We will address actions required at the national and individual levels in the next chapter. The subject areas that will be addressed in the following sections are listed below (Rogers 2023):

- Cohesive action from the global community
  - Global environmental regulations
  - Obtaining a sustainable human population
  - Modifying the human diet
  - Financial fairness and equity distribution
- Improving environmental regulation at national levels
- Understanding the natural setting of our urban and industrial areas
  - Protecting sensitive ecological areas
  - Smart urban development
  - Improving urban transportation
  - Improving building design
  - Incentivizing multi-family housing
- Knowing the chemicals that are used and how they might cause harm to the environment and humans if they are released
- Knowing what alternative chemicals are available
- How to reduce chemical use
- Preventing chemical releases
- Reducing energy use
- Using alternative sources of energy
- Minimizing solid waste, air emissions, and water discharges
- Reducing water use
- Minimizing stormwater runoff
- Repairing environmental damage to the extent possible
- Attaining environmental stewardship

15.4 COHESIVE POLICY ACTIONS FROM THE GLOBAL COMMUNITY

At least seven policy-oriented actions at the global level are required if sustainability will be effective. These actions include the following (Rogers 2023):

- Building a global community
- Cohesive environmental policy, regulations, and enforcement
- Controlling the human population
- Modifying the human diet

- Financial fairness and equity distribution
- Modifying business models
- Improving our relationship with nature

15.4.1 Building a Global Community

Building a global community is possible and is achieved periodically. An example is the Olympic Games which bring the world together to celebrate sport every 2 years. In addition, banning or limiting the use of CFC which degrade the ozone layer is another example where the global community cooperated to address an environmental threat. Building a global community on sustainability issues will require significant cooperation on several actions for an extended period of time, and also must include removing nationalistic political, social, and religious boundaries (Harari 2018). Patriotism for a nation, which is common, must graduate to a global level.

With the publishing of a recent report on GHG emissions by the IPCC (2021), which described the observed climate change impacts, there is a positive cooperative dialogue between almost all countries of the world to reduce GHG emissions. Hopefully, this recent turn of events will finally wake the world up and respond to climate change directly and with scientific vigor.

15.4.2 Cohesive Environmental Policy, Regulations, and Enforcement

Pollution does not respect boundaries. Like all matter, pollution follows the laws of physics and will disperse from a point of origin and given enough time will be detected worldwide. One country can adopt and implement effective sustainability measures but, in the end, will eventually become polluted at similar levels as other countries that do not implement the same measures regardless of financial resources or technology. Therefore, we must develop a global community and enact cohesive environmental policies and enforcement at the international level. Climate change is a good example that action must be taken at a global level as GHGs and a plethora of other pollutants migrate the globe and become ubiquitous. Just in the last few decades, there is much less pack ice in the arctic and glaciers are shrinking at a significant pace.

15.4.3 Controlling the Human Population

There are at least three factors common to countries that have significant pollution issues and those include (1) lack of financial resources, (2) over-population, and (3) lack of effective environmental regulations (Rogers 2019a, 2019b).

We have discussed in the previous section that capitalism currently relies on growth for itself to be sustainable. Much of the economic growth that fuels capitalism is achieved through an endless and continued increase in the human population and a growing standard of living, especially in underdeveloped countries. In 1900, the human population was approximately 1 billion and currently is estimated at more than 7 billion (United Nations 2024c; Harari 2015). A continued growing human population is clearly not sustainable and has grabbed the attention of world and religious leaders (Pope Francis 2015).

The question is what is a sustainable population of humans on Earth? Clearly, we have exceeded that balance with our current practices and disregard for our plants. The answer lies in reducing adverse impacts and balancing those achievements with a human population that does not adversely impact future generations and other life on Earth.

Many human technological innovations have been developed through our history including (1) the agricultural revolution approximately 12,000 years ago, (2) the Industrial Revolution in the last 300 years, and (3) the information and computer revolution just recently. These technological marvels have enabled humans to establish large cities, have time for recreation, and greatly increase our population (Harari 2015). If humans can solve complex technological unknowns, humans can also solve the human population issue before nature does it for us.

15.4.4 Modifying the Human Diet

Much of the farming in the United States is concentrated in the production of meat (USDA 2011). Meat production in the United States also consumes huge volumes of water and creates significant biological pollution and is also a significant contributor of methane which is a powerful GHG (USEPA 2024b). Slight modifications in the diet of Americans that include reducing meat and animal fat consumption combined with modifying farming techniques through the addition of urban farms and other measures can have a significant positive impact on the following (USEPA 2024c):

- Improving the health of Americans by improving our diet and reducing exposure to contaminants such as pesticides and herbicides
- Reducing GHG emissions
- Reducing fugitive emissions
- Significantly increasing the amount of natural land
- Decreasing water use
- Decreasing fertilizer, pesticide and herbicide use
- Decreasing exposure to harmful chemicals
- Decreasing biological pollution
- Improving surface water quality

15.4.5 Financial Fairness and Equity Distribution

There is a correlation between effective environmental regulations and the standard of living (Rogers 2019a, 2019b). Those countries without financial resources have more significant environmental uses than countries with deeper financial resources. Fortunately, the United Nations and The World Bank are very aware of this situation and have been working diligently to improve the environment of undeveloped nations (United Nations 2024d; World Bank 2024a). In addition, the USEPA is also aware and has been assisting numerous other nations in improving environmental quality by implementing several environmental initiatives worldwide (USEPA 2024d).

15.4.6 Modifying Business Models

A significant potential impediment to achieving sustainability is the current capitalistic economic system. Every country in the world is unique but also shares commonalities with respect to basic economic systems. The current capitalistic economic system must change from growth-minded to sustainability-minded. Sustainability incentives must be integrated at the international, national, community, and individual economic levels.

Thousands of businesses worldwide are adjusting their business models by incorporating sustainability measures and incentives into their culture and using these modifications to promote themselves by meeting the highest verified standards of social and environmental performance, public transparency, and legal accountability to balance profit and purpose. Two widely used methods include ISO 14001 and Certified B Corporations.

15.4.6.1 International Organization for Standardization

The International Organization for Standardization (ISO) 14000 is a family of standards originating in 1992 and has now grown into detailed international standards that specify requirements for an effective environmental management system (EMS) (American Society for Quality 2024).

ISO 14001 is a voluntary standard that examines every aspect of a particular business and evaluates how each aspect potentially impacts the environment. ISO 14001 then examines how to reduce environmental impact by modifying aspects. ISO 14001 examines every step of the manufacturing process from each raw material, energy consumption and type, chemical use, packaging, space, location, transportation, and each type of waste generated (American Society for Quality 2024). ISO 14001 essentially states that everything that is not included in the manufactured product eventually becomes a waste.

ISO 14001 is widely used throughout the world, especially in the automobile manufacturing industry. Automobile manufacturers required suppliers to be ISO certified as early as 2002 (Thornton 2001).

ISO 14001 is an environmental management system that has also been used as a tool to evaluate status of environmental compliance but has found more purpose in reducing waste and generating manufacturing efficiencies that translate into business advantages by lowering costs and liabilities and other factors.

Certified B Corporations is an additional method that can steer businesses toward sustainability. Certified B Corporations believe (B Corporations 2024):

- That they must be the change that people seek in the world
- That people and places matter
- That businesses should aspire to do no harm and benefit all through their products, practices, and profits
- That businesses should act with the understanding that each person depends upon another including future generations

Currently, there are hundreds of Certified B Corporations throughout the world with 470 just in the United States (B Corporations 2024).

15.4.6.2 Environmental, Social, and Governance

Environmental, Social, and Governance (ESG) refer to a set of standards used to measure an organization's environmental and social impact. ESG, at its core, is a means by which companies can be evaluated and compared to a broad range of socially desirable standards. ESG describes a set of factors used to measure the non-financial impacts of particular investments and companies.

ESG has an origin from investment philosophies associated with sustainability and socially responsible investing. Early efforts concentrated on identified companies from portfolios largely because of environmental, social, or governance concerns. ESG has three guiding objectives are the following (Yeoh 2022):

- The "E" captures energy efficiencies, GHG emissions, deforestation, biodiversity, climate change, and pollution mitigation, waste management, and water use.
- The "S" addresses labor standards, wages and benefits, workplace and board diversity, racial justice, pay equity, human rights, talent management, and other human financial and social justice parameters.
- The "G" represents governance over environmental and social aspects of ESG. It addresses board composition and structure, strategic sustainability oversight and compliance, executive compensation, political contributions, lobbying, bribery, and corruption.

Even though ESG is focused on companies, it tracks closely with the United Nations goals on sustainability and is a mechanism to involve businesses through the world at setting and achieving sustainability objectives and treating humans and the environment more fairly.

15.4.7 Improving Our Relationship with Nature

In general, humans have not been kind to the Earth. For the most part, humans have modified the planet to suit their perceived needs and for profit. Some of the numerous negative results include species extinction and global pollution. However, there are actions that can be implemented to improve the health of the planet. Many will be described in the next section and in the two remaining chapters of this book. In basic terms, actions at the global level to improve our relationship with nature should focus on the several items discussed in this section that promote building a global community, which include the following:

- Cohesive environmental policy, regulations, and enforcement
- Controlling the human population
- Modifying the human diet
- Financial fairness and equity distribution
- Modifying business models
- Addressing climate change

In addition, there are methods and examples of protecting those natural areas that include national parks and heritage sites throughout the world that can be improved to prevent degradation in their many forms. In addition, wildlife corridors should be expanded so that the health of existing species can be improved by increasing the range and genetic diversity of wildlife (Tallamy 2019). Currently, many national parks operate as genetic islands which threaten the long-term health of species within those islands. By establishing wildlife corridors that could be as small as a vegetated bridge over a highway is an example and provides a means by which wildlife can overcome anthropogenic barriers.

Humans now know that mastering nature cannot occur through force, which may have been a belief as recent as 50 years ago, but by understanding. This concept is not new; many naturalists throughout the history have written and documented the destruction of the natural areas worldwide. One response to this destruction, the United States established national parks so that at least some of the most significant natural areas would essentially be persevered for future generations to study and enjoy.

Coordination of several efforts to understand nature and improve the quality of our air, water, land, and biosphere should be conducted at the highest hierarchy global level.

15.4.8 Addressing Climate Change

As discussed in Chapter 8, the United Nations has played perhaps the most significant role in educating and promoting actions for reducing GHGs throughout the world. The IPCC was established by the United Nations Environment Programme (UNEP) and the World Meteorological Organization in 1988 (IPCC 2021). The IPCC is the United Nations body for assessing the science related to climate change. The IPCC is widely considered the world leader in setting the standards for assessing, mitigating, and reducing GHG emissions (IPCC 2021).

In 1998, a multi-stakeholder group that included IPCC, businesses, nongovernmental organizations (NGOs), and governments formed with the goal of developing internationally accepted GHG accounting and reporting standards. The first standard was published in 2001 and subsequently updated in 2015 (World Resource Institute 2015; USEPA 2005; 2020).

The standard, now commonly known as the GHG Protocol, developed a methodology that measures GHG from a company or other organization that involved three steps termed Scopes. Scope 1 is direct emissions of GHG. Scope 2 is upstream GHG-producing activities such as electrical generation. Scope 3 involves downstream activities such as transportation, product use, waste disposal, travel, and others.

Targets for GHG emission reduction have focused on a 50% reduction by 2030 and to become carbon neutral by 2050. The IPCC must continue to assess the science of climate change and take a lead role in reducing GHG emissions at the global level.

15.5 COORDINATING AND PRIORITIZING SUSTAINABILITY INITIATIVES AT THE GLOBAL LEVEL

From a historical perspective, the global community came together and realized that the degradation of the ozone layer in the upper atmosphere and rising carbon dioxide being responsible for climate change were both from anthropogenic sources. This realization stimulated action that has resulted in coordination and cooperation at the global level to address these negative anthropogenic impacts on our planet from the United Nations. Now 195 nations have signed up to find solutions and keep our planet from overheating and restoring the ozone layer, which we covered in previous chapters. Therefore, the United Nations, World Health Organization (WHO), World Bank, and every nation must play a role in addressing other potentially catastrophic issues facing our species that threaten our existence including protecting the air, the water, and the land. If this can be accomplished, then protection of the biosphere can be achieved.

The following sections list specific subject areas where either improvement must be instituted or be created to address sustainability at the global level. These subject areas discussed include the following (Rogers 2020a, 2020b):

- Science-Based Regulation Improvements
- Modifying Environmental Enforcement Emphasis and Policy
- Ecological Restoration and Preservation
- Managing Natural Resources
- Improving Water Quality
- Improving Water Quality of the Oceans
- Modifying Agricultural Practices
- Limitations on Herbicides and Pesticides
- Controlling the Use of Fertilizers
- Modify Forest Management Techniques
- Improve Cooking Fuels
- Improving Urban Air Quality
- Banning Harmful Chemicals
- Addressing Emerging Contaminants
- Addressing Invasive Species
- Limiting Species Extinction
- Improving Pollution Prevention Measures
- Modify Urban Land-Use Policies
- Addressing Noise Pollution
- Regulating Household Waste
- Addressing Non-Point Source Pollution
- Improve Consumer Products and Packaging
- Sustainability Legislation
- Measuring Sustainability Progress

15.5.1 Science-Based Regulation Improvements

We discovered in earlier chapters that contaminants behave differently and are more prone to release in different geologic, hydrogeologic, geographic, and climatic regions. For instance, regulations specific to geologic risk factors such as earthquakes (e.g., California) or climate change (eastern seaboard and the coast along the Gulf of Mexico) should better reflect those areas where there is the highest risk.

15.5.2 Modify Environmental Enforcement Emphasis and Policy

At this point, it is not apparent that a need for modifying USEPA's approach to enforcement is in need of change, especially as it has been seemingly effective up to this point. However, it is time to move beyond a litigation-heavy and punishment-based enforcement model and enter into more of a reform- and repair-based enforcement policy that emphasizes sustainability.

We will learn in the next chapter that some countries are more advanced compared to the United States in how they have chosen to organize and implement their environmental regulations and where they focus internal resources. In addition, the governments of some countries have different attitudes and relationships with industry. The regulatory agencies of some countries have chosen to form working partnerships with industry with the intent to work more closely together to solve complex environmental issues rather than an adversarial approach which much too often results in polarizing points of view and expensive litigation.

The result of working together saves time, effort, and resources and, in the long run, has resulted in the fact that some countries are better at protecting human health and the environment than the United States.

15.5.3 Ecological Restoration and Preservation

Much has been documented on the destruction of habitat throughout the world from the Amazon rainforest to wildfires in California and wetlands throughout the United States. In addition, almost every country in the world can point to an anthropogenic-induced environment that inflicted significant harm to the environment (Rogers 2019b, 2020a).

We will discuss this topic at each hierarchy level because action is required at each level. At the global level, we focus sustainability efforts on preserving and expanding National Parks throughout the world. In addition, establishing wildlife corridors that connect National Parks has major advantages in preserving species and also expanding the genetic gene pool. This is accomplished such that unconnected National Parks do not become genetic islands (Tallamy 2019).

The United Nations already has a program in place that designates areas as World Heritage Sites. Currently, there are well over 1,000 World Heritage Sites worldwide (United Nations 2024e). The purpose of World Heritage Sites is to appreciate and learn about our legacy from the past, what we live with today, and what we pass on to future generations. Our cultural and natural heritage is both irreplaceable sources of life and inspiration (United Nations 2024a).

15.5.4 Managing Natural Resources

Many of the countries where there has been an environmental incident that has caused enormous harm to human health and the environment were or are related to natural resource exploitation. Examples include the following (Rogers 2020a):

- Exxon Valdez oil spill in Alaska
- Deepwater Horizon oil spill in the Gulf of Mexico
- Poor mining techniques throughout the world
- Poor forest management techniques throughout the world
- Hydroelectric power generation throughout the world
- Farming

These examples highlight the need for common regulatory treatment on a global scale for resource exploitation.

15.5.5 Improving Water Quality

According to the United Nations (2015), the most prevalent water quality problem globally is eutrophication, which is a result of high-nutrient loads (mainly phosphorus and nitrogen). The most significant source of water pollution on Earth is agriculture. Other sources include domestic sewage, industrial effluents, and atmospheric inputs from fossil-fueled burning and wildfires. A significant issue related to water is sanitation and the spread of disease through untreated water with human excrement and from livestock and wildlife called **zoonotic pathogens** (WHO 2013). In addition, the WHO (2013, 2015a, 2015b) and the United Nations (2015, 2016a, 2016b, 2016c) estimate that 2.8 billion people do not have access to proper sanitation, while Baum et al. (2013) estimate that the actual number could be as high as 4.1 billion because most evaluations do not account for all biological disease-causing pathogens.

The WHO lists the top ten cities with the most water pollution in the world as the following (WHO 2024):

1. Sumgavit, Azerbaijan	2. Linfen, China
3. Tianjing, China	4. Sukinda, India
5. Vapi, India	6. La Oroya, Peru
7. Dzerzhinsk, Russia	8. Norilsk, Russia
9. Chernobyl, Russia	10. Kabwa, Zambia

Contaminants that have affected the cities with the most water pollution in the world include the following (WHO 2024):

• Volatile organic compounds (VOCs)	• Biological contaminants including *Escherichia coli*
• Oils and fuel	• Pesticides
• Radioactive materials	• Fertilizers
• Garbage	• Plastic
• Zoonotic pathogens	• Other pathogens

In 2011, the WHO published minimum standards for water quality. The WHO Water Quality Standards are intended as a benchmark for those countries that have not established standards of their own and are considered minimum standards (WHO 2015a). However, much needs to be done with respect to improving water quality throughout the planet. Much of the needed action is installation of infrastructure which many poor countries cannot afford because the population is increasing at a pace that is faster than infrastructure construction. The causes in many cases are exacerbated by social issues, corruption, war, and poverty which may be widespread in some regions (Rogers 2019a).

15.5.6 Improving the Quality of the Oceans

Evaluating the environmental health of the oceans and ocean fisheries is much more difficult because the oceans are large, have many different jurisdictions, and counting the number of fish is difficult even with advanced technologies. However, there is evidence that suggests that ocean fisheries are under stress from over-exploitation. This is evidenced by the fact that the numbers of fish caught by species are declining even though the number of fishing vessels has increased and fishing technologies have improved. For example, in 1974, only 10% of ocean fisheries were considered stressed whereas nearly 30% were stressed in 2010 (World Ocean Review 2024). However, the National Oceanic and Atmospheric Administration (NOAA) of the United States released a report in 2017 that concluded that only 15% of fisheries were stressed, and represents a decrease over previous evaluations (NOAA 2021).

China catches more than double the amount of fish at nearly 14 million metric tons in 2010 than any other nation. Indonesia and the United States rank second and third, respectively, at just over 4 million tons each (World Ocean Review 2024).

Actions required to improve the health of the oceans will require more study because the oceans are so large and the anthropogenic effects are not fully understood.

15.5.7 Modifying Agriculture Practices

Agriculture is now the biggest polluter in the United States and the world (United Nations 2016c). The United Nations estimates that most land pollution is caused by agricultural activities, such as grazing, use of pesticides, fertilizers, irrigation, plowing, and confined animal facilities (United Nations 2016c). Agricultural activities release significant amounts of greenhouse gases from plowing, decaying vegetation, and from livestock. Fugitive emissions and erosion are also significant and release significant amounts of chemicals into the air and the water. In addition, we have also learned that fertilizers are a significant source of pollution to the oceans of the world.

As environmental regulations have been so successful at reducing the harmful effects of industry in the United States, it is now time to reduce the harmful effects on the environment from agriculture.

A method of farming that reduces environmental impacts is termed **regenerative farming.** Regenerative farming is a philosophy of common principles that include re-establishing relationships between humans and land through improving soil health, reducing or eliminating the use of harmful chemicals, growing diverse crops, holistic livestock management, innovative and efficient use of resources, and equitable labor (Natural Resource Defense Council (NRDC) 2024). This type of farming is not new. Indigenous cultures, and even those in the United States in the 19th century, practiced this type of farming before the widespread use of pesticides, herbicides, and extensive use of fertilizers. Current farm and food production in the United States is not designed to prioritize climate, ecosystem's water quality, human health, and establishing a healthy and productive relationship with nature (NRDC 2024).

15.5.8 Limits on Herbicide and Pesticide Use

Simply put, pesticides and herbicides are manufactured to kill things. USEPA (2024e) defines a **pesticide** as preventing, destroying, repelling, or mitigating any pest. An **herbicide** is defined as a substance that is used to kill unwanted plants commonly referred to as weeds. **Pests** are defined as living organisms occurring where they are not wanted or causing damage to crops, humans, or other animals (USEPA 2024e). Examples include the following:

- Insects
- Mice or other animals
- Unwanted plants (weeds)
- Fungi
- Microorganisms, such as bacteria and viruses
- Invasive species that include many plants and animals

Perhaps the most famous of all pesticides is the now-banned substance commonly referred to as **DDT** (dichlorodiphenyltrichloroethane). It became famous as an environmental contaminant after the book *Silent Spring* written by Rachel Carson was published in 1962 (Carson 1962). The book cataloged the environmental impacts of the indiscriminate spraying of DDT in the United States and questioned the logic of releasing large amounts of chemicals into the environment without fully understanding their effects on ecology or human health. DDT was widely used in agriculture and by consumers in the 1940s and 1950s. DDT (CAS Registry Number 50-29-3) is now considered a probable carcinogen (Group: A2) by USEPA (2024f) and is no longer used as a pesticide in the United States after it was banned in 1972. The 2001 United Nations Environmental Program meeting held in Stockholm, Sweden (and put into effect in 2004) permitted its use for "vector control" – organisms that produce pathogens, such as mosquitoes) (USEPA 2024f).

Currently, approximately 1 billion pounds of conventional pesticides and herbicides are used to control weeds, insects, and other pests just in the United States (USEPA 2017; USGS 2024).

15.5.9 Controlling the Use of Fertilizers

Fertilizers are chemical compounds designed to promote plant and fruit growth when applied (USEPA 2024g). The most common fertilizers are nitrogen (N), phosphorus (P), and potassium (K). Nitrogen is found primarily in an organic form in soils, but it can also occur as nitrate. Because nitrate is very soluble and mobile, it is transported by surface water to rivers, lakes, and streams where it can promote algal growth. In many cases the algal growth is extensive. Nitrate can also contaminate drinking water. Phosphorus occurs in the soil in organic and inorganic forms, but being more soluble than nitrate it can also be depleted in the soil through surface water runoff. Phosphorus can also promote algal growth in rivers, lakes, and streams, because it is a limiting nutrient in freshwater. Potassium (K) in fertilizers is commonly incorporated as potash – an oxide form of potassium that includes the compounds potassium chloride, potassium sulfate, potassium nitrate, and potassium carbonate. The term "potash" comes from the historical practice of extracting potassium carbonate (K_2CO_3) by leaching wood ashes and evaporating the solution in large iron pots. Fertilizers containing potassium generally do not promote algal growth (USEPA 2024g).

The United States Department of Agriculture (USDA) has tracked fertilizer use in the United States since 1960. According to the USDA, approximately 63.5 kg (140 pounds) of fertilizer containing N–P–K are applied each year per acre of land farmed and this amount has increased more than 200% since 1960 (USDA 2019). In 2018, USDA estimated that 1.41 million km^2 (350 million acres) were planted for crops which means that approximately 22.2 trillion kg (49 trillion pounds) of fertilizers were applied to farm land in the United States (USDA 2019).

In urban areas, fertilizers containing N–P–K are common and routinely applied to residential lawns and golf courses to help maintain green healthy grass and gardens (USEPA 2024g). According to USEPA (2024h), nitrogen, phosphorus, and potassium fertilizers are not currently known to cause cancer. Exposure to high concentrations may cause nausea and vomiting (USEPA 2024h).

The most common fertilizers include nitrogen (N), phosphorus (P), and potassium (K). A major sink for fertilizers is surface water because they are applied to the soil surface and are considered soluble in water and mobile – especially in the case of nitrates and phosphorus. Once in surface water, nitrate and phosphorus can promote excessive algal growth. Significant algal growth can deplete the dissolved oxygen in surface water and cause suffocation and death to aquatic organisms. The solubility of some fertilizers, combined with the relationship between surface water and groundwater, may lead to groundwater contamination. The natural process of enrichment of surface waters with plant nutrients is termed **eutrophication**. When anthropogenic activities such as fertilization or sewage discharges accelerate this natural process, **cultural eutrophication** occurs (McGucken 2000).

15.5.10 Forest Management

Forest management throughout the planet is considered poor at best. Worldwide, forests provide habitats for 80% of amphibians, 75% of bird species, and 68% of mammal species (United Nations

2024f). Approximately one-third of the world's forests have been lost due to multiple anthropogenic effects. Most of the loss of forests has been due to land-use conversion to agriculture.

From 2011 to 2020, Africa has experienced the most forest loss of 3.9 million hectares with South America next at an estimated loss of 2.6 million hectares (United Nations 2024f). The United Nations has suggested that actions needed to reduce the loss of forests worldwide include modifying human diets, improved governance, and improved and upgraded regulations (United Nations 2024f).

15.5.11 Cooking Fuels

Improving access to affordable and reliable energy services for cooking is essential for poor and developing countries in reducing the harmful health effects from using traditional biomass, including manure and dung, charcoal, coal, and dead vegetation. As of 2011, the United Nations has estimated that 2.64 billion people rely on this cooking method. Besides the obvious increased human health effects, cooking in this manner also emits large quantities of GHGs into the atmosphere (World Bank 2024b).

15.5.12 Urban Air Quality

The CAA has made a major impact in improving the air quality throughout the United States, especially with respect to automobile exhaust, ozone, particulate matter, and other criteria pollutants. According to USEPA (2018), compared to emissions in 1980, carbon monoxide emissions have been reduced by 72%, lead by 99%, nitrogen oxides by 61%, VOCs by 54%, particulate matter (PM10) by 61% and, sulfur dioxide by 89%. However, these improvements achieved in the United States have not had the same effect when applied to many other countries of the world.

Improvements in exhaust from diesel trucks and buses have been slowed and are especially noticeable in urban areas. In addition, other sources of particulate matter are of high concern in urban areas as well and need improvement. This need is in part the result of an increasing population and that urban areas continue to grow as population centers that will inevitably increase automobile congestion and emissions if not addressed as quickly as possible.

Even though the United States is in need of improving urban air, many locations in the rest of the world are far worse when it comes to urban air pollution.

15.5.13 Banning Harmful Chemicals

Consistency in banning certain chemicals known to be harmful worldwide is in need of improvement. Knowing that chemical contaminants do not respect country borders begs for a global approach because even if one country bans certain contaminants, the country still may become contaminated by those same contaminants if they migrate from a nearby country that has not banned the same contaminants. The Toxic Substance Control Act (TSCA) is a set of regulations in the United States that provides a mechanism to ban certain harmful contaminants worldwide.

15.5.14 Emerging Contaminants

Emerging contaminants are chemicals or substances that have been recently discovered in the environment and, in some instances, have caused harm or have impacted resources to the extent that they can no longer be used for fear of health risks. Emerging contaminants include many compounds including per- and polyfluoroalkyl substances (PFAS), PIP, many pharmaceutical compounds, 1,4-dioxane, and others. Many of these emerging compounds are now being detected in the environment and have not been adequately evaluated. This highlights a need for improved prevention measures and the need for improving handling and disposal practices.

15.5.15 Addressing Invasive Species

Some consider invasive species a type of emerging contaminant. For clarity, we treat them separately because there are so many invasive species and their behavior in the environment to so different compared to chemicals and other substances because their mode of migration is through breeding. Invasive species have been present for the last few hundred years, but environmental damage is increasing and in need of further regulatory protections at the global level.

Worldwide, there are thousands of invasive species (United Nations 2018). Invasive species are characterized as aggressive when they are transported into areas where they are not naturally occurring usually because they do not have a natural predator or other organisms have not evolved defenses (United Nations 2018, 2024g; USEPA 2024i).

15.5.16 Limiting Species Extinction

Species extinction is a global issue that is getting worse. Worldwide, there are thousands of species of animals and plants that are in danger of extinction because of direct human involvement that includes development, habitat loss and destruction, poaching, pollution, and other factors. The World Wildlife Fund maintains a general list of those animals that are in danger and also includes an additional category different from that of the United States by the addition of another category called "critically endangered." Some of the animals currently listed by the World Wildlife Fund as critically endangered or endangered include the following (World Wildlife Fund 2024):

- Amur Leopard
- Cross River Gorilla
- Malayan Tiger
- South China Tiger
- Sumatran Rhino
- Asian Elephant
- Bluefin Tuna
- Chimpanzee
- Hectors Dolphin
- Indochinese Tiger
- Sea Lion
- Tiger
- Black Rhino
- Eastern Lowland Gorilla
- Sumatran Orangutan
- Sumatran Elephant
- Sumatran Tiger
- Western Lowland Gorilla
- Blue Whale
- Borneo Pygmy Elephant
- Ganges River Dolphin
- North Atlantic Wright Whale
- Snow Leopard
- Black Spider Monkey
- Bornean Orangutan
- Javan Rhino
- Orangutan
- Mountain Gorilla
- Amur Tiger
- African Elephant
- Bonobo
- Galapagos Penguin
- Indian Elephant
- Red Panda
- Sri Lanka Elephant
- African Giraffe

The above list is just to name a few of the more familiar animals that are now considered threatened with extinction. Unless major efforts from the world community are made immediately, the loss of these animals and countless others is likely inevitable. The list below includes some North American species that have gone extinct since 1500 AD and include the following (International Union for Conservation of Nature (IUCN) 2024):

- Great Auk – 1852
- California Golden Bear – 1922
- Labrador Duck – 1878
- Cascade Mountain Wolf – 1940
- Mexican Grizzly Bear – 1964
- Sea Mink – 1860
- South Rocky Mountain Wolf – 1935
- Carolina Parakeet – 1918
- Heath Hen – 1932
- Bachman's Warbler – 1988
- Passenger Pigeon – 1914
- Ivory-Billed Woodpecker – 1942
- Eastern Cougar – 2011 declared
- Eastern Elk – 1887
- Newfoundland Wolf – 1911
- Caribbean Monk Seal – 1952
- Southern California Kit Fox – 1903
- Dusky Seaside Sparrow – 1987
- Eskimo Curlew – 1981
- Tacoma Pocket Gopher – 1970

According to the World Wildlife Fund (2024) and the IUCN (2024), hundreds of species of animals are now extinct directly due to humans in several different ways including hunting, habitat destruction, poisoning, development, and pollution. According to the United Nations (2024g), more than 1 million species of animals and plants are threatened with extinction as a direct result of human activities.

15.5.17 Improving Pollution Prevention

The Pollution Prevention Act (PPA) and Oil Spill Prevention Act (OSPA), both of 1990, do not get enough credit in the history of environmental regulations of the United States (USEPA 2024a). The PPA and OSPA were the first real efforts to turn from regulating the environment at the "end of the pipe" to promoting the prevention of pollution from entering the environment. One of the driving factors behind the PPA and OSPA was the realization of how much it costs to clean up the environment after a release.

It was a sobering lesson that just one incident could cause so much harm, cost billions of dollars, and, in some instances, could not be cleaned up to acceptable levels because technologies simply did not exist. It was during this time that USEPA and many others discovered that sometimes you can throw money at a contaminated site or incident and make it go away and then there are those where all the money and technology in the world would not be able to fix the problem before enormous harm was a reality. Therefore, out of this realization, USEPA began to focus efforts on pollution prevention, in part because it made common sense, in that it is proactive at environmental protection and lowers response costs much like fire prevention. As we shall see in the next chapter, it was out of these and similar efforts in the European Union that the term "sustainability" took hold.

15.5.18 Urban Land Use

One important item that should be added to pollution prevention efforts is urban land use. There is much that we can accomplish to protect human health and the environment by not locating high-risk activities in areas that are especially sensitive to pollution.

Currently, urban planning for industrial activities is near transportation routes and is generally not located near residential areas. This is a good practice. However, there is often a complete disregard for evaluating geologic and hydrogeologic risk factors that can greatly influence the migration of contaminants through air, water, and soil that ultimately lead to human and ecological exposure from these areas. The examples for Woburn Massachusetts, Hinkley Site in California, Toms River, Love Canal, and Times Beach are just a few of hundreds of examples that highlight this risk but have not been adequately addressed with effective legislation for land-use planning.

15.5.19 Noise Pollution

An additional type of pollution that is not traditionally considered pollution that must be addressed is noise pollution. Noise pollution is regulated in some countries, such as the European Union, China, and Mexico (Rogers 2019a). One may ask how noise pollution relates to a sustainability issue. Recent research indicates that many animal species react to noise and, in fact, noise that travels in the oceans is harmful to whales and dolphins that depend on echolocation to survive (National Geographic Society 2024; United States National Park Service 2024).

15.5.20 Household Waste

Some local efforts have been initiated at educating the public on the potential dangers of some household wastes that include cleaners, solvents, electrical equipment, waste paints, oil and grease, and many other substances. However, much more effort and incentives are needed. The landfills where much of the waste from households is disposed are not as adequately constructed to prevent releases to groundwater from many of the substances that are located within the majority of municipal landfills. In addition, septic tanks are an additional source of pollution to the ground and groundwater and highlight the need for expanding local POTWs so that this untreated waste is not directly discharged into the environment.

15.5.21 Non-Point Source Pollution

As described in the summary of the CWA, non-point source pollution is in need of regulation, especially from rural and agricultural land to minimize the build-up and discharge of pollutants that include fertilizers, pesticides, herbicides, bacteria, and other pollutants into the waters of the United States and causing massive marine deaths in the Gulf of Mexico and perhaps other coastal locations (e.g., Florida). This will be more difficult to regulate in part because it affects areas that are not within the geographical boundary of the United States. The source of much of this pollution originates from non-point sources located in rural and agricultural land within the interior portions of the country.

Environmental regulations concentrate on urban land which only accounts for 2.5% of developed land in the United States (USEPA 2024a). Other areas that include crop land, range land, and managed forest account for 75% of developed land in the United States and are generally not regulated by the USEPA. USEPA has been successful at improving the environmental quality in urban areas to such a degree that pollution effects from non-point sources, which account for 75% of other developed land in the United States, are much more noticeable and in need of regulation and protection.

15.5.22 Consumer Products and Packaging

Plastic has become one of the more important current pollution issues and was the focus of Earth Day 2018 (Earth Day 2021). Plastic has many valuable uses. However, humans have become addicted to single use of disposable plastic. Around the world, 1 million plastic drinking bottles are purchased every minute, and up to 5 trillion single-use plastic bags are used worldwide every year. In total, half of all plastic produced is designed to be used only once and then thrown away. Plastic waste is now so common in the natural environment that some scientists have even suggested that plastic could serve as a geological indicator of the Anthropocene Era (United Nations 2024h). From the 1950s through the 1970s, only a small amount of plastic was produced, which was not a significant issue. By the 1990s, plastic waste had more than tripled. In the early 2000s, plastic output increased more than in the previous 40 years. As of the end of 2018, more than 300 million metric tons of plastic waste is generated per year, which is equivalent to the weight of the entire human population (United Nations 2024h). Although plastic containers and packaging are nearly impossible to regulate by USEPA, there is a clear need to address this issue.

15.5.23 Sustainability Legislation

The United States is in need of a clearer picture of what it means to be sustainable. To many, it is a buzzword that lacks specific goals or even a basic plan for achieving sustainability. The European Union and other countries have established a clearer picture and have set goals for sustainability. When we discuss sustainability later in this book, we will discuss the need to move beyond reactive measures to proactive sustainability measures to improve our environment in the future. If there are delays in enacting sustainability measures, it may be too late. A whole chapter is dedicated to this discussion point.

15.5.24 Measuring Sustainability Progress

A measurable sustainability program for any location in the world may seem impossible because of the sheer number of potential variables that greatly increase the complexity and because of the many differences between countries. However, this is just not the case. Sustainability is defined as creating and maintaining conditions under which humans can exist in productive harmony to support present and future generations (United Nations 2024a; USEPA 2024j). The term "productive harmony" applies to the environment. Therefore, sustainability can be reduced to three variables that include (1) understanding the environment, (2) understanding how aspects of human activities impact the environment, and (3) developing and implementing measures to mitigate unacceptable risk (Rogers 2019a, 2019b).

Understanding the geological, hydrological, and ecological environment is the first step in building a sustainability model. Building a sustainability model requires employing two fundamental equations, each containing three steps. The first term of the first equation is represented by the general term Geologic/Environmental/Anthropogenic-induced hazard. The second step is evaluating the magnitude and risk occurrence of the hazard. The third term is evaluating the sensitivity of the exposed target population or ecological environment The output is termed **impact severity**. The first term in the second equation then utilizes the impact severity term and combines this term with anthropogenic behavioral aspects that affect the environment either through negative or positive outcomes. The third variable in the second equation is evaluating the effectiveness of risk reduction measures. The output of the second equation is termed the sustainability index (see Equations 15.1 and 15.2). The Sustainability Index represents a measure of sustainability for any particular location with a higher value representing increased risk and harm to human health or the environment and, therefore, less sustainable (Rogers 2020b):

$$\text{Geologic/Environmental/Anthropogenic-Induced Hazard} \times \text{Magnitude and Risk of Occurrence} \times \text{Sensitivity} = \text{Impact Severity} \tag{15.1}$$

$$\frac{\text{Impact Severity} \times \text{Behavioral Aspects}}{\text{Risk Reduction Measures}} = \text{Sustainability Index} \tag{15.2}$$

The Sustainability Index can be scaled to evaluate environmental health of a single location or whole ecosystem depending on system inputs and evaluation criteria described in the following sections.

15.5.25 Summary of Improving or Need for New Regulations

The sections highlighted above describe the majority of subject areas where environmental regulations in the United States are either in need of improvement or where new regulations are needed. For reference purposes and to list those areas that have the greatest need for new or improved environmental regulations, the following is provided (Rogers 2023):

- Modify environmental enforcement emphasis and policy
- Science-based improvements to environmental regulations to account for regional risks and climate change
- Agriculture
- Pesticides and herbicides
- Fertilizers
- Septic tanks

- Invasive species
- Urban planning and land use
- Residential and household waste
- Urban air
- Further or new restrictions on harmful chemicals, especially those that are persistent and mobile in the environment
- Emerging contaminants, such as those described in Chapter 2
- Noise
- Plastic
- Sustainability

Many of the items listed above fall into the exceptions scenario that we first touched on and explained in the introduction of this chapter. The main theme with the exceptions is that environmental regulations should address all identified risks for a given practice, situation, or exposure scenario. It makes little sense if human health and the environment are protected from 9 out of 10 exposure risks when the remaining risk factor causes harm. This example is realized when considering the lack of regulations for the items listed above. Another item worthy of discussion is that there are political factors and areas where there are overlapping jurisdictions that may have negative outcomes that must be addressed when discussing the need for further environmental regulations in the United States. For instance, the United States Department of Agriculture would almost certainly interject an interest in the environmental regulation of farm land. In addition, the Department of Interior would also be interested in sustainability regulations as it pertains to national forests and parks and environmental regulations on rangeland.

This will become more evident as we discuss the challenges that are faced by many of the countries of the world in the next chapter.

15.6 SUMMARY AND CONCLUSION

The global-level actions outlined in the chapter are necessary. Achieving succes will require cooperation not often experienced in our short history. We must act as a global community to more fully understand the impacts humans have made on our planet and act together and diligently through focused efforts to limit the negative effects and restore ecosystems where possible. These actions must be undertaken before it is too late not just for the survival of other species but for the survival of our own species as well.

REFERENCES

American Society for Quality (AQS). 2021. *ISO 14001: Environmental Management Systems Standard*. https://asq.org/quality-resources/iso-14001. (Accessed May 5, 2024).

Baum, R. Luh, J., and Bartman, J. 2013. Sanitation: A Global Estimate of Sewerage Connections without Treatment and the Resulting Impact. *Journal of Environmental Science and Technology*. Vol. 47, no. 4, pp. 1994–2000.

B Corporations. 2024. *About B Corporations*. https://bcorporation.net/about-b-corps. (Accessed May 5, 2024).

Cadman, T., Hurlbert, M., and Simonelli, A. C. 2022. *Earth System Law: Standing on the Precipice of the Anthropocene*. Routledge Books. London, United Kingdom. 320pp.

Carson, R. 1962. *Silent Spring*. Houghton Mifflin. Boston, MA.

Earth Day. 2021. *Earth Day Network Campaign: End Plastic Pollution*. https://earthday.or/campaigns/plastic-campaign. (Accessed May 5, 2021).

Harari, Y. N. 2015. *Sapiens*. Harper Collins Publishers. New York, NY. 443pp.

Harari, Y. N. 2018. *21 Lessons for the 21st Century*. Random House Books. New York, NY. 372pp.

Intergovernmental Panel on Climate Change (IPCC). 2021. *Climate Change 2021: The Physical Science Basis*. IPCC AR6 WGI. IPCC. 3949pp.

International Union for Conservation of Nature (IUCN). 2024. *Extinct Species in North America since 1500AD.* www.iucn.org/resources/spatial-data-download. (Accessed February 6, 2025).).

McGucken, W. 2000. *Lake Erie Rehabilitated: Controlling Cultural Eutrophication, 1960s–1990s.* University of Akron Press. Akron, OH.

National Academies of Sciences, Engineering, and Medicine. 2024. *Science and Religion.* www.nationalacademies.org/evolution/science-and-religion. (Accessed June 2, 2024).

National Geographic Society. 2024. *Noise Pollution.* www.nationalgeographicsociety.org/noise-pollution/. (Accessed May 5, 2024).

National Oceanic and Atmospheric Administration (NOAA). 2021. *Status of Fish Stocks 2017.* www.fisheries.noaa.gov/status-stocks-2017. (Accessed May 5, 2024).

Natural Resource Defense Council (NRDC). 2024. *Regenerative Agriculture.* www.nrdc.org/resources/regenerative-agriculture. (Accessed May 5, 2024).

Ohio State University. 2024. *Earth Day 1971. Walt Kelly, Pogo Collection.* OSU Library. https://library.osu.edu/site/40stories "We Have Met the Enemy and He Is Us" – Tales from the Vault: 40 Years / 40 Stories (osu.edu). (Accessed May 5, 2024).

Pope Francis. 2015. Environmental Encyclical and Care for Our Common Home. *The Vatican.* 184p. www.w2.vatican.va/content/francesco/en/encyclicals/documents/papa-francesco-2015024-enciclica-laudato-si.html. (Accessed May 5, 2024).

Rogers, D. T. 2019a. *Environmental Compliance and Sustainability: Global Challenges and Perspectives.* CRC Press. Boca Raton, FL. 583p.

Rogers, D. T. 2019b. Key Policy Changes Required to Reduce Environmental Degradation of the United States and Beyond. In: *International Association of Hydrogeologists 46th Congress.* Vol. 1. Malaga, Spain. International Association of Hydrogeologists (IAH), London, UK. 529p.

Rogers, D. T. 2020a. *Urban Watersheds: Geology, Contamination, Environmental Regulations, and Sustainability.* CRC Press. Boca Raton, FL. 606pp.

Rogers, D. T. 2020b. Geological- and Chemical-Based Environmental Risk Factor Sustainability Model. *European Journal of Sustainability Development.* Vol. 9, no. 4, pp. 303–316.

Rogers, D. T. 2023. *Environmental Compliance Handbook: Sustainability and Future Environmental Regulations.* Vol. 4. CRC Press. Boca Raton, FL. 548pp.

Tallamy, D. W. 2019. *Nature's Best Hope.* Timber Press. Portland, OR 254pp.

Thornton, R. V. 2001. ISO 14001 Certification Mandate Reaches the Automobile Industry. *Journal of Environmental Quality Management.* Vol. 10, no. 1, pp. 89–96.

United Nations. 2015. *International Decade for Action "Water for Life".* United Nations Department of Economic and Social Affairs (UNDESA). New York, NY. www.un.org/waterforlifedecade/quality.shtml. (Accessed May 5, 2024).

United Nations. 2016a. *Global Drinking Water Quality Index and Development and Sensitivity Analysis Report.* UNEP Water Programme Office. Burlington, ON, Canada. 58 pages.

United Nations. 2016b. *World Air Pollution Status.* United Nations News Center. New York, NY. www.un.org/sustainabledevelopment/2016/09. (Accessed May 5, 2024).

United Nations. 2016c. *Human Development Report.* United Nations Development Programme. www.hdr.undr.org/sites/2017. (Accessed May 5, 2024).

United Nations. (2018). *Biodiversity: Invasive Species.* United Nations System-Wide Earthwatch. www.un.org/earthwatch/biodiversity/invasivespecies.html. (Accessed May 5, 2024).

United Nations. 2024a. *What Is Sustainable Development.* www.un.org/sustainabledevelopment. (Accessed May 5, 2024).

United Nations. 2024b. *Global Sustainability Goals.* www.undp.org/content/undp/en/home/sustainable-development-goals.html. (Accessed May 5, 2024).

United Nations. 2024c. *World Population Prospects.* United Nations Department of Economic and Social Affairs. Population Division. https://esa.un.org/unpd/wpp/data. (Accessed May 5, 2024).

United Nations. 2024d. *Funds, Programmes, Specialized Agencies and Others.* www.un.org/en/sections/about-un/funds-programmes-specialized-agencies-and-others/. (Accessed May 5, 2024).

United Nations. 2024e. *Description and Statistics about United Nations World Heritage Sites.* https://whc.unesco.org/en/about/. (Accessed May 5, 2024).

United Nations 2024f. *IAS Network. Current State of Forests of the World.* www.un.org/stateofforests. (Accessed May 5, 2024).

United Nations. 2024g. *Intergovernmental Science-Policy Platform on Biodiversity and Ecosystems Services (IPBES) Summary for Policymakers of the Methodological Assessment Report of the IPBES Scenarios and Models of Biodiversity and Ecosystems Services.* www.ipbes.net/system/tdf/downloads/pdf/spm_deliverable_3c_scenarios_20161124.pdf?file=1&type=node&id=15245. (Accessed May 5, 2024).

United Nations. 2024h. *The Facts on Plastic Pollution.* www.unenvironment.org/beat-plastic-pollution. (Accessed May 5, 2024).

United States Department of Agriculture (USDA). 2011. *Land Use in the United States.* www.ers.usda.gov/data-products/major-land-use. (Accessed May 5, 2024).

United States Department of Agriculture (USDA). 2019. *United States Fertilizer Use and Cost.* USDA. Washington, DC.

United States Environmental Protection Agency (USEPA). 2005. *Climate Leaders Greenhouse Gas Inventory Protocol Design Principles.* Climate leaders greenhouse gas inventory protocol (epa.gov). https://nepis.epa.gov/climateleadership.inventory-management-plan-guidance. (Accessed May 5, 2024).

United States Environmental Protection Agency (USEPA). 2017. *Pesticide Industry Usage.* USEPA Biological and Economic Analysis Division. Office of Pesticide Programs. Washington, DC. 24pp.

United States Environmental Protection Agency (USEPA). 2018. *Air Quality – 2018 National Summary.* www.epa.gov/air-trends/air-quality-national-summary. (Accessed May 5, 2024).

United States Environmental Protection Agency (USEPA). 2020. *GHG Inventorying and Target Setting.* www.epa.gov/GHG-inventorying-an-target-setting=self -assessment/v1.0. (Accessed May 5, 2024).

United States Environmental Protection Agency (USEPA). 2024a. *Summary of Environmental Regulations of the United States.* epa.gov/summary-environmental-laws. (Accessed May 5, 2024).

United States Environmental Protection Agency (USEPA). 2024b. *Regulations for Greenhouse Gas (GHG) Emissions.* www.epa.gov/regulations-emissions-vehicles-and-engines/regulations-greenhouse-gas-ghg-emissions. (Accessed May 5, 2024).

United States Environmental Protection Agency (USEPA). 2024c. *What Is Sustainability?* www.epa.gov/sustainability. (Accessed May 5, 2024).

United States Environmental Protection Agency (USEPA). 2024d. *International Cooperation.* www.epa.gov/international-cooperation. (Accessed May 5, 2024).

United States Environmental Protection Agency (USEPA). 2024e. *Pesticides.* www.epa.gov/pesticides. (Accessed May 5, 2024).

United States Environmental Protection Agency (USEPA). 2024f. *DDT – A brief History and Status.* www.epa.gov/ingredients-used-pesticide-products/ddt-brief-history-and-status. (Accessed May 5, 2024).

United States Environmental Protection Agency (USEPA). 2024g. *Agriculture Nutrient Management and Fertilizer.* www.epa.gov/agriculture/agriculture-nutrient-management-and-fertilizer. (Accessed May 5, 2024).

United States Environmental Protection Agency (USEPA). 2024h. *Integrated Risk Information System.* www.epa.gov/iris. (Accessed May 5, 2024).

United States Environmental Protection Agency (USEPA). 2024i. *Invasive Species.* www.epa.gov/greatlakes/invasive-species. (Accessed May 5, 2024).

United States Environmental Protection Agency (USEPA). 2024j. *National Environmental Protection Act.* www.epa.gov/NEPA. (Accessed May 5, 2024).

United States Geological Survey (USGS). 2024. *Pesticide and Herbicide Use in the United States.* // www.usgs.gov/centers/oki-water/science/pesticides. (Accessed May 5, 2024).

United States National Park Service. 2024. *Can You Hear Me Now.* Managing Underwater Noise Pollution. https://npsorg/articles/canyouhearmenow.htm. (Accessed May 5, 2024).

World Bank. 2024a. *The Fight to Improve Lives of the World's Poor.*www.worldbank.org/en/news/press-release/2013/07/23/improve-lives-world-poor-world-bank-group-delivers-nearly-53-billion-support-developing-countries-fy13. (Accessed May 5, 2024).

World Bank. 2024b. *Household Cooking Fuel Choice and Adoption of Improved Cookstoves in Developing Countries.* Policy Research Working Paper No. 6903. https://worldbank.org/curated/en/wps6903. (Accessed May 5, 2024).

World Health Organization (WHO). 2013. *Water Quality and Health Strategy 2013–2020.* www.who.int/water_sanitation_health/dwq/en/. (Accessed May 5, 2024).

World Health Organization (WHO). 2015a. *Guidelines for Drinking-Water Quality.* Geneva, Switzerland. 516pp.

World Health Organization (WHO). 2015b. *Progress on Drinking Water and Sanitation*. World Health Organization. New York, NY. 90pp.

World Health Organization (WHO). 2024. *United Nations Global Analysis and Assessment of Sanitation and Drinking Water: Indonesia*. www.who.org/drinkingwater/Indonesia. (Accessed May 5, 2024).

World Ocean Review. 2024. *The Global Hunt for Fish*. https://worldoceanreview.com. (Accessed May 5, 2024).

World Resource Institute. 2015. *The Greenhouse Gas Protocol*. World Resource Institute. Washington, DC. 114pp.

World Wildlife Fund (WWF). 2024. *Species List of Endangered, Vulnerable, and Threatened Animals*. www.wwf.org/species/endangered-list. (Accessed May 5, 2024).

Yeoh, P. 2022. *Environmental, Social, and Governance (ESG) Laws, Regulations, and Practices in the Digital Era*. Wolter Kluwer Publishers. 424pp.

16 Risk Evaluation, Building a Sustainability Model, and Mitigation Alternatives

16.1 INTRODUCTION

In this last chapter, we combine what has been presented throughout this book into three discussion points. The first is a review of the geologic, environmental, and anthropogenic risk evaluation processes. The second discussion point focuses on building a sustainability model. The third discussion point summarizes geologic, environmental, and anthropogenic-induced risk impact mitigation alternatives.

Creating an effective sustainability model begins with in-depth knowledge of geology and geologic hazards, environmental hazards, and anthropogenic-induced hazards, especially in urban areas. This information is then integrated within a comprehensive range of anthropogenic activities. Once these two inputs are understood, then an effective strategy can be developed to target specific high-risk hazards and minimize negative impacts to humans and the environment if the risk of exposure to the hazard is unacceptable.

For purposes of this book, a **hazard** refers to a natural geologic, environmental, or anthropogenic-induced event. **Risk** is a measure of the potential magnitude of an event and the frequency of occurrence. Risk can be applied by evaluating the probability of a negative outcome or impact from a geologic, environmental, or anthropogenic-induced hazard source. **Sensitivity** is a measure of how resilient a target population or ecological sector is to the hazard. To put it in terms that perhaps can be better understood, we do not always have control over the hazard, but we can lower our sensitivity of experiencing a negative outcome as a result of knowing and quantifying the risk and implementing appropriate mitigation measures.

16.2 RISK EVALUATION

In this section, we will review and provide clarity on the risk evaluation process. One overriding factor in evaluating any kind of risk is to collect the appropriate data. Using multiple lines of evidence, verifying observations, and collecting additional data to address data gaps are all critical steps. In addition, a robust quality assurance and quality control commitment is key to ensuring that any data set can be relied upon.

Natural geological, environmental, and anthropogenic-induced hazard identification and associated impacts are of key concern to lives and livelihoods across the Earth. Understanding and monitoring geological and environmental processes and using this information to identify and quantify risk potential is critical to improve risk reduction methods so that sustainability potential for survival is greatly improved (British Geological Survey 2024).

Natural geologic, environmental, and anthropogenic-induced hazards can be thought of as three sets of hazards that are related and are linked to one another through geology. One set are geologic hazards and include earthquakes, landslides, tsunamis, volcanic eruptions, coastal hazards, and subsidence and

DOI: 10.1201/9781003463153-19

sinkholes. The second set are environmental risks that are influenced by geology and include extreme atmospheric events such as flooding, tornados, hurricanes, snowstorms, drought, severe thunderstorms, extreme heat and cold, and wildfires. Environmental risks are influenced by the geology of a particular area including topography, specific location, rock types, geomorphology, drainage, vegetation, elevation, and other geographical features. A third set is termed anthropogenic-induced hazards and occur when geological and environmental hazards are ignored or not fully understood or when human activities exacerbate or create risk. In this chapter, we will address each type of hazard.

Equation 16.1, which is also presented in Chapter 5 and Chapter 15, presents the concept of determining the risk of a geologic, environmental, or anthropogenic-induced hazard in more mathematical terms. Each hazard may exist, but the hazard must be realized and must occur within a certain magnitude for the realization of impact. Sensitivity is an expression of exposure to the risk. The outcome is then a measure of the impact severity (Rogers 2020a, 2020b).

$$\text{Geologic/environmental/anthropogenic-induced hazard} \times \text{magnitude and risk of occurrence} \times \text{sensitivity} = \text{impact severity} \quad (16.1)$$

Step 1 of Equation 16.1 is identification of the geologic, environmental, or anthropogenic-induced hazard. For purposes of this example, each hazard should be evaluated independently. Step 2 evaluates the risk of occurrence of the hazard and the range in magnitude. Typically, the magnitude is expressed as the amount of energy released, if measurable and appropriate. The third step represents sensitivity. For most hazards, being within the hazard zone increases sensitivity and the highest risk is usually associated with proximity to the location where the most energy is released.

For example, being at a location near the epicenter of a large earthquake rather than a few hundred kilometers typically would result in increased vulnerability due to the close proximity of the hazard. This also holds true for many other hazards such as tornadoes, hurricanes, volcanic eruptions, and others. However, there are some complicating factors that must be considered and that is exposure to multiple hazards at or about the same time. For instance, exposure to a hurricane does not just include wind speed. It also includes flying debris, flooding due to extreme precipitation and storm surge, and potential tornadoes. Each of these hazards should be considered individually and then should be presented as a value of accumulated impact severity.

16.3 BUILDING A SUSTAINABILITY MODEL

A sustainability model for any location on Earth may seem impossible because of the sheer number of potential variables that greatly increase the complexity and because of the many differences in geology, geography, and urban development variables. However, this is just not the case. Sustainability is defined as creating and maintaining conditions under which humans can exist in productive harmony to support present and future humans generations (United Nations 2020; USEPA 2024a). The term **productive harmony** applies to the environment. Therefore, sustainability can be reduced to three variables that include (1) understanding the environment, (2) understanding how aspects of human activities impact the environment, and (3) developing and implementing measures to mitigate unacceptable risk (Rogers 2019a, 2019b, 2020b).

16.3.1 Sustainability Model Elements and Basic Equation

Understanding the geological, hydrological, and ecological environment is the first step in building a sustainability model and is represented by the general term **impact severity**. Impact severity is obtained from Equation 16.1. The second step is evaluating aspects of human behavior that affect the environment either through negative or positive outcomes. The third variable is evaluating the effectiveness of risk reduction measures. An equation for sustainability is created by combining these fundamental concepts through which a sustainability index as the output (see Equation 16.2). The Sustainability Index represents a measure of sustainability for any particular location with a higher value representing increased risk and potential harm to human health or the environment and therefore, less sustainable (Rogers 2020a, 2020b):

$$\frac{\text{Impact Severtiy} \times \text{Behavioral Aspects}}{\text{Risk Reduction Measures}} = \text{Sustainability Index} \tag{16.2}$$

The Sustainability Index can be scaled to evaluate environmental health of a single location, household, or a whole ecosystem depending on system inputs and evaluation criterion described in the following sections.

As the focus of this book is on evaluating geologic and environmental hazards primarily affecting urban areas, the focus of the sustainability model example presented in the next section is on exposure of environmental contaminants that are released into the environment from anthropogenic sources. Exposure to contaminants in urban areas was chosen because, according to World Health Organization (2024) and the United Nations (2024a, 2024b), exposure to environmental contaminants affects every human and is responsible for the premature deaths of millions of humans every year.

16.3.2 Sustainability Model Methods

The methods to develop and examine geologic sensitivity, climate change impacts, aspects of behavior, and risk reduction and adaptation measures will each be addressed separately and then will be combined into a sustainability model that is tested to evaluate its accuracy and effectiveness at reducing environmental risk caused by anthropogenic activities (Rogers 2020b).

16.3.2.1 Geologic Sensitivity to an Anthropogenic-Induced Risk

Despite centuries of attempting to modify the places where we live and work, humans do not have complete control over the geologic and hydrogeologic environment. Therefore, an increased understanding of the geologic environment is required to minimize or eliminate the potential harmful effects of geologic hazards and contaminants upon human health and the environment. To achieve this goal, an extensive understanding of the geology and hydrogeology is required. This must be followed up by a detailed understanding of the regional and local influence that the geology and hydrogeology play in the influence of contaminant fate and transport (Rogers 2019a).

The largest cities on Earth are dominated by a geologic environment of unconsolidated sedimentary deposits and are located near water. Most of those sedimentary deposits are saturated with water very near the surface, and function as a source of transportation as many of the water bodies are large enough to permit the transport of goods. More importantly for our survival, most of these water sources also provide potable water, including drinking water. However, the presence of water can also enhance the transport of contaminants because water acts as a solvent and can dissolve many pollutants and potentially transport them a long distance. This means that a chemical contaminant released from an anthropogenic source can potentially migrate undetected to a location where an exposure can occur and cause harm to humans and the environment. The basic scientific factors that control the severity of harm to the environment from pollutant releases are (1) the environment and geology where the release occurred, (2) the amount of chemical release and over what period of time, and (3) the physical chemistry and toxicity of compounds released (Rogers 1996; Murray and Rogers 1999a; Kaufman et al. 2005; Rogers 2018).

To evaluate the environment and geology at the locations where a contaminate release has occurred requires a detailed understanding of the geology and hydrogeology in great detail so that an accurate assessment of contaminate fate and transport can be conducted. This type of assessment is commonly referred to as geologic vulnerability analysis. A logical first step in conducting geologic vulnerability analysis is to understand the near-surface geologic and hydrogeologic environment on a local and regional scale.

This information is commonly presented as a geologic map and a geologic sensitivity map. A geologic map provides detailed information of the subsurface with a perspective and concentration of effort toward providing information helpful to determine how contaminants may migrate in the subsurface environment so that a vulnerability map can be constructed. Geologic sensitivity analysis and mapping have been developed by Murray and Rogers (1999a) and Kaufman et al. (2003, 2005).

Central to this method is an objective numerical rating system using weighting coefficients for numerous geologic and hydrogeologic parameters (Rogers 1992, 1996, 2002; Murray and Rogers 1999b; Kaufman et al. 2005). These factors provide detailed subsurface information relevant to construct a geologic sensitivity map within any urban area and is presented as a matrix of sensitivity factors that are rated through a linear scoring scale as shown in Table 16.1 (Rogers 2016a, 2016b, 2020a).

TABLE 16.1
Geologic Sensitivity Matrix and Scoring Scale

Parameter Identification	Parameter Description	Rating Strength
1	**Depth to groundwater**	
	Less than 10 feet below the ground surface	10
	10 to 30 feet	5
	Greater than 30 feet	1
2	**Composition, extent, and thickness of soil units in the vadose zone**	
	Thick and extensive sequence of sand and gravel	10
	Interbedded sands and clay deposits	5
	Thick and extensive sequence of clay	1
3	**Composition, areal extent, and thickness of saturated zone**	
	Thick and extensive sequence of sand and gravel	10
	Interbedded sands and clay deposits	5
	Thick and extensive sequence of clay	1
4	**Occurrence and relative abundance of groundwater**	
	25% or less likelihood before encountering an aquiclude	10
	25% to 74% likelihood	5
	Greater than 75% likelihood	1
5	**Area of groundwater recharge**	
	Significant area of recharge	10
	Moderate area of recharge	5
	Not a significant area of recharge	1
6	**Areas of groundwater discharge**	
	Significant area of discharge	10
	Moderate area of discharge	5
	Not a significant area of discharge	1
7	**Travel time and distance to point of potential exposure**	
	Less than 10 years	10
	10 to 25 years	5
	Greater than 25 years	1
8	**Source of potable water**	
	Current source of potable water	10
	Potential source of potable water	5
	Not a potential source of potable water	1

Source: Rogers, D. T. 2020a. *Urban Watersheds: Geology, Contamination, Environmental Regulations, and Sustainability.* CRC Press, Boca Raton, FL. 615p.

16.3.2.2 Climate Change Sensitivity Risk

Climate change sensitivity can be evaluated within the framework of the sustainability model, but it is recommended to be evaluated separately from other hazards so that specific impacts from climate change are highlighted. In this sustainability model, site and regional specific information is then compared to future potential climate change scenarios to evaluate potential outcomes of climate change impacts. Potential climate change outcomes can then be used to develop adaptive alternatives to lower the severity of risk from climate change impacts. The following sections present a framework to evaluate future potential climate change impacts at any location on Earth.

16.3.2.2.1 Climate Change Data Inputs

Data inputs from each location to be evaluated should include the following information (Rogers and Rogers 2022a):

- Site size
- Maximum building height
- Material of construction
- Age of construction
- Elevation (meters above mean sea level)
- Location and coordinates (longitude and latitude)
- Site history
- Geography
- Geology
- Ecological considerations
- Hydrogeology
- Hydrology
- Nearest surface water bodies and distance
- Local and regional population and demographics
- Urban infrastructure considerations
- Water source
- Wastewater disposal
- Chemical inventory
- Environmental compliance history
- Other historical considerations

16.3.2.2.2 Climate Change Modeling

Global climate models represent Earth as millions of grid boxes and then solve mathematical equations to calculate how energy is transferred between those boxes using the laws of thermodynamics. These models of how energy is cycled through all parts of the planet can be used to estimate dozens of environmental variables, such as wind, temperature, moisture, just to name a few. The models are tested by simulating historical conditions and then matching the results to historical observations. It the models can adequately recreate the past, they are then run forward in time to predict what may happen in the future (National Oceanic and Atmospheric Administration (NOAA) 2024).

The current models of choice for climate change are from the Coupled Model Intercomparison Project Phase 5 (CMIP5) (World Climate Research Programme 2024). Global climate models from CMIP5 are used to evaluate the exposure of any land parcel on Earth. The goal of the CMIP5 is to better understand past, present, and future climate changes arising from natural, unforced variability or in response to changes in radiative forcing in a multi-model context. The understanding includes assessments of model performance during a selected historical period and quantifications of the causes of the spread in future projections. Idealized experiments are also used to increase understanding of the model responses. In addition, to these long-time scale responses, experiments are performed to investigate the predictability of the climate system on various time and space scales as well as making predictions from observed climate data.

To model climate projections of any one location, several models available under the CMIP5 program should be utilized and should be referenced against the ensemble mean results and inter-model variance from ten or more statistically downscaled climate models and two or more climate projection scenarios. Future projections can be calibrated for 20, 40, 60, or 80 into the future for each climate projections scenarios such as reasonable best case and reasonable worst-case climate simulation using a Representative Concentration Pathway (RCP) from 4.5 to 8.5 (NOAA 2024).

16.3.2.2.3 Climate Change Modeling Results

Figure 16.1 presents the results of a model simulation using a single model run for temperature variation compared to the 20th century average worldwide. Two model run scenarios were simulated, one using an RCP of 4.5 (represented as the image on the left) and the other using an RCP of 8.5 (represented as the image on the right). The figure shows that global temperature rise under the reasonable best-case scenario is moderate, whereas under the reasonable worst-case scenario, the temperate rise globally is much more significant.

Model runs can be simulated for several climate change variables the include the following (NOAA 2024):

- Temperature extremes
- Hurricane intensity
- Tornado intensity
- Drought
- Flooding
- Extreme precipitation events
- Sea-level rise
- Sea-surface temperatures

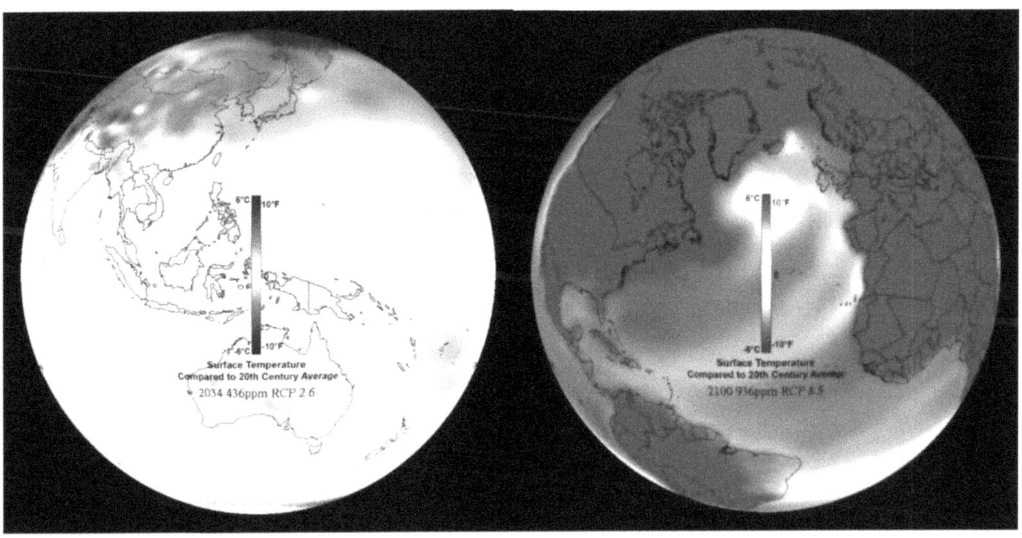

FIGURE 16.1 Model simulation results for a single model run for temperature variation compared to the 20th century average worldwide using RCP-4.5 on the left and RCP-8.5 on the right. (From National Oceanic and Atmospheric Administration (NOAA). 2024. *Climate Model: Temperature Change (RCP 4.5)- 2006-2100 and RCP-8.5 -2006-2100*. https://sos.noaa.gov/datasets/climate-models. (Accessed May 7, 2024).)

16.3.2.3 Behavioral Aspects

Aspects of behavior are divided into two separate activities. The first is evaluating the potential risk posed by contaminants used or stored at any location and the second includes aspects of human activities that may increase or decrease the potential for a release to the environment.

16.3.2.3.1 Contaminant Risk Factors

A required variable in understanding how a specific contaminant will behave in the environment originates from the physical chemistry of each individual chemical. The physical chemistry of a specific chemical can vary greatly and can translate into a very different behavior when released into the environment. Therefore, the risk posed by each different chemical contaminant can vary by more than eight orders of magnitude in some circumstances. With so much chemical variability, there is nearly an infinite number of outcomes and in many circumstances an extremely toxic contaminant may not present as much risk as a moderately toxic contaminant if it is released at the right location, in the right amount, and at the right time. This scenario commonly occurs if a more toxic chemical does not migrate and degrades quickly, and the less toxic chemical has higher mobility and persistence and lasts for decades before degrading. However, if an extremely toxic chemical were released in a heavily populated area, it can have the potential to inflict enormous harm as evidenced by the incident in Bhopal, India, where a chemical was released and inflicted casualties in the thousands in a short period of time (Rogers 2019b).

An additional factor when evaluating chemical behavior in the environment is to consider how each contaminant behaves in air, water, and soil. The behavior of a chemical in the environment is influenced by factors that include solubility, vapor pressure, density, chemical stability, persistence, and adsorption potential (USEPA 1989, 1996; United States Geological Survey (USGS) 2006; Rogers 2020b).

Toxicity is an important factor when examining risk but should not be the only factor when assessing the risk of a chemical released into the environment. Mobility and persistence are very often just as important, if not more important, because these two factors directly influence a chemical's ability to migrate in the geologic environment to a point of exposure. The environmental risk posed by a specific chemical compound to contaminate the air, the water, or the land, termed **Contaminate Risk Factor** (CRF), is a function of toxicity, mobility, and persistence (Kaufman et al. 2005; Rogers et al. 2007). In general terms, the risk to humans or the environment only exists if a contaminate pathway is completed. In general terms, the CRF is expressed in Equation 16.3 (Kaufman et al. 2005; Rogers et al. 2007).

$$\text{CRF} = \text{Toxicity} \times \text{mobility} \times \text{persistence} \tag{16.3}$$

Toxicity values are continually updated by the United Stated Environmental Protection Agency (USEPA) Integrated Risk Information System (IRIS) (2024b, 2024c). To guard against a false-negative response, the toxicity value selected should reflect the highest toxicity value for any exposure pathway—ingestion, dermal adsorption, and inhalation regardless of whether its carcinogenic or chronic. Mobility is calculated by multiplying Henry's Law constant and the retardation factor. Calculating mobility is presented in Equation 16.4 (Kaufman et al. 2005; Rogers et al. 2007):

$$M = (H)\,(R) \tag{16.4}$$

where M = mobility
H = Henry's Law constant
R = retardation factor

Henry's Law constant (H) (atm. mol^{-1} m^{-3}) is very useful in evaluating the mobility of a specific chemical as it is a measure of the tendency of a chemical to evaporate in the atmosphere and also is used to evaluate water solubility. It is a measure of the tendency for substances to volatilize (Sander 1999). Henry's law is related to vapor pressure and is expressed as Equation 16.5 (Kaufman et al. 2005; Rogers et al. 2007):

$$H = (\text{VP})\,(\text{MW})\,(W_s) \tag{16.5}$$

where VP = vapor pressure
MW = molecular weight
W_s = water solubility

The retardation factor is presented in Equation 16.6.

$$R = 1 + \frac{(\rho\text{b})(\text{Kd})}{\eta} \tag{16.6}$$

where R = retardation factor
ρb = bulk density of aquifer matrix (g/cm^3)
Kd = distribution coefficient (mL•g^{-1})
η = effective porosity (calculated as a percent value)

The distribution coefficient is presented in Equation 16.7.

$$\text{Kd} = (\text{Foc})\,(\text{Koc}) \tag{16.7}$$

where Kd = distribution coefficient
Foc = organic carbon partition coefficient (kg/kg)
Koc = fraction of total organic carbon in soil (l/kg)

Values for the fraction of organic carbon should be collected in the field. If field collection is not possible, standard values and ranges can be obtained from USEPA (1996), Wiedemeier et al. (1999), USEPA (2002a, 2002b), and Suthersan and Payne (2005). The retardation factor represents the ratio between groundwater and contaminant migration in groundwater or other media. Therefore, if the value calculated equals 1, this implies that both are traveling at the approximate same rate (USEPA 1989, 2002b). The CRF for water is calculated in Equation 16.8 by multiplying the inverse of the chemical compound's Toxicity (T) by the inverse of its Mobility (M) and its Persistence (P):

$$\text{CRF}_{\text{WATER}} = \frac{1}{(\text{T})} \times \frac{1}{(\text{M})} \times (\text{P}) \tag{16.8}$$

where $\text{CRF}_{\text{water}}$ = Contaminant Risk Factor for Groundwater, and
T = Toxicity
M = Mobility
P = Persistence

The inverse of the toxicity value must be used because the integer values assigned for toxicity decrease with increasing carcinogenicity (USEPA 2024b, 2024c). The inverse of the mobility values must also be used because the calculated values of retardation decrease with increasing mobility.

The CRF for soil is calculated in Equation 16.9 by multiplying the inverse of the chemical compound's Toxicity (*T*) by its Mobility and its persistence:

$$CRF_{soil} = \frac{1}{(T)} \times (M) \times (P) \quad (16.9)$$

Atmospheric contaminants are released to the atmosphere as a gas and as particulate matter, and sometimes as a water droplet (USEPA 2024d).

The CRF_{air} is expressed as Equation 16.10 (Kaufman et al. 2021; Rogers et al. 2012, Rogers 2018).

$$CRF_{air} = Toxicity \times [(mobility_{gas} \times persistence_{gas}) + (mobility_{particulate} \times persistence_{particulate})] \quad (16.10)$$

Mobility for air is represented separately for contaminants in the gas phase and for solid particulate matter and is combined with persistence. The mobility of atmospheric contaminants in the gas phase is presented in Equation 16.11:

$$M_{AIR} = (H) \quad (16.11)$$

where M = mobility
H = Henry's Law constant

The mobility of a contaminant sorbed to particulate matter is represented by Equation 16.12:

$$M_{particulate} = \frac{1}{SpG} \times Koc \quad (16.12)$$

where $M_{particulate}$ = mobility of a contaminant sorbed to particulate matter
Spg = specific gravity
Koc = partitioning coefficient

The CRF for air is presented in Equation 16.13:

$$CRF_{AIR} = \frac{1}{(T)} \times \left[\left(M_{gas} \times P_{gas}\right) + \left(M_{particulate} \times P_{particulate}\right)\right] \quad (16.13)$$

where CRF_{air} = Air Contaminant Risk Factor
T = Toxicity
M_{gas} = Mobility of a contaminant in the gas phase
P_{gas} = Persistence of a contaminant in the gas phase
$M_{particulate}$ = Mobility of a contaminant sorbed to particulate matter
$P_{particulate}$ = Persistence of a contaminant sorbed to particulate matter

CRF analysis provides valuable information for assessing and predicting contaminate behavior anywhere in the environment as demonstrated by examining Table 16.2, which lists CRFs for many common contaminants of the world. Analysis of the values listed in Table 16.2 demonstrates the variability of risks posed by contaminants released in different media. For instance, the variability of polychlorinated biphenyls (PCBs) in the environment ranges by more than 7 orders of magnitude.

TABLE 16.2
Contaminant Risk Factors for Water, Soil, and Air for Each Contaminant Group

Contaminant Type	Groundwater CRF (CRF_{water})	Soil CRF (CRF_{soil})	Air CRF (CRF_{air})
DNAPL	978.00	1,508.00	67.52
LNAPL	0.47	10.16	20.23
PAHs	0.001	142.00	124.82
Total PCBs	0.0004	5,264.00	37.00
Chlordane	0.0001	7,501.00	173.00
Arsenic	27.24	3,723.00	44.00
Chromium VI	1,926.00	5.17	210.00
Lead	0.55	52.44	3.45
Mercury	2.56	22,013.00	166.00

DNAPL, volatile organic compounds heavier that than water; LNAPL, volatile organic compounds lighter than water; PAHs, polynuclear aromatic hydrocarbons, PCBs, polychlorinated biphenyls.

16.3.2.3.2 Aspect Risks of Behavior

Risks can be characterized into two measured variables: aspect risks of behavior and risk reduction actions. Much of this information can be obtained from data contained in an environmental audit. The recommended point values for each attribute are presented within a scaled range. Higher point values indicate higher relative risk. Categories included in the checklist follow the environmental audit guidelines and are as follows (Rogers 2018, 2019a, 2019b):

- Air (0 to 7 points)
- Solid and Hazardous Waste and Landfills (0 to 7 points)
- Water (0 to 7 points)
- Spills (0 to 5 points)
- PCBs (0 to 5 points)
- Toxic Substance Control Act or Equivalent (CRFs) (0 to 5 points)
- Community Right to Know or Equivalent (0 to 5 points)
- Tanks (0 to 3 points each)
- Asbestos (0 to 3 points)
- Regulatory Inspections (0 to 5 points each)
- Site Inspection (0 to 3 points on each item noted)

USEPA defines conducting an environmental audit as a systematic evaluation to determine the conformance to quantitative specification to environmental laws, regulations, standards, permits, or other legally required documents (USEPA 2024c). This definition is consistent with environmental audits in most countries of the world including China, the EU, and others.

Conducting an environmental audit is the first step on a long and winding road toward environmental compliance and ultimately can lead to the path of sustainability and environmental stewardship. Most large companies with several manufacturing locations conduct environmental audits on a regular basis. Some choose to conduct environmental audits more often than others, but most seem to choose conducting an environmental audit every year or every 2 years. Some companies conduct environmental audits with in-house staff to retain institutional knowledge. Some choose to retain an environmental consulting firm to conduct environmental audits because it's conducted by an independent third party.

Some environmental audits are conducted without any prior notice and some are conducted with prior notice. Typically, an audit is conducted by an environmental professional and much of the information from an environmental can be used to evaluate the environmental risks posed at any facility or property. The objectives of an environmental audit include (1) verifying compliance with environmental requirements, (2) evaluating the effectiveness of in-place environmental management systems, and (3) assessing risks posed from regulated and unregulated materials, chemicals, and practices. From a fundamental point of view, the typical environmental audit encompasses evaluating subject areas that include evaluating how behavior at a facility impacts the environment through the land, water, and air.

If the environmental audit is announced prior the audit date, a request for applicable documents to be available to be reviewed in advance audit will save time. As applicable to the facility being reviewed, the following list of documents should be available for review during the site visit:

1. Current site map
2. Current organization chart
3. Current process flow diagrams depicting inputs, process units, waste streams, and other outputs
4. List of processes that have been shut down since the last audit
5. List of processes that are expected to start up with the next 12 to 24 months
6. Spill Prevention Control and Countermeasures (SPCC) Plan
7. Spill reports for the last 3 years
8. All Resource Conservation and Recovery Act (RCRA) manifests since the last audit
9. All special waste notifications and shipping documents since the last audit
10. All analytical tests results and determination for special and hazardous waste
11. All analytical tests for each non-hazardous waste stream
12. Contingency plan
13. Personnel training records since the last audit
14. List of arrangements with local agencies
15. National Pollution Discharge Elimination Plan (NPDES) permit and most recent application
16. Discharge monitoring reports and any associated letters of explanation of permit limit exceedances since the last audit
17. Stormwater plan, intent to comply documents, and all monitoring reports and sample data
18. All notices of violation, noncompliance letters, or consent orders including facility's noncompliance explanations and, if in significant noncompliance, the facility's plan to return to compliance
19. Water flow diagrams and water balances
20. Sketches of all wastewater treatment systems within the facility
21. Names of wastewater analytical laboratory, a list of all analytical methods used for wastewater analysis, and the quality assurance/quality control procedures used for wastewater sample collection and analysis
22. Air permit application modifications
23. All construction and operating permits for air sources
24. Copies of all visible emission observations taken by facility personnel or contractors for the past 3 years
25. Environmental management plan
26. Any other environmental permit to operate (associated with air, land, or water)
27. Waste shipment summary for the past 2 years
28. Toxic Release Inventory Reports (Form R) for the past 2 years

For facilities not in the United States, and as applicable to the specific country, province or state, additional documents to be made available to the reviewers include the following:

1. Noise studies and noise compliance (e.g., Mexico, China, etc.)
2. Environmental Impact Assessment
3. Operating Permits
4. European Union Registration, Evaluation, Authorization, and Restriction of Chemicals (REACH) certification status for those locations within the European Union.

As demonstrated by the list above, an environmental audit, if conducted correctly, should provide enough technical information to conduct an environmental risk evaluation.

16.3.2.4 Constructing the Sustainability Model

Three distinct elements have now been developed that can be combined into a model for sustainability that includes the following:

1. Understanding the natural environment
2. A detailed analysis of the aspects of behavior
3. Evaluating the effectiveness or ineffectiveness of risk reduction actions

Combining these three elements we arrive at an equation with an output that is termed the Sustainability Index and is presented in Equation 16.14 (Rogers 2018, 2020a, 2020b). Note that productive actions to reduce risk lower the overall Sustainability Index by using the inverse.

$$\frac{\text{Impact Severtiy} \times \text{Behavioral Aspects} \times 1}{\text{Risk Reduction Measures}} = \text{Sustainability Index} \qquad (16.14)$$

Examining the method for calculating the Sustainability Index, a higher Sustainability Index value indicates higher risk. The CRF is a critical value and appears in many places within the Operational Aspects variable. The CRF appears at several locations that include air, water, solid and hazardous waste, Emergency Planning and Community Right-to-Know Act (EPCRA), Toxic Substances Control Act (TSCA), and pollution prevention efforts. Critical in this evaluation are chemicals and their behavior when released into the environment that includes each chemical's toxicity, persistence, and mobility. Each of these factors is important especially when combined with geologic vulnerability because it can be used as a predictor to evaluate whether a release of any particular chemical will cause harm and at what scale.

CRFs are also part of the evaluation for the risk reduction measures. As CRFs are part of two of the three elements of the sustainability model (operational aspects and risk reduction measures), it ensures that the Sustainability Index that is calculated results in a higher risk if chemicals are used at a specific facility that have high CRFs. In contrast, eliminating chemicals with high CRFs will have a significant positive effect at lowering risks and greatly reducing the Sustainability Index.

16.3.2.5 Sustainability Model Application

To evaluate whether the Sustainability Index can actually be effective at identifying hazard risk and subsequently lowering hazard risk through mitigation efforts, increasing sustainability, and achieving environmental stewardship, it was implemented and put through rigorous testing. Two scenarios were selected to test the model. One scenario evaluated environmental contaminant hazard risk at 67 manufacturing sites in 12 countries over a period of 14 years. The other evaluated hazard risks due to future climate change impacts. The following section describes the findings of each scenario.

16.3.2.5.1 Contaminant Hazard Risk Assessment

For a period of 14 years that included 7 two-year evaluation periods, as many as 67 manufacturing facilities in 12 countries were tracked using the Sustainability Model outlined in this chapter. Some

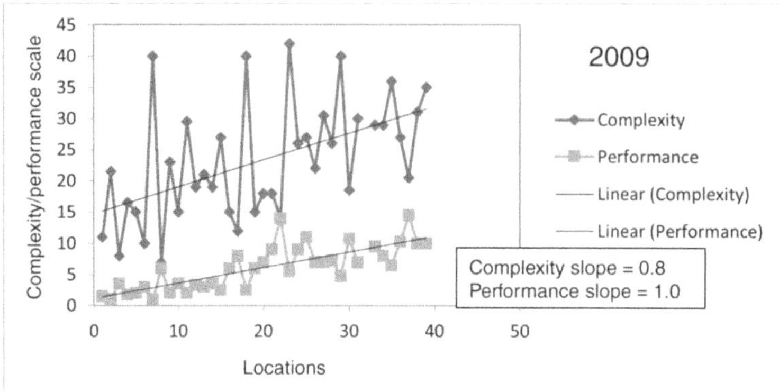

FIGURE 16.2 Complexity and performance results in 2009.

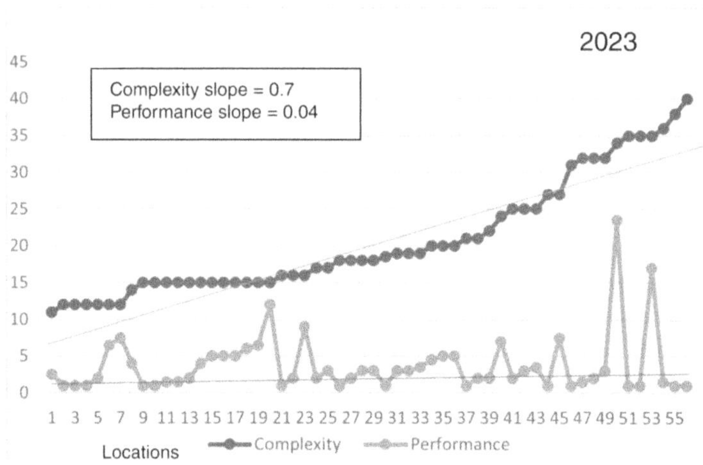

FIGURE 16.3 Complexity performance results in 2023.

facilities were not evaluated during all seven periods because of mergers, acquisitions, divestitures, new construction, or were shut down.

An environmental audit was used to gather the necessary information to score each facility's Operational Aspects and Risk Reduction Measures and was conducted at least every 2 years. Figures 16.2 and 16.3 show a graphical representation of Aspects of Behavior and Risk Reduction Measures for each facility during each of the 2 years evaluation periods for 2009 and 2010 and then from 2018 and 2019, respectively.

Upon examination of Figures 16.2 and 16.3, the Aspects of Behavior did not vary and only decreased slightly during the 14-year evaluation period. However, increased performance and attention to Risk Reduction Measures were significant.

The main reason that Behavioral Aspects did not appear to improve significantly is because many Behavioral Aspect improvements were offset by expanding operations which tended to increase complexity. Another measure of Risk Reduction Measures is to compare the number of regulatory inspections with the number of Notice of Violations (NOVs) issued by regulatory authorities. The number of inspections remained relatively constant, but the number of NOVs decreased by approximately 75%. This positive outcome is due to a significant increase in Risk Reduction Measures that also reduced Behavioral Aspects. The Risk Reduction Action also represents initial efforts to create and improve sustainability initiatives. Some manufacturing locations have achieved nearly a zero waste-generating status of solid or hazardous waste and have also achieved decreases in water use by more than 95%.

The Sustainability Model has undergone 14 years of testing at as many at 67 manufacturing locations in 12 different countries of the world. Over the 14-year period, improvements in Risk Reduction Measures have been realized by an average of 80% resulting in a significant reduction in environmental noncompliance and releases of hazardous substances to the environment and greatly improve the sustainability of each facility (Rogers 2019a, 2019b; Rogers and Rogers 2020b). Efforts to reduce Operational Aspects were largely offset by new or more strict environmental regulations, facility expansion, new equipment, or increased production that required additional permitting and regulatory limitations. However, even with the offsets, slightly more than 50% of facilities reduced Operational Aspects.

The most significant challenge during the 14-year implementation and evaluation period was changing cultural attitudes and behaviors. This highlights the difficult actions that must be addressed to change cultural attitudes and behaviors toward Earth (Rogers 2019b).

The success of implementing a sustainability program could not have occurred without acknowledging and overcoming cultural attitudes and behaviors ingrained in most western capitalist organizations which is a constant drive for growth and profit first and everything else second. The issues and challenges surrounding and weaved within the solutions for overcoming pollution will take commitment and sacrifice from each of us. One person may not make much of a difference but collectively we can and must (Rogers 2019b). Changing attitude and culture requires several steps and involves all organization levels. Those steps include (Rogers 2019b, Rogers and Rogers 2022a, 2022b) the following:

1. A corporate culture of respect and commitment to establishing and maintaining environmental stewardship for the Earth
2. Education and training
3. A clear plan with stated purpose and goals
4. Incentives, rewards, and penalties
5. Measuring and recording performance
6. Data analysis
7. Plan adaptations including additions or subtractions and adjusting timelines

16.3.2.5.2 Climate Change Hazard Risk Assessment

Global climate models from the Coupled Model Intercomparison Project Phase 5 (CMIP5) were used to evaluate the exposure manufacturing operations of a United States-based multinational manufacturing company to future climate change impacts.

Regional geographic and site-specific data from 110 manufacturing sites in 18 countries were collected and referenced against the ensemble mean results and intermodel variance from an ensemble of 10 statistically downscaled climate models and 2 climate projection scenarios (RCP 4.5 and RCP 8.5) for four 20-year periods that were calibrated to the years 2040, 2060, 2080, and 2100. Each of the 10 models used simulated a reasonable best-case (RCP 4.5) and reasonable worst-case (RCP 8.5) scenario. Therefore, a total of 80 separate model runs were conducted. Results were calibrated and compared to the 30-year climate average at each location (Rogers and Rogers 2022b).

The climate aspects evaluated included the following (Rogers and Rogers 2022b):

- Mean atmospheric temperature
- Mean precipitation
- Sea-level fluctuations
- Drought potential
- Severe atmospheric weather events
- Wildfire potential

Data collected from each location included (Rogers and Rogers 2022b) the following:

- Site size
- Maximum building height
- Material of construction
- Age of construction
- Elevation (meters above mean sea level)
- Location and coordinates (longitude and latitude)
- Site history
- Geography
- Geology
- Ecological considerations
- Hydrogeology
- Hydrology
- Nearest surface water bodies and distance
- Local and regional population and demographics
- Urban infrastructure considerations
- Water source
- Wastewater disposal
- Chemical inventory
- Environmental compliance history
- Other historical considerations

The results of the collective model runs indicated the following (Rogers and Rogers 2022b):

- Every location experienced a projected increase in both maximum and minimum daily temperatures
- Thirty five of the 110 experienced a projected a significant increase in both maximum and minimum daily temperatures
- Four of the 110 locations experienced increased risk from sea-level rise
- Twelve of the 110 locations experienced increased risk of drought
- Seventy five of the 110 locations experienced increased risk of extreme precipitation events
- Forty five of the 110 locations experienced risk of increased severity of weather events that included severe thunderstorms, tornadoes, and hurricanes
- None of the location indicated a significant increases risk from wildfires

The model findings for the climate change risk hazard assessment of future potential impacts on these 110 locations modeled were then used as severity inputs to the sustainability model represented in Equation 16.14 in Section 16.3.2.4. Sustainability model results revealed important considerations for climate adaptation plans for each site. The most significant identified parameters that impacted climate change sensitivity included location, elevation, local and regional geology, topography, geography, hydrology, materials of construction, density of local populations, and infrastructure considerations. The results were used for long-term strategic planning to employ climate change mitigation efforts for each location.

16.4 MITIGATION ALTERNATIVES

The following subsections will review mitigation alternatives for each geologic and environmental large-scale hazard. They include the following (Rogers 2019a, 2023):

- Extreme atmospheric events
- Species extinction, abundance, and habitat destruction
- Ocean pollution
- Deforestation
- Energy and climate change
- Agriculture

16.4.1 Mitigating Impacts from Extreme Atmospheric Events

Extreme atmospheric risks include drought, hurricanes, tornadoes, extreme precipitation events, extreme temperatures, flooding, severe thunderstorms, severe winter weather, lightning, hail, high winds, and wildfires. In this section, we will collectively address each of the list listed extreme atmospheric events. To minimize impacts the following are recommended (FEMA 2025):

- Construct maps that identify area of highest risk and potential impacts
- Monitor the occurrence of extreme atmospheric events and modify hazards maps as necessary
- Monitor water supplies
- Consider water conservation measures
- Prevent or minimize overgrazing
- Enhance landscaping
- Reduce urban heat island effects by planting more trees, installing vegetated roofs, use cool roofing products
- Limit urban development in hazard areas
- Incorporate mitigation impact measures into building codes
- Improve stormwater management planning
- Adopt policies to reduce stormwater runoff
- Elevate to retrofit structures and utilities
- Protect infrastructure
- Protect critical facilities, such as hospitals and emergency response structures
- Consider the construction of flood and sea walls, dams, dikes, levees, and other engineering measures
- Restore natural flood mitigation measures such as wetlands and estuaries
- Increase awareness and education
- Prepare emergency and medical supplies
- Protect critical facilities and equipment from lightning
- Conduct lightning awareness programs
- Protect power lines from wind and storm damage potential
- Improve land-use planning and regulations from severe atmospheric events including wildfires
- Encourage the construction of safe rooms
- Encourage fire-resistant construction techniques and materials
- Create defensible space around structure for wildfires and other severe atmospheric events

16.4.2 Addressing Species Extinction, Abundance, and Habitat Destruction

Creating a living landscape with indigenous species instead of a grass lawn along with other actions will create more habitat, increase biodiversity, and improve ecosystem productivity (Darke and Tallamy 2019). Many ecologists, entomologists, wildlife biologists, and other scientists have a term for grass lawns, and they refer to them as "The Green Desert." This is largely because the grass used for lawns in the United States is not indigenous. It arrived in the United States from Europe in the early period of colonization of North America. Indigenous insect species have not adapted to these grasses and generally do not consider them a food source, so they tend to ignore them. If the insects ignore them, then so do the birds that feed off the insects. This this begins a cycle of abandonment and ecosystem decline (Darke and Tallamy 2019).

Restoring habitat for ecosystem diversity requires many considerations to be examined to construct the richest habitat possible. A way to think of building a healthy diverse habitat is to think in layers. The layers include the canopy, understory trees, shrubs, herbaceous plants, the ground layer, the dynamic edge, wet edges, wetlands, and below-ground considerations. The next step after an individual living landscape is created is to invite neighbors to construct their own living landscapes and connect them. This creates wildlife corridors and will attract a multitude of birds and other mammals (Tallamy 2019; Darke and Tallamy 2019).

The following presents a list of action that the individual and local community can undertake to increase species habitat and improve biodiversity, and ecosystem productivity (Rogers 2023):

- Establish plant and wildlife corridors on individual home sites and coordinate with neighbors to establish larger and connected corridors
- Plant a variety of native tress and understory plants to increase food and shelter options for insects, birds, and other wildlife
- Plant milk weed
- Compost yard waste and food waste
- Limit or eliminate pesticide and herbicide application and use
- Limit fertilizer use
- Consider collecting rainwater from roofs to water plants and vegetation during dry periods
- Join a local garden club
- Plant a vegetable garden
- Plant additional flower gardens or other native species
- Consider installing rain barrels on gutter downspouts for watering gardens and plants
- Purchase organically grown and ecofriendly food
- Become more ecologically educated
- Plant a butterfly garden
- Plant indigenous or native species in gardens
- Plant indigenous or native species rather than grass
- Leaf and understory plant and vegetative matter should remain during late fall and winter to improve the number and health of insect populations in the spring.

16.4.3 Addressing Ocean Pollution

At the individual or local level, addressing ocean pollution is a monumental task. Currently, the biggest threat to the oceans is generally considered plastic pollution. Historically, the oceans were used as the dumping ground and human privy for centuries (Rogers 2023). With plastic production increasing, low levels of recycling, and poor waste management, and between 4 and 12 million metric tons of plastic dumped into the oceans yearly, the outlook is not anticipated to improve (Rogers 2023). Plastic pollution impacts sea turtles, whales, seabirds, fish, coral reefs, and countless other marine species and habitats. Scientists estimate that more than half of the world's sea turtles and near every seabird on Earth have plastic in their bodies (Ocean Society 2024).

The following presents a list of action items that the individual and local community can undertake for preventing ocean pollution (Rogers 2023):

- Donate to charitable organizations that focus on cleaning the ocean of pollution
- Reduce use of single-use plastic products

- Support legislation to curb plastic production and waste
- Recycle properly
- Participate in beach and river clean-up activities
- Avoid products with microplastics
- Spread the word
- Support organizations addressing plastic pollution
- Recycle plastic, paper, glass, and aluminum (contact for municipality for additional information and recycling opportunities)
- Utilize reusable shopping bags
- Become more educated
- Contact your local and state-elected officials and encourage them to support sustainable legislation
- Avoid purchasing food in individual serving sizes
- Avoid purchasing items that are considered single use
- Plant trees

16.4.4 Deforestation

As discussed in Chapter 7, deforestation is one of the primary causes of habitat loss on Earth. Deforestation is a leading cause of reduction in species abundance, endangering species, and species extinction. The following presents a list of action items that the individual and local community can undertake for preventing ocean pollution (Rogers 2023):

- Plant trees
- Consider plant native vegetation rather than grass
- Do not purchase products that contain palm oil
- Do not burn wood excessively
- Consider downsizing your home
- Utilize reusable shopping bags
- Reduce meat consumption
- Print documents only when required
- Use recycled paper
- Choose to purchase items with minimal packaging material (remember if you don't consume or use it, it becomes a waste)
- Recycle plastic, paper, glass, and aluminum (contact for municipality for additional information and recycling opportunities)
- Receive and pay bills online
- Call companies that participate in junk mail advertisements
- Contact your local and state-elected officials and encourage them to support sustainable legislation

16.4.5 Agriculture

According to the United Nations (2024a), agriculture is the largest land use on Earth and now occupies 40% of the land surface on the planet. In addition, agriculture and associated activities constitute the largest pollution source on Earth and is also the largest single factor responsible for species population decline. In addition, agriculture affects ocean health in many ways that include excessive sedimentation, pollution from multiple sources and types of pollution, including fertilizers, pesticides, and herbicides. Agriculture also contributes significantly to climate change and ocean acidification (United Nations 2024b).

The following presents a list of action items that the individual and local community can undertake for preventing ocean pollution (Rogers 2023):

- Plant a vegetable garden
- Consume less meat
- Consume more plant food
- Purchase organically grown and ecofriendly food
- Become more educated
- Encourage and inspire others
- Contact your local and state-elected officials and encourage them to support sustainable legislation
- Limit fertilizer and pesticide and herbicide use

16.4.6 Reducing Energy and Climate Change Impacts

Local and individual actions that can be undertaken to reduce energy and climate change impacts include the following (Rogers 2023):

- Bike to work and to school
- Walk
- Carpool
- Consider living closer to work
- Drive a more fuel-efficient automobile, hybrid, or electric automobile
- Use mass transit
- Consider an electric lawn mower and snow blower
- Install solar house panels
- Purchase green electricity, if possible, from your utility company
- Lower water heater temperature
- Dry clothes by hanging them outside
- Upgrade appliances with improved energy efficiency
- Connect your outside lights to a timer
- Install a heat blanket on the hot water heater and lower its temperature 10% to 20%
- Use low-flow shower heads
- Take shorter showers
- Wash clothes in cold water
- Purchase "Energy Star" products
- Plant deciduous trees in locations around your home to provide shade in the summer and light into you home during the winter
- Turn your computer off rather than using sleep mode or screen saver
- Conserve energy – turn off appliances and lights when you leave a room
- Keep woodstoves and fireplaces well maintained. Consider replacing old wood burning stoves with an EPA-certified model.
- Keep vehicle tires at recommended air pressure
- Attempt to limit or avoid drive thru lines
- Get regular engine tune ups and regular maintenance checks
- Request to work from home, if possible
- Do full loads of laundry
- Run the dishwasher only when full
- Turn down the heat or turn air conditioning when not at home

- Lower the thermostat during winter and especially at night and increase the thermostat during the summer
- Unplug electrical devices such as toasters, printers, chargers, computers and other items when not in use
- Improve insulation of your home

16.4.7 Reducing Chemical Impacts

There are many things that can be done to reduce chemical impacts to the air, water, and land. Some of the most important include the following (Rogers 2020a):

- Eliminate or reduce the use of pesticides and herbicides
- Limit the amount of fertilizers
- Increase regulatory authority on agriculture
- Ban harmful chemicals
- Evaluate emerging contaminants
- Reduce noise pollution of the air and water
- Improve household waste regulations
- Address non-point source pollution improvements
- Improve geologic maps of urban areas

16.5 SUMMARY AND CONCLUSION

To be successful at achieving the goal of coexisting in harmony with nature will require cooperation not often experienced in our short history. Humans must respond individually, as a community, and globally to understand and repair the impacts we have made to our planet. These actions must be undertaken before it is too late, not just for the survival of other species, but for the survival of our own species as well.

Human innovation and our success as a species have been remarkable. One of the many significant lessons we have learned is that the speed of human technological advance has greatly outpaced the capacity of nature's ability to adapt. We look for clues in our past that can help us predict a future outcome. However, we have no point of reference. We are moving into an uncertain future, and we are moving so quickly that even scientific research has difficulty digesting the possible realities of our future. Scientists have documented the negative impacts of our past that have resulted in enormous harm to nature and our planet. Once we accept responsibility for those actions, we can innovate yet again with our collective minds and repair our home so that nature and we can begin to heal.

REFERENCES

British Geological Survey. 2024. *Global Geological Risk*. www.bgs.ac.uk/geology-projects/global-geological-risk. (Accessed March 11, 2024).

Darke, R., and Tallamy, D. 2019. *The Living Landscape*. Timber Press. Portland, Oregon. 392p.

Federal Emergency Management Agency (FEMA), 2025. Guidance on Extreme Temperatures for State, Local, Tribal and Territorial Leaders. www.fema.gov/sites/default/files/documnets/fema_guidance-extreme-temperature-for-state-local-tribal-territorial-leaders.pdf. (Accessed February 6, 2025).

Kaufman, M. M., Rogers, D. T., and Murray, K. S. 2003. Surface and Subsurface Geologic Risk Factors to Ground Water Affecting Brownfield Redevelopment Potential. *Journal of Environmental Quality*. Vol. 32, pp. 490–499.

Kaufman, M. M., Rogers, D. T., and Murray, K. S. 2005. An Empirical Model for Estimating Remediation Costs at Contaminated Sites. *Journal of Water, Air and Soil Pollution*. Vol. 167, pp. 365–386.

Kaufman, M. M., Rogers, D. T., and Murray, K. S. 2021. Urban Watersheds. CRC Press, Boca Raton, FL. 583p.

Murray, K. S., and Rogers, D. T. 1999a. Groundwater Vulnerability, Brownfield Redevelopment and Land Use Planning. *Journal of Environmental Planning and Management*. Vol. 42, no. 6, pp. 801–810.

Murray, K. S., and Rogers, D. T. 1999b. *Evaluation of Groundwater Vulnerability in an Urban Watershed*. Proceedings of the 2nd International Congress on Water Resources and Environmental Research. Brisbane, Australia. pp. 877–883.

National Oceanic and Atmospheric Administration (NOAA). 2024. *Climate Model: Temperature Change (RCP 4.5) - 2006-2100*. https://sos.noaa.gov/catalog/datasets/climate-model-temperature-change-rcp-85-2006-2100. (Accessed May 7. 2024).

Ocean Society. 2024. *The Ocean Plastic Pollution Problem*. www.oceansociety.org/resources/ways-to-reduce-ocean-plastic. (Accessed May 10, 2024).

Rogers, D. T. 1992. *The Importance of Site Observation and Follow-up Environmental Site Assessment – A Case Study. Groundwater Management Book 12*. National Groundwater Association. Dublin, OH. pp. 563–572.

Rogers, D. T. 1996. *Environmental Geology of Metropolitan Detroit*. Clayton Environmental Consultants. Novi, MI.

Rogers, D. T. 2002. The Development and Significance of a Geologic Sensitivity Map of the Rouge River Watershed in Southeastern Michigan, USA. In: Bobrowsky, P. T. editor. *Geoenvironmental Mapping: Methods, Theory, and Practice*. A. A. Balkema Publishers. The Netherlands. pp. 295–319.

Rogers, D. T. 2016a. Next Generation of Urban Hydrogeologic Investigations. *Journal of the Italian Geological Society*. Rome, Italy. Vol. 39, no. 1, pp. 349–353.

Rogers, D. T. 2016b. *Scientific Advancements that Improve the Conceptual Site Model in Urban Hydrogeological Site Investigations*. 35th International Geological Congress. Paper 3019. Cape Town, South Africa.

Rogers, D. T. 2018. *Derivation of a Comprehensive Environmental Risk Model for Urban Groundwater Protection*. International Association of Hydrogeologists Congress. Vol. 1. Daejeon, Korea.

Rogers, D. T. 2019a. *Key Policy Changes Required to Reduce Environmental Degradation in the United States and Beyond*. International Association of Hydrogeologists 46th Congress. Vol. 1 . Malaga, Spain. p. 328.

Rogers, D. T. 2019b. *Environmental Compliance and Sustainability: Global Challenges and Perspectives*. CRC Press. Boca Raton, FL. 543p.

Rogers, D. T. 2020a. *Urban Watersheds: Geology, Contamination, Environmental Regulations, and Sustainability*. CRC Press. Boca Raton, FL. 615p.

Rogers, D. T. 2020b. Geological- and Chemical-based Environmental Risk Factor Sustainability Model. *International Journal of Sustainable Development*. Vol. 9, no. 4, pp. 303–316. Rome, Italy.

Rogers, D. T. 2023. *Environmental Compliance Handbook: Sustainability and Future Environmental Regulations*. CRC Press. Boca Raton, FL. 548p.

Rogers, D. T., Murray, K. S., and Kaufman, M. M. 2007. Assessment of Groundwater Contaminant Vulnerability in an Urban Watershed in Southeast Michigan, USA. In: Howard, K. W. F. editor. *Urban Groundwater – Meeting the Challenge*. Taylor & Francis, London, England.

Rogers, D. T., Murray, K. S., and Kaufman, M. M. 2012. *Environmental Risk Analysis through Integration of Geologic Vulnerability and Air, Water, and Soil Contaminant Risk Factor Derivation*. International Geological Congress. Vol. 1. Brisbane, Australia. p. 3003.

Rogers, D. T., and Rogers, M. H. 2022a. *Integrating Future Anthropogenic Climate Impacts into A Geological and Chemical-based Sustainability Model*. Geological Society of America. Paper No. 381709. Denver, Colorado.

Rogers, M. H., and Rogers, D. T. 2022b. Incorporating Climate Change Modeling to Parcel Level Sustainability Efforts Globally. *International Journal of Sustainable Development*. Vol. 11, No. 4, pp. 211–220. Rome, Italy.

Sander, R. 1999. *Henry's Law Constants for Inorganic and Organic Species of Potential Importance in Environmental Chemistry (Version 3)*. http://www.henrys-law.org. (Accessed May 20, 2018).

Suthersan, S. S., and Payne, F. C. 2005. *In Situ Remediation Engineering*. CRC Press. Boca Raton, FL.

Tallamy, D. W. 2019. *Nature's Best Hope*. Timber Press. Portland, Oregon. 254p.

United Nations. 2020. *What is Sustainable Development*. www.un.org/sustainabledevelopment. (Accessed April 15, 2020).

United Nations. 2024a. *World Population Prospects*. United Nations Department of Economic and Social Affairs. Population Division. https://esa.un.org/unpd/wpp/data. (Accessed May 4, 2024).

United Nations. 2024b. *Human Development Report*. United Nations Development Programme. https://hdr.undp.org. (Accessed April 11, 2024).

United States Environmental Protection Agency (USEPA). 1989. *Transport and Fate of Contaminants in the Subsurface*. USEPA. EPA/625/4-89/019. Washington, DC.

United States Environmental Protection Agency (USEPA). 1996. *Bioscreen. Natural Attenuation Decision Support System, Version 1.4*. USEPA. Office of Research and Development. Washington, DC.

United States Environmental Protection Agency (USEPA). 2002a. *Supplemental Guidance for Developing Soil Screening Levels for Superfund Sites*. USEPA. Office of Solid Waste and Emergency Response. OSWER 9355.4-24. Washington, DC.

United States Environmental Protection Agency (USEPA). 2002b. *Health assessment of 1,3-butadiene*. National Center for Environmental Assessment, USEPA/600/P-98/001F. Washington, DC.

United States Environmental Protection Agency (USEPA). 2024a. *What Is Sustainability?* www.epa.gov/sustainability. (Accessed April 30, 2024).

United States Environmental Protection Agency (USEPA). 2024b. *Integrate Risk Information System (IRIS)*. www.epa.gov/ncea/iris/intro.htm. (Accessed May 11, 2024).

United States Environmental Protection Agency (USEPA). 2024c. *EPA Online Tools for Site Assessment Calculation*. Ecosystems Research Division. USEPA. www.epa.gov/athens/learn2model/part-two/onsite/retard.html. (Accessed Ma 11, 2024).

United States Environmental Protection Agency (USEPA). 2024d. *National Air Quality: Status and Trends of Key Air Pollutants*. www.epa.gov/air-trends. (Accessed May 11, 2024).

United States Geological Survey (USGS). 2006. *Volatile Organic Compounds in Nation's Ground Water and Drinking – Water Supply Wells*. USGS Circular 1292. Washington, DC.

Wiedemeier, T. H., Rifai, H. S., Newell, C. J., and Wilson, T. J. 1999. *Natural Attenuation of Fuels and Chlorinated Solvents in the Subsurface*. John Wiley & Sons. New York, NY.

World Climate Research Programme (WCRP). 2024. *WCRP Coupled Model Intercomparison Project*. https://wcrp-climate.org/wgcm-cmip. (Accessed May 7, 2024).

World Health Organization. 2024. *Nine Out of Ten People Breathe Unhealthy Air*. Geneva. Switzerland. www.who.int/news-room/detail/02-02-2018. (Accessed April 13, 2024).

Index

Note: Figures and tables are denoted by *italic* and **bold** text respectively.

F

K

L

M

N

O

P

R

V

W

Z